Electromagnetic Field Theory and Transmission Lines

Electromagnetic Field Theory and Transmission Lines

G.S.N. Raju
M.E., Ph.D (IIT-KGP), FIE, FIETE
Department of Electronics and Communications Engineering
College of Engineering, Andhra University
Visakhapatnam, India

Pearson

ISBN: 978-81-317-0171-3

First Impression, 2006
Sixth Impression,
Seventh Impression, 2011

Published by Pearson India Education Services Pvt. Ltd, CIN: U72200TN2005PTC057128.

Head Office: 1st Floor, Berger Tower, Plot No. C-001A/2, Sector 16B, Noida - 201 301, Uttar Pradesh, India.

Registered Office: 6th Floor, Tower A, Unit A and B, International Tech Park, CapitaLand Chennai, 200 Feet Radial Road,Zamin Pallavaram, Old Pallavaram Chennai – 600117, Tamil Nadu, India

Website: in.pearson.com; Email: companysecretary.india@pearson.com

Digitally printed in India by Trinity Academy for Corporate Training Ltd, New Delhi in the year of 2026

To

my brother,

Prof. Krishnam Raju *[MSc (Math.Physics), PhD],*

a man of academic excellence with impeccable character,

and a great administrator

who has made me what I am today.

PREFACE

Electromagnetic Field Theory and Transmission Lines is a core subject for all the students of BE/BTech in Electronics and Communication Engineering, Electronics Engineering, Electrical Engineering, Electronic Instrumentation Engineering, MSc (Electronics), MSc (Applied Physics), AMIETE, AMIE, and other courses throughout the world. EMF theory is essential for the design and analysis of all communication and radar systems. Moreover, it has numerous applications in all fields of life. It is a universal theory and has many advantages over the circuit theory which has limited applications. It is also useful in bio-medical engineering in connection with radiation therapy. It is extremely useful to interpret electromagnetic interference in the systems for compatibility studies. The behaviour of electromagnetic waves between the transmitter and receiver can be understood only with the concepts of the electromagnetic field (EMF) theory.

I have been teaching this important subject for several years, referring to books written by several experts like Kraus, Jordan, Hayt, Kreyszig and Narayana Rao. However, the general belief is that no single book available caters to the needs of a complete course required for undergraduate and post graduate programmes. In view of this, an attempt is being made to bring out a simplified book on the subject of Electromagnetic Field Theory and Transmission Lines.

I hope that this book will be extremely useful for students, teachers, professionals, engineers, technicians, designers and also for short-term course organisers.

Any suggestions to improve the book will be welcome.

G.S.N. RAJU

ACKNOWLEDGEMENT

I am grateful to my gurus, Prof. B.N. Das and Prof. Ajoy Chakraborty of IIT Kharagpur, who taught me fundamentals of Applied Electromagnetics.

I thank all my colleagues, Prof. Madhusudhan Rao, Prof. Venkata Rao, Prof. Veena Kumari, Dr. Hari Babu, Prof. Raja Rajeswari, Prof. Satyanarayana Reddy, Dr. Mallikarjuna Rao, Mrs. Santa Kumari, Dr. Gopala Rao, Mr. Sampath Kumar and Dr. Panduranga Reddy for their cooperation throughout.

It is a privilege to convey my gratitude with warm regards and respects to Sri K. Raghu, Chairman, Raghu Engineering College, and Sri S.V.S.S. Rama Chandra Raju, former PF Commissioner and Director, Raghu Engineering College, for their kind appreciation of my efforts in teaching and research and for giving me unmatched support in all my academic endeavors. I also thank Sri N.V. Rama Raju for extending support in my academic and professional efforts.

I am most grateful to Prof. K.C. Reddy, a renowned economist and Chairman, Council for Higher Education, Government of Andhra Pradesh for releasing the International edition of the book on *Antennas and Wave Propagation* and for his kind appreciation of the book.

I convey my gratitude to Prof. Y.C. Simhadri, Hon'ble Vice-Chancellor of Andhra University for his full support with encouragement and also for sending an excellent message of appreciation for *Antennas and Wave Propagation*.

I am thankful to Prof. D. Ganapathi Rao, who is kind enough to release the Indian edition of the book.

It is a privilege to thank Radm. S. Mohapatra, former Director, Dr. K. Satyanarayana, Director, Dr. V. Bhujanga Rao, Associate Director, Sri B. Vara Prasada Rao, Deputy Director and Senior Scientist, Cdr. T.K. Guha Roy, NSTL for their keen interest and moral support to my work in electromagnetics.

I am grateful to Prof. P.S. Rao, Principal, and Prof. P. Soma Raju, Secretary and Correspondent, GVP College of Engineering for their kind appreciation of *Antennas and Wave Propagation*.

I cherish the association of all my research scholars Mallikarjuna Rao, Sridevi, Gopala Rao, Misra, Sudhakar, Baba, Padma Raju, Chandra Bhushana Rao, Prasad,

Habibullah Khan, Subrahmanyam, Narayana, Sadasiva Rao, Gutti, Mallikarjuna Prasad, K.V.S.N. Raju, Murali Krishna, Rajesh and Swamy in putting their efforts for their Ph.D. degrees. I thank my friends, Prof. Appa Rao, Prof. K.V.L.P. Raju, Prof. Paramahamsa and Gopala Krishnam Raju for the moral support given to me throughout.

I thank Prof. Sarvarayudu, Prof. Raghuram, Prof. Prabhakara Rao, Prof. P.V.D.S. Rao of JNTU and Prof. Subbaiah of Nagarjuna University for their keen interest and appreciation for the book.

I thank Mr. Narayana, a sincere and hardworking scholar for assisting in proof reading with interest and devotion.

I love all my students who are always fond of my teaching, guidance and discipline.

I extend my regards to all senior authors and experts in Electromagnetic Field Theory for their invaluable contributions.

I am extremely grateful to the Department of Electronics and Communication Engineering, College of Engineering, Andhra University, for the encouragement given through out my teaching and research career.

I am grateful to the reviewers of this book for their excellent positive feedback.

I thank the most intelligent and Pro-poor Doctor Dr. G. Prasad, M.D., D.M., for his support and happiness given to me during preparation of the book. I also thank Dr. B. Dharani Priya for her appreciation of the book.

I express my thanks for the support extended by our technical and office staff, Koteswara Rao, Somayajulu, Sahu, Prasada Reddy, Appala Raju, Srinivas, Babji, Ramesh, Adilakshmi, Nara Hari, Ramana, Tavudu and Kondamma. The assistance given by Sankar is acknowledged with thanks for preparing the manuscript.

I take this opportunity to convey my respects to my parents Venkatrama Raju and Sitama and thanks to my wife Kanaka Durga, daughter Narmada Devi (B.Tech-CSE) and son Venkata Krishna Varma (B.Tech) IIT-Delhi for their love, affection, help and patience throughout the preparation of the book.

Finally, I thank all those who helped me directly and indirectly in bringing out this book.

G.S.N. RAJU

CONTENTS

INTRODUCTION

This introduction provides

- a list of some applications of EMF theory
- a comparison between EMF and Circuit theories: the reasons why EMF theory is superior and why it is essential for engineers and scientists
- small and large value representation, frequency ranges of TV channels and so on for general information
- information about the notations used in this book
- comprehensive background of the parameters, symbols, units and meanings

Electromagnetic Field (EMF) Theory is often called Electromagnetics. It is a subject which deals with electric field, magnetic field and also electromagnetic fields and phenomena.

EMF Theory is essential to design and analyse all communication and radar systems. In fact, it is also used in Bio-systems and in this context it is called Bio-electromagnetics.

APPLICATIONS OF ELECTROMAGNETIC FIELD THEORY

Electromagnetic Field Theory is used in

1. Wireless communications
2. Satellite communications
3. TV communications
4. Cellular communications
5. Radio broadcast
6. Mobile communications
7. Microwave communications
8. All types of antenna analysis and design
9. All types of transmission lines and waveguides
10. Bio-medical systems
11. Electrical machines
12. Speed-trap radars
13. Weather forecast radars
14. Remote sensing radars
15. Radio astronomy radars
16. Ground mapping radars
17. Meteorological radars
18. Plasmas
19. Electromagnetic interference and compatibility
20. Electromechanical energy conversion systems
21. Electric motors

22. Cyclotrons
23. Nuclear research
24. Radiation therapy
25. Heating deep tissues
26. Stimulating Bio-physiological responses
27. Relieving some pathological conditions
28. Induction heating
29. Melting
30. Forging
31. Surface hardening
32. Annealing
33. Soldering
34. Dielectric heating
35. Joining plastic sheets
36. Sealing plastic sheets
37. Agriculture
38. Reducing acidity in vegetables to improve taste
39. Telephones
40. Fibre optic communications
41. Lasers
42. Masers
43. Electric relays
44. Transformers
45. Microwave ovens

The design and analysis of a system, device or circuit requires the use of some theory or the other. The analysis of a system is universally defined as one by which the output is obtained from the given input and system details. On the other hand, the design of a system is one by which the system details are obtained, from the given input and output. These two important tasks are executed by two most popular theories, namely, circuit and electromagnetic theories. The differences between them are listed on next page.

DIFFERENCES BETWEEN CIRCUIT THEORY AND ELECTROMAGNETIC FIELD THEORY

Circuit theory	*Field theory*
1. Deals with voltage (V) and current (I)	Deals with Electric (**E**) and Magnetic (**H**) fields
2. V and I are scalars	**E** and **H** are vectors
3. V and I are produced from **E** and **H** respectively	**E** and **H** are produced from V and I respectively
4. V and I are functions of time (t)	**E** and **H** are functions time (t) and space variables (x, y, z) or (ρ, ϕ, z) or (r, θ, ϕ)
5. Radiation effects are neglected	Radiation effects can be considered
6. Using circuit theory, transmitter and receiver circuits can be analysed and designed. But it cannot be used to design or analyse a medium like free space	Using field theory, the medium also can be designed and analysed
7. This is simplified approximation of field theory	This is a more accurate theory
8. The variables of circuit theory, V and I are integrated effects of variables of field theory **E** and **H**	The variables of field theory, **E** and **H** are integrated effects of variables of circuit theory V and I
9. Circuit theory cannot be used to analyse or design a complete communication system	Field theory can be used where circuit theory fails to hold good for the analysis and design of a communication system
10. Is useful at low frequencies	Is useful at all frequencies, particularly at high frequencies
11. At low frequencies the length of connecting wires is very much smaller than λ	At high frequencies the length of connecting components are of the order of λ
12. Cannot be applied in free space	Is applicable in free space
13. Is simple	Is complex but it is simplified by using appropriate mathematics
14. Basic laws are Ohms law, Kirchoff's laws	Basic laws are Coulomb's law, Gauss's law, Ampere's circuit law
15. Basic theorems are Thevenin's, Nortan's, Reciprocity, Superposition, Maximum power transfer theorems	Basic theorems are Reciprocity, Helmholtz, Stoke's, Divergence and Poynting theorems
16. Basic equations are Mesh/Loop equations	Basic equations are Maxwell, Poission, Laplace and Wave

NOTATION OF SCALAR PARAMETERS

Parameter	*Notation /symbol*	*Definition*	*Unit name*
Frequency	f	It is the reciprocal of one time period of a periodic waveform	Hertz (Hz) 1 Hz = 1 cycle/sec
Energy	W	It is the work done when force is exerted through a distance of one metre	Joule (J) 1 Joule = 10^7 ergs
Power	P	It is the time rate of energy	Watt (W) 1 W = 1 Joule/sec or 1 W = 1 volt × 1 amp
Charge	Q	It is the product of current and time	Coulomb (C) 1 C = 1 A-sec
Resistance	R	It is the ratio of voltage and current	Ohm (Ω) 1 Ω = 1 volt/1 amp
Conductance	G	It is the reciprocal of R	Mho $1 \text{ Mho} = \dfrac{1 \text{ Amp}}{1 \text{ Volt}}$
Resistivity	ρ	It is the resistance measured between two parallel faces of a unit cube	Ohm-metre
Conductivity	σ	It is the reciprocal of resistivity	Mho/metre
Electromotive force	Emf, V	It is the ratio of power to current	Volt 1 Volt = 1 J/C or watt/amp
Electric flux	ψ	It is displaced charge	Coulomb (C)
Magnetic flux	φ	$\phi = -\int_0^t V\,dt$	weber (wb) 1 wb = 1 volt-sec
Magnetomotive force	V_m (mmf)	$V_m = \int_A^B \mathbf{H}.d\mathbf{L}$	Amp (A)
Capacitance	C	$C = \dfrac{Q}{V}$	Farads (F) 1 F = 1 C/1 Volt

Parameter	*Notation /symbol*	*Definition*	*Unit name*
Inductance	L	$L = \frac{N\phi}{I}$	Henry (H)
Mutual inductance	M	$M = N_2 \phi_{12}/I_1$	Henry (H) 1 H = 1 wb/amp
Permittivity	$\in$	$\in = \frac{D}{E}$	Farad/metre (F/m)
Permeability	μ	$\mu = \frac{B}{H}$	Henry/metre (H/m)
Permittivity of free space	$\in_0$	8.854×10^{-12} F/m	F/m
Permeability of free space	μ_0	$4\pi \times 10^{-7}$ H/m	H/m
Relative permittivity of a medium	$\in_r$	$\in_r = \frac{\in}{\in_0}$ $\in_r = 1$ for free space	No unit
Relative permeability of a medium	μ_r	$\mu_r = \frac{\mu}{\mu_0}$ $\mu_r = 1$ for free space	No unit
Electric susceptibility	χ_e	$\chi_e = \in_r - 1$ χ is pronounced as Chi	No unit
Magnetic susceptibility	χ_m	$\chi_m = \mu_r - 1$	No unit
Steradian	Str	It is a measure of solid angle	Steradian
Intrinsic impedance of free space	η_0	$\eta_0 = \sqrt{\frac{\mu_0}{\in_0}}$	$120\ \pi\Omega$
Differential length	dL	Small length	m
Area	S	Product of two lengths	m^2
Differential area	dS	Product of two differential lengths	m^2

Parameter	*Notation /symbol*	*Definition*	*Unit name*
Differential volume	dv	Product of differential length, width and breadth	m^3
Volume	v	Product of length, width and breadth	m^3
Angular frequency	ω	$2\pi f$	rad/sec
Wavelength	λ	v/f	m
Electric potential	V	$-\int \mathbf{E.dL}$	Volt
Magnetic scalar potential	V_m	$-\int \mathbf{H.dL}$	Ampere
Surface charge density	ρ_s	Q/S	c/m^2
Line charge density	ρ_L	Q/L	c/m
Volume charge density	ρ_v	Q/v	c/m^3
Propagation constant	γ	$\alpha + j\beta$	dB/m
Attenuation constant	α	It is a measure of reduction of EM wave as it progresses	dB/m
Depth of penetration or skin depth	δ	It is the depth at which an EM wave is attenuated to 37 per cent of original value	$\delta = 1/\alpha$ (m)
Phase constant	β	It is a measure of phase shift of EM wave	rad/m
Group velocity	v_g	It is the velocity with which energy propagates in a guided structure	m/sec
Phase velocity	v_p	It defines a point of constant phase	m/sec
VSWR	S	V_{max}/V_{min}	No unit

Parameter	*Notation /symbol*	*Definition*	*Unit name*
Reflection coefficient	ρ	$\frac{\text{Reflected wave}}{\text{Incident wave}} = \frac{V_r}{V_i}$	No unit
Transmission coefficient	T	$\frac{\text{Transmitted wave}}{\text{Incident wave}}$	No unit
Surface impedance	z_S	$z_S = \frac{E_{\tan}}{J_S}$	Ω (Ohm)
Capacitance	C	$C = \frac{Q}{V}$	Farad
Velocity of EM wave in free space	V_0	λf	3×10^8 m/s
Guide wavelength	λ_g	$\lambda_g = \frac{2\pi}{\beta_g}$	Metres
Phase constant in waveguide	β_g	$\beta_g = \frac{2\pi}{\lambda_g}$	Rad/m
Wavelength in parallel plates	λ_p	$\lambda_p = \frac{2\pi}{\beta_p}$	Metres

NOTATION OF VECTOR PARAMETERS

Parameter	*Notation /symbol*	*Definition*	*Unit name*
Force	**F**	It is the product of mass and acceleration	Newton (N) 1 Newton $= \frac{\text{Kg-m}}{\text{sec}^2}$
Electric field strength	**E**	It is the force per one Coulomb	Volt/m or Newton/C
Conduction current density	$\mathbf{J}_c$	It is defined as the ratio of current to area	A/m^2

Parameter	*Notation /symbol*	*Definition*	*Unit name*
Displacement electric flux density	**D**	$\mathbf{D} = \in \mathbf{E}$	C/m^2
Displacement current density	$\mathbf{J}_d$	$\mathbf{J}_d = \frac{\partial \mathbf{D}}{\partial t}$	A/m^2
Magnetic flux density	**B**	$\mathbf{B} = \mu\mathbf{H}$	Wb/m^2 or Tesla
Magnetic field strength	**H**	It is the current per metre width	A/m
Velocity	**V**	Rate of displacement	m/sec
Vector	Bold faced letter	-	-
Unit vector	Bold faced small letter	-	-
Electric dipole moment	**p**	Q**d**	Coulomb-m
Magnetic dipole moment	**m**	I**A**	A-m^2
Polarisation	**P**	$\chi_e \in_0 \mathbf{E}$	Coulomb/m^2
Magnetisation	**M**	$\chi_m \mathbf{H}$	Ampere/m
Torque	**T**	$\mathbf{R} \times \mathbf{F}$	N-m
Surface current density	$\mathbf{J}_s$	Current per metre	Ampere/m
Poynting vector	**P**	$\mathbf{E} \times \mathbf{H}$	Watt/m^2
Tangential component of **E**	$\mathbf{E}_t$	Tangential component of **E**	Volt/m
Tangential component of **H**	$\mathbf{H}_t$	Tangential component of **H**	Ampere/m

Parameter	*Notation /symbol*	*Definition*	*Unit name*
Normal component of **D**	D_n	Normal component of **D**	C/m^2
Normal component of **B**	B_n	Normal component of **B**	Wb/m^2

SMALL VALUE REPRESENTATION

Value	*Prefix*	*Symbol*	*Length*
10^{-1}	Deci	d	dm
10^{-2}	Centi	c	cm
10^{-3}	milli	m	mm
10^{-6}	micro	μ	μm
10^{-9}	nano	n	nm
10^{-12}	pico	p	pm
10^{-15}	femto	f	fm
10^{-18}	atto	a	am

LARGE VALUE REPRESENTATION

Value	*Prefix*	*Symbol*	*Length*	*Frequency*
10	deka	da	dam	daHz
10^2	hecto	h	hm	hHz
10^3	Kilo	K	Km	kHz
10^6	Mega	M	Mm	MHz
10^9	Giga	G	Gm	GHz
10^{12}	Tera	T	Tm	THz

FREQUENCY RANGES OF TV CHANNELS

Channel no.	*Frequency band (MHz)*
2	54-60
3	60-66
4	66-72
5	76-82
6	82-88
7	174-180
8	180-186
9	186-192
10	192-198
11	198-204
12	204-210
13	210-216
UHF Band	470-806

SOME GREAT CONTRIBUTORS TO ELECTROMAGNETIC FIELD THEORY

Benjamin Franklin (1706–1790)

American scientist-statesman, established the law of conservation of charge and determined that there are both positive and negative charges. He invented the lightning rod in 1750.

Charles Augustin De Coulomb (1736–1806)

French colonel, formulated Coulomb's law in 1785.

Andre Marie Ampere (1775–1836)

French physicist, invented the Solenoidal Coil for producing magnetic fields. He developed Oersted's discovery and introduced the concept of current element and force between current elements in 1825.

Hans Christian Oersted (1777–1851)

Danish professor of physics. In 1820 he discovered that a magnetic needle is deflected by a current in a wire.

KARL FRIENDRICH GAUSS (1777–1855)

German mathematician, developed Divergence theorem and Gauss's law. Gauss was the first scientist who measured electric and magnetic quantities in absolute units.

MICHAEL FARADAY (1791–1861)

Director of the Royal Society in London. He formulated Faraday's law in 1837.

HEINRICH FRIEDRICH EMIL LENZ (1804–1865)

Professor of physics, discovered Lenz's law.

HERMANN LUDWIG FERDINAND VON HELMHOLTZ (1821–1894)

Professor at Berlin, working in the field of Physiology, Electrodynamics and Optics. He formulated the wave equations in Electromagnetics. He established Helmholtz theorem.

JAMES CLERK MAXWELL (1831–1879)

Scottish physicist, was a professor at Cambridge University, England. He published the first Unified Theory of Electricity and Magnetism. Founded the science of electromagnetism and formulated relations between electric and magnetic fields. These are the laws of Gauss, Faraday and Ampere.

JEAN-BAPTISTE BIOT (1774–1862) AND FELIX SAVART (1791–1841)

They were professors of physics at the College de France. The Biot–Savart law was proposed in 1820.

HEINRICH HERTZ (1857–1894)

Professor at the Karlsruhe Polytechnic. In 1886 he assembled the apparatus for a complete radio system with an end-loaded dipole as transmitting antenna and a resonant square loop antenna as receiver.

GUGLIELMO MARCONI (1874–1937)

He repeated Hertz's experiments in 1901 and startled the world by announcing that he had received radio signals from across the Atlantic.

CHAPTER

1

MATHEMATICAL PRELIMINARIES

Mathematics is the backbone of all Sciences, Economics as well as all branches of Engineering.

The main objective of this chapter is to provide a brief and necessary background of the following main topics of mathematics:

- fundamentals of scalar and vector coordinate systems
- operations with del
- determinants and matrices
- series and identities
- trigonometric functions
- differentiation and integration formulae
- integral theorems and so on
- points/formulae to remember, solved problems, objective questions and exercise problems.

Do you know?

The analysis and design of any system or device is impossible without mathematics.

Mathematics is the backbone of science and engineering. For all analytical or computational purposes, mathematical background is essential. Mathematical modeling of systems is a common practice. The design and analysis of any antenna problem is possible only with mathematical concepts. No engineering system can be designed without using mathematics. Moreover, the solutions of antenna problems are simplified by mathematical approach.

In view of this, basics of mathematics is presented in this chapter.

1.1 FUNDAMENTALS OF SCALARS AND VECTORS

A **scalar** has magnitude and an algebraic sign. For example, temperature, mass, charge and work.

A **vector** has both magnitude and direction. For example, velocity, force, electric field and magnetic field.

In this book, a scalar is represented by simple letters like A, B and a vector is represented by bold-faced letters like **A**, **B**. Unit vectors are represented by small bold-faced letters like **a**, **b**.

The vector **A** is expressed in two forms:

(i) $\mathbf{A} = (A_x, A_y, A_z)$. A_x, A_y, A_z are known as the components of vector **A**.

(ii) $\mathbf{A} = A_x \mathbf{a}_x + A_y \mathbf{a}_y + A_z \mathbf{a}_z$. $\mathbf{a}_x, \mathbf{a}_y, \mathbf{a}_z$ are unit vectors along the coordinate axes.

The magnitude of **A** is written as A,

that is, $A = |\mathbf{A}|$

The unit vector of **A** is **a** and it is given by

$$\mathbf{a} = \frac{\mathbf{A}}{A}$$

The sum and difference of two vectors are given by

$$\mathbf{A} + \mathbf{B} = (A_x + B_x)\,\mathbf{a}_x + (A_y + B_y)\,\mathbf{a}_y + (A_z + B_z)\,\mathbf{a}_z$$

$$\mathbf{A} - \mathbf{B} = (A_x - B_x)\,\mathbf{a}_x + (A_y - B_y)\,\mathbf{a}_y + (A_z - B_z)\,\mathbf{a}_z$$

The dot product is denoted by

$$\mathbf{A}\,.\,\mathbf{B} \text{ or } \mathbf{B}\,.\,\mathbf{A}$$

$$\mathbf{A}\,.\,\mathbf{B} = \mathbf{B}\,.\,\mathbf{A} = AB\cos\theta$$

$$= A_x B_x + A_y B_y + A_z B_z$$

Here θ is the angle between the vectors **A** and **B**.

Dot product of two vectors is a scalar.

The cross product is denoted by $\mathbf{A} \times \mathbf{B}$.

$$\mathbf{A} \times \mathbf{B} = AB \sin \theta \, \mathbf{a}_n.$$

where $\mathbf{a}_n$ is the unit vector perpendicular to **A** and **B**

or
$$\mathbf{A} \times \mathbf{B} = \begin{vmatrix} \mathbf{a}_x & \mathbf{a}_y & \mathbf{a}_z \\ A_x & A_y & A_z \\ B_x & B_y & B_z \end{vmatrix}$$

$$= \mathbf{a}_x [A_y B_z - A_z B_y] + \mathbf{a}_y [A_z B_x - A_x B_z] + \mathbf{a}_z [A_x B_y - A_y B_x]$$

where $\mathbf{A} = A_x \mathbf{a}_x + A_y \mathbf{a}_y + A_z \mathbf{a}_z$

and $\mathbf{B} = B_x \mathbf{a}_x + B_y \mathbf{a}_y + B_z \mathbf{a}_z$

The cross product of two vectors is a vector.

1.2 COORDINATE SYSTEMS

Coordinate system is defined as a system which is used to represent a point in space.

Coordinate systems, basically, are of three types:

1. Cartesian coordinate system
2. Cylindrical coordinate system
3. Spherical coordinate system

Cartesian Coordinate System

Here a point, P is represented by $P(x, y, z)$. The variables are x, y and z.

A point obtained by the intersection of three planes given by

$$x = k_1 \text{ (constant)}$$
$$y = k_2 \text{ (constant)}$$
$$z = k_3 \text{ (constant)}$$

The unit of x, y and z is metre.

The Cartesian coordinates are represented in Fig. 1.1.

The three axes, x, y and z are mutually perpendicular to each other. These are said to be orthogonal to each other.

The unit vectors along the coordinate axes are represented by $\mathbf{a}_x$, $\mathbf{a}_y$ and $\mathbf{a}_z$. Their magnitude is unity and they are in the increasing directions of x, y and z axes respectively.

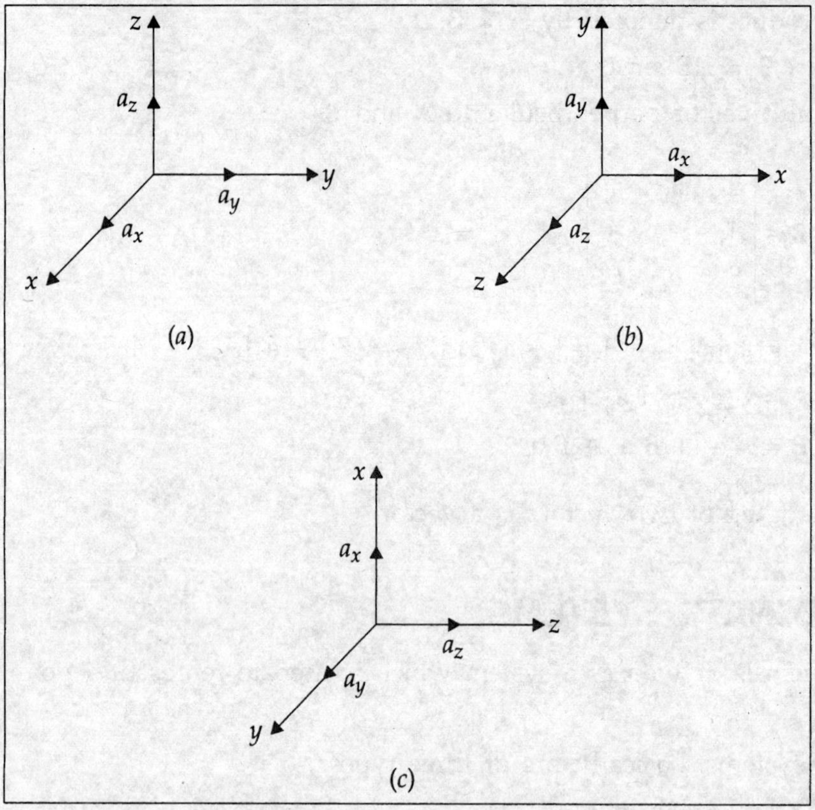

Fig. 1.1 Cartesian coordinate system

Properties of unit vectors

$\mathbf{a}_x \cdot \mathbf{a}_x = 1$	$\mathbf{a}_y \cdot \mathbf{a}_z = 0$
$\mathbf{a}_y \cdot \mathbf{a}_y = 1$	$\mathbf{a}_x \times \mathbf{a}_y = \mathbf{a}_z$
$\mathbf{a}_z \cdot \mathbf{a}_z = 1$	$\mathbf{a}_y \times \mathbf{a}_z = \mathbf{a}_x$
$\mathbf{a}_x \times \mathbf{a}_x = 0$	$\mathbf{a}_z \times \mathbf{a}_x = \mathbf{a}_y$
$\mathbf{a}_y \times \mathbf{a}_y = 0$	$\mathbf{a}_y \times \mathbf{a}_x = -\mathbf{a}_z$
$\mathbf{a}_z \times \mathbf{a}_z = 0$	$\mathbf{a}_z \times \mathbf{a}_y = -\mathbf{a}_x$
$\mathbf{a}_x \cdot \mathbf{a}_y = 0$	$\mathbf{a}_x \times \mathbf{a}_z = -\mathbf{a}_y$
$\mathbf{a}_x \cdot \mathbf{a}_z = 0$	

Cylindrical Coordinate System

Here, a point, P is represented by $P(\rho, \phi, z)$. ρ represents the radius of a cylinder, ϕ is called the azimuthal angle and z is the same as in Cartesian coordinate system.

The unit of ρ is metre.

The unit of ϕ is degree or radian.

The unit of z is metre.

In cylindrical coordinate system, a point is obtained by the intersection of three surfaces, namely,

a cylindrical surface, $\rho = k_1$ (constant), metre

a plane, $\phi = \alpha$ (constant), radian, and

another plane, $z = k_2$ (constant), metre

All the three surfaces are mutually perpendicular to each other. These are said to be mutually orthogonal.

In this book, the term Cylindrical Coordinate system is used to indicate circular cylindrical coordinate system. However, a point in cylindrical coordinate system is shown in Fig. 1.2. The coordinate ρ is the radius of the cylinder. ϕ is measured from x-axis and z is the same as in Cartesian system.

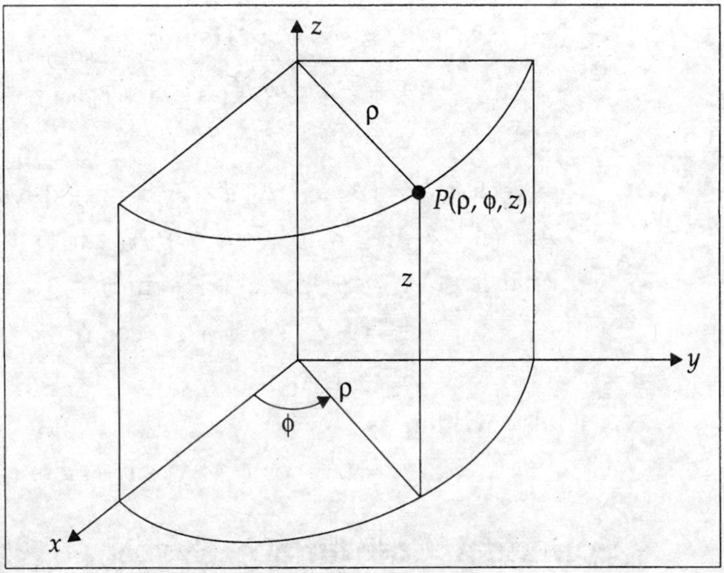

Fig. 1.2 A point in cylindrical coordinates

Here $\mathbf{a}_\rho$, $\mathbf{a}_\phi$, $\mathbf{a}_z$ represent the unit vectors along the coordinates ρ, ϕ, z. Their magnitude is unity and they are in the increasing directions of ρ, ϕ, z respectively.

It is obvious that increase in ρ results in cylinders of higher radius. ϕ increases in anti-clockwise direction. z is the same as in Cartesian system.

The relations between x, y, z and ρ, ϕ, z:

$$x = \rho \cos \phi$$
$$y = \rho \sin \phi$$
$$z = z$$

and $\rho = \sqrt{x^2 + y^2}, \quad 0 \le \rho < \infty$

$\phi = \tan^{-1} x, \quad 0 \le \phi < 2\pi$

$z = z, \quad 0 \le z < \infty$

Dot products of $\mathbf{a}_x$, $\mathbf{a}_y$ and $\mathbf{a}_z$ with $\mathbf{a}_\rho$, $\mathbf{a}_\phi$ and $\mathbf{a}_z$ are given by

$$\mathbf{a}_x . \mathbf{a}_\rho = \cos\phi$$

$$\mathbf{a}_x . \mathbf{a}_\phi = -\sin\phi$$

$$\mathbf{a}_y . \mathbf{a}_\rho = \sin\phi$$

$$\mathbf{a}_y . \mathbf{a}_\phi = \cos\phi$$

$$\mathbf{a}_z . \mathbf{a}_\rho = 0$$

$$\mathbf{a}_z . \mathbf{a}_\phi = 0$$

The unit vectors of cylindrical coordinates in terms of Cartesian coordinates are given by

$$\mathbf{a}_\rho = \cos\phi\ \mathbf{a}_x + \sin\phi\ \mathbf{a}_y$$

$$\mathbf{a}_\phi = -\sin\phi\ \mathbf{a}_x + \cos\phi\ \mathbf{a}_y$$

$$\mathbf{a}_z = \mathbf{a}_z$$

A vector, $\mathbf{A} = (A_x\, \mathbf{a}_x + A_y\, \mathbf{a}_y + A_z\, \mathbf{a}_z)$ is expressed in cylindrical coordinates as

$\mathbf{A} = (A_\rho, A_\phi, A_z) = [(A_x \cos\phi + A_y \sin\phi), (-A_x \sin\phi + A_y \cos\phi), A_z]$

That is $\quad A_\rho = A_x \cos\phi + A_y \sin\phi$

$$A_\phi = -A_x \sin\phi + A_y \cos\phi$$

$$A_z = A_z$$

A is also written as

$$\mathbf{A} = (A_x \cos\phi + A_y \sin\phi)\, \mathbf{a}_\rho + (-A_x \sin\phi + A_y \cos\phi)\, \mathbf{a}_\phi + A_z\, \mathbf{a}_z.$$

Spherical Coordinate System

Here $P(r, \theta, \phi)$ represents a point. r represents the radius of a sphere, θ is the angle of elevation measured from z-axis and ϕ is the azimuthal angle measured from x-axis.

A point is obtained by the intersection of three surfaces, namely,

a spherical surface, $r = k$ (constant), metre

a cone, $\theta = \alpha$ (constant), radian, and

a plane, $\phi = \beta$ (constant), radian

All these three surfaces are mutually perpendicular to each other.

These are said to be orthogonal.

$\mathbf{a}_r$, $\mathbf{a}_\theta$, $\mathbf{a}_\phi$ represent unit vectors along the coordinate axes. Their magnitude is unity and they are in the increasing directions of r, θ and ϕ axes.

A point in spherical coordinate system is shown in Fig. 1.3.

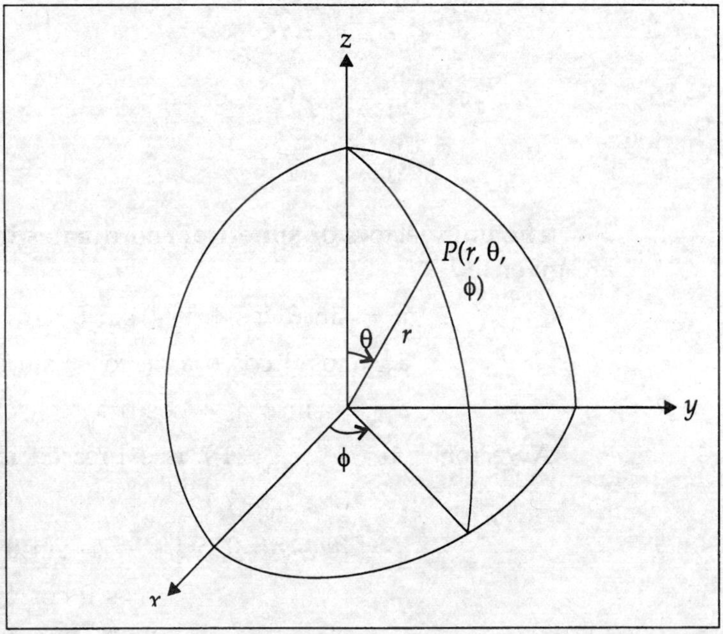

Fig. 1.3 A point in spherical coordinate system

Increase of r results in spheres of larger radii. θ increases in the clockwise direction and ϕ increases in the anti-clockwise direction.

The variables of Cartesian and spherical coordinates are related by

$$x = r \sin\theta \cos\phi, \qquad -\infty < x < \infty$$

$$y = r \sin\theta \sin\phi, \qquad -\infty < y < \infty$$

$$z = r \cos\theta, \qquad -\infty < z < \infty$$

and

$$r = \sqrt{x^2+y^2+z^2} \qquad 0 \le r \le \infty$$

$$\theta = \cos^{-1}\frac{z}{\sqrt{x^2+y^2+z^2}} \qquad 0 \le \theta \le \pi$$

$$\phi = \tan^{-1}\frac{y}{x} \qquad 0 \le \phi \le 2\pi$$

The relations between the variables of cylindrical and spherical coordinates are given by

$$\rho = r \sin \theta$$
$$\phi = \phi$$
$$z = r \cos \theta$$
$$r = \sqrt{\rho^2 + z^2}$$
$$\theta = \tan^{-1} \frac{r}{z}$$
$$\phi = \phi$$

The unit vectors of spherical coordinates in terms of Cartesian coordinates are given by

$$\mathbf{a}_r = \sin\theta \cos\phi \, \mathbf{a}_x + \sin\theta \sin\phi \, \mathbf{a}_y + \cos\theta \, \mathbf{a}_z$$
$$\mathbf{a}_\theta = \cos\theta \cos\phi \, \mathbf{a}_x + \cos\theta \sin\phi \, \mathbf{a}_y - \sin\theta \, \mathbf{a}_z$$
$$\mathbf{a}_\phi = -\sin\phi \, \mathbf{a}_x + \cos\phi \, \mathbf{a}_y$$

A vector, $\mathbf{A} = (A_x, A_y, A_z)$, is expressed in spherical coordinates as

$$\mathbf{A} = (A_r, A_\theta, A_\phi)$$
$$= [(A_x \sin\theta \cos\phi + A_y \sin\theta \sin\phi + A_z \cos\theta),$$
$$(A_x \cos\theta \cos\phi + A_y \cos\theta + \sin\phi - A_z \sin\theta),$$
$$(-A_x \sin\phi + A_y \cos\phi)]$$

The dot products of $\mathbf{a}_x$, $\mathbf{a}_y$ and $\mathbf{a}_z$ with $\mathbf{a}_r$, $\mathbf{a}_\theta$ and $\mathbf{a}_\phi$ are given by

$$\mathbf{a}_x \cdot \mathbf{a}_r = \sin\theta \cos\phi$$
$$\mathbf{a}_x \cdot \mathbf{a}_\theta = \cos\theta \cos\phi$$
$$\mathbf{a}_x \cdot \mathbf{a}_\phi = -\sin\phi$$
$$\mathbf{a}_y \cdot \mathbf{a}_r = \sin\theta \sin\phi$$
$$\mathbf{a}_y \cdot \mathbf{a}_\theta = \cos\theta \sin\phi$$
$$\mathbf{a}_y \cdot \mathbf{a}_\phi = \cos\phi$$
$$\mathbf{a}_z \cdot \mathbf{a}_r = \cos\theta$$
$$\mathbf{a}_z \cdot \mathbf{a}_\theta = -\sin\theta$$
$$\mathbf{a}_z \cdot \mathbf{a}_\phi = 0$$

Here,

$$A_r = A_x \sin\theta \cos\phi + A_y \sin\theta \sin\phi + A_z \cos\theta$$
$$A_\theta = A_x \cos\theta \cos\phi + A_y \cos\theta \sin\phi - A_z \sin\theta$$
$$A_\phi = -A_x \sin\phi + A_y \cos\phi$$

The point $A(x, y, z) = A(\rho, \phi, z) = A(r, \theta, \phi)$ in Cartesian, cylindrical and spherical coordinate systems is shown in a single Fig. 1.4 to understand the concept at a glance.

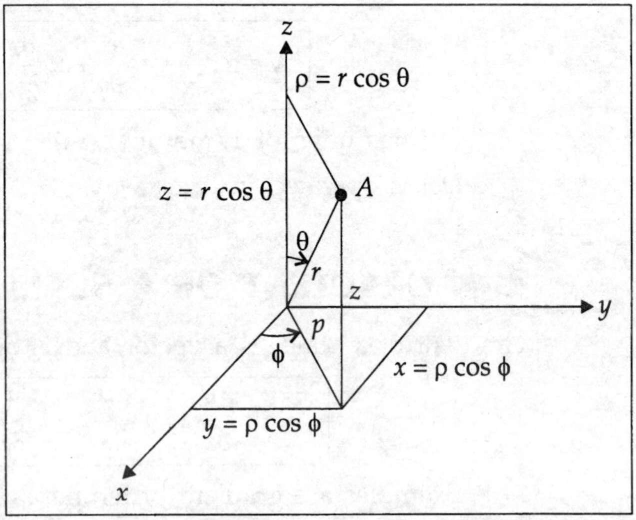

Fig. 1.4 Coordinates of a point in all the three systems

The relations for differential length, area and volume are given in Table 1.1.

Table 1.1 Differential Quantities in Different Coordinates

Coordinate system	*dL*	*dS*	*dv*
Cartesian	$dx\,\mathbf{a}_x + dy\,\mathbf{a}_y + dz\,\mathbf{a}_z$	$dx\,dy\,\mathbf{a}_z$ or $dy\,dz\,\mathbf{a}_x$ or $dz\,dx\,\mathbf{a}_y$	$dxdydz$
Cylindrical	$d\rho\,\mathbf{a}_\rho + \rho\,d\phi\,\mathbf{a}_\phi + dz\,\mathbf{a}_z$	$\rho\,d\rho\,d\phi\,\mathbf{a}_z$ or $\rho\,d\phi\,dz\,\mathbf{a}_\rho$ or $d\rho\,dz\,\mathbf{a}_\phi$	$\rho\,d\rho\,d\phi\,dz$
Spherical	$dr\,\mathbf{a}_r + r\,d\theta\,\mathbf{a}_\theta + r\sin\theta\,d\phi\,\mathbf{a}_\phi$	$r\,dr\,d\theta\,\mathbf{a}_\phi$ or $r\sin\theta\,dr\,d\phi\,\mathbf{a}_\theta$ or $r^2\sin\theta\,d\theta\,d\phi\,\mathbf{a}_r$	$r^2\sin\theta\,dr\,d\theta\,d\phi$

Relations between the polar coordinates (ρ, ϕ) and Cartesian coordinates (x, y) are:

$$x = \rho\cos\phi$$
$$y = \rho\sin\phi$$
$$\rho = \sqrt{x^2 + y^2}$$
$$\phi = \tan^{-1}\left(\frac{y}{x}\right)$$
$$dx\,dy = \rho\,d\rho\,d\phi$$

1.3 DEL (∇) OPERATOR

Definition The del or nabla is known as differential vector operator and is defined as

$$\nabla \equiv \mathbf{a}_x \frac{\partial}{\partial x} + \mathbf{a}_y \frac{\partial}{\partial y} + \mathbf{a}_z \frac{\partial}{\partial z}$$

Del has units of 1/metre (1/m)

Del is operated in three ways.

1.4 GRADIENT OF A SCALAR, V (=∇V)

Gradient of a scalar is a vector and is defined as

$$\nabla V \equiv \frac{\partial V}{\partial x}\mathbf{a}_x + \frac{\partial V}{\partial y}\mathbf{a}_y + \frac{\partial V}{\partial z}\mathbf{a}_z$$

Examples are gradient of temperature, gradient of electric potential and so on.

It gives the maximum space rate of change of the scalar. The scalar can be temperature, potential and so on.

1.5 DIVERGENCE OF A VECTOR, A (= ∇.A)

Divergence of a vector is a scalar and is defined as

$$\nabla .\, \mathbf{A} \equiv \text{div}\, \mathbf{A} \equiv \frac{\partial A_x}{\partial x} + \frac{\partial A_y}{\partial y} + \frac{\partial A_z}{\partial z}$$

Divergence means the spreading or diverging of a quantity from a point. It is applicable to vectors only. The divergence of a vector indicates the net flow of quantities like gas, fluid, vapour, electric and magnetic flux lines. In other words, it is a measure of the difference between outflow and inflow.

The divergence of a vector is positive if the net flow is outward.

It is negative if the net flow is inward.

The fluid is said to be incompressible if the divergence is zero, that is, $\nabla . \mathbf{A} = 0$ is the condition of incompressibility.

Examples and Features of Divergence

1. Leaking air from a balloon yields positive divergence.
2. Rushing of air into the drum under the carriage of a train yields negative divergence.

3. Divergence of water or oil is almost zero and hence they are incompressible.
4. Divergence of electric flux density is equal to volume charge density,

or, $$\nabla \cdot \mathbf{D} = \rho_v$$

5. Divergence of magnetic flux density is equal to zero,

or, $$\nabla \cdot \mathbf{B} = 0$$

6. Divergence of gradient of scalar electric potential is equal to the laplacian of the scalar,

or, $$\nabla \cdot \nabla V = \nabla^2 V.$$

1.6 CURL OF A VECTOR ($\equiv \nabla \times \mathbf{A}$)

Curl of a vector is a vector and is defined as

$$\text{Curl } \mathbf{A} = \nabla \times \mathbf{A} = \begin{vmatrix} \mathbf{a}_x & \mathbf{a}_y & \mathbf{a}_z \\ \frac{\partial}{\partial x} & \frac{\partial}{\partial y} & \frac{\partial}{\partial z} \\ \mathbf{A}_x & \mathbf{A}_y & \mathbf{A}_z \end{vmatrix}$$

$$= \mathbf{a}_x\left[\frac{\partial}{\partial y}\mathbf{A}_z - \frac{\partial}{\partial z}\mathbf{A}_y\right] + \mathbf{a}_y\left[\frac{\partial}{\partial z}\mathbf{A}_x - \frac{\partial}{\partial x}\mathbf{A}_z\right] + \mathbf{a}_z\left[\frac{\partial}{\partial x}\mathbf{A}_y - \frac{\partial}{\partial y}\mathbf{A}_x\right]$$

It is a measure of the tendency of a vector quantity to rotate or twist or curl. In other words, the rate of rotation or angular velocity at a point is the measure of curl.

As the curl of a vector represents rotation, it is also written as

$$\text{curl } \mathbf{A} = \text{rot } \mathbf{A} = \nabla \times \mathbf{A}$$

It may be noted that curl (gradient of a scalar) = $\nabla \times (\nabla V)$ is zero.

This means that the gradient of fields is irrotational. Also div (curl) = 0.

Examples:

- When a leaf floats in sea water and its rotation is about the *z*-axis, curl of velocity **V** is in the *z*-direction. When $(\nabla \times V)_z$ is positive, it represents rotation from x to y.
- For a rotating rigid body, the curl of velocity is in the direction of the axis of rotation. Its magnitude is equal to twice the angular speed of rotation.

In Cartesian coordinates,

$$\nabla V = \frac{\partial V}{\partial x}\mathbf{a}_x + \frac{\partial V}{\partial y}\mathbf{a}_y + \frac{\partial V}{\partial z}\mathbf{a}_z \quad \text{is a vector.}$$

$$\nabla . \mathbf{A} = \frac{\partial A_x}{\partial x} + \frac{\partial A_y}{\partial y} + \frac{\partial A_z}{\partial z} \text{ is a scalar.}$$

$$\nabla \times \mathbf{A} = \left(\frac{\partial A_z}{\partial y} - \frac{\partial A_y}{\partial z} \right) \mathbf{a}_x + \left(\frac{\partial A_x}{\partial z} - \frac{\partial A_z}{\partial x} \right) \mathbf{a}_y + \left(\frac{\partial A_y}{\partial x} - \frac{\partial A_x}{\partial y} \right) \mathbf{a}_z \text{ is a vector.}$$

In cylindrical coordinates,

$$\nabla V = \frac{\partial V}{\partial \rho} \mathbf{a}_\rho + \frac{1}{\rho} \frac{\partial V}{\partial \phi} \mathbf{a}_\phi + \frac{\partial V}{\partial z} \mathbf{a}_z \text{ is a vector.}$$

$$\nabla . \mathbf{A} = \frac{1}{\rho} \frac{\partial (\rho A_\rho)}{\partial \rho} + \frac{1}{\rho} \frac{\partial A_\phi}{\partial \phi} + \frac{\partial A_z}{\partial z} \text{ is a scalar.}$$

$$\nabla \times \mathbf{A} = \left(\frac{1}{\rho} \frac{\partial A_z}{\partial \phi} - \frac{\partial A_\phi}{\partial z} \right) \mathbf{a}_\rho + \left[\frac{\partial A_\rho}{\partial z} - \frac{\partial A_z}{\partial \rho} \right] \mathbf{a}_\phi + \frac{1}{\rho} \left[\frac{\partial (\rho A_\phi)}{\partial \rho} - \frac{\partial A_\rho}{\partial \phi} \right] \mathbf{a}_z$$

is a vector.

In spherical coordinates,

$$\nabla V = \frac{\partial V}{\partial r} \mathbf{a}_r + \frac{1}{r} \frac{\partial V}{\partial \theta} \mathbf{a}_\theta + \frac{1}{r \sin \theta} \frac{\partial V}{\partial \phi} \mathbf{a}_\phi \text{ is a vector.}$$

$$\nabla . \mathbf{A} = \frac{1}{r^2} \frac{\partial}{\partial r} (r^2 A_r) + \frac{1}{r \sin \theta} \frac{\partial}{\partial \theta} (\sin \theta A_\theta) + \frac{1}{r \sin \theta} \frac{\partial A_\phi}{\partial \phi} \text{ is a scalar.}$$

$$\nabla \times \mathbf{A} = \frac{1}{r \sin \theta} \left[\frac{\partial}{\partial \theta} (A_\phi \sin \theta) - \frac{\partial A_\theta}{\partial \phi} \right] \mathbf{a}_r + \frac{1}{r} \left[\frac{1}{r \sin \theta} \frac{\partial A_r}{\partial \phi} - \frac{\partial (r A_\phi)}{\partial r} \right] \mathbf{a}_\theta$$

$$+ \frac{1}{r} \left[\frac{\partial}{\partial r} \left(r A_\theta - \frac{\partial A_r}{\partial \theta} \right) \right] \mathbf{a}_\phi \text{ is a vector.}$$

Vector identities

$$\mathbf{B} . \mathbf{B} = B^2$$

$$\mathbf{B} . \mathbf{B}^* = B^2$$

$$\mathbf{B} + \mathbf{C} = \mathbf{C} + \mathbf{B}$$

$$\mathbf{B} . \mathbf{C} = \mathbf{C} . \mathbf{B}$$

$$\mathbf{C} \times \mathbf{B} = -\mathbf{B} \times \mathbf{C}$$

$$(\mathbf{D} + \mathbf{B}) . \mathbf{C} = \mathbf{D} . \mathbf{C} + \mathbf{B} . \mathbf{C}$$

$$(\mathbf{D} + \mathbf{B}) \times \mathbf{C} = \mathbf{D} \times \mathbf{C} + \mathbf{B} \times \mathbf{C}$$

$$\mathbf{D} . \mathbf{B} \times \mathbf{C} = \mathbf{B} . \mathbf{C} \times \mathbf{D} = \mathbf{C} . \mathbf{D} \times \mathbf{B}$$

$$\mathbf{D} \times (\mathbf{B} \times \mathbf{C}) = (\mathbf{D} . \mathbf{C}) \mathbf{B} - (\mathbf{D} . \mathbf{B}) \mathbf{C}$$

$$(\mathbf{E} \times \mathbf{B}) . (\mathbf{C} \times \mathbf{D}) = \mathbf{E} . \mathbf{B} \times (\mathbf{C} \times \mathbf{D})$$

$$= (\mathbf{E} \cdot \mathbf{C})(\mathbf{B} \cdot \mathbf{D}) - (\mathbf{E} \cdot \mathbf{D})(\mathbf{B} \cdot \mathbf{C})$$

$$(\mathbf{E} \times \mathbf{B}) \times (\mathbf{C} \times \mathbf{D}) = (\mathbf{E} \times \mathbf{B} \cdot \mathbf{D})\,\mathbf{C} - (\mathbf{E} \times \mathbf{B} \cdot \mathbf{C})\,\mathbf{D}$$

$$\nabla (V_m V) = V_m \nabla V + V \nabla V_m$$

$$\nabla \cdot (\nabla \times \mathbf{C}) = 0$$

$$\nabla \times \nabla V = 0$$

$$\nabla (V_m + V) = \nabla V_m + \nabla V$$

$$\nabla \cdot (\mathbf{C} + \mathbf{B}) = \nabla \cdot \mathbf{C} + \nabla \cdot \mathbf{B}$$

$$\nabla \times (\mathbf{C} + \mathbf{B}) = \nabla \times \mathbf{C} + \nabla \times \mathbf{B}$$

$$\nabla \cdot (V\mathbf{C}) = \mathbf{C} \cdot \nabla V + V \nabla \cdot \mathbf{C}$$

$$\nabla \times (V\mathbf{C}) = \nabla V \times \mathbf{C} + V \nabla \times \mathbf{C}$$

$$\nabla (\mathbf{C} \cdot \mathbf{B}) = (\mathbf{C} \cdot \nabla)\,\mathbf{B} + (\mathbf{B} \cdot \nabla)\,\mathbf{C} + \mathbf{C} \times (\nabla \times \mathbf{B}) + \mathbf{B} \times (\nabla \times \mathbf{C})$$

$$\nabla \cdot (\mathbf{C} \times \mathbf{B}) = \mathbf{B} \cdot \nabla \times \mathbf{C} - \mathbf{C} \cdot \nabla \times \mathbf{B}$$

$$\nabla \times (\mathbf{C} \times \mathbf{B}) = \mathbf{C}\,(\nabla \cdot \mathbf{B}) - \mathbf{B}\,(\nabla \cdot \mathbf{C}) + (\mathbf{B} \cdot \nabla)\,\mathbf{C} - (\mathbf{C} \cdot \nabla)\,\mathbf{B}$$

$$\nabla \times \nabla \times \mathbf{C} = \nabla (\nabla \cdot \mathbf{C}) - \nabla^2 \mathbf{C}$$

$$\oint_L \mathbf{C} \cdot d\mathbf{L} = \int_s (\nabla \times \mathbf{C}) \cdot d\mathbf{s}$$

$$\oint_s \mathbf{C} \cdot d\mathbf{s} = \int_v (\nabla \cdot \mathbf{C})\, dv$$

$$\oint_s (\mathbf{a}_n \times \mathbf{C})\, ds = \int_v (\nabla \times \mathbf{C})\, dv$$

$$\oint_s V\, d\mathbf{s} = \int_v \nabla v\, dv$$

$$\oint_L V\, d\mathbf{L} = \int_s \mathbf{a}_n \times \nabla v\, ds$$

1.7 LAPLACIAN OPERATOR (∇^2)

It is defined as $\nabla \cdot \nabla$. Its unit is $\frac{1}{\text{m}^2}$. It is a scalar differential operator. It is operated on a scalar as well as a vector.

$$\nabla^2 \equiv \frac{\partial^2}{\partial x^2} + \frac{\partial^2}{\partial y^2} + \frac{\partial^2}{\partial z^2}$$

Laplacian of a scalar electric potential ($\nabla^2 V$), is a scalar.

Laplacian of an electric field ($\nabla^2 \mathbf{E}$), is a vector.

Laplacian of a scalar and a vector in different coordinate systems.

$$\nabla^2 V = \frac{\partial^2 V}{\partial x^2} + \frac{\partial^2 V}{\partial y^2} + \frac{\partial^2 V}{\partial z^2} \quad \text{(Cartesian)}$$

$$\nabla^2 V = \frac{1}{\rho}\frac{\partial}{\partial \rho}\left(\rho \frac{\partial V}{\partial \rho}\right) + \frac{1}{\rho^2}\frac{\partial^2 V}{\partial \phi^2} + \frac{\partial^2 V}{\partial z^2} \quad \text{(Cylindrical)}$$

$$\nabla^2 V = \frac{1}{r^2}\frac{\partial}{\partial r}\left(r^2 \frac{\partial V}{\partial r}\right) + \frac{1}{r^2 \sin\theta}\frac{\partial}{\partial \theta}\left(\sin\theta \frac{\partial V}{\partial \theta}\right) + \frac{1}{r^2 \sin^2\theta}\frac{\partial^2 V}{\partial \phi^2} \quad \text{(Spherical)}$$

$$\nabla^2 \mathbf{A} = \frac{\partial^2 \mathbf{A}}{\partial x^2} + \frac{\partial^2 \mathbf{A}}{\partial y^2} + \frac{\partial^2 \mathbf{A}}{\partial z^2} \quad \text{(Cartesian)}$$

$$= \frac{\partial}{\partial x^2}(A_x \mathbf{a}_x + A_y \mathbf{a}_y + A_z \mathbf{a}_z) + \frac{\partial}{\partial y^2}(A_x \mathbf{a}_x + A_y \mathbf{a}_y + A_z \mathbf{a}_z) + \frac{\partial}{\partial z^2}(A_x \mathbf{a}_x + A_y \mathbf{a}_y + A_z \mathbf{a}_z).$$

1.8 DIRAC DELTA

The properties of Dirac Delta are

$$\delta(t - t_0) = \infty \quad \text{if } t = t_0 \text{ and}$$
$$= 0 \quad \text{otherwise}$$

Direct Delta is used to represent very short pulse of high amplitude. Sometimes it is also called unit impulse function.

At $t = 0$, it is represented by $\delta(t)$

At $t = t_0$, it is represented by $\delta(t - t_0)$

$$\int_0^\infty \delta(t - t_0)\, dt = 1$$

Dirac Delta is shown in Fig. 1.5.

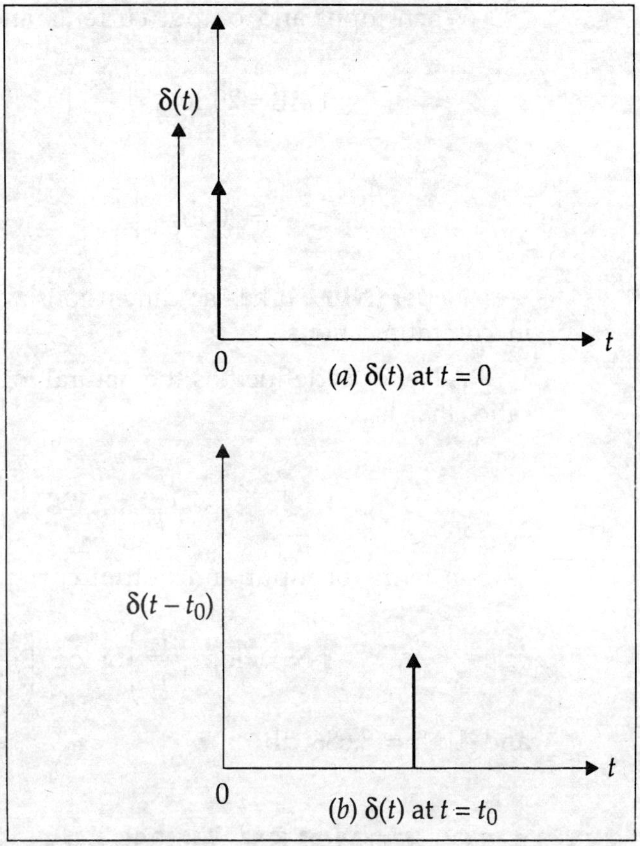

Fig. 1.5 Dirac Delta

1.9 DECIBEL AND NEPER CONCEPTS

Decibel (dB) is defined as ten times the common logarithm of the power ratio, that is,

$$1 \text{ dB} \equiv 10 \log_{10}\left(\frac{P_1}{P_2}\right)$$

If P_1 and P_2 are input and output powers respectively of an electric circuit, then dB is negative if $P_2 > P_1$. This indicates power loss. If $P_1 > P_2$, dB is positive and this indicates power gain.

Decibel has no dimensions and it is used to express the ratio of two powers, voltages, currents or sound intensities.

One Bel (B) is equal to ten decibels.

When input and output currents and voltages are known, we have

$$1 \text{ dB} = 20 \log_{10}\left(\frac{V_2}{V_1}\right)$$

$$= 20 \log_{10}\left(\frac{I_2}{I_1}\right)$$

Neper (NP) It has no dimensions and is used to express the ratio of powers in communications.

A Neper is defined as the natural logarithm of the square root of the power ratio, that is,

$$1 \text{ Np} \equiv \log_e \sqrt{\frac{P_2}{P_1}} = \frac{1}{2} \log_e \left(\frac{P_2}{P_1}\right)$$

In terms of input and output currents or voltages, it is given by

$$1 \text{ Np} = \log_e \left(\frac{I_2}{I_1}\right) \text{ or } \log_e \left(\frac{V_2}{V_1}\right)$$

and 1 Np= 8.686 dB.

1.10 COMPLEX NUMBERS

By definition, a complex number is an ordered pair represented by

$$A \equiv (x, y)$$

where x is the real part and y is the imaginary part of A. Hence it is also represented as $A = x + jy$.

Here, $j \equiv \sqrt{-1} \equiv (0, 1)$ is called imaginary unit.

Properties of Complex Numbers

1. Two complex numbers are equal if their real parts are equal and their imaginary parts are also equal.
2. If $A_1 = x_1 + jy_1$, $A_2 = x_2 + jy_2$ the sum of two complex numbers A_1 and A_2 is given by

   $$A = A_1 + A_2 = (x_1 + x_2) + j\,(y_1 + y_2)$$

3. Subtraction of A_2 from A_1 is given by

   $$B = A_1 - A_2 = (x_1 - x_2) + j\,(y_1 - y_2)$$

4. The product of A_1 and A_2 is

$$C = A_1 A_2 = (x_1 + jy_1)(x_2 + jy_2)$$
$$= x_1 x_2 + jx_1 y_2 + jx_2 y_1 - y_1 y_2$$
$$= (x_1 x_2 - y_1 y_2) + j(x_1 y_2 + x_2 y_1)$$

5. The division of A_1 by A_2 is given by

$$D = \frac{A_1}{A_2} = \frac{x_1 + jy_1}{x_2 + jy_2}$$

Multiplying numerator and denominator by $(x_2 - jy_2)$, we get

$$D = \frac{A_1}{A_2} = \frac{(x_1 + jy_1)}{(x_2 + jy_2)} \frac{(x_2 - jy_2)}{(x_2 - jy_2)}$$
$$= \frac{x_1 x_2 - jx_1 y_2 + jx_2 y_1 + y_1 y_2}{(x_2^2 + y_2^2)}$$

So
$$D = \frac{(x_1 x_2 + y_1 y_2) + j(x_2 y_1 - x_1 y_2)}{(x_2^2 + y_2^2)}$$

1.11 LOGARITHMIC SERIES AND IDENTITIES

$$\log(1+x) = x - \frac{x^2}{2} + \frac{x^3}{3} - \frac{x^4}{4} + \ldots\ldots$$

This series converges for $-1 < x < 1$.

It diverges for $x = -1$

$$\log(1+x) = -\left(x - \frac{x^2}{2} + \frac{x^3}{3} - \frac{x^4}{4} + \ldots\ldots\right)$$

$$\log_a(xy) = \log_a x + \log_a y$$

$$\log_a\left(\frac{x}{y}\right) = \log_a x - \log_a y$$

$$\log_a\left(\frac{1}{y}\right) = -\log_a y$$

$$\log_a(x^n) = n \log_a x$$

$$\log_a(x^{1/n}) = \frac{1}{n} \log_a x$$

$$\log_a x = \log_b x . \log_a b$$

$$= \log_b x / \log_b a$$

$$\log_e x = \log_{10} x . \log_e 10 = 2.30 \log_{10} x$$

$$\log_{10} x = \log_e x . \log_{10} e = 0.434 \log_e x$$

1.12 QUADRATIC EQUATIONS

A quadratic equation is represented by $ax^2 + bx + c = 0$ where a, b and c are constants. Its roots are given by

$$x_1 = \frac{-b+\sqrt{b^2-4ac}}{2a}$$

and $$x_2 = \frac{-b-\sqrt{b^2-4ac}}{2a}$$

The roots may be real or imaginary and may be equal or unequal depending on whether the quantity $(b^2 - 4ac)$ is positive, zero or negative.

1.13 CUBIC EQUATIONS

It is given by

$$f(x) = x^3 + ax^2 + bx + c = 0$$

By substituting $\left(y-\frac{a}{3}\right)$ for x, $f(x)$ becomes

$$y^3 + dy + e = 0$$

Here, $$d = \frac{1}{3}(3b-a^2) \text{ and}$$

$$e = \frac{1}{27}(2a^2-9ab+27c)$$

If $$A = \left(-\frac{e}{2}+\sqrt{\frac{e^2}{4}+\frac{d^3}{27}}\right)^{\frac{1}{3}} \text{ and}$$

$$B = \left(-\frac{e}{2} - \sqrt{\frac{e^2}{4} + \frac{d^3}{27}}\right)^{\frac{1}{3}}, \text{ then}$$

the three values of y are

$$y = A + B$$

$$y = -\frac{A+B}{2} + \frac{(A-B)\,(-3)^{\frac{1}{2}}}{2} \text{ and}$$

$$y = -\frac{A+B}{2} - \frac{(A-B)\,(-3)^{\frac{1}{2}}}{2}$$

If $\left(\frac{e^2}{4} + \frac{d^2}{27}\right) < 0$, the trigonometric solution is found by evaluating the cube roots of complex quantities. Then the angle ϕ is evaluated from

$$\cos\phi = -\frac{e}{2} \Big/ \left(-\frac{d^3}{27}\right)^{\frac{1}{2}}$$

The values of y will be

$$y = 2\left(-\frac{d}{3}\right)^{\frac{1}{2}} \cos\frac{\phi}{3}$$

$$y = 2\left(-\frac{d}{3}\right)^{\frac{1}{2}} \cos\left(\frac{\phi}{3} + 120^{\circ}\right) \text{ and}$$

$$y = 2\left(-\frac{d}{3}\right)^{\frac{1}{2}} \cos\left(\frac{\phi}{3} + 240^{\circ}\right).$$

1.14 DETERMINANTS

A quantity $\begin{vmatrix} a_1 & a_2 \\ b_1 & b_2 \end{vmatrix}$ is defined as a determinant.

It is said to be a second order determinant. Its value is given by

$$\begin{vmatrix} a_1 & a_2 \\ b_1 & b_2 \end{vmatrix} = a_1 b_2 - a_2 b_1$$

Example The value of a determinant

$$\begin{vmatrix} 1 & 2 \\ 3 & 5 \end{vmatrix} \text{ is } 1\times 5 - 2\times 3 = 5 - 6 = -1$$

Similarly, $\begin{vmatrix} a_1 & a_2 & a_3 \\ b_1 & b_2 & b_3 \\ c_1 & c_2 & c_3 \end{vmatrix}$ is called a determinant of the third order.

Example The value of a third order determinant $\begin{vmatrix} 1 & -2 & -3 \\ -4 & 5 & -6 \\ -7 & 8 & -9 \end{vmatrix}$

$$= 1\times\begin{vmatrix} 5 & -6 \\ 8 & -9 \end{vmatrix} - (-2)\times\begin{vmatrix} -4 & -6 \\ -7 & -9 \end{vmatrix} + (-3)\times\begin{vmatrix} -4 & 5 \\ -7 & 8 \end{vmatrix}$$

$$= 3 - 12 - 9 = -18$$

Application of Determinants

They are widely used in solving simultaneous equations and are also useful in developing the theory of matrices.

Minor of a Determinant

In a determinant $\begin{vmatrix} a_1 & a_2 & a_3 \\ b_1 & b_2 & b_3 \\ c_1 & c_2 & c_3 \end{vmatrix}$, the minor of b_3 is given by $\begin{vmatrix} a_1 & a_2 \\ c_1 & c_2 \end{vmatrix}$ and the minor of c_2 is given by $\begin{vmatrix} a_1 & a_3 \\ b_1 & b_3 \end{vmatrix}$

that is, the minor of an element in a determinant is a determinant obtained by deleting the row and the column which intersect at that element.

The Cofactor The cofactor of an element in a determinant is its minor with a suitable sign. The sign in the ith row and j^{th} column is given by $(-1)^{i+j}$. Then the cofactor, B_3 of b_3 is $(-1)^{2+3}$ × minor of b_3.

$$B_3 = -\begin{vmatrix} a_1 & a_2 \\ c_1 & c_2 \end{vmatrix}$$

Properties of Determinants

1. Value of the determinant remains the same after changing its rows into columns and columns into rows.
2. Value of the determinant remains the same if any two rows or columns are interchanged but the sign changes.
3. If two rows or columns are identical, the value of the determinant is zero.
4. If each element of a row or column is multiplied by a factor, the determinant is multiplied by the same factor.
5. If each element of a row consists of n terms, the determinant can be expressed as the sum of n determinants.
6. When equi-multiples of the corresponding elements of one or more parallel lines are added to each element of a line, the determinant value remains the same.

1.15 MATRICES

A **matrix** is defined as a rectangular array of numbers or functions enclosed in brackets. These numbers or functions are known as the elements of the matrix.

Examples $\begin{bmatrix} 1 \\ 3 \end{bmatrix}, \begin{bmatrix} 2 & 1 & 4 \\ 3 & 0 & 1 \end{bmatrix}, [a\ b\ c], \begin{bmatrix} a & b \\ c & d \end{bmatrix}, \begin{bmatrix} e^{-x} & x \\ x^2 & e^x \end{bmatrix}$

$\begin{bmatrix} 1 \\ 3 \end{bmatrix}$ is a matrix containing two rows and one column.

Similarly, $\begin{bmatrix} 2 & 1 & 4 \\ 3 & 0 & 1 \end{bmatrix}$ is a matrix containing two rows and three columns.

Here, a row means a horizontal line and a column means a vertical line.

Applications of Matrices

Matrices are widely used to solve linear systems of equations. They often appear as models of different problems. For example, in electrical circuits and numerical methods they are used for solving differential equations.

Types of Matrices

1. Row Matrix—Example [1 2 3].

2. Column Matrix—Example $\begin{bmatrix} 1 \\ 2 \\ 3 \end{bmatrix}$

3. Square Matrix—Example $\begin{bmatrix} 1 & 2 \\ 3 & 4 \end{bmatrix}$

4. Diagonal Matrix—Example $\begin{bmatrix} 1 & 0 & 0 \\ 0 & -2 & 0 \\ 0 & 0 & 3 \end{bmatrix}$ and $\begin{bmatrix} 2 & 0 & 0 \\ 0 & 1 & 0 \\ 0 & 0 & 3 \end{bmatrix}$

5. Unit Matrix—Example $\begin{bmatrix} 1 & 0 & 0 \\ 0 & 1 & 0 \\ 0 & 0 & 1 \end{bmatrix}$

6. Null Matrix—Example $\begin{bmatrix} 0 & 0 & 0 \\ 0 & 0 & 0 \end{bmatrix}$

7. Symmetric Matrix—Example $\begin{bmatrix} 1 & 2 & 3 \\ 2 & 1 & 6 \\ 3 & 6 & 7 \end{bmatrix}$

8. Skew Symmetric Matrix—Example $\begin{bmatrix} 0 & 1 & -2 \\ -1 & 0 & 3 \\ 2 & -3 & 0 \end{bmatrix}$

9. Upper Triangular Matrix—Example $\begin{bmatrix} 1 & 3 & 4 \\ 0 & 2 & 5 \\ 0 & 0 & 6 \end{bmatrix}$

10. Lower Triangular Matrix—Example $\begin{bmatrix} 1 & 0 & 0 \\ 3 & 5 & 0 \\ 4 & 6 & 7 \end{bmatrix}$

Properties of Matrices

1. Two matrices A and B are equal if and, only if, they are of the same order and each element of A is equal to the corresponding element of B.
2. The sum of A and B exists only when they are of the same order.

3. The difference of A and B exists only when they are of the same order.

$$\text{If } A=\begin{bmatrix} a_1 & b_1 \\ a_2 & b_2 \\ a_3 & b_3 \end{bmatrix}, B=\begin{bmatrix} c_1 & d_1 \\ c_2 & d_2 \\ c_3 & d_3 \end{bmatrix}$$

$$A+B=\begin{bmatrix} a_1+c_1 & b_1+d_1 \\ a_2+c_2 & b_2+d_2 \\ a_3+c_3 & b_3+d_3 \end{bmatrix}$$

$$A-B=\begin{bmatrix} a_1-c_1 & b_1-d_1 \\ a_2-c_2 & b_2-d_2 \\ a_3-c_3 & b_3-d_3 \end{bmatrix}$$

4. Addition of matrices is commutative, that is, $A + B = B + A$.
5. Addition and subtraction of matrices are associative, that is,

$$(A + B) - C = A + (B - C)$$
$$= B + (A - C)$$

6. The multiplication of a matrix, A by a scalar, K gives a matrix each element of which is K times the corresponding elements of A. The distributive law is valid for such a product,

   that is, $$K(A + B) = KA + KB$$
7. All the laws of ordinary algebra hold good for the addition or subtraction of matrices and their multiplication by scalars.
8. Two matrices can be multiplied only when the number of columns in the first is equal to the number of rows in the second. Such matrices are said to be conformable.
9. $AB \neq BA$

   Sometimes if AB exists, BA may not exist at all.
10. $IA = AI = A$ if A is a square matrix which has the same order as that of I.
11. $OA = AO = 0$, where O is a null matrix.
12. If $AB = 0$, it does not mean that A or B is a null matrix.
13. Multiplication of matrices is associative, that is, $(AB)C = A(BC)$ provided A, B are conformable for the product AB and B, C are conformable for the product BC.
14. Multiplication of matrices is distributive, that is, $A(B + C) = AB + AC$ provided A, B are conformable for AB and A, C are conformable for AC.
15. If A is a square matrix, then $A \, . \, A = A^2$.

16. The matrix obtained from a given matrix A by interchanging rows and columns is called Transpose of A denoted by A'.
17. For a symmetric matrix, $A' = -A$
18. $(AB)' = B'.A'$
19. Adjoint of a matrix A is the transposed matrix of cofactors of A.
20. If A is any matrix, then a matrix, B if it exists such that $AB = BA = I$, is called the Inverse of A. Both the matrix and its inverse must be non-singular. Inverse of a matrix, A is denoted by A^{-1} so that $A \cdot A^{-1} = A^{-1}. A = I$.

   Also, $$A^{-1} = \frac{Adj\,A}{|A|}$$
21. Inverse of a matrix is unique.
22. $(AB)^1 = B^{-1} A^{-1}$
23. A matrix is said to be of rank r when it has at least one non-zero minor of order r and every minor of order higher than r vanishes. It means that the rank of a matrix is the largest order of any non-vanishing minor of the matrix.

1.16 FACTORIAL

Factorial of n is defined as the product of the first n natural numbers.

Factorial of n is represented by $n!$ or $\angle n$

$$\text{Factorial of } n = n! = 1 \times 2 \times 3 \times \ldots\ldots\ldots \times n$$

Using Stirling's approximation, $n!$ is evaluated from

$$n! = e^{-n}\, n^n\, (2\pi n)^{\frac{1}{2}}$$

$$0! = 1$$

1.17 PERMUTATIONS

Permutations are defined as different possible arrangements which can be formed with a given number of items, taking some or all of them at a time.

The number of permutations of n different things taken r at a time is given by np_r and $np_r = \dfrac{n!}{(n-r)!}$

If p, q, r types of elements exist among n elements, then the number of permutations are given by

$$n!/(p! + q! + r! + \ldots\ldots\ldots), n = p + q + r$$

The number of permutations of n different things taken all at a time is np_n.

1.18 COMBINATIONS

Combinations are defined as the number of possible groups which can be formed by taking some or all the number of things at a time irrespective of the order.

The number of combinations or groups of n different elements taken r at a time is given by nc_r

where $$nC_r = \frac{n!}{r!\,(n-r)!}$$

$$(n+1)\,C_r = nC_r + nC_{r-1}$$

$$nC_1 + nC_2 + nC_3 + \ldots + nC_n = 2^n - 1$$

1.19 BASIC SERIES

Binomial Series These are given by

$$(a \pm b)^n = a^n \pm na^{n-1}b + \frac{n(n-1)}{2!}a^{n-2}b^2 \pm \frac{n(n-1)(n-2)}{3!}a^{n-3}b^3$$

$$+ (\pm 1)^r \frac{n\,(n-1)\ldots\ldots\ldots(n-r+1)}{r!}a^{n-r}b^r$$

Here the last term shown is the $(r + 1)$th term.

If n is a positive integer, the series is finite and the last term is b^n. If n is less than unity, the series is infinite and converges only when $b < a$. For any value of n, and convergent, when x^2 is less than unity, we have

$$(1 \pm x)^n = 1 \pm nx + \frac{n\,(n-1)}{2!}x^2 \pm \frac{n\,(n-1)\,(n-2)}{3!}x^3 +$$

$$\frac{n\,(n-1)\,(n-2)\,(n-3)}{4!}x^4 \pm \ldots$$

$$(1 \pm x)^{-n} = 1 \mp nx + \frac{n\,(n-1)}{2!}x^2 \mp \frac{n\,(n-1)\,(n+2)}{3!}x^3 + \frac{n\,(n-1)\,(n+2)\,(n-3)}{4!}x^4 + \ldots$$

$$(1 \pm x)^{-1} = 1 \mp x + x^2 \mp x^3 + x^4 \mp x^5 \ldots$$

$$(1 \pm x)^{\frac{1}{2}} = 1 \pm \frac{1}{2}x - \frac{1}{2.4}x^2 \pm \frac{1.3}{2.4.6}x^3 - \ldots$$

$$(1 \pm x)^{\frac{1}{2}} = 1 \mp \frac{1}{2}x + \frac{1.3}{2.4}x^2 \pm \frac{1.3.5}{2.4.6}x^3 + \ldots$$

$$\left.\begin{aligned}(1+x)^{-1} &\approx 1-x\\ (1-x)^{-1} &\approx 1+x\end{aligned}\right\} \text{ if } |x| << 1$$

1.20 EXPONENTIAL SERIES

$$\sum (n^2) = 1^2 + 2^2 + \ldots\ldots\ldots\ldots + n^2 = \frac{n\,(n+1)\,(2n+1)}{6}$$

$$\sum (n^3) = 1^3 + 2^3 + \ldots\ldots\ldots\ldots + n^3 = \frac{n^2(n+1)^2}{4}$$

$$e^x = 1 + \frac{x}{1!} + \frac{x^2}{2!} + \frac{x^2}{3!} + \ldots$$

This series converges for all values of x.

For $x = 1$,

$$e^x = e^1 = 1 + \frac{1}{1!} + \frac{1}{2!} + \frac{1}{3!} + \ldots\ldots\ldots\ldots = 2.718$$

$$e^x\, e^y = e^{x+y}, \quad e^x / e^y = e^{x-y}, \quad (e^x)^y = e^{xy}$$

1.21 SINE AND COSINE SERIES

If x is a variable,

$$\sin x = x - \frac{x^3}{3!} + \frac{x^5}{5!} - \ldots\ldots\ldots\ldots\ldots\infty$$

$$\cos x = 1 - \frac{x^2}{2!} + \frac{x^4}{4!} - \ldots\ldots\ldots\ldots\infty$$

1.22 SINH AND COSH SERIES

$$\sinh x = x + \frac{x^3}{3!} + \frac{x^5}{5!} + \ldots\ldots\ldots\ldots\infty$$

$$\cosh x = 1 + \frac{x^2}{2!} + \frac{x^4}{4!} + \ldots\ldots\ldots\ldots\infty$$

1.23 HYPERBOLIC FUNCTIONS

Hyperbolic sine of $x = \sinh x = \dfrac{(e^x - e^{-x})}{2}$

Hyperbolic cosine of $x = \cosh x = \dfrac{(e^x + e^{-x})}{2}$

Hyperbolic tan of $x = \dfrac{(e^x - e^{-x})}{(e^x + e^{-x})} = \tanh x$

Hyperbolic cot of $x = \dfrac{1}{\tanh x} = \dfrac{(e^x + e^{-x})}{(e^x - e^{-x})} = \coth x$

Similarly, $\operatorname{sech} x = \dfrac{1}{\cosh x} = \dfrac{2}{(e^x + e^{-x})}$

$$\operatorname{cosech} x = \frac{1}{\sinh x} = \frac{2}{(e^x - e^{-x})}$$

It may be noted that

$$\sinh 0 = 0,\ \cosh 0 = 0,\ \text{and}\ \tanh 0 = 0$$

$$\cosh x + \sinh x = e^x$$

$$\cosh x + \sinh x = e^{-x}$$

$$\cosh^2 x - \sinh^2 x = 1$$

$$\sinh^2 x = \frac{1}{2}(\cosh 2x - 1)$$

$$\cosh^2 x = \frac{1}{2}(\cosh 2x + 1)$$

$$\sinh(x \pm y) = \sinh x \cosh y \pm \cosh x \sinh y$$

$$\cosh(x \pm y) = \cosh x \cosh y \pm \sinh x \sinh y$$

$$\tanh(x \pm y) = \frac{\tanh x \pm \tanh y}{1 \pm \tanh x \tanh y}$$

1.24 SINE, COSINE, TAN AND COT FUNCTIONS

For all values of θ,

$$\sin\theta = \frac{e^{j\theta} - e^{-j\theta}}{2j}$$

$$\cos\theta = \frac{(e^{j\theta} + e^{-j\theta})}{2}$$

If $\qquad \theta = jx,$

$$\sin jx = \frac{(e^{-x} - e^{x})}{2j} = -\frac{(e^{x} - e^{-x})}{2j}$$

$$= j^2\frac{(e^{x} - e^{-x})}{2j} = j\frac{(e^{x} - e^{-x})}{2}$$

$$= j\sinh x$$

and $\qquad \cos jx = \dfrac{e^{-x} + e^{x}}{2} = \cosh x$

Similarly,

$$\tan jx = j\tanh x$$

$$\sinh jx = j\sin x$$

$$\cosh jx = \cos x$$

$$\tanh jx = j\tan x$$

$$\sin(-x) = -\sin x$$

$$\cos(-x) = \cos x$$

$$1 \text{ degree} = 0.01745 \text{ radian}$$

$$1 \text{ radian} = 57.29577 \text{ degrees}$$

$$\sin^2 x + \cos^2 x = 1$$

$$\sin(x + y) = \sin x \cos y + \cos x \sin y$$

$$\sin(x - y) = \sin x \cos y - \cos x \sin y$$

$$\cos(x + y) = \cos x \cos y - \sin x \sin y$$

$$\cos(x - y) = \cos x \cos y + \sin x \sin y$$

$$\sin 2x = 2 \sin x \cos x$$

$$\cos 2x = \cos^2 x - \sin^2 x$$

$$\sin x = \cos\left(x - \frac{\pi}{2}\right) = \cos\left(\frac{\pi}{2} - x\right)$$

$$\cos x = \sin\left(x + \frac{\pi}{2}\right) = \sin\left(\frac{\pi}{2} - x\right)$$

$$\sin(\pi - x) = \sin x$$

$$\cos(\pi - x) = -\cos x$$

$$\cos^2 x = \frac{1}{2}(1 + \cos 2x)$$

$$\sin^2 x = \frac{1}{2}(1 - \cos 2x)$$

$$\sin x \sin y = \frac{1}{2}[-\cos(x + y) + \cos(x - y)]$$

$$\cos x \cos y = \frac{1}{2}[\cos(x + y) + \cos(x - y)]$$

$$\sin x \cos y = \frac{1}{2}[\sin(x + y) + \sin(x - y)]$$

$$\sin a + \sin b = 2 \sin\frac{a + b}{2} \cos\frac{a - b}{2}$$

$$\cos a + \cos b = 2 \cos\frac{a + b}{2} \cos\frac{a - b}{2}$$

$$\cos a - \cos b = -2 \sin \frac{a+b}{2} \sin \frac{a-b}{2}$$

$$A \cos x + B \sin x = \sqrt{A^2 + B^2} \cos(x \pm \delta)$$

$$\tan \delta = \frac{\sin \delta}{\cos \delta} = \frac{B}{A}$$

$$A \cos x + B \sin x = \sqrt{A^2 + B^2} \sin(x \pm \delta)$$

$$\tan \delta = \frac{\sin \delta}{\cos \delta} = \pm \frac{A}{B}$$

$$= \frac{\sin x}{\cos x}$$

$$\cot x = \frac{\cos x}{\sin x}$$

$$\sec x = \frac{1}{\cos x}$$

$$\operatorname{cosec} x = \frac{1}{\sin x}$$

$$\tan(x+y) = \frac{\tan x + \tan y}{1 - \tan x \tan y}$$

$$\tan(x-y) = \frac{\tan x - \tan y}{1 + \tan x \tan y}$$

1.25 SOME SPECIAL FUNCTIONS

Gamma function

$\Gamma(\alpha)$ is defined as

$$\Gamma(\alpha) \equiv \int_0^\infty e^{-t} t^{\alpha-1} dt$$

This is true if $\alpha > 0$

$$\Gamma\left(\frac{1}{2}\right)=\sqrt{\pi}.$$

Beta function

$$B(x,y) \equiv \int_0^1 t^{x-1}(1-t)^{y-1}\,dt, \qquad (x>0,\ y>0)$$

$$= \frac{\Gamma(x)\,\Gamma(y)}{\Gamma(x+y)}$$

Error function

$$\operatorname{erf} x = \frac{2}{\sqrt{\pi}} \int_0^x e^{-t^2}\,dt$$

$$\operatorname{erf}(\infty) = 1.$$

Bessel function

Bessel functions are a set of solutions for differential equations of a certain type. These are useful to obtain propagation characteristics of electromagnetic waves in circular waveguides.

Considering circular waveguides, solutions of the axial component is written conveniently in terms of Bessel function. For transverse magnetic wave, the solution is

$$E_{z,\,nm} = J_n\,(K_c r)\,(A \cos n\phi + B \sin m\phi) \text{ and}$$

for transverse electric wave

$$H_{z,\,nm} = J_n\,(K_c\,r)\,(A^1 \cos n\phi + B^1 \sin m\phi)$$

where $J_n\,(K_c\,r)$ = Bessel function of the first kind

r = radius of the guide

K_c = cut-off wave number

$A, B, A^1, B^1,$ = constants

The solutions for the Bessel function are obtained for certain values of K_c. These values of K_c are known as eigen values. If K_c is to produce solution of the Bessel function, $(K_c\,r)$ must be the roots of the Bessel function. Then

$$J_n(K_c\,r) = 0$$

The propagation parametres for nm^{th} mode of TM waves are

Phase constant,

$$\beta_{nm} = (K^2 - K_{c,nm}^2)^{1/2}$$

where $$K_{c,nm} = \frac{p_{nm}}{r}$$

Also, $$\beta_{nm} = \left[K^2 - \left(\frac{p_{nm}}{r}\right)^2\right]^{1/2}$$

where p_{nm} = the roots of the Bessel function

K= free space wave number $\frac{2\pi}{\lambda}$

The cut-off wavelength for the TM wave,

$$\lambda_{c,nm} = \frac{2\pi}{K_{c,nm}} = \frac{2\pi r}{p_{nm}}$$

The roots of the Bessel function for TM mode are shown in Table 1.2.

Table 1.2 Roots of Bessel Function (TM)

Order n	*First order* p_{n1}	*Second order* p_{n2}	*Third order* p_{n3}
0	2.405	5.520	8.654
1	3.832	7.016	10.174
2	5.135	8.417	11.620

The roots of the Bessel function for TE mode are shown in Table 1.3.

Table 1.3 Roots of Bessel Function (TE)

Order n	*First order* p'_{n_1}	*Second order* p'_{n_2}	*Third order* p'_{n_3}
0	3.832	7.016	10.174
1	1.841	5.331	8.536
2	3.054	6.706	9.970

The propagation parametres for TE_{nm} mode are

$$\beta_{nm} = \left[K^2 - K'^2_{c,nm}\right]^{1/2}$$

where
$$K'^2_{c,nm} = \frac{p'_{nm}}{r}$$

So,
$$\beta_{nm} = \left[K^2 - \left(\frac{p'_{nm}}{r}\right)^2\right]^{1/2}$$

$$\lambda_{c,nm} = \frac{2\pi r}{p'_{nm}}$$

Guide wavelength,
$$\lambda_g = \frac{\lambda}{\left[1 - \left(\frac{\lambda}{\lambda_{c,nm}}\right)^2\right]^{1/2}}$$

The variation of the first three orders of the Bessel function and their roots are shown in Fig. 1.6.

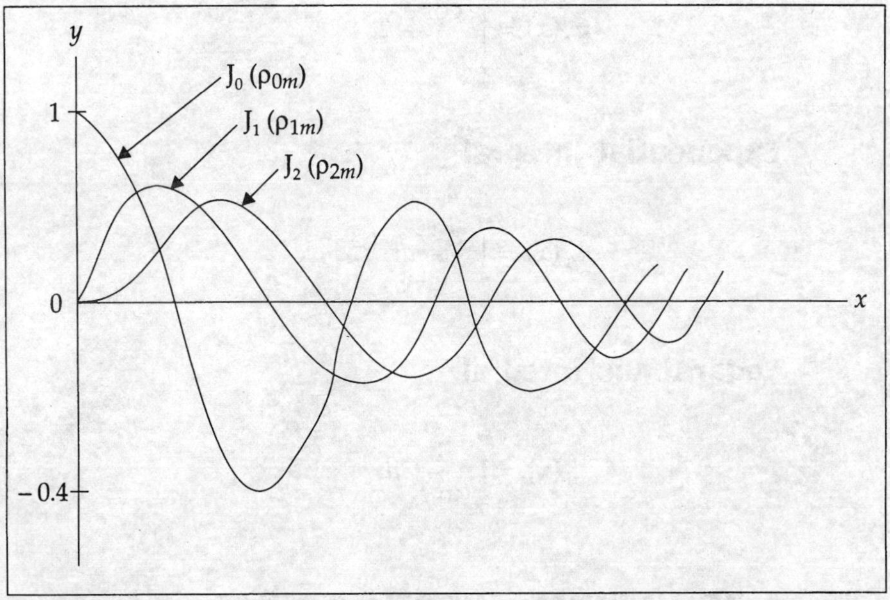

Fig. 1.6 Bessel function

Fresnel integral

$$c(x) = \int_0^x \cos(t^2)\, dt$$

$$s(x) = \int_0^x \sin(t^2)\, dt$$

$$C(\infty) = \sqrt{\pi/8}$$

$$S(\infty) = \sqrt{\pi/8}$$

Sine integral

$$s_i(x) = \int_0^x \frac{\sin t}{t}\, dt$$

$$s_i(\infty) = \pi/2$$

Cosine integral

$$C_i(x) = \int_x^\infty \frac{\cos t}{t}\, dt$$

Exponential integral

$$E_i(x) = \int_x^\infty \frac{e^{-t}}{t}\, dt$$

Logarithmic integral

$$l_i(x) = \int_0^\infty \frac{1}{\ln t}\, dt.$$

1.26 PARTIAL DERIVATIVE

Assume $z = f(x, y)$ is a real function of two independent real variables, x and y. If x is kept constant, say $x = x_1$ and y is a variable, then $f(x_1, y)$ depends on y only. If the derivative of $f(x, y_1)$ with y for a value $y = y_1$ exists, then the value of this derivative is called partial derivative of $f(x, y)$ with respect to y at the point (x_1, y_1). It is represented by

$$\frac{\partial f}{\partial x}\Big/_{(x_1,y_1)} \text{ or } \frac{\partial z}{\partial x}\Big/_{(x_1,y_1)}$$

1.27 SOME DIFFERENTIATION FORMULAE

Assume u, v to be functions of x and let a, n be the constants. Then,

$$\frac{d(a)}{dx} = 0$$

$$\frac{d(u+v)}{dx} = \frac{du}{dx} + \frac{dv}{dx}$$

$$\frac{d(au)}{dx} = a\frac{du}{dx}$$

$$\frac{d(u.v)}{dx} = u\frac{dv}{dx} + v\frac{du}{dx}$$

$$\frac{d}{dx}\left(\frac{u}{v}\right) = \frac{v\frac{du}{dx} - u\frac{dv}{dx}}{v^2}$$

$$\frac{du^n}{dx} = nu^{n-1}\frac{du}{dx}$$

$$\frac{d}{dx}\log_e(u) = \frac{1}{u}\frac{du}{dx}$$

$$\frac{d}{dx}(x^n) = n.x^{n-1}$$

$$\frac{d}{dx}(e^x) = e^x$$

$$\frac{d}{dx}(a^x) = a^x \ln(a)$$

$$\frac{d}{dx}(\sin x) = \cos x$$

$$\frac{d}{dx}(\cos x) = -\sin x$$

$$\frac{d}{dx}(\tan x) = \sec^2 x$$

$$\frac{d}{dx}(\cot x) = -\operatorname{cosec}^2 x$$

$$\frac{d}{dx}(\sinh x) = \cosh x$$

$$\frac{d}{dx}(\cosh x) = \sinh x$$

$$\frac{d}{dx}(\ln (x)) = \frac{1}{x}$$

$$\frac{d}{dx}(\log_a x) = \frac{\log_a e}{x}$$

$$\frac{d}{dx}(\sin^{-1} x) = \frac{1}{\sqrt{1-x^2}}$$

$$\frac{d}{dx}(\cos^{-1} x) = -\frac{1}{\sqrt{1-x^2}}$$

$$\frac{d}{dx}(\tan^{-1} x) = \frac{1}{1+x^2}$$

$$\frac{d}{dx}(\cot^{-1} x) = -\frac{1}{1+x^2}$$

$$\frac{d}{dx}(\log_e (u)) = \frac{1}{u}\frac{du}{dx}$$

$$\frac{d}{dx}(\sin u) = \cos u \frac{du}{dx}$$

$$\frac{d}{dx}(\cos u) \; -\sin u \frac{du}{dx}$$

$$\frac{d}{dx}(\tan u) = \sec^2 u \frac{du}{dx}$$

$$\frac{d}{dx}(\cot u) = -\csc^2 u \frac{du}{dx}$$

$$\frac{d}{dx}(\sec u) = \sec u \tan u \frac{du}{dx}$$

$$\frac{d}{dx}(\csc u) = -\csc u \cot u \frac{du}{dx}$$

$$\frac{d}{dx}(\sinh^{-1} u) = \frac{1}{\sqrt{u^2+1}}\frac{du}{dx}$$

$$\frac{d}{dx}(\cosh^{-1} u) = \frac{1}{\sqrt{u^2-1}}\frac{du}{dx}$$

$$\frac{d}{dx}(\tanh^{-1} u) = \frac{1}{1-u^2}\frac{du}{dx}$$

$$\frac{d}{dx}(\coth^{-1} u) = \frac{1}{1-u^2}\frac{du}{dx}$$

$$\frac{d}{dx}(\text{csch}^{-1} u) = -\frac{1}{u\sqrt{1+u^2}}\frac{du}{dx}$$

$$\frac{d}{dx}(\text{sech}^{-1} u) = -\frac{1}{u\sqrt{1-u^2}}\frac{du}{dx}$$

1.28 SOME USEFUL INTEGRATION FORMULAE

$$\int a\,dx = ax, \text{ } a \text{ being a constant}$$

$$\int (u+\upsilon)\,dx = \int u\,dx + \int \upsilon\,dx, \text{ } u \text{ and } v \text{ are variables}$$

$$\int u\,d\upsilon = u\upsilon - \int \upsilon\,du$$

$$\int \log_e x\,dx = x\log_e x - x$$

$$\int x^n\,dx = \frac{x^{n+1}}{n+1} \qquad (n \neq -1)$$

$$\int \frac{dx}{x} = \log_e x + c \qquad (c = \text{a constant})$$

$$\int e^{ax}dx = \frac{1}{a}\,e^{ax}$$

$$\int a^x\,dx = \frac{a^x}{\log_e a}$$

$$\int a^x \log_e a \, dx = a^x$$

$$\int \sin ax \, dx = -\frac{1}{a} \cos ax$$

$$\int \cos ax \, dx = \frac{1}{a} \sin ax$$

$$\int \tan ax \, dx = -\frac{1}{a} \log_e \cos ax$$

$$\int \cot ax \, dx = \frac{1}{a} \log_e \sin ax$$

$$\int \sec ax \, dx = \frac{1}{a} \log_e (\sec ax + \tan ax)$$

$$\int \csc ax \, dx = \frac{1}{a} \log_e (\csc ax - \cot ax)$$

$$\int \frac{dx}{x^2 + a^2} = \frac{1}{a} \tan^{-1} \frac{x}{a}$$

$$\int \frac{dx}{\sqrt{a^2 - x^2}} = \sin^{-1} \frac{x}{a}$$

$$\int \frac{dx}{\sqrt{x^2 + a^2}} = \sinh^{-1} \frac{x}{a}$$

$$\int \frac{dx}{x^2 - a^2} = \cosh^{-1} \frac{x}{a}$$

$$\int \sin^2 x \, dx = \frac{1}{2} x - \frac{1}{4} \sin 2x$$

$$\int \cos^2 x \, dx = \frac{1}{2} x + \frac{1}{4} \sin 2x$$

$$\int \tan^2 x \, dx = \tan x - x$$

$$\int \cot^2 x \, dx = -\cot x - x$$

$$\int \ln x \, dx = x \ln x - x$$

$$\int e^{ax} \sin bx \, dx = \frac{e^{ax}}{a^2 + b^2} (a \sin bx - b \cos bx)$$

$$\int e^{ax}\cos bx\,dx = \frac{e^{ax}}{a^2+b^2}(a\cos bx + b\sin bx)$$

$$\int \cos^3 ax\,dx = \frac{1}{a}\sin ax - \frac{1}{3a}\sin^3 ax$$

$$\int \sin^3 ax\,dx = -\frac{1}{a}\cos ax + \frac{1}{3a}\cos^3 ax$$

1.29 RADIAN AND STERADIAN

The **radian** is the measure of a plane angle. One radian is defined as a plane angle with its vertex at the centre of a circle of radius r which is subtended by an arc whose length is equal to r.

This is shown in Fig. 1.7. It is well known that the circumference of a circle is $2\pi r$ and there are 2π radians in a full circle.

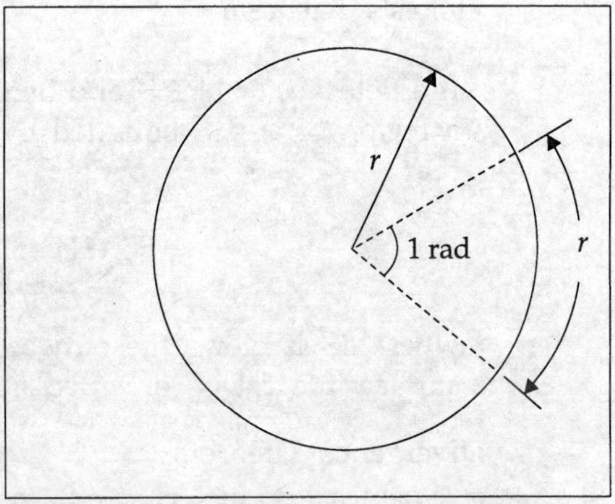

Fig. 1.7 Radian

Steradian is the measure of a solid angle. It is defined as the solid angle with its vertex at the centre for a sphere of radius r which is subtended by a spherical surface area equal to the area of a square, whose side length is r.

The differential area ds on the surface of a sphere of radius r is

$$ds = r^2 \sin\theta\, d\theta\, d\phi \ (m^2)$$

The element of solid angle $d\Omega$ of the sphere is given by

$$d\Omega = \frac{ds}{r^2} \quad \text{(Str)}$$

This is illustrated in Fig. 1.8.

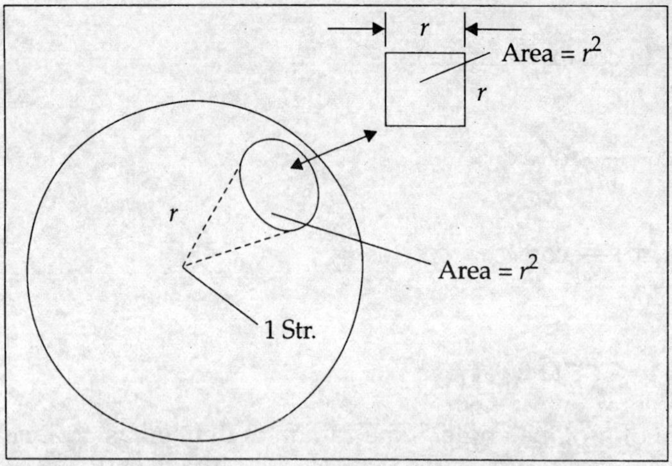

Fig. 1.8 Steradian

1.30 INTEGRAL THEOREMS

Stokes's Theorem

If a field vector like **H** and its first derivatives are continuous at all points in a region of area S bounded by a closed curve, then Stokes's theorem states that

$$\int_S (\nabla \times \mathbf{H}).d\mathbf{S} = \oint_L \mathbf{H}.d\mathbf{L}$$

where $d\mathbf{S}$ is along the outward normal to the surface. By Stokes theorem, surface integral is converted into line integral and vice versa.

Divergence Theorem

If a field vector like **D** and its first derivatives are continuous at all points in a region of volume V bounded by a closed elementary surface $d\mathbf{S}$, then Divergence theorem states that

$$\oint_s \mathbf{D}.d\mathbf{S} = \int_\upsilon (\nabla.\mathbf{D})\, d\upsilon$$

This means a surface integral is converted into volume integral and vice versa by Divergence theorem.

POINTS/FORMULAE TO REMEMBER

- Unit vector of a vector **A** is $\mathbf{a} = \frac{\mathbf{A}}{A}$
- The coordinate axes are perpendicular to each other.
- A point in Cartesian coordinate system is obtained by the intersection of three planes.
- A point in cylindrical coordinate system is obtained by the intersection of two planes and one cylindrical surface.
- A point in spherical coordinate system is obtained by the intersection of a plane, a spherical surface and a cone.
- Del, ∇ is a vector differential operator.
- The unit of ∇ is 1/m.
- Gradient of a scalar is a vector.
- Divergence of a scalar is a vector.
- Curl of a vector is a vector.
- 1 dB is defined as $10 \log_{10} \frac{P_1}{P_2}$
- 1 Neper is defined as $\log_e \left(\frac{V_1}{V_2} \right)$
- 1 Neper = 8.686 dB.
- A determinant is represented by $\begin{vmatrix} a_1 & a_2 \\ b_1 & b_2 \end{vmatrix}$
- A matrix is defined as a rectangular array of numbers or functions enclosed in brackets.
- $\cos jx = \cosh x$
- $\sin jx = j \sinh x$
- 1 degree = 0.01745 radians
- 1 radian = 57.29 degrees
- The unit of magnetic scalar potential is ampere.
- The unit of magnetic vector potential is wb/m.

- Reflection coefficient is a complex quantity.
- Electric field, magnetic field, electric flux density, magnetic flux density, conductor current density, displacement current density, magnetic current density are vector quantities.
- The unit of Poynting vector is watts/m^2.
- Torque is a vector and has unit of Newton-metre.
- The unit of solid angle is steradian.
- Stokes' theorem gives the relation between a closed line integral and a surface integral.
- Divergence theorem gives the relation between a closed surface integral and a volume integral.
- The unit of magnetic current density is volts/m^2.
- ∇^2 is Laplacian scalar operator.

SOLVED PROBLEMS

Problem 1.1 If a vector $\mathbf{A} = \mathbf{a}_x + 2\mathbf{a}_y + 3\mathbf{a}_z$, find its magnitude and direction.

Solution $\mathbf{A} = \mathbf{a}_x + 2\mathbf{a}_y + 3\mathbf{a}_z$,

The magnitude of $\mathbf{A} = A = \sqrt{1^2+2^2+3^2} = \sqrt{1+4+9} = \sqrt{14} = 3.742$

$$\boxed{A = 3.742}$$

Its direction is given by unit vector, $\mathbf{a} = \dfrac{\mathbf{A}}{A} = \dfrac{(\mathbf{a}_x + 2\mathbf{a}_y + 3\mathbf{a}_z)}{3.742}$

$$\boxed{\mathbf{a} = 0.267\mathbf{a}_x + 0.534\mathbf{a}_y + 0.802\mathbf{a}_z}$$

Problem 1.2 If $\mathbf{A} = 2\mathbf{a}_x + 5\mathbf{a}_y + 6\mathbf{a}_z$ and $\mathbf{B} = \mathbf{a}_x - 3\mathbf{a}_y + 6\mathbf{a}_z$, find $\mathbf{A}+\mathbf{B}$ and $\mathbf{A}-\mathbf{B}$.

Solution

$$\mathbf{A} + \mathbf{B} = (A_x + B_x)\,\mathbf{a}_x + (A_y + B_y)\,\mathbf{a}_y + (A_z + B_z)\,\mathbf{a}_z$$

$$= (2+1)\,\mathbf{a}_x + (5-3)\,\mathbf{a}_y + (6+6)\,\mathbf{a}_z$$

$$\boxed{\mathbf{A}+\mathbf{B} = 3\mathbf{a}_x + 2\mathbf{a}_y + 12\mathbf{a}_z}$$

$$\mathbf{A} - \mathbf{B} = (2-1)\,\mathbf{a}_x + (5+3)\,\mathbf{a}_y + (6-6)\,\mathbf{a}_z$$

$$\boxed{\mathbf{A}-\mathbf{B} = \mathbf{a}_x + 8\mathbf{a}_y}$$

Problem 1.3 If $\mathbf{A} = \mathbf{a}_x + \mathbf{a}_y + 2\mathbf{a}_z$ and $\mathbf{B} = 2\mathbf{a}_x + \mathbf{a}_y + \mathbf{a}_z$, find **A.B**.

Solution

$$\mathbf{A.B} = A_x B_x + A_y B_y + A_z B_z$$

$$= 1.2 + 1.1 + 2.1 = 2 + 1 + 2 = 5$$

Problem 1.4 Given $\mathbf{A} = 2\mathbf{a}_x + \mathbf{a}_y + 2\mathbf{a}_z$ and $\mathbf{B} = \mathbf{a}_x + 2\mathbf{a}_y + \mathbf{a}_z$, find $\mathbf{A} \times \mathbf{B}$.

Solution

$$\mathbf{A} \times \mathbf{B} = \begin{vmatrix} \mathbf{a}_x & \mathbf{a}_y & \mathbf{a}_z \\ 2 & 1 & 2 \\ 1 & 2 & 1 \end{vmatrix}$$

$$= \mathbf{a}_x (1-4) + \mathbf{a}_y (2-2) + \mathbf{a}_z (4-1)$$

$$\boxed{\mathbf{A} \times \mathbf{B} = -3\mathbf{a}_x + 3\mathbf{a}_z}$$

Problem 1.5 For $\mathbf{A} = (1, 3, 4)$ and $\mathbf{B} = (1, 0, 2)$, find **A.B**.

Solution Here $A_x = 1, \quad A_y = 3, \quad A_z = 4$

$B_x = 1, \quad B_y = 0, \quad B_z = 2$

$$\mathbf{A.B} = A_x B_x + A_y B_y + A_z B_z$$

$$= 1 \times 1 + 3 \times 0 + 4 \times 2$$

$$= 1 + 0 + 8 = 9$$

Problem 1.6 In Cartesian coordinates, a point is described by $P(1, 2, 4)$. What are the orthogonal planes whose intersection give this point?

Solution The planes which intersect at the point $P(1, 2, 4)$ are:

$x = 1$

$y = 2$ and

$z = 4$

Problem 1.7 Represent the point $P(0.5, 0.2, 0.1)$ in terms of unit vectors along the coordinate axes.

Solution At the point, $P(0.5, 0.2, 0.1)$, the coordinates are $x = 0.5$, $y = 0.2$ and $z = 0.1$. Therefore, P is represented by a vector

$$\mathbf{P} = 0.5\mathbf{a}_x + 0.2\mathbf{a}_y + 0.1\mathbf{a}_z$$

Problem 1.8 How is a point $P(2.0\,\text{m}, \pi/4, 1.0\,\text{m})$ obtained in cylindrical coordinates?

Solution The point $P(2.0\,\text{m}, \pi/4, 1.0\,\text{m})$ is obtained by the intersection of a cylindrical surface, $\rho = 2.0\,\text{m}$ a plane, $\phi = \pi/4$ and a plane, $z = 1.0\,\text{m}$.

Problem 1.9 A point in Cartesian coordinates is given by $P(1, 2, 3)$. Express it in cylindrical coordinates.

Solution The coordinates of P in Cartesian system are

$$x = 1, y = 2, z = 3$$

The coordinates of P in cylindrical system are given by

$$\rho = \sqrt{x^2 + y^2} = \sqrt{1^2 + 2^2} = \sqrt{5} = 2.236$$

$$\phi = \tan^{-1}\frac{y}{x} = \tan^{-1}\frac{2}{1} = \tan^{-1} 2 = 63.43°$$

$$z = 3$$

In cylindrical coordinates P is (2.236, 63.43°, 3).

Problem 1.10 If a vector, **A** is $4\mathbf{a}_x + 2\mathbf{a}_y + \mathbf{a}_z$, express it in cylindrical coordinate system.

Solution $\mathbf{A} = 4\mathbf{a}_x + 2\mathbf{a}_y + \mathbf{a}_z$

We have $A_x = 4, A_y = 2, A_z = 1$

$$\phi = \tan^{-1}\frac{y}{x} = \tan^{-1}\frac{2}{4} = 26.56°$$

$$A_\rho = A_x \cos\phi + A_y \sin\phi$$

$$= 4\cos 26.56° + 2\sin 26.56°$$

that is, $A_\rho = 3.57 + 0.894 = 4.46$

$$A_\phi = -A_x \sin\phi + A_y \cos\phi$$

$$= -4 \times 0.447 + 2 \times 0.894$$

$$= -1.789 + 1.789$$

$$= 0$$

$$A_z = 1.0$$

A in cylindrical coordinates is

$$\mathbf{A} = 4.46\mathbf{a}_\rho + \mathbf{a}_z$$

Problem 1.11 How is a point P (1.0 m, 10°, 30°) obtained in spherical coordinates?

Solution The point P (1.0 m, 10°, 30°) is obtained by the intersection of a sphere, $r = 1.0$ m a cone, $\theta = 10°$ and a plane, $\phi = 30°$.

Problem 1.12 If a point in Cartesian coordinates is given by $P(1, 2, 3)$, express it in spherical coordinates.

Solution The Cartesian coordinates of P are

$$x = 1, y = 2, z = 3$$

The coordinates of P in spherical system are given by

$$r = \sqrt{x^2+y^2+z^2} = \sqrt{1^2+2^2+3^2} = \sqrt{1+4+9} = \sqrt{14} = 3.741$$

Problem 1.13 A scalar function, V is given by $V = xyz^2$. Find the gradient of V.

Solution Grad $V = \nabla V = \dfrac{\partial V}{\partial x}\mathbf{a}_x + \dfrac{\partial V}{\partial y}\mathbf{a}_y + \dfrac{\partial V}{\partial z}\mathbf{a}_z$

$$= \frac{\partial}{\partial x}(xyz^2)\,\mathbf{a}_x + \frac{\partial}{\partial y}(xyz^2)\,\mathbf{a}_y + \frac{\partial}{\partial y}(xyz^2)\,\mathbf{a}_z$$

$$\boxed{\nabla V = yz^2\mathbf{a}_x + xz^2\mathbf{a}_y + 2xyz\mathbf{a}_z}$$

Problem 1.14 If a vector, $\mathbf{B} = 4xy^2\mathbf{a}_x + 2y^3\mathbf{a}_y + xyz\mathbf{a}_z$, find the divergence of $\mathbf{B}$.

Solution Div $\mathbf{B} = \nabla.\mathbf{B}$

$$= \frac{\partial}{\partial x}B_x + \frac{\partial}{\partial y}B_y + \frac{\partial}{\partial z}B_z$$

$$= \frac{\partial}{\partial x}(4xy^2) + \frac{\partial}{\partial y}(2y^3) + \frac{\partial}{\partial z}(xyz)$$

$$= 4y^2 + 6y^2 + xy = 10y^2 + xy$$

$$\boxed{\nabla.\mathbf{B} = 10y^2 + xy}$$

Problem 1.15 Given a vector, $\mathbf{A} = 3x\mathbf{a}_x + y\mathbf{a}_y + 5z\mathbf{a}_z$, find the curl of $\mathbf{A}$.

Solution Curl $\mathbf{A} = \nabla \times \mathbf{A}$

$$= \begin{vmatrix} \mathbf{a}_x & \mathbf{a}_y & \mathbf{a}_z \\ \dfrac{\partial}{\partial x} & \dfrac{\partial}{\partial y} & \dfrac{\partial}{\partial z} \\ 3x & y & 5z \end{vmatrix}$$

$$= \mathbf{a}_x\left[\frac{\partial}{\partial y}(5z) - \frac{\partial}{\partial z}(y)\right] + \mathbf{a}_y\left[\frac{\partial}{\partial z}(3x) - \frac{\partial}{\partial x}(5z)\right] + \mathbf{a}_z\left[\frac{\partial}{\partial x}(y) - \frac{\partial}{\partial y}(3x)\right]$$

$$\boxed{\nabla \times \mathbf{A} = 0}$$

Problem 1.16 If the scalar potential is given by $V = x^2 - y^2 - z^2$ volts, find the Laplacian of V.

Solution $V = x^2 - y^2 - z^2$ volts

$$\nabla^2 V = \frac{\partial^2 V}{\partial x^2} + \frac{\partial^2 V}{\partial y^2} + \frac{\partial^2 V}{\partial z^2}$$

$$= 2 - 2 - 2 = -2$$

Problem 1.17 If power gain, A_p is 22, find its value in decibels.

Solution Power gain = 22

A_p in dB $= 10 \log_{10} 22$

$$= 13.424 \text{ dB}$$

Problem 1.18 When the ratio of output and input voltages of an amplifier is 95, find its gain in dB.

Solution A_V = voltage gain = 95

A_V in dB $= 20 \log_{10} 95$

$$= 39.55 \text{ dB}$$

Problem 1.19 If the power gain in a communication system is 16, what is its value in nepers?

Solution A_p = power gain = 16

$$A_p \text{ in Np} = \log_e \sqrt{A_p} = \log_e \sqrt{16} = 1.386 \text{ Np}$$

Problem 1.20 If the current gain in a circuit is 34, what is its value in nepers?

Solution A_I Current gain = 34

A_I in Np $= \log_e 34$

$$= 3.526$$

Problem 1.21 If a complex number, A is given by $2 + j4$, find its magnitude and phase.

Solution $A = 2 + j4$

Magnitude of $A = \sqrt{2^2 + 4^2} = \sqrt{4 + 16} = \sqrt{20} = 4.472$

Phase of A, $\phi = \tan^{-1}\frac{4}{2} = \tan^{-1} 2 = 63.43°$

Problem 1.22 If a complex number is given by $A = 1 + j3$, find its complex conjugate, magnitude of A and its phase.

Solution The complex conjugate of A is

$$A^* = 1 - j3$$

The magnitude of $A = \sqrt{1^2 + 3^2} = \sqrt{1+9} = \sqrt{10} = 3.1622$

The phase of $A = \tan^{-1}\left(\dfrac{y}{x}\right) = \tan^{-1}\left(\dfrac{3}{1}\right) = 71.56°$

Problem 1.23 If the amplitude of a complex number, A is 5.0 and its phase is 45°, find its real and imaginary parts.

Solution The given complex number

$$A = 5.0 \angle 45°$$

Its magnitude, $\rho = 5.0$

Its phase, $\phi = 45°$

So the real part of $A = \rho \cos\phi = 5 \cos 45° = 3.535$

The imaginary part of $A = \rho \sin\phi = 3.535$

A is represented as $3.535 + j3.535$.

Problem 1.24 If $A_1 = 2 + j3$ and $A_2 = 4 + j5$, find the sum of A_1 and A_2.

Solution $A_1 = 2 + j3,\ A_2 = 4 + j5$

$$A = A_1 + A_2 = 2 + j3 + 4 + j5 = 6 + j8$$

Problem 1.25 If $A_1 = j6$ and $A_2 = 1 - j2$, find $A_1 - A_2$.

Solution $A_1 = j6,\ A_2 = 1 - j2$

$$A_1 - A_2 = j6 - 1 + j2 = -1 + j8$$

Problem 1.26 Given two complex numbers, $A = 0.4 + j5$, $B = 2 + j3$, find the product of A and B.

Solution $A = 0.4 + j5$

$$B = 2 + j3$$

The product, $AB = (0.4 + j5)(2 + j3)$

$$= 0.4 \times 2 + j0.4 \times 3 + j5 \times 2 + j5 \times j3$$

$$\boxed{AB = -14.2 + j11.2}$$

Problem 1.27 Given two complex numbers, $A = 10 + j6$ and $B = 2 - j3$, find the ratio of A and B.

Solution

$$A = 10 + j6$$

$$B = 2 - j3$$

$$\frac{A}{B} = \frac{(10 + j6)}{(2 - j3)} = \frac{(10 + j6)}{(2 - j3)} \times \frac{(2 + j3)}{(2 + j3)}$$

$$= \frac{20 + j30 + j12 - 18}{4 + 9}$$

$$\boxed{\frac{A}{B} = 0.153 + j3.23}$$

Problem 1.28 Find the roots of the quadratic equation $x^2 + 2x + 4 = 0$.

Solution For $x^2 + 2x + 4 = 0$, the roots are

$$x_{1,2} = \frac{-2 \pm \sqrt{4 - 4 \times 1 \times 4}}{2 \times 1} = \frac{-2 \pm \sqrt{-12}}{2}$$

$$\boxed{\begin{aligned} x &= -1.0 + j1.732 \\ x &= -1.0 - j1.732 \end{aligned}}$$

Problem 1.29 For the determinant, $\begin{vmatrix} 1 & 2 & 3 \\ 4 & 5 & 6 \\ 7 & 8 & 9 \end{vmatrix}$ find the minor of 8 and 9.

Solution The minor of the element, 8 is $\begin{vmatrix} 1 & 3 \\ 4 & 6 \end{vmatrix}$

The minor of the element, 9 is $\begin{vmatrix} 1 & 2 \\ 4 & 5 \end{vmatrix}$

Problem 1.30 For the determinant, $\begin{vmatrix} 1 & 3 & 2 \\ 6 & 1 & 5 \\ 7 & 9 & 8 \end{vmatrix}$ find the cofactor of 6 and 5.

Solution The minor of 6 is $\begin{vmatrix} 3 & 2 \\ 9 & 8 \end{vmatrix}$

Its cofactor $=(-1)^{i+j}\begin{vmatrix}3 & 2\\ 9 & 8\end{vmatrix}$

$$=(-1)^{2+1}\begin{vmatrix}3 & 2\\ 9 & 8\end{vmatrix}$$

$$=-\begin{vmatrix}3 & 2\\ 9 & 8\end{vmatrix}$$

The minor of 5 is $\begin{vmatrix}1 & 3\\ 7 & 9\end{vmatrix}$

Its cofactor $=(-1)^{i+j}\begin{vmatrix}1 & 3\\ 7 & 9\end{vmatrix}$

$$=(-1)^{2+3}\begin{vmatrix}1 & 3\\ 7 & 9\end{vmatrix}$$

$$=\begin{vmatrix}1 & 3\\ 7 & 9\end{vmatrix}$$

Problem 1.31 Find the factorial of 4 and 6.

Solution Factorial of $4 = 4\times 3\times 2\times 1 = 24$

Factorial of $6 = 6\times 5\times 4\times 3\times 2\times 1 = 720$

$$\boxed{\begin{aligned}4! &= 24\\ 6! &= 720\end{aligned}}$$

OBJECTIVE QUESTIONS

1. The wavelength of an EM wave depends on its velocity. (Yes/No)
2. The velocity of an EM wave depends on the medium in which it propagates. (Yes/No)
3. The unit vectors along the Cartesian coordinate axes are not mutually perpendicular to each other. (Yes/No)
4. In cylindrical coordinates, ρ has the unit of degrees. (Yes/No)
5. In cylindrical coordinates, ρ has the units of metres. (Yes/No)
6. In spherical coordinates, θ increases in anticlockwise direction. (Yes/No)
7. In cylindrical coordinates, ϕ increases in anticlockwise direction. (Yes/No)
8. In spherical coordinates, θ increases in clockwise direction. (Yes/No)
9. Del is a scalar operator. (Yes/No)
10. ∇^2 is a vector operator. (Yes/No)
11. ∇^2 can be operated on scalar and vector. (Yes/No)
12. $\sin jx$ is equal to $j \sinh x$. (Yes/No)
13. $\sin jx$ is equal to $j \sin x$. (Yes/No)
14. $\cosh jx$ is $\cos x$. (Yes/No)
15. $\tanh jx$ is $j \tan x$. (Yes/No)
16. Fresnel integrals $C(\infty) = S(\infty)$. (Yes/No)
17. Fresnel integral, $C(\infty) = \sqrt{\pi/8}$. (Yes/No)
18. Error function, $erf(\infty) = 1$. (Yes/No)
19. In cylindrical coordinates, $\phi =$ a constant is a plane. (Yes/No)
20. In spherical coordinates, $\theta =$ a constant is a cone. (Yes/No)
21. $\mathbf{a}_x \cdot \mathbf{a}_\rho = \sin\phi$. (Yes/No)
22. $\mathbf{a}_x \cdot \mathbf{a}_\phi = \sin\phi$. (Yes/No)
23. $\mathbf{a}_y . \mathbf{a}_\rho = 0$. (Yes/No)
24. $\mathbf{a}_z . \mathbf{a}_\rho = 0$. (Yes/No)

25. $\mathbf{a}_z . \mathbf{a}_\phi = 0$. (Yes/No)

26. $\mathbf{a}_y . \mathbf{a}_\phi = \cos \phi$. (Yes/No)

27. The gradient of a vector does not exist. (Yes/No)

28. Divergence of a scalar does not exist. (Yes/No)

29. Curl of a scalar exists. (Yes/No)

30. Del (∇) has no units. (Yes/No)

31. Divergence of water or oil is almost zero. (Yes/No)

32. Divergence of gradient of scalar electric potential is the Laplacian of the potential. (Yes/No)

33. $j = (0, 1)$. (Yes/No)

34. Complex conjugate of $A = 2 + j5$ is $2 - j5$. (Yes/No)

35. A complex number $B = (x, y)$ is nothing but $(x + jy)$. (Yes/No)

36. $\int_0^\infty \delta(t - t_0)\, dt = 1$ (Yes/No)

37. 1 Neper = 8.686 dB (Yes/No)

38. $\log_e x = 0.434 \log_{10} x$ (Yes/No)

39. $\mathbf{A} . \mathbf{A} = \mathbf{A} . \mathbf{A}^* = A^2$. (Yes/No)

40. Radar X-band is ________________.

41. The frequency range of radar mm-band is ________________.

42. The frequency range UHF is ________________.

43. The wavelength range of VHF band is ________________.

44. In cylindrical coordinates, ρ is the ________________.

45. In spherical coordinates, r is ________________.

46. dv in cylindrical coordinates is ________________.

47. dL in spherical coordinates is ________________.

48. Dot product of $\mathbf{a}_z$ and $\mathbf{a}_\phi$ is ________________.

49. Dot product of $\mathbf{a}_x$ and $\mathbf{a}_r$ is ________________.

50. Dot product of $\mathbf{a}_x$ and $\mathbf{a}_y$ is ________________.

51. In spherical coordinates, θ varies between ______________.

52. In spherical coordinates, ϕ varies between ____________.

53. The divergence of an incompressible fluid is ___________.

54. Del has ______________ units.

55. A point is obtained by the intersection of ___________________ in spherical coordinates.

56. 1 NP is equal to ____________.

57. 1 dB is defined as ______________.

58. 1 NP is defined as ______________.

59. 0! is _______.

60. For $x << 1, (1+x)^{-1} =$ __________.

61. For $x << 1, (1+x)^{+1} =$ ___________.

62. sinh 0 is ___________.

63. tanh 0 is _____________.

64. cosh 0 is ______________.

65. Gamma function, $\Gamma\left(\frac{1}{2}\right)$ is __________.

Answers

1. Yes	2. Yes	3. No	4. No	5. Yes
6. No	7. Yes	8. Yes	9. No	10. No
11. Yes	12. Yes	13. Yes	14. Yes	15. Yes
16. Yes	17. Yes	18. Yes	19. Yes	20. Yes
21. No	22. No	23. No	24. Yes	25. Yes
26. Yes	27. Yes	28. Yes	29. No	30. No
31. Yes	32. Yes	33. Yes	34. Yes	35. Yes
36. Yes	37. Yes	38. Yes	39. Yes	40. 8-12 GHz
41. 40-300 GHz	42. 300 – 3,000 MHz		43. 10 – 1 m	
44. Radius of cylinder		45. Radius of sphere		46. $\rho\ d\rho\ d\phi\ dz$

47. $dr\,\mathbf{a}_r + r\,d\theta\,\mathbf{a}_\theta + r\sin\theta\,d\phi\,\mathbf{a}_\phi$			48. Zero	49. $\sin\theta\cos\phi$
50. Zero	51. 0 and π	52. 0 and 2π	53. Zero	54. 1/m
55. A plane, a cone and a sphere			56. 8.686 dB	
57. $10\log_{10}\left(\frac{P_1}{P_2}\right)$		58. $\log_e\left(\frac{I_2}{I_1}\right)$		59. 1
60. $1-x$	61. $1+x$	62. Zero	63. Zero	64. One
65. $\sqrt{\pi}$				

EXERCISE PROBLEMS

1. Find the magnitude and direction for the vectors $\mathbf{A} = 5\mathbf{a}_x + 3\mathbf{a}_y + 2\mathbf{a}_y$ and $\mathbf{B} = 4\mathbf{a}_y + 0.5\mathbf{a}_z$.

2. If $\mathbf{A} = (2, 4, 1)$ and $\mathbf{B} = (0, 0.1, 5)$, find $\mathbf{A} + \mathbf{B}$, $\mathbf{A} - \mathbf{B}$.

3. If $\mathbf{A} = 6\mathbf{a}_x + 3\mathbf{a}_z$ and $\mathbf{B} = 2\mathbf{a}_y + 5\mathbf{a}_z$, find $\mathbf{A}.\mathbf{B}$.

4. When $\mathbf{A}$ = (2, 0, 7), $\mathbf{B}$ = (3, 2, 0), what is $\mathbf{A} \times \mathbf{B}$?

5. If a point in Cartesian coordinates is given by (2, 0, 4), express it in cylindrical and spherical coordinate systems.

6. The electric scalar potential is given by $V = 3x + 5y^2 + 4z^3$. Find its gradient.

7. For a vector $A = 12a_x + 2a_y + 3a_z$, find the divergence and curl.

8. If a voltage gain is 22 dB, express it in nepers.

9. Find the roots of the quadratic equation $5x^2 + 2x + 4 = 0$.

10. What are the roots of the cubic equation given by $x^3 - 2x^2 + 4x - 8 = 0$?

11. Find the value of the determinant $\begin{vmatrix} 1 & 2 & 3 \\ 2 & 1 & 3 \\ 3 & 0 & 2 \end{vmatrix}$.

12. Find the minor of 0 in Problem 11.

13. What is the cofactor of 4 and 3 in the determinant $\begin{matrix} 4 & 0 & 1 \\ 5 & 2 & 6 \\ 0 & 3 & 2 \end{matrix}$?

14. Find the value of $(1 + x)^{1/2}$ if $x = 0.01$, using the series expansion.

15. Find $\sin x$ if $x = 0.2$, using the series expansion.

CHAPTER 2

ELECTROSTATIC FIELDS

Electrostatic fields are produced by charges at rest.

The main objective of this chapter is to provide detailed concepts of electrostatics. They include:

- applications of electrostatics
- charge distributions
- Coulomb's law, applications and limitations
- Gauss's law, applications and limitations
- potential functions
- Poisson's and Laplace's equations
- conductors, dielectrics and capacitors
- energy stored in electrostatic fields and capacitors
- fields in dielectrics and boundary conditions
- solved problems, points/formulae to remember, objective and multiple choice questions and exercise problems.

Do you know?

An electrostatic field of 200 V/m exists outdoor on a normal day.

2.1 INTRODUCTION

Electrostatic fields are also called static electric fields or steady electric fields. These fields are not variant with time. **They are produced by static charges or charge distributions**. These fields have a wide range of applications.

In this chapter, the important laws, theorems and equations, with their applications are presented to provide a thorough understanding of electrostatics. The important areas are Coulomb's and Gauss's laws, Stoke's and divergence theorems and Poisson's and Laplace's equations.

2.2 APPLICATIONS OF ELECTROSTATIC FIELDS

Electrostatic fields are used

1. in cathode ray oscilloscopes to obtain the electron beam deflection
2. in ink-jet printers to obtain speed of printing and quality of print
3. to sort out minerals in ore separators, to sort out seeds in agriculture and for spraying plants and trees
4. in electrostatic generators
5. to produce potential
6. to produce force on charges for their mobility
7. in electric power transmission
8. in lightning protection
9. to measure moisture content
10. to spin cotton
11. in field-effect transistors
12. in capacitors
13. in LCDs
14. in touch pads
15. in capacitance keyboards
16. in fast baking of bread
17. in ECGs, EEGs, ERG, EMG, EOG in medical applications
18. in spray painting
19. in electrodeposition
20. in electrochemical machining
21. X-ray machines
22. computer peripherals

2.3 DIFFERENT TYPES OF CHARGE DISTRIBUTIONS

Charges at rest produce electrostatic field. The charges are basically of two types: positive and negative.

Properties and Functions of Charges

(i) Charge is conserved. It can neither be created nor destroyed.

(ii) Charge is quantisised that is, it comes in discrete quantities and in integer multiples of the basic unit of charge. Electron charge or proton charge is the basic charge.

(iii) When a charge is accelerated, electromagnetic field is produced. A fraction of the field is detached and propagates at the speed of light. The detached field carries energy, momentum and angular momentum. This is called **electromagnetic radiation**.

(iv) Charges are surrounded by electric and magnetic fields.

(v) A charge experiences a force in the presence of a field.

(vi) Charges mediate the integration of fields.

Charge density and charge distribution indicate the same property. In practice, there are four types of charge distributions.

(i) **Point charges, Q (Coulomb)** These are the charges which do not occupy any space, that is, the volume of the point charge is zero. For example, an electron is considered to be a point charge and has a charge of 1.6×10^{-19} Coulombs (C).

(ii) **Line charge distribution, ρ_L (c/m)** This is a charge distribution in which the charge is distributed along a line like a filament, that is, this has only length but no width or breadth. ρ_L is defined as the charge per unit length, that is,

$$\boxed{\rho_L \equiv \underset{\Delta L \to 0}{\text{Lt}} \frac{\Delta Q}{\Delta L} = \frac{dQ}{dL}}$$

where ΔQ is small charge, ΔL is small length, dQ is differential charge and dL is differential length.

Sometimes, ρ_L is simply considered as Q/L.

An example is the electron beam in CRT.

Problem 2.1 If there is a charge of $10\,\mu C$ over a filament length of 0.5 m, find its line charge density.

Solution Here, $Q = 10\mu C = 10 \times 10^{-6} C$

$L = 0.5 \text{m}$

Line charge density, $\rho_L = Q/L = 10/0.5 = 20 \,\mu\text{C/m}$

(iii) **Surface charge distribution, ρ_s (c/m²)** When a charge is confined to the surface of a conductor, it is said to be surface charge distribution. Such a surface has both length and width but no breadth.

Surface density is defined as the charge per unit area, that is,

$$\boxed{\rho_s \equiv \underset{\Delta s \to 0}{\text{Lt}} \frac{\Delta Q}{\Delta S} = \frac{dQ}{dS}}$$

where ΔS is small area and dS is differential area.

Sometimes, ρ_s is simply considered as Q/S.

An example is the conductor surface of a capacitor.

Problem 2.2 If there is a total charge of 10 pC over a surface area of 0.2 m², find the surface charge density.

Solution Here, $Q = 10 \text{ pC} = 10 \times 10^{-12} \text{C}$

$S = 0.2 \text{ m}^2$

Surface charge density, $\rho_s = Q/S = (10 \times 10^{-12})/0.2$

$= 50 \text{pC/m}^2$

(iv) **Volume charge distribution, ρ_υ (c/m³)** Volume charge density is defined as the charge per unit volume, that is,

$$\boxed{\rho_\upsilon \equiv \underset{\Delta \upsilon \to 0}{\text{Lt}} \frac{\Delta Q}{\Delta \upsilon} = \frac{dQ}{d\upsilon}}$$

where $d\upsilon$ is differential volume.

Sometimes, ρ_υ is simply considered as $\frac{Q}{\upsilon}$.

Examples are ionospheric region, electron cloud in vacuum tube.

Problem 2.3 If there exists a total charge of 12 nC in a spherical volume of 0.1 m³, find the volume charge density.

Solution Here $Q = 12 \text{ nC} = 12 \times 10^{-9} \text{C}$

$$\upsilon = 0.1 \text{ m}^3$$

The volume charge density,

$$\rho_\upsilon = Q/\upsilon$$
$$= 12/0.1$$
$$= 120 \text{ nC/m}^3$$

2.4 COULOMB'S LAW

Charles Augustin De Coulomb (1736–1806) was a French physicist. After conducting several experiments on charged bodies, he concluded that there exists a force between them. He formulated a law known as Coulomb's law.

The law appears in the following forms:

(i) Gaussian form

$$\text{Force,} \quad \mathbf{F} = \frac{Q_1 Q_2}{r^2} \mathbf{a}_r$$

(ii) SI form

$$\text{Force,} \quad \mathbf{F} = \frac{1}{4\pi \epsilon_0} \frac{Q_1 Q_2}{r^2} \mathbf{a}_r$$

(iii) Heavside-Lorentz form

$$\text{Force,} \quad \mathbf{F} = \frac{1}{4\pi} \frac{Q_1 Q_2}{r^2} \mathbf{a}_r$$

Law statement Coulomb's law states that there exists a force between charged bodies and it is:

1. Proportional to the product of the two charges,
2. Inversely proportional to the square of the distance between the charges. The force also depends on the medium in which the charges are located.

The force is a vector quantity and it is attractive if the charges are unlike and repulsive if the charges are alike. It acts along the straight line joining the two charges.

Mathematically, Coulomb's law is given by,

$$\mathbf{F} \,\alpha\, \frac{Q_1 Q_2}{r^2} \mathbf{a}_r \text{ Newton or } \mathbf{F} = k \frac{Q_1 Q_2}{r^2} \mathbf{a}_r$$

where k = a constant of proportionality

$$= \frac{1}{4\pi \in} \text{ (metre/farad-radian)}$$

Q_1, Q_2 are two point charges (C)

= force between the charges (Newton)

r = distance between the charges (m)

$\in$ = permittivity of the medium in which the charges are located (F/m)

$= \in_0 \in_r$

$\in_0$ = permittivity of vacuum or free space

$$= \frac{1}{36\pi} \times 10^{-9} = 8.854 \times 10^{-12} \text{ F/m}$$

$\in_r$ = relative permittivity of the medium with respect to free space (has no unit)

= 1 for free space

k in free space

$$= \frac{1}{4\pi \in_0} = 9 \times 10^9 \text{ m/F}$$

$\mathbf{a}_r$ = unit vector along the line joining the two charges

The force on Q_2 due to Q_1 in free space is written in the form of

$$\mathbf{F}_2 = \frac{Q_1 Q_2}{4\pi \in_0} \frac{\mathbf{a}_{21}}{r_{21}^2}$$

where,
$$\mathbf{a}_{21} = \frac{\mathbf{r}_2 - \mathbf{r}_1}{|\mathbf{r}_2 - \mathbf{r}_1|}$$

$\mathbf{r}_1$ = location of Q_1

$\mathbf{r}_2$ = location of Q_2

$r_{21} = |\mathbf{r}_2 - \mathbf{r}_1|$

$\in_0$ = permittivity of free space $= 8.854 \times 10^{-12}$ F/m or $\left(\dfrac{1}{36\pi \times 10^9}\right)$ F/m

$$\frac{1}{4\pi \in_0} = 9 \times 10^9$$

The force on Q_1 due to Q_2 is

$$\boxed{\mathbf{F}_1 = -\mathbf{F}_2}$$

Problem 2.4 If Q_1 and Q_2 are two point charges and are located at $\mathbf{r}_1$ and $\mathbf{r}_2$ respectively in free space, find (a) the force on Q_2 due to Q_1, (b) the force on Q_1 due to Q_2.

Solution Q_1 = first charge

Q_2 = second charge

$\mathbf{r}_1$ = location of Q_1

$\mathbf{r}_2$ = location of Q_2

Distance vector along the line joining two charges,

$$\mathbf{r}_{21} = \mathbf{r}_2 - \mathbf{r}_1$$

The magnitude of $\mathbf{r}_{21}$ is

$$|\mathbf{r}_{21}| = r_{21} = |\mathbf{r}_2 - \mathbf{r}_1|$$

The direction of $\mathbf{r}_{21} = \dfrac{\mathbf{r}_{21}}{r_{21}} = \mathbf{a}_{21}$

(a) The force on Q_2 due to Q_1

$$\boxed{\mathbf{F}_2 = \frac{Q_1 Q_2}{4\pi \in_0 r_{21}^2}\mathbf{a}_{21}, \text{N}}$$

(b) The force on Q_1 due to Q_2,

$$\boxed{\mathbf{F}_1 = \frac{Q_1 Q_2}{4\pi \in_0 r_{12}^2}\mathbf{a}_{12}, \text{N}}$$

Here
$$\mathbf{r}_{12} = \mathbf{r}_1 - \mathbf{r}_2 = -(\mathbf{r}_2 - \mathbf{r}_1) = -\mathbf{r}_{21}$$

$$r_{12} = r_{21} = |\mathbf{r}_1 - \mathbf{r}_2| = |\mathbf{r}_2 - \mathbf{r}_1|$$

$$\mathbf{a}_{12} = -\mathbf{a}_{21}$$

$$\mathbf{F}_1 = -\mathbf{F}_2$$

Problem 2.5 A charge of $Q_1 = -1.0\,\mu\text{C}$ is placed at the origin of a rectangular coordinate system and a second charge, $Q_2 = -10$ mC is placed on the x-axis at a distance of 50 cm from the origin. Find the force on Q_1 due to Q_2 if they are in free space.

Solution $Q_1 = -1.0\,\mu\text{C}$ at (0, 0, 0)

$Q_2 = -10$ mC at (0.5, 0, 0)

By Coulomb's law,

$$\mathbf{F}_{12} = \frac{Q_1 Q_2}{4\pi \epsilon_0 r^2}\mathbf{a}_{12}$$

$$\mathbf{r} = (0, 0, 0) - (0.5, 0, 0) = -0.5\mathbf{a}_x$$

$$r = |\mathbf{r}| = 0.5$$

$$\mathbf{a}_{12} = -\mathbf{a}_x, \frac{1}{4\pi \epsilon_0} = 9\times 10^9$$

$$\mathbf{F}_{12} = \frac{(-1.0\times 10^{-6})\times(-10\times 10^{-3})\times 9\times 10^9(-\mathbf{a}_x)}{0.25}$$

$$\boxed{\mathbf{F}_{12} = -360\mathbf{a}_x \text{ Newton}}$$

Problem 2.6 A point charge, $Q_1 = 2\mu C$ is at (2, 3, 6) and another charge, $Q_2 = 5\,\mu C$ is at (0, 0, 0) in free space. Find the force on Q_1 due to Q_2.

Solution We have

$$\mathbf{F}_{12} = \frac{Q_1 Q_2}{4\pi \epsilon_0 r^2}\mathbf{a}_{12}$$

$$\frac{1}{4\pi \epsilon_0} = 9\times 10^9$$

$$Q_1 = 2\mu C = 2\times 10^{-6}\,C$$

$$Q_2 = 5\mu C = 5\times 10^{-6}\,C$$

$$\mathbf{r} = \mathbf{r}_1 - \mathbf{r}_2$$

$$\mathbf{r}_1 = 2\mathbf{a}_x + 3\mathbf{a}_y + 6\mathbf{a}_z$$

$$\mathbf{r}_2 = 0\mathbf{a}_x + 0\mathbf{a}_y + 0\mathbf{a}_z$$

$$\mathbf{r} = 2\mathbf{a}_x + 3\mathbf{a}_y + 6\mathbf{a}_z$$

$$r = |\mathbf{r}| = \sqrt{4+9+36} = \sqrt{49} = 7$$

$$\mathbf{a}_{12} = \frac{\mathbf{r}}{r} = \frac{2\mathbf{a}_x + 3\mathbf{a}_y + 6\mathbf{a}_z}{7}$$

$$\mathbf{F}_{12} = \frac{9\times 10^9 \times 2\times 10^{-6}\times 5\times 10^{-6}}{7^2\times 7}(2\mathbf{a}_x + 3\mathbf{a}_y + 6\mathbf{a}_z)$$

$$= 0.26\,(2\mathbf{a}_x + 3\mathbf{a}_y + 6\mathbf{a}_z) \times 10^{-3}$$

$$\boxed{\mathbf{F}_{12} = (0.52\mathbf{a}_x + 0.78\mathbf{a}_y + 1.56\mathbf{a}_z)\ \text{mN}}$$

Problem 2.7 Three equal point charges of $2\mu C$ are in free space at (0, 0, 0), (2, 0, 0) and (0, 2, 0), respectively. Find net force on $Q_4 = 5\mu C$ at (2, 2, 0).

Solution

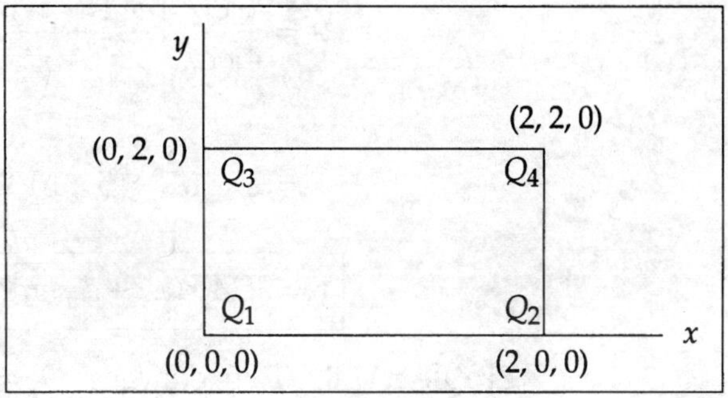

Fig. 2.1 Location of charges

$$Q_1 = Q_2 = Q_3 = 2\mu C = Q$$

$$Q_4 = 5\mu C$$

$$\mathbf{r}_1 = (0,\ 0,\ 0)$$

$$\mathbf{r}_2 = (2,\ 0,\ 0) = 2\mathbf{a}_x$$

$$\mathbf{r}_3 = (0,\ 2,\ 0) = 2\mathbf{a}_y$$

$$\mathbf{r}_4 = (2,\ 2,\ 0) = 2\mathbf{a}_x + 2\mathbf{a}_y$$

$$\mathbf{r}_{41} = \mathbf{r}_4 - \mathbf{r}_1 = 2\mathbf{a}_x + 2\mathbf{a}_y$$

$$\mathbf{r}_{42} = \mathbf{r}_4 - \mathbf{r}_2 = 2\mathbf{a}_x + 2\mathbf{a}_y - 2\mathbf{a}_x = 2\mathbf{a}_y$$

$$\mathbf{r}_{43} = \mathbf{r}_4 - \mathbf{r}_3 = 2\mathbf{a}_x + 2\mathbf{a}_y - 2\mathbf{a}_y = 2\mathbf{a}_x$$

$$r_{41} = \sqrt{4+4} = \sqrt{8} = 2.828$$

$$r_{42} = 2,\ r_{43} = 2$$

The net force on Q_4 due to Q_1, Q_2, Q_3 is

$$\mathbf{F} = \mathbf{F}_{41} + \mathbf{F}_{42} + \mathbf{F}_{43}$$

$$\mathbf{F}_{41} = \frac{QQ_4}{4\pi \in_0 r_{41}^2}\mathbf{a}_{41}$$

$$\mathbf{a}_{41} = \frac{2\mathbf{a}_x + 2\mathbf{a}_y}{2.828}$$

$$\mathbf{F}_{42} = \frac{QQ_4}{4\pi \in_0 r_{42}^2}\mathbf{a}_{42}$$

$$\mathbf{a}_{42} = \mathbf{a}_y$$

$$\mathbf{F}_{43} = \frac{QQ_4}{4\pi \in_0 r_{43}^2} \mathbf{a}_{43}$$

$$\mathbf{a}_{43} = \mathbf{a}_x$$

$$\mathbf{F} = \frac{QQ_4}{4\pi \in_0} \left[\frac{\mathbf{a}_{41}}{r_{41}^2} + \frac{\mathbf{a}_{42}}{r_{42}^2} + \frac{\mathbf{a}_{43}}{r_{43}^2} \right]$$

$$= 2\times 10^{-6} \times 5\times 10^{-6} \times 9\times 10^9 \left[\frac{2\mathbf{a}_x + 2\mathbf{a}_y}{(2.828)^3} + \frac{\mathbf{a}_y}{4} + \frac{\mathbf{a}_x}{4} \right]$$

$$\boxed{\mathbf{F} = 30.45\,(\mathbf{a}_x + \mathbf{a}_y)\text{ mN}}$$

Problem 2.8 Two charges $Q_1 = 2\mu C$ and $Q_2 = 5\mu C$ are located at $(-3, 7, -4)$ and $(2, 4, -1)$, respectively. Determine the force on Q_2 due to Q_1 and the force on Q_1 due to Q_2.

Solution The force on Q_2 due to Q_1 is given by

$$\mathbf{F}_{21} = \frac{Q_1 Q_2}{4\pi \in_0 r_{21}^2} \mathbf{a}_{21}$$

where $\mathbf{r}_{21} = \mathbf{r}_2 - \mathbf{r}_1$

$$\mathbf{a}_{21} = \frac{\mathbf{r}_2 - \mathbf{r}_1}{|\mathbf{r}_2 - \mathbf{r}_1|}$$

$$\mathbf{r}_2 = 2\mathbf{a}_x + 4\mathbf{a}_y - \mathbf{a}_z$$

$$\mathbf{r}_1 = -3\mathbf{a}_x + 7\mathbf{a}_y - 4\mathbf{a}_z$$

$$\mathbf{r}_{21} = \mathbf{r}_2 - \mathbf{r}_1 = 5\mathbf{a}_x - 3\mathbf{a}_y + 3\mathbf{a}_z$$

$$|\mathbf{r}_{21}| = \sqrt{25+9+9} = \sqrt{43} = 6.557$$

$$\mathbf{F}_{21} = \frac{2\times 10^{-6} \times 5\times 10^{6} \times 9\times 10^9}{(43)^{3/2}} (5\mathbf{a}_x - 3\mathbf{a}_y + 3\mathbf{a}_z)$$

$$\boxed{\mathbf{F}_{21} = (1.596\mathbf{a}_x - 0.957\mathbf{a}_y + 0.957\mathbf{a}_z)\text{ mN}}$$

Now the force on Q_1 due to Q_2

$$\boxed{\mathbf{F}_{12} = -\mathbf{F}_{21} = (-1.596\mathbf{a}_x + 0.957\mathbf{a}_y - 0.957\mathbf{a}_z)\text{ mN}}$$

2.5 APPLICATIONS OF COULOMB'S LAW

Coulomb's law is used to:

1. find the force between a pair of charges
2. find the potential at a point due to a fixed charge
3. find the electric field at a point due to a fixed charge
4. find the displacement flux density indirectly
5. find the potential and electric field due to any type of charge distribution
6. find the charge if the force and the electric field are known.

2.6 LIMITATION OF COULOMB'S LAW

It is difficult to apply the law when charges are of arbitrary shape. Here, the distance, r cannot be determined accurately as the centres of arbitrarily shaped charged bodies cannot be identified accurately.

2.7 ELECTRIC FIELD STRENGTH DUE TO POINT CHARGE

Electrostatic field is produced by a charge at rest. It is defined by Coulomb's law. Electric field strength or electric field intensity or electric field indicate the same property in this book.

Definition 1 Electric field due to a charge is defined as the Coulomb's force per unit charge. It is a vector and has the unit of Newton per Coulomb or volt per metre, that is,

$$\text{Electric field, } \mathbf{E} \equiv \frac{\mathbf{F}}{Q}, \text{ N/C}$$

where **F** is Coulomb's force, Newton and Q is charge, Coulomb.

Definition 2 Electric field is also defined as a negative gradient of a potential due to a charge, that is,

$$\mathbf{E} \equiv -\nabla V, \text{ volts / metre}$$

where V is potential due to the charge, volts.

Let Q_f = fixed point charge, C

Q_t = test point charge, C

$\mathbf{r}_f$ = location of fixed charge

$\mathbf{r}_t$ = location of test charge

Then the force on Q_t due to a fixed charge in free space, Q_f is given by

$$\mathbf{F}_{tf} = \frac{Q_f\, Q_t}{4\pi \epsilon_0\, r_{tf}^2} \mathbf{a}_{tf}, \text{ N/C}$$

The electric field, **E** at the location of Q_t due to Q_f is defined as the ratio of force on Q_t due to Q_f and the test charge, Q_t, that is,

$$\mathbf{E} \equiv \frac{\mathbf{F}_{tf}}{Q_t}$$

$$\mathbf{E} = \frac{Q_f}{4\pi \epsilon_0\, r_{tf}^2} \mathbf{a}_{tf}, \text{ N/C}$$

If there are N point charges, the field at a point is the vectorial sum of fields due to N charges, that is,

$$\mathbf{E} = \mathbf{E}_1 + \mathbf{E}_2 + \ldots + \mathbf{E}_N, \text{ N/C}$$

2.8 SALIENT FEATURES OF ELECTRIC INTENSITY

1. It has units of Newton/Coulomb or volts/metre.
2. It is a vector.
3. It has both direction and magnitude.
4. Its direction is the same as that of Coulomb's force.
5. Its magnitude depends on the magnitude of Coulomb's force and the charge on which the force is acting.
6. It depends on the medium.
7. It depends on the permittivity of the medium.
8. It depends on the distance of the charge from another charge which produces Coulomb's force.
9. It depends on the location of the charges.
10. It originates from a positive charge and terminates on a negative charge.
11. When a unit charge at a distance is moved around a fixed charge, the field lines and force appear as in Fig. 2.2.

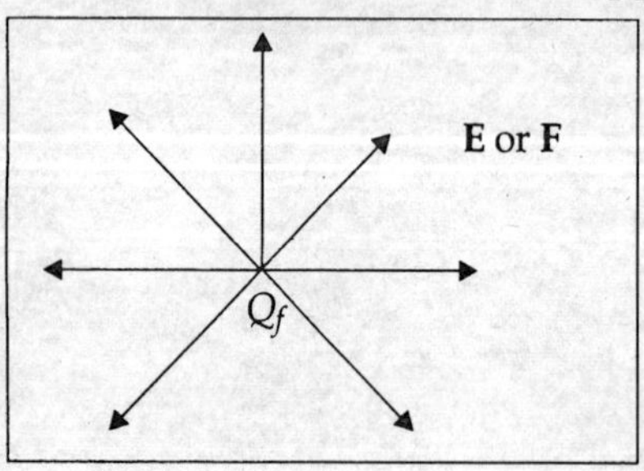

Fig. 2.2 Coulomb's force and electric field

Problem 2.9 If Coulomb's force, $\mathbf{F} = 2\mathbf{a}_x + \mathbf{a}_y + \mathbf{a}_z$ N, is acting on a charge of 10C, find the electric field intensity, its magnitude and direction.

Solution Force,

$$\mathbf{F} = 2\mathbf{a}_x + \mathbf{a}_y + \mathbf{a}_z, \text{N}$$

$$Q = 10\,\text{C}$$

$$\mathbf{E} = \frac{\mathbf{F}}{Q}$$

$$= \frac{2\mathbf{a}_x + \mathbf{a}_y + \mathbf{a}_z}{10}$$

$$\boxed{\mathbf{E} = 0.2\mathbf{a}_x + 0.1\mathbf{a}_y + 0.1\mathbf{a}_z, \text{N/C}}$$

The magnitude of **E** is

$$E = |\mathbf{E}|$$

$$= \sqrt{(0.2)^2 + (0.1)^2 + (0.1)^2}$$

$$= \sqrt{0.04 + 0.01 + 0.01} = \sqrt{0.06}$$

$$= 0.2449 \text{ V/m}$$

The direction of **E** is

$$\mathbf{a}_E = \frac{\mathbf{E}}{E}$$

$$= \frac{0.2\mathbf{a}_x + 0.1\mathbf{a}_y + 0.1\mathbf{a}_z}{0.2449}$$

$$\boxed{\mathbf{a}_E = 0.816\mathbf{a}_x + 0.408\mathbf{a}_y + 0.408\mathbf{a}_z}$$

Problem 2.10 If a charge of 2μC is located at $P_1(1, 0, 0)$ and another charge of 1μC is located at $P_2(0, 1, 0)$ in free space, find the electric field at P_2.

Solution Let

$$Q_f = 2\mu\text{C} = 2 \times 10^{-6}\,\text{C}$$

$$Q_t = 1\mu\text{C} = 1 \times 10^{-6}\,\text{C}$$

$$\mathbf{r}_f = (1, 0, 0) = \mathbf{a}_x$$

$$\mathbf{r}_t = (0, 1, 0) = \mathbf{a}_y$$

$$\mathbf{r}_{tf} = \mathbf{r}_t - \mathbf{r}_f = \mathbf{a}_y - \mathbf{a}_x$$

$$r_{tf} = \sqrt{1^2 + 1^2} = \sqrt{2}$$

$$\mathbf{a}_{tf} = \frac{\mathbf{r}_{tf}}{r_{tf}} = \frac{(\mathbf{a}_y - \mathbf{a}_x)}{\sqrt{2}}$$

Electric field strength at P_2 is

$$\mathbf{E} = \frac{Q_f}{4\pi \epsilon_0 \, r_{tf}^2} \mathbf{a}_{tf}$$

$$= \frac{2 \times 10^{-6} \, (\mathbf{a}_{tf})}{4\pi \epsilon_0 \, (\sqrt{2})^2}$$

$$= \frac{2 \times 10^{-6}}{4\pi \, \epsilon_0 \, (\sqrt{2})^2} \frac{(\mathbf{a}_y - \mathbf{a}_x)}{\sqrt{2}}$$

As
$$\frac{1}{4\pi \, \epsilon_0} = 9 \times 10^9$$

$$\mathbf{E} = 9 \times 10^9 \times 10^{-6} \times \frac{2}{2} \left(\frac{\mathbf{a}_y - \mathbf{a}_x}{\sqrt{2}} \right)$$

$$\boxed{\mathbf{E} = 6.364 \, (\mathbf{a}_y - \mathbf{a}_x) \text{ kN/C}}$$

Problem 2.11 There are three charges which are given by $Q_1 = 1\mu\text{C}$, $Q_2 = 2\mu\text{C}$ and $Q_3 = 3\mu\text{C}$. The field due to each charge at a point, P in free space is $\mathbf{a}_x + 2\mathbf{a}_y - \mathbf{a}_z$, $\mathbf{a}_y + 3\mathbf{a}_z$ and $2\mathbf{a}_x - \mathbf{a}_y$ N/C. Find the total field at the point, P due to all the three charges.

Solution The field, $\mathbf{E}_1$ at P due to $Q_1 \, (1\mu\text{C})$

$$= \mathbf{a}_x + 2\mathbf{a}_y - \mathbf{a}_z, \text{ N/C}$$

The field, $\mathbf{E}_2$ due to $Q_2 \, (2\mu\text{C})$

$$= \mathbf{a}_y + 3\mathbf{a}_z, \text{ N/C}$$

The field, $\mathbf{E}_3$ due to $Q_3 \, (3\mu\text{C})$

$$= 2\mathbf{a}_x - \mathbf{a}_y, \text{ N/C}$$

The total field at P,

$$\mathbf{E} = \mathbf{E}_1 + \mathbf{E}_2 + \mathbf{E}_3$$

$$= \mathbf{a}_x + 2\mathbf{a}_y - \mathbf{a}_z + \mathbf{a}_y + 3\mathbf{a}_z + 2\mathbf{a}_x - \mathbf{a}_y$$

$$\boxed{\mathbf{E} = 3\mathbf{a}_x + 2\mathbf{a}_y + 2\mathbf{a}_z, \text{ N/C}}$$

Problem 2.12 A charge, $Q_1 = -10\,\text{nC}$ is at the origin in free space. If the x-component of $\mathbf{E}$ is to be zero at the point (3, 1, 1), what charge, Q_t should be kept at the point (2, 0, 0)?

Solution Consider Fig. 2.3.

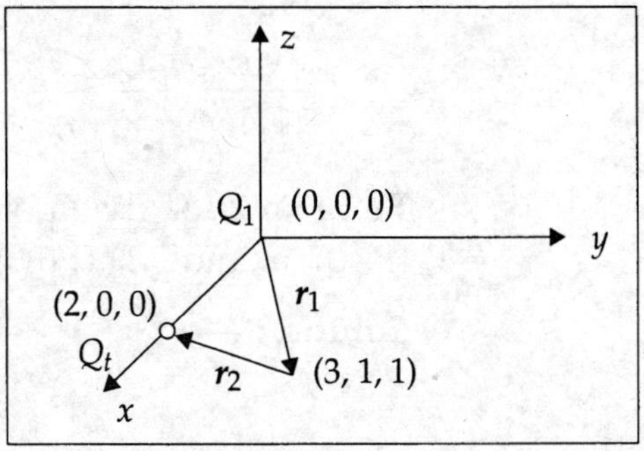

Fig. 2.3

$$Q_1 = -10 \text{ nC at the origin}$$

$$\mathbf{r}_1 = (3, 1, 1) - (0, 0, 0)$$

$$\mathbf{r}_1 = (3-0)\,\mathbf{a}_x + (1-0)\,\mathbf{a}_y + (1-0)\,\mathbf{a}_z$$

$$= 3\mathbf{a}_x + \mathbf{a}_y + \mathbf{a}_z$$

$$r_1 = \sqrt{9+1+1} = \sqrt{11}$$

$$\mathbf{r}_2 = (3, 1, 1) - (2, 0, 0)$$

$$\mathbf{r}_2 = (3-2)\,\mathbf{a}_x + (1-0)\,\mathbf{a}_y + (1-0)\,\mathbf{a}_z$$

$$= \mathbf{a}_x + \mathbf{a}_y + \mathbf{a}_z$$

$$r_2 = \sqrt{1+1+1} = \sqrt{3}$$

$$\mathbf{E} = \mathbf{E}_1 + \mathbf{E}_2$$

$$= \frac{Q_1}{4\pi \epsilon_0 r_1^2}\mathbf{a}_{r_1} + \frac{Q_t}{4\pi \epsilon_0 r_2^2}\mathbf{a}_{r_2}$$

$$\mathbf{a}_{r1} = \frac{\mathbf{r}_1}{r_1} = \frac{3\mathbf{a}_x + \mathbf{a}_y + \mathbf{a}_z}{\sqrt{11}}$$

$$\mathbf{a}_{r2} = \frac{\mathbf{r}_2}{|r_2|} = \frac{\mathbf{a}_x + \mathbf{a}_y + \mathbf{a}_z}{\sqrt{3}}$$

E at (3, 1, 1)

$$= \frac{-10 \times 10^{-9} \times 9 \times 10^9}{(11)^{3/2}}(3\mathbf{a}_x + \mathbf{a}_y + \mathbf{a}_z) + \frac{Q_t \times 9 \times 10^9}{(3)^{3/2}}(\mathbf{a}_x + \mathbf{a}_y + \mathbf{a}_z)$$

$$\mathbf{E} = \frac{-90}{(11)^{3/2}}(3\mathbf{a}_x + \mathbf{a}_y + \mathbf{a}_z) + \frac{Q_t \times 9 \times 10^9}{(3)^{3/2}}(\mathbf{a}_x + \mathbf{a}_y + \mathbf{a}_z)$$

If E_x is to be zero, we should have

$$\frac{-90\times 3}{(11)^{3/2}}+\frac{Q_t\times 9\times 10^9}{(3)^{3/2}}=0 \text{ or } \boxed{Q_t = 4.27\,\text{nC}}$$

Problem 2.13 Two point charges $Q_1 = 5.0\,\text{C}$ and $Q_2 = 1.0\text{ nC}$ are located at $(-1, 1, -3)$ m and $(3, 1, 0)$ m, respectively. Determine the electric field at Q_1.

Solution $Q_1 = 5.0$ C is at $(-1, 1, -3)$ m

$Q_2 = 1.0$ nC is at $(3, 1, 0)$ m

Electric field, **E** at Q_1 $(-1, 1, -3)$ m is

$$\mathbf{E} = \frac{Q_2}{4\pi \epsilon_0 r^2}\mathbf{a}_r$$

$$\mathbf{r} = (-1-3)\,\mathbf{a}_x + (1-1)\,\mathbf{a}_y(-3-0)\,\mathbf{a}_z$$

$$= -4\mathbf{a}_x - 3\mathbf{a}_z$$

$$r = |\mathbf{r}_1| = \sqrt{16+9} = 5$$

$$\mathbf{a}_r = \frac{-4\mathbf{a}_x - 3\mathbf{a}_z}{5}$$

$$\mathbf{E} = \frac{1.0\times 10^{-9}\times 9\times 10^9}{25\times 5}(-4\mathbf{a}_x - 3\mathbf{a}_z)$$

$$\boxed{\mathbf{E} = -0.288\mathbf{a}_x - 0.216\mathbf{a}_z \text{ V/m}}$$

2.9 ELECTRIC FIELD DUE TO LINE CHARGE DENSITY

By definition, line charge density is given by

$$\rho_L = \frac{dQ}{dL},\ \text{C/m}$$

$$dQ = \rho_L\, dL$$

or, $$Q = \int \rho_L dL$$

Here, Q is the total charge.

But electric field due to Q at a distance of r is given by

$$\mathbf{E} = \frac{Q}{4\pi \epsilon_0 r^2}\mathbf{a}_r$$

$$\mathbf{E} = \int \frac{\rho_L \, dL}{4\pi \epsilon_0 \, r^2} \mathbf{a}_r$$

Problem 2.14 If a fine filament carries a uniform charge distribution of ρ_L c/m, find the electric field at a point, $P_2 (x_1, y_1, z_1)$.

Solution Consider a differential length, dL on the line carrying the uniform charge. Let it be at a point $P_1 (x_2, y_2, z_2)$ on the line.

The position $P_1 = \mathbf{r}_1 = x_1 \, \mathbf{a}_x + y_1 \, \mathbf{a}_y + z_1 \, \mathbf{a}_z$

The position $P_2 = \mathbf{r}_2 = x_2 \, \mathbf{a}_x + y_2 \, \mathbf{a}_y + z_2 \, \mathbf{a}_z$

$$\mathbf{r} = \mathbf{r}_2 - \mathbf{r}_1$$

$$= (x_2 - x_1) \, \mathbf{a}_x + (y_2 - y_1) \, \mathbf{a}_y + (z_2 - z_1) \, \mathbf{a}_z$$

$$|\mathbf{r}| = r = |\mathbf{r}_2 - \mathbf{r}_1|$$

$$\mathbf{a}_r = \frac{\mathbf{r}}{r}$$

The electric field strength at P_2 is

$$\mathbf{E} = \frac{1}{4\pi \epsilon_0} \int \frac{\rho_L dL}{r^2} \mathbf{a}_r$$

or,

$$\mathbf{E} = \frac{1}{4\pi \epsilon_0} \int \frac{\rho_L (\mathbf{r})}{r^3} dL$$

Problem 2.15 Find out electric field due to a uniformly charged short line (refer Fig. 2.4).

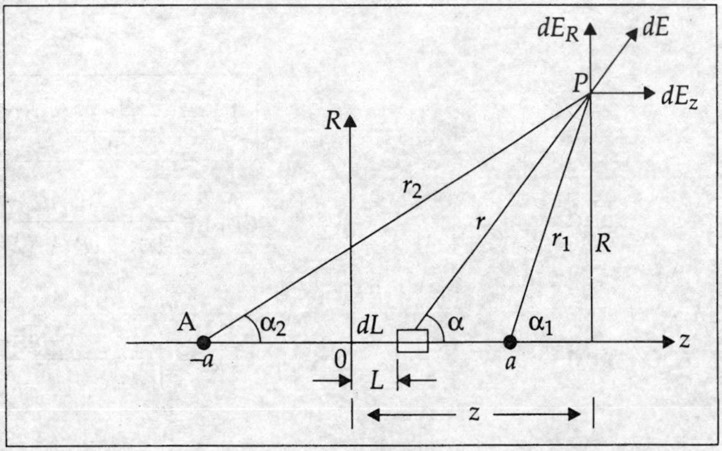

Fig. 2.4 Field due to a finite length of line charge

Solution It may be noted that there exists symmetry with the line charge along z-axis. Assume that the origin is at the centre of the line.

Consider dL at a distance of L from the origin. The magnitude of the charge, dQ of dL length is given by

$$dQ = \rho_L \, dL$$

Now $$r = \sqrt{R^2 + (z-L)^2}$$

$d\mathbf{E}$ at $$P = \frac{\rho_L \, dL}{4\pi \in_0 [R^2 + (z-L)^2]} \mathbf{a}_r$$

or, $$dE_z = \frac{\rho_L \, dL}{4\pi \in_0 [R^2 + (z-L)^2]} \cos\alpha$$

$$E_z = \int_{\alpha_2}^{\alpha_1} \frac{\rho_L \cos\alpha}{4\pi \in_0 [R^2 + (Z-L)^2]} \frac{R^2 + (Z-L)^2}{R} d\alpha$$

Let $$\tan\alpha = \frac{R}{(z-L)}$$

or, $$(z-L) = R\cot\alpha$$

$$-dL = -R \operatorname{cosec}^2 \alpha \, d\alpha$$

$$= \frac{-R[R^2 + (z-L)^2]}{R^2} d\alpha$$

$$= \frac{\rho_L}{4\pi \in_0 R} \int_{\alpha_2}^{\alpha_1} \cos\alpha \, d\alpha$$

$$\boxed{E_z = \frac{\rho_L}{4\pi \in_0 R} (\sin\alpha_1 - \sin\alpha_2)}$$

and $$dE_R = \frac{\rho_L dL}{4\pi \in_0 [R^2 + (z-L)^2]} \sin\alpha$$

and $$E_R = \int_{\alpha_2}^{\alpha_1} \frac{\rho_L \sin\alpha}{4\pi \in_0 [R^2 + (z-L)^2]} \frac{[R^2 + (z-L)^2]}{R} d\alpha$$

$$= \frac{\rho_L}{4\pi \in_0 R} \int_{\alpha_2}^{\alpha_1} \sin\alpha \, d\alpha$$

$$E_R = \frac{\rho_L}{4\pi\epsilon_0 R}(\cos\alpha_2 - \cos\alpha_1)$$

2.10 ELECTRIC FIELD STRENGTH DUE TO INFINITE LINE CHARGE

The electric field due to a uniform infinite line charge is given by

$$\mathbf{E} = \frac{\rho_L}{2\pi\epsilon_0 \rho}\mathbf{a}_\rho$$

Proof For a line charge extending from $-\infty$ to ∞ along z-axis, the electric field does not vary with z when ϕ and ρ are constants. It also does not vary with ϕ when z and ρ are constants due to symmetry.

The position of infinite line charge is shown in Fig. 2.5.

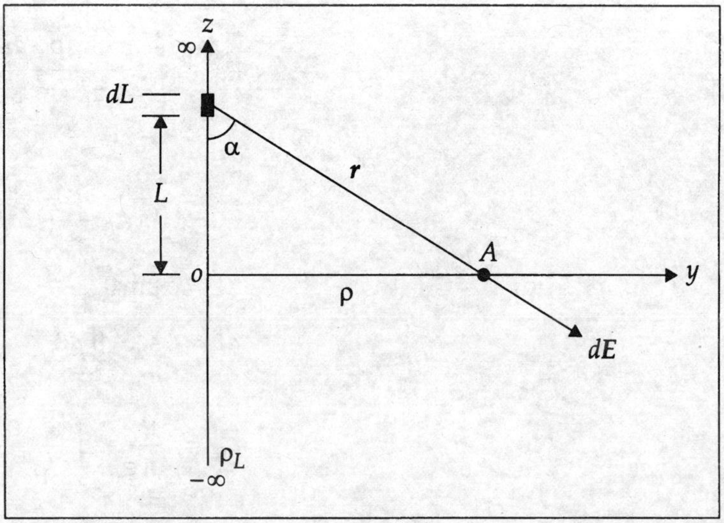

Fig. 2.5 Infinite line charge

Consider a differential length, dL along the line charge. Let it be at a distance of L from the origin. By definition, $\rho_L \equiv \frac{dQ}{dL}$.

The charge dQ of $dL = \rho_L\, dL$.

Consider a point A on y-axis at a distance of ρ from the origin. Let the field at A due to dQ be $d\mathbf{E}$.

From Coulomb's law

$$d\mathbf{E} = \frac{dQ}{4\pi\epsilon_0 r^2}\mathbf{a}_r$$

$$= dE_\rho \mathbf{a}_\rho + dE_z \mathbf{a}_z$$

$$dE_\rho = \frac{\rho_L \, dL \sin\alpha}{4\pi\epsilon_0 r^2}$$

$$= \frac{\rho_L \, dL}{4\pi\epsilon_0 r^2} \cdot \frac{\rho}{r}$$

$$= \frac{\rho_L \, dL\rho}{4\pi\epsilon_0 r^3}$$

Here, $r = |\mathbf{r}| = \sqrt{L^2 + \rho^2}$

So, $$dE_\rho = \frac{\rho\rho_L \, dL}{4\pi\epsilon_0 (L^2 + \rho^2)^{3/2}}$$

$$E_\rho = \int_{-\infty}^{\infty} \frac{\rho\rho_L \, dL}{4\pi\epsilon_0 \, (L^2 + \rho^2)^{3/2}}$$

$$= \frac{\rho_L}{4\pi\epsilon_0} \int_{-\infty}^{\infty} \frac{\rho \, dL}{(L^2 + \rho^2)^{3/2}}$$

Put $$L = \rho \tan\theta$$

$$dL = \rho \sec^2\theta \, d\theta$$

Now $$E_\rho = \frac{\rho_L}{4\pi\epsilon_0} \int_{-\pi/2}^{\pi/2} \frac{\rho^2 \sec^2\theta \, d\theta}{(\rho^2 \tan^2\theta + \rho^2)^{3/2}}$$

$$= \frac{\rho_L}{2\pi\epsilon_0 \rho}$$

As $E_z = 0$ due to symmetry, the field strength is given by

$$\boxed{\mathbf{E} = \frac{\rho_L}{2\pi\epsilon_0 \rho} \mathbf{a}_\rho}$$

Problem 2.16 A charge density of $\rho_L = 10 \text{ PC/m}$ is uniformly distributed along an infinite line. Determine electric field at a point $(\sqrt{2}, \sqrt{2}, 1)$.

Solution The field at $(\sqrt{2}, \sqrt{2}, 1)$ is

$$\mathbf{E} = \frac{\rho_L}{2\pi\epsilon_0 \rho}\mathbf{a}_\rho$$

where $\rho_L = 10\,\text{PC/m}$

$$\rho = \sqrt{x^2 + y^2} \text{ in cylindrical coordinates}$$

$$= \sqrt{2+2} = 2$$

But $\dfrac{1}{4\pi\epsilon_0} = 9\times 10^9$

or $\dfrac{1}{2\pi\epsilon_0} = 18\times 10^9$

So, $\mathbf{E} = \dfrac{18\times 10^9 \times 10\times 10^{-12}}{2}\mathbf{a}_\rho$

$$= 9\times 10^{-2}\mathbf{a}_\rho$$

$$\boxed{\mathbf{E} = 90\mathbf{a}_\rho \text{ mV/m}}$$

Problem 2.17 Find **E** at (2, 0, 2) if a line charge of 10 PC/m lies along the y-axis.

Solution We have $\mathbf{E} = \dfrac{\rho_L}{2\pi\epsilon_0 \rho}\mathbf{a}_\rho$

where $\boldsymbol{\rho} = (2, 0, 2) - (0, 0, 0)$

$$= 2\mathbf{a}_x + 2\mathbf{a}_z$$

$$|\rho| = \rho = \sqrt{4+4} = \sqrt{8}$$

$$\mathbf{a}_\rho = \frac{\boldsymbol{\rho}}{\rho} = \frac{2\mathbf{a}_x + 2\mathbf{a}_z}{\sqrt{8}}$$

$$\frac{1}{2\pi\epsilon_0} = 18\times 10^9$$

$$\mathbf{E} = \frac{18\times 10^9 \times 10\times 10^{-12}}{8}(2\mathbf{a}_x + 2\mathbf{a}_z)$$

or,
$$\mathbf{E} = \frac{18\times10^{-2}}{4}(\mathbf{a}_x + \mathbf{a}_z)$$

$$\boxed{\mathbf{E} = 45.0\,(\mathbf{a}_x + \mathbf{a}_z)\text{ mV/m}}$$

Problem 2.18 Three parallel line charges, $\rho_{L1} = 5\text{ nC/m}$, $\rho_{L2} = 4\text{ nC/m}$ and $\rho_{L3} = -6\text{ nC/m}$ are located at (0, 0), (3, 0) and (0, 4) m, respectively. Find **D** and **E** at (3, 4).

Solution We have
$$\rho_{L1} = 5\text{ nC/at } (0,0)$$
$$\rho_{L2} = 4\text{ nC/m at } (3,0)$$
$$\rho_{L3} = -6\text{ nC/m at } (0,4)$$
$$\mathbf{D} = \mathbf{D}_1 + \mathbf{D}_2 + \mathbf{D}_3$$

where
$$\mathbf{D}_1 = \frac{\rho_{L1}}{2\pi\rho_1}\mathbf{a}_1$$
$$r_1 = (0,0)$$
$$\mathbf{r} = 3\mathbf{a}_x + 4\mathbf{a}_y$$
$$\mathbf{r} - \mathbf{r}_1 = 3\mathbf{a}_x + 4\mathbf{a}_y$$
$$\rho_1 = |\mathbf{r} - \mathbf{r}_1| = \sqrt{9+16} = 5$$
$$\mathbf{a}_1 = \frac{\mathbf{r} - \mathbf{r}_1}{|\mathbf{r} - \mathbf{r}_1|} = \frac{(3\mathbf{a}_x + 4\mathbf{a}_x)}{5}$$
$$\mathbf{D}_1 = \frac{5\times10^{9}}{2\pi\times5}\,\frac{3\mathbf{a}_x + 4\mathbf{a}_y}{5}$$
$$= \frac{10^{-9}}{\pi}(0.3\mathbf{a}_x + 0.4\mathbf{a}_y)$$
$$\mathbf{D}_2 = \frac{\rho_{L2}}{2\pi\rho_2}\mathbf{a}_2$$
$$r_2 = 3\mathbf{a}_x$$
$$\rho_2 = |\mathbf{r} - \mathbf{r}_2|$$
$$\rho_2 = |4\mathbf{a}_y| = 4$$
$$\mathbf{a}_2 = \frac{\mathbf{r} - \mathbf{r}_2}{|\mathbf{r} - \mathbf{r}_2|} = \frac{4\mathbf{a}_y}{4} = \mathbf{a}_y$$

$$\mathbf{D}^2 = \frac{4\times10^{-9}}{2\pi}\mathbf{a}_y = \frac{10^{-9}}{\pi}(0.5\mathbf{a}_y)$$

$$\mathbf{D}_3 = \frac{\rho_{L3}}{2\pi\rho_3}\mathbf{a}_3$$

$$\mathbf{r}_3 = 4\mathbf{a}_y$$

$$\rho_3 = |\mathbf{r}-\mathbf{r}_3|$$

or,

$$\rho_3 = |3\mathbf{a}_x + 4\mathbf{a}_y - 4\mathbf{a}_y| = 3$$

$$\mathbf{a}_3 = \frac{\mathbf{r}-\mathbf{r}_3}{|\mathbf{r}-\mathbf{r}_3|} = \frac{3\mathbf{a}_x}{3} = \mathbf{a}_x$$

$$\mathbf{D}_2 = \frac{4\times10^{-9}}{2\pi\times4}\mathbf{a}_y = \frac{10^{-9}}{\pi}(0.5\mathbf{a}_y)$$

$$\mathbf{D}_3 = \frac{-6\times10^{-9}}{2\pi\times3}\mathbf{a}_x = \frac{-10^{-9}}{\pi}\mathbf{a}_x$$

$$\mathbf{D} = \mathbf{D}_1 + \mathbf{D}_2 + \mathbf{D}_3 = \frac{10^{-9}}{\pi}(0.3\mathbf{a}_x + 0.4\mathbf{a}_y + 0.5\mathbf{a}_y - \mathbf{a}_x)$$

or,

$$\boxed{\mathbf{D} = (-0.222\mathbf{a}_x + 0.286\mathbf{a}_y)\ \text{nC/m}^2}$$

$$\mathbf{E} = \frac{\mathbf{D}}{\epsilon_0} = \frac{1}{8.854\times10^{-12}}\mathbf{D}$$

$$\boxed{\mathbf{E} = (-25\mathbf{a}_x + 32\mathbf{a}_y)\ \text{N/C}}$$

Problem 2.19 Three infinitely long lines charged uniformly are parallel to the *z*-axis. They are separated by a distance of *b* m. The charge density of each is $\rho_L = 2.0\,\text{PC/m}$. Find the electric field **E** at a point *P* on the *y*-axis at $y = a$ m. If $a = b = 1$ m, what is the electric field, **E**?

Solution Consider Fig. 2.6.

Expression for the electric field at a distance of *P* due to an infinitely long charged line, **E** is

$$\mathbf{E} = \frac{\rho_L}{2\pi\epsilon_0\rho}\mathbf{a}_\rho, \text{V/m}$$

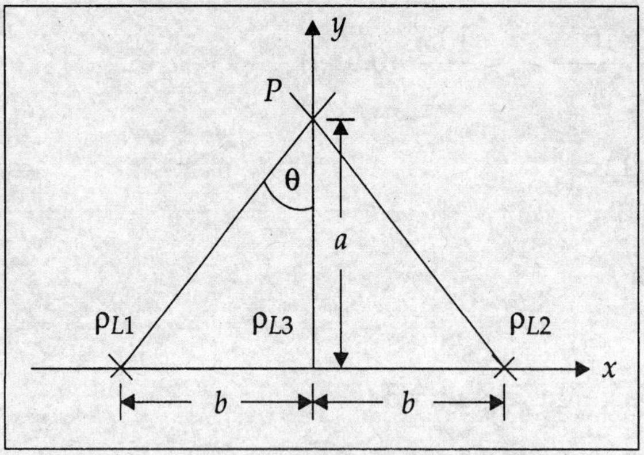

Fig. 2.6 A system of line charges

$\mathbf{E}_{y_1}$ at the point P due to ρ_{L_1}

$$= \frac{1}{2\pi\epsilon_0}\left(\frac{\rho_{L_1}\cos\theta}{\sqrt{a^2+b^2}}\right)$$

$\mathbf{E}_{y_3}$ at the point P due to ρ_{L3}

$$= \frac{1}{2\pi\epsilon_0}\left(\frac{\rho_{L_3}}{a}\right)$$

$\mathbf{E}_{y_2}$ at the point P due to ρ_{L2}

$$= \frac{1}{2\pi\epsilon_0}\left(\frac{\rho_{L_2}\cos\theta}{\sqrt{a^2+b^2}}\right)$$

As $\rho_{L1} = \rho_{L2} = \rho_{L3} = \rho_L$

Total $$E_y = \frac{\rho_L}{2\pi\epsilon_0}\left(\frac{1}{a} + \frac{2\cos\theta}{\sqrt{a^2+b^2}}\right) \text{V/m}$$

$$\cos\theta = \frac{a}{\sqrt{a^2+b^2}}$$

$$E_y = \frac{\rho_L}{2\pi\epsilon_0}\left(\frac{1}{a} + \frac{2a}{a^2+b^2}\right)$$

$$= \frac{\rho_L}{2\pi\epsilon_0}\left(\frac{a^2+b^2+2a^2}{a\,(a^2+b^2)}\right)$$

$$\mathbf{E} = \frac{\rho_L}{2\pi\epsilon_0 a}\left[\frac{(3a^2+b^2)}{(a^2+b^2)}\right]\mathbf{a}_y, \text{V/m}$$

If $a = b = 1\,\text{m}$, $\rho_L = 2.0\,\text{PC/m}$

$$\mathbf{E} = \frac{2\times10^{-12}}{2\pi\epsilon_0 1}\left(\frac{3.1+1}{1+1}\right)\mathbf{a}_y$$

$$= \frac{2\times10^{-12}\times2}{2\pi} = \frac{2\times10^{-12}\mathbf{a}_y}{\pi\epsilon_0}$$

$$\boxed{\mathbf{E} = 0.0719\mathbf{a}_y, \text{V/m}}$$

Problem 2.20 An infinite length of uniform line charge has $\rho_L = 10\,\text{PC/m}$ and it lies along the *z*-axis. Determine the electric field **E** at (4, 3, 3) m.

Solution In cylindrical coordinates,

$$\rho = \sqrt{x^2+y^2} = \sqrt{16+9} = 5\,\text{m}$$

Due to symmetry, **E** is constant with *z*. Hence

$$\mathbf{E} = \frac{\rho_L}{2\pi\epsilon_0\rho}\mathbf{a}_\rho$$

or,

$$\mathbf{E} = 10\times10^{-12}\times18\times10^9\,\mathbf{a}_\rho$$

$$= 180\times10^{-3}\,\mathbf{a}_\rho\ \text{V/m}$$

$$\boxed{\mathbf{E} = 180\mathbf{a}_\rho, \text{mV/m}}$$

Problem 2.21 An infinitely long uniform line distribution of $\rho_L = 3.0\,\text{nC/m}$ is at $y = 3, z = 5$. Determine **E** at (a) (0, 0, 0), (b) (0, 2, 1), (c) (3, 2, 1).

Solution $\rho_L = 3.0\times10^{-9}\ \text{C/m}$ is at $(x, 3, 5)$.

Let us take a general point (x, y, z). Then the radial vector from the location of ρ_L, that is, $(x, 3, 5)$ to (x, y, z) is

$$\mathbf{r} = (y-3)\,\mathbf{a}_y + (z-5)\,\mathbf{a}_z$$

$$r = \sqrt{(y-3)^2+(z-5)^2}$$

$$\mathbf{a}_r = \mathbf{a}_\rho = \frac{(y-3)\,\mathbf{a}_y + (z-5)\,\mathbf{a}_z}{\sqrt{(y-3)^2+(z-5)^2}}$$

$$\mathbf{E} \text{ at } (x, y, z) = \frac{\rho_L}{2\pi\epsilon_0 \rho}\mathbf{a}_\rho$$

$$= 3.0\times10^{-9}\times18\times10^{9}\frac{[(y-3)\,\mathbf{a}_y+(z-5)\,\mathbf{a}_z]}{(y-3)^2+(z-5)^2}$$

(a) **E** at (0, 0, 0)

$$= 54\,\frac{[-3\mathbf{a}_y-5\mathbf{a}_z]}{9+25}$$

$$\boxed{\mathbf{E}=(-4.76\mathbf{a}_y-7.94\mathbf{a}_z)\ \text{V/m}}$$

(b) **E** at (0, 2, 1)

$$= 54\,\frac{[(2-3)\,\mathbf{a}_y+(1-5)\,\mathbf{a}_z]}{1+16},\ \text{V/m}$$

$$\boxed{\mathbf{E}=(-3.17\mathbf{a}_y-12.7\mathbf{a}_z)\ \text{V/m}}$$

(c) **E** at (2, 3, 1)

$$= 54\,\frac{[-4\mathbf{a}_z]}{16}$$

$$\boxed{\mathbf{E}=-13.5\mathbf{a}_z\ \text{V/m}}$$

2.11 FIELD DUE TO SURFACE CHARGE DENSITY, ρ_S (C/m^2)

Field due to surface charge density,

$$\boxed{\mathbf{E} = \frac{\rho_s}{2\epsilon_0}\mathbf{a}_n}$$

Consider an infinitely charged sheet lying in *x-y* plane. Assume that the sheet is uniformly charged.

Consider a strip of a differential width of dx as in Fig. 2.7.

The sheet extends from $-\infty$ to ∞ in both x and y directions. It is obvious that the field does not vary with x or y due to symmetry. Now, there is only z-component.

By definition, $\rho_s = \dfrac{dQ}{dS}$

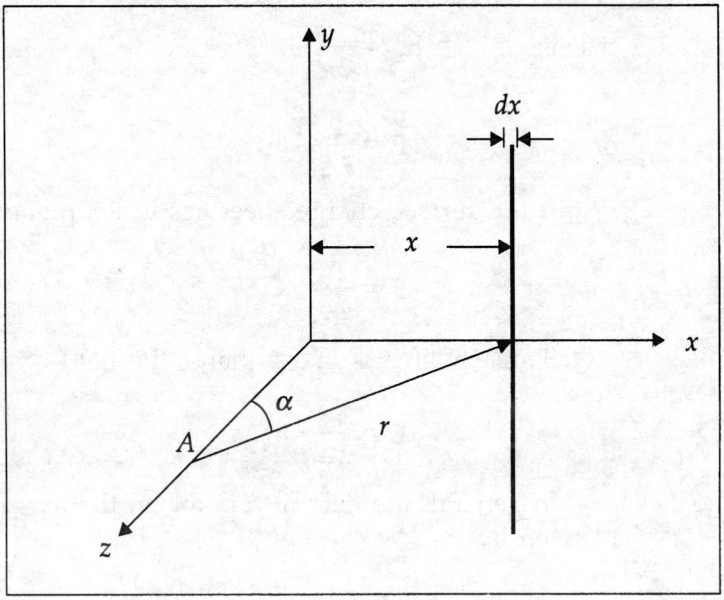

Fig. 2.7 A differential strip in an infinite surface charged sheet

or, $$dQ = \rho_s \, dS$$

$$= \rho_s \, dx \, dy$$

that is, $$\frac{dQ}{dy} = \rho_s \, dx$$

For a differential strip of width dx, we have

$$\rho_L = \rho_S \, dx$$

The field at a point, A on z-axis is given by

$$d\mathbf{E} = \frac{\rho_S \, dx}{2\pi\epsilon_0 \, r} \, \mathbf{a}_r$$

$$\therefore \quad dE_z = \frac{\rho_S \, dx}{2\pi\epsilon_0 \, r} \cos\alpha$$

Here, $$r = \sqrt{z^2 + x^2}$$

$$\cos\alpha = \frac{z}{r}$$

$$\therefore \quad E_z = \frac{\rho_S}{2\pi\epsilon_0} \int_{-\infty}^{\infty} \frac{z^2}{(z^2 + x^2)} \, dx$$

$$= \frac{\rho_S}{2\pi\epsilon_0} \left[\tan^{-1} \frac{x}{z} \right]_{-\infty}^{\infty}$$

that is, $$E_z = \frac{\rho_S}{2\epsilon_0}$$

or, $$E = \frac{\rho_S}{2\epsilon_0}\mathbf{a}_z$$

If the surface charge sheet lies in y-z plane, the field at a point on x-axis is

$$\mathbf{E} = \frac{\rho_S}{2\epsilon_0}\mathbf{a}_x$$

Similarly, if it is in x-z plane, the field at a point on y-axis is

$$\mathbf{E} = \frac{\rho_S}{2\epsilon_0}\mathbf{a}_y$$

In general, the field at a point on the axis normal to the plane of the sheet is given by

$$\mathbf{E} = \frac{\rho_S}{2\epsilon_0}\mathbf{a}_n$$

Problem 2.22 An infinite sheet in x-y plane extending from $-\infty$ to ∞ in both directions has a uniform charge density of 10 nC/m^2. Find the electric field at $z = 1.0\,\text{cm}$.

Solution $$\rho_s = 10\,\text{nC/m}^2$$

For a sheet of charge lying in x-y plane, the field at any point on z-axis is given by

$$\mathbf{E} = \frac{\rho_s}{2\epsilon_0}\mathbf{a}_z$$

$$= \frac{10\times10^{-9}}{2\times 8.854\times10^{-12}}\mathbf{a}_z$$

$$= 0.5647\times10^3\,\mathbf{a}_z$$

$$\boxed{\mathbf{E} = 564\mathbf{a}_z \text{ V/m}}$$

Problem 2.23 A sheet of charge lies in y-z plane at $x = 0$ and has uniform surface charge density of 5.0 PC/m^2. Find the electric field at a point, $P(-5, 0, 0)$ on x-axis.

Solution $$\rho_s = 5.0\,\text{PC/m}^2$$

$$= 5.0\times10^{-12}\,\text{C/m}^2$$

E at $P(-5, 0, 0)$

$$= \frac{-\rho_s}{2\epsilon_0} \mathbf{a}_x$$

$$= -\frac{5\times10^{-12}}{2\times 8.854\times10^{-12}} \mathbf{a}_x$$

$$\boxed{\mathbf{E} = -0.282\mathbf{a}_x, \text{V/m}}$$

Problem 2.24 A point charge, Q is at the centre of a neutral spherical conducting shell. Find the surface charge density at the inner surface and at the outer surface (Fig. 2.8).

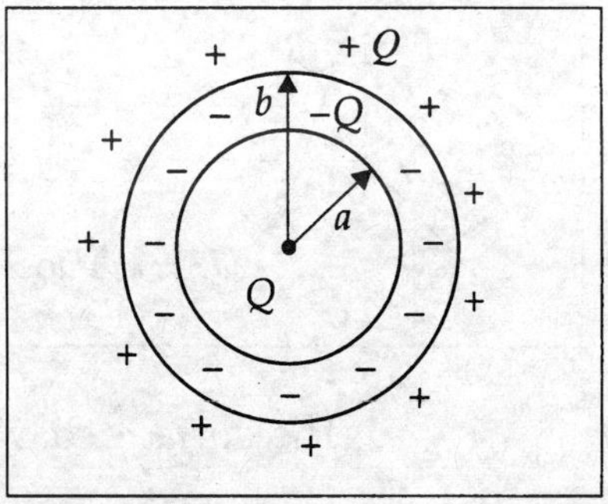

Fig. 2.8 Neutral spherical conducting shell

Solution The surface density at the inner surface is

$$\rho_s \,/\, \text{at } a = -\frac{Q}{4\pi a^2}$$

The surface density at the outer surface is

$$\rho_s \,/\, \text{at } b = \frac{Q}{4\pi b^2}$$

Problem 2.25 A plane $z = 1.0$ m has a uniform charge density of $\rho_s = 2.0\,\text{PC/m}^2$. Find the electric field **E** above the plane.

Solution $\rho_s = 2.0\times10^{-12}\ \text{C/m}^2$

$$\mathbf{E} = \frac{\rho_s}{2\epsilon_0} \mathbf{a}_z$$

$$\frac{1}{4\pi\epsilon_0} = 9\times10^9$$

$$\frac{1}{2\epsilon_0} = 18\pi\times10^9$$

$$\mathbf{E} = 2.0\times10^{-12}\times18\pi\times10^9\,\mathbf{a}_z$$

or,
$$\boxed{\mathbf{E} = 36\pi\mathbf{a}_z \text{ mV/m}}$$

Problem 2.26 Determine the force on a point charge of 5 nC at (0, 0, 5) m due to uniformly distributed charge of 5 mC over a circular disc of radius $r \le 1$ m in $z = 0$ plane.

Solution Consider Fig. 2.9.

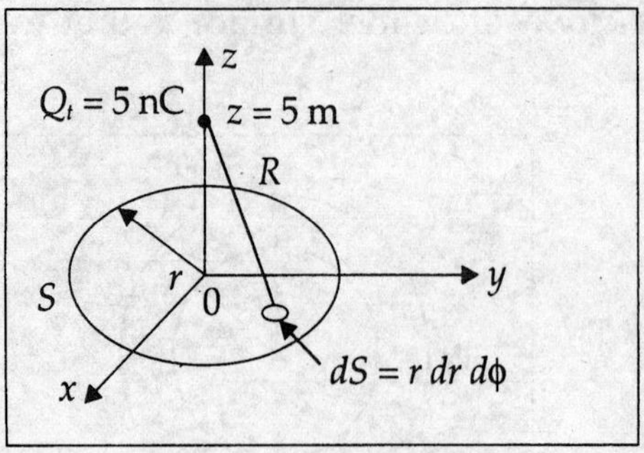

Fig. 2.9 Uniformly charged disc

The surface charge density, ρ_s is

$$\rho_s = \frac{Q}{S} = \frac{5\times10^{-3}}{\pi r^2} = 1.59\text{ mC/m}^2$$

From Fig. 2.9

$$\mathbf{R} = -r\mathbf{a}_r + 5\mathbf{a}_z$$

The differential charge on dS is

$$dQ = \rho_s\, r\, dr\, d\phi$$

Then the differential force due to differential charge is

$$d\mathbf{F} = \frac{dQ\times Q_t}{4\pi\epsilon_0 R^2}\mathbf{a}_R$$

$$= \frac{(\rho_s\, r\, dr\, d\phi)\times5\times10^{-9}}{4\pi\epsilon_0 R^2}\mathbf{a}_R$$

$$\mathbf{a}_R = \frac{-r\mathbf{a}_r + 5\mathbf{a}_z}{\sqrt{r^2+25}}$$

as
$$R = \sqrt{r^2+25}$$

$$d\mathbf{F} = \frac{1.59\times10^{-3}\times5\times10^{-9}}{4\pi\epsilon_0\,(r^2+25)^{3/2}}\, r\,dr\,d\phi\,(-r\mathbf{a}_r + 5\mathbf{a}_z)$$

From Fig. 2.9, it is obvious that the radial components will be cancelled out due to symmetry.

The force on Q_t is

$$\mathbf{F} = \int_0^{2\pi}\int_0^{5} \frac{7.95\times10^{-12}\times9\times10^{9}\times5\,r\,dr\,d\phi}{(r^2+25)^{3/2}}\,\mathbf{a}_z$$

$$= 357.75\times10^{-3}\int_0^{2\pi}\int_0^{5}\frac{r\,dr\,d\phi\,\mathbf{a}_z}{(r^2+25)^{3/2}}$$

$$= 2246.67\times10^{-3}\int_0^{5}\frac{r\,dr}{(r^2+25)^{3/2}}\,\mathbf{a}_z$$

$$= 2246.67\times10^{-3}\left[\frac{-1}{\sqrt{r^2+25}}\right]_0^5\mathbf{a}_z$$

$$= 2246.67\times10^{-3}\times0.0585\,\mathbf{a}_z$$

$$\boxed{\mathbf{F} = 0.1314\,\mathbf{a}_z\ \text{N}}$$

2.12 FIELD DUE TO VOLUME CHARGE DENSITY, ρ_υ (C/m^3)

Volume charge density is defined as

$$\boxed{\rho_\upsilon \equiv \underset{\Delta\upsilon\to0}{\text{Lt}}\ \frac{\Delta Q}{\Delta\upsilon} \equiv \frac{dQ}{d\upsilon}}$$

or,

$$\boxed{Q \equiv \int_\upsilon dQ = \int_\upsilon \rho_\upsilon\, d\upsilon}$$

Here, determination of field due to volume charge density simply involves the estimation of total charge, Q from ρ_υ.

As $$\mathbf{E} = \frac{Q}{4\pi\epsilon_0 r^2}\,\mathbf{a}_r,$$

$$\boxed{\mathbf{E} = \int_{\upsilon} \frac{\rho_\upsilon}{4\pi\epsilon_0 r^2}\, d\upsilon\, \mathbf{a}_r \left(\frac{N}{C}\right)}$$

Problem 2.27 A sphere of volume 0.1 m^3 has a charge density of 8.0 PC/m^3. Find the electric field at a point (2, 0, 0) if the centre of the sphere is at (0, 0, 0).

Solution The field, **E** due to volume charge density is

$$\mathbf{E} = \int \frac{\rho_\upsilon\, d\upsilon}{4\pi\epsilon_0 r^2}\,\mathbf{a}_r$$

$$= \frac{1}{4\pi\epsilon_0}\int \frac{\rho_\upsilon\, d\upsilon}{r^2}\,\mathbf{a}_r$$

$$= \frac{Q}{4\pi\epsilon_0 r^2}\,\mathbf{a}_r$$

$$\rho_\upsilon = 8.0\,\text{PC/m}^3$$

$$\upsilon = 0.1\,\text{m}^3$$

$$Q = \rho_\upsilon \upsilon$$

$$= 8 \times 0.1 = 0.8\,\text{PC}$$

$$\mathbf{r} = (2, 0, 0) - (0, 0, 0)$$

$$= 2\mathbf{a}_x$$

$$r = |\mathbf{r}| = 2$$

$$\mathbf{a}_r = \frac{\mathbf{r}}{r} = \frac{2\mathbf{a}_x}{2} = \mathbf{a}_x$$

$$\mathbf{E} = \frac{9 \times 10^9 \times 0.8 \times 10^{-12}}{4}\,\mathbf{a}_x$$

$$\mathbf{E} = 1.8 \times 10^{-3}\,\mathbf{a}_x$$

$$\boxed{\mathbf{E} = 1.8\mathbf{a}_x,\ \text{mV/m}}$$

Problem 2.28 Find the electric field at a point on the axis of a charged disc (Fig. 2.10).

Solution Consider a differential area shown by the dotted line.

The differential area $= 2\pi r\, dr$

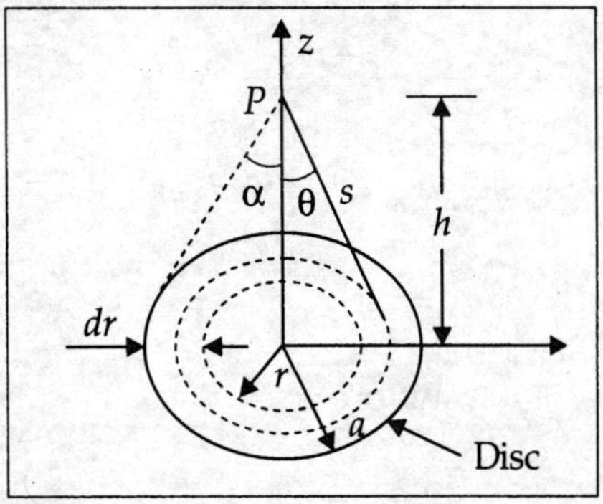

Fig. 2.10 Uniformly charged disc

The field at P $$d\mathbf{E} = \frac{\rho_s 2\pi r\, dr}{4\pi\epsilon_0 s^2}\mathbf{a}_r$$

The vertical component of **E** is

$$dE_z = \frac{\rho_s (2\pi r) dr}{4\pi\epsilon_0 s^2}\cos\theta$$

$$E_z = \int_{r=0}^{a} \frac{\rho_s (2\pi r)\, dr}{4\pi\epsilon_0 s^2}\cos\theta$$

at $r = 0,\ \theta = 0$

at $r = a,\ \theta = \alpha$

Also $r = h\tan\theta$

that is, $dr = h\sec^2\theta\, d\theta$

$$dE_z = \frac{\rho_s \sin^2\theta}{4\pi\epsilon_0 r^2} 2\pi r\, dr \cos\theta$$

$$= \frac{\rho_s \sin^2\theta}{4\pi\epsilon_0 r^2}(2\pi r)\, h\sec^2\theta\cos\theta\, d\theta$$

$$= \frac{\rho_s \sin\theta\, d\theta}{2\epsilon_0}$$

$$\mathbf{E} = \int_{\theta=0}^{\alpha} dE_z \mathbf{a}_n$$

$$= \int_{\theta=0}^{\alpha} \frac{\rho_s}{2\epsilon_0} \sin\theta \, d\theta \, \mathbf{a}_n$$

$$\boxed{\mathbf{E} = \frac{\rho_s}{2\epsilon_0}(1-\cos\alpha)\, \mathbf{a}_n \text{ V/m}}$$

where $$\cos\alpha = \frac{h}{(h^2 + a^2)^{1/2}}$$

Problem 2.29 Determine the charge enclosed in a cylinder (Fig. 2.11) when the volume charge density is $\rho_v = 1.0 e^{-z}(x^2 + y^2)^{-1/4}$ C/m^3.

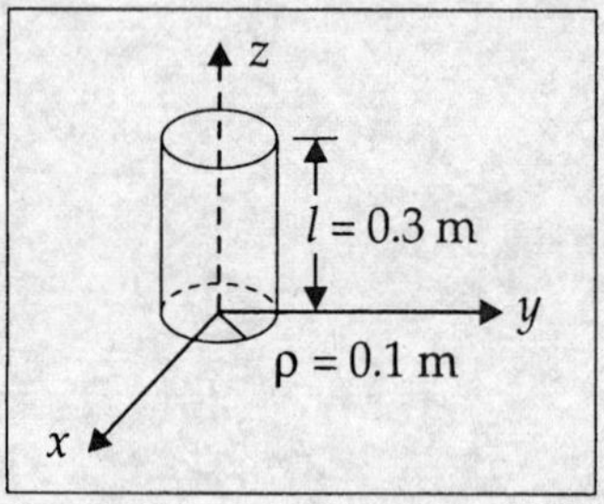

Fig. 2.11 Charge in a cylinder

Solution In cylindrical coordinates,

$$\rho = (x^2 + y^2)^{1/2}$$

$$\rho_v = 1.0 e^{-z} \rho^{-1/2}$$

Hence the total charge

$$Q = \iiint \rho_v \, dv$$

$$dv = \rho \, d\rho \, d\phi \, dz$$

$$Q = \int_0^{0.3}\int_0^{2\pi}\int_0^{0.1} e^{-z} \rho^{-1/2} \rho \, d\rho \, d\phi \, dz$$

$$= \int_0^{0.3}\int_0^{2\pi} e^{-z} \frac{2}{3} [\rho^{3/2}]_0^{0.1} \, d\phi \, dz$$

$$\boxed{Q = 34.3 \text{ mC}}$$

Problem 2.30 In a spherical region, the electric displacement is given by $\mathbf{D} = 10r^2 \, \mathbf{a}_r$ mC/m^2. Find the total charge enclosed by the volume specified by $r = 40$ cm, $\theta = \pi/4$ and $\phi = 2\pi$.

Solution The point form of Gauss's law is

$$\nabla . \mathbf{D} = \rho_\upsilon$$

and

$$Q = \iiint \rho_\upsilon \, d\upsilon$$

$$= \iiint (\nabla . \mathbf{D}) \, d\upsilon$$

$$(\nabla . \mathbf{D})_{\text{spherical}} = \frac{1}{r^2} \frac{\partial}{\partial r} (r^2 D_r) + \frac{1}{r \sin\theta} \frac{\partial}{\partial \theta} (\sin\theta \, D_\theta) + \frac{1}{r \sin\theta} \frac{\partial D_\phi}{\partial \phi}$$

But in the present problem, **D** is only a function of r, or, $D_\phi = 0$, $D_\theta = 0$.

So,
$$\nabla . \mathbf{D} = \frac{1}{r^2} \frac{\partial}{\partial r} (r^2 D_r)$$

$$= \frac{1}{r^2} \frac{\partial}{\partial r} (r^2 \times 10 r^2)$$

$$= \frac{1}{r^2} \frac{\partial}{\partial r} (10 r^4)$$

$$= \frac{1}{r^2} 40 r^3 = 40 r$$

$$Q = \iiint (\nabla . \mathbf{D}) \, d\upsilon$$

$$d\upsilon = r^2 \sin\theta \, dr \, d\theta \, d\phi$$

$$Q = \int_0^{2\pi} \int_0^{\pi/4} \int_0^{0.4} (40 r) \, r^2 \sin\theta \, dr \, d\theta \, d\phi$$

$$= \left[40 \frac{r^4}{4} \right]_0^{0.4} \times [-\cos\theta]_0^{\pi/4} \times [\phi]_0^{2\pi}$$

$$\boxed{Q = 4.7111 \, \text{mC}}$$

2.13 POTENTIAL

A charge at rest produces potential at a specified point. It is a scalar quantity.

Potential at a point due to a fixed charge is defined as the work done in bringing one Coulomb of charge from infinity to the point against the force created by the fixed charge, that is, the potential is the work done per unit charge.

The potential, V at a point due to a fixed charge, Q_f is given by

$$V \equiv \frac{\text{work done to bring a charge, } Q \text{ from } \infty \text{ to the point towards } Q_f}{Q}$$

Simply $$V \equiv \frac{\text{work done}}{Q}, \text{ J/C or volt}$$

2.14 POTENTIAL AT A POINT

The potential at a point due to a point charge is given by

$$\boxed{V = \frac{Q}{4\pi \epsilon_0 r}}$$

Proof Consider a fixed charge, Q and 1 C of charge at an infinite distance (Fig. 2.12). There exists a force on 1 C due to Q. If 1 C of charge is moved against the force of repulsion, some work has to be done.

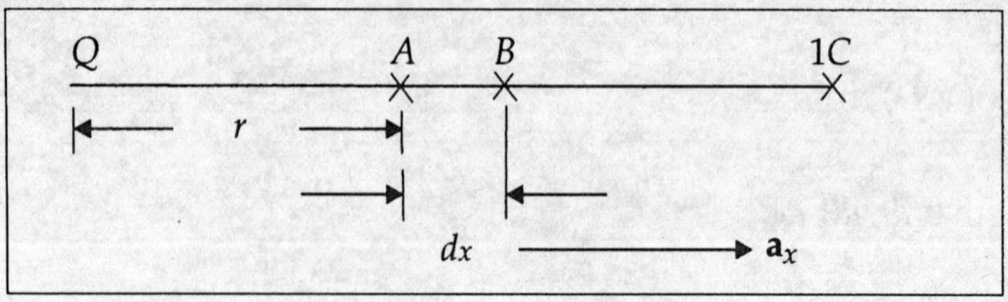

Fig. 2.12 Determination of potential

If 1 C is at point B which is at a distance of x from Q, then force on 1 C due to Q is

$$\mathbf{F} = \frac{Q}{4\pi \epsilon_0 x^2} \mathbf{a}_x$$

If 1 C is moved through a distance of dx in the opposite direction of $\mathbf{a}_x$, then the differential work done is

$$dW = -\mathbf{F}.dx\mathbf{a}_x$$

$$= -\frac{Q}{4\pi \epsilon_0 x^2}$$

Total work done in moving 1 C from ∞ to r is

or,
$$W = -\int_{\infty}^{r} \frac{Q}{4\pi\epsilon_0 x^2} dx$$

$$= -\left[\frac{-Q}{4\pi\epsilon_0 x}\right]_{\infty}^{r}$$

$$W = \frac{Q}{4\pi\epsilon_0 r}$$

This work done is the potential. The potential at a distance of r from Q is

$$\boxed{V = \frac{Q}{4\pi\epsilon_0 r} \text{Joules/C}}$$

Problem 2.31 A charge of 10 PC is at rest in free space. Find the potential at a point, A 10 cm away from the charge.

Solution The potential at A is

$$V = \frac{Q}{4\pi\epsilon_0 r}$$

$$Q = 10\,\text{PC} = 10 \times 10^{-12}\,\text{C}$$

$$r = 10\,\text{cm} = 0.1\,\text{m}$$

$$\frac{1}{4\pi\epsilon_0} = 9 \times 10^9$$

$$V_A = \frac{9 \times 10^9 \times 10 \times 10^{-12}}{0.1}$$

$$= 9 \times 10^{11} \times 10^{-12}$$

$$\boxed{V_A = 0.9\,\text{volts}}$$

Problem 2.32 The potential at a point A is 10 volts and at B it is 15 volts. If a charge, $Q = 10\mu\text{C}$ is moved from A to B, what is the work required to be done?

Solution
$$V_A = 10\,\text{V}$$

$$V_B = 15\,\text{V}$$

$$V_{BA} = 15 - 10 = 5\,\text{volt}$$

By definition $$V_{BA} = \frac{\text{work done}}{Q}$$

$$Q = 10\,\mu\text{C}$$

Work done $$= V_{BA} \times Q$$

$$= 5 \times 10\,\mu\text{C}$$

$$\text{W} = 50\,\mu\text{J}$$

Problem 2.33 Two point charges $Q_1 = 2\,\text{nC}$ and $Q_2 = 4\,\text{nC}$ are located at (1, 1, 1) and (1, 0, 0) respectively. Determine the potential at P (1, 1, 0) due to the point charge.

Solution $$Q_1 = 2\,\text{nC}$$

$$\mathbf{r}_1 = (1,1,1) = \mathbf{a}_x + \mathbf{a}_y + \mathbf{a}_z$$

$$\mathbf{r}_p = (1,1,0) = \mathbf{a}_x + \mathbf{a}_y$$

$$\mathbf{r}_p - \mathbf{r}_1 = \mathbf{a}_x + \mathbf{a}_y - \mathbf{a}_x - \mathbf{a}_y - \mathbf{a}_z$$

$$= -\mathbf{a}_z$$

$$|\mathbf{r}_p - \mathbf{r}_1| = 1.0$$

The potential at P due to Q_1 is

$$V_1 = \frac{Q_1}{4\pi\epsilon_0 |\mathbf{r}_p - \mathbf{r}_1|}$$

$$= \frac{9 \times 10^9 \times 2 \times 10^{-9}}{1.0} = 18\,\text{V}$$

$$Q_2 = 4\,\text{nC}$$

$$\mathbf{r}_2 = (1,0,0) = \mathbf{a}_x$$

$$\mathbf{r}_p - \mathbf{r}_2 = \mathbf{a}_x + \mathbf{a}_y - \mathbf{a}_x = \mathbf{a}_y$$

$$|\mathbf{r}_p - \mathbf{r}_2| = 1.0$$

The potential at P due to Q_2 is

$$V_2 = \frac{Q_2}{4\pi\epsilon_0 |\mathbf{r}_p - \mathbf{r}_2|}$$

$$= \frac{9 \times 10^9 \times 4 \times 10^{-9}}{1} = 36\,\text{V}$$

The total potential is $V = V_1 + V_2 = 18 + 36$

$$\boxed{V = 54 \text{ volts}}$$

Problem 2.34 An electric field is given by $\mathbf{E} = 10y\,\mathbf{a}_x + 10x\,\mathbf{a}_y$, V/m. Find the potential function, V. Assume $V = 0$ at the origin.

Solution We have

$$\mathbf{E} = -\nabla V$$

$$= \frac{-\partial V}{\partial x}\mathbf{a}_x - \frac{\partial V}{\partial y}\mathbf{a}_y - \frac{\partial V}{\partial z}\mathbf{a}_z$$

or,

$$V = -\int \mathbf{E}.d\mathbf{L},\ d\mathbf{L} = dx\,\mathbf{a}_x + dy\,\mathbf{a}_y + dz\,\mathbf{a}_z$$

$$V = -\int 10y\,dx - \int 10x\,dy$$

$$V = -10xy + B - 10xy + C$$

At $x = 0$ and $y = 0$, $V = 0$

$$B + C = 0$$

$$\boxed{V = -20xy \text{ volts}}$$

$$V/(0,0,0) = 0$$

$$\boxed{V = 0 \text{ volts}}$$

2.15 POTENTIAL DIFFERENCE

The potential difference between two points A and B is defined as the work done by an applied force in moving a unit positive charge from A to B in electric field.

The work done,

$$W = -Q\int_A^B \mathbf{E}.d\mathbf{L}$$

or,

$$V_{AB} \equiv \frac{W}{Q} \equiv -\int_A^B \mathbf{E}.d\mathbf{L}$$

where Q = charge that is being moved from A to B.

Potential difference between A and B is also defined as the difference between the potentials at A and B.

Let V_A be the potential at A. V_B be the potential at B. Then

$$V_A = -\int_{\infty}^{A} \mathbf{E}.d\mathbf{L}$$

$$V_B = -\int_{\infty}^{B} \mathbf{E}.d\mathbf{L}$$

$$\boxed{V_{AB} = V_A - V_B}$$

2.16 SALIENT FEATURES OF POTENTIAL DIFFERENCE

1. Potential difference depends only on the initial and final points.
2. It does not depend on the path between the points.
3. It is zero around a closed path,

   that is, $$V = -\oint \mathbf{E}.d\mathbf{L} = 0$$

   or, $$V = -\int_{A}^{A} \mathbf{E}.d\mathbf{L} = 0$$

4. Negative value of V_{AB} represents loss in potential energy in moving Q from A to B.
5. Positive value of V_{AB} represents gain in potential energy.
6. V_{AB} has the units of Joules/Coulombs or Volt.
7. V_{AB} depends on the distance between A and B.
8. V_{AB} is created by a fixed charge.
9. If B is reference at ∞ from the fixed charge, V_{AB} is the potential of A itself.
10. The potential at ∞ is zero.

Problem 2.35 A point charge, Q = 10 nC is at the origin. Determine the potential difference at A (1, 0, 0) with respect to B (2, 0, 0).

Solution

$$Q = 10\,\text{nC} = 10 \times 10^{-9}\,\text{C}$$

$$r_A = 1\,\text{m}$$

$$r_B = 2\,\text{m}$$

The potential difference,

$$V_{AB} = \frac{Q}{4\pi\epsilon_0}\left(\frac{1}{r_A} - \frac{1}{r_B}\right)$$

$$= 10\times10^{-9} \times 9\times10^{9}\left(\frac{1}{1} - \frac{1}{2}\right)$$

$$\boxed{V_{AB} = 45\,\text{V}}$$

2.17 POTENTIAL GRADIENT

Potential gradient is defined as the gradient of potential, that is,

Potential gradient $\equiv \nabla V$

where ∇ is vector differential operator and V is scalar potential.

2.18 SALIENT FEATURES OF POTENTIAL GRADIENT

1. Potential gradient is a vector.
2. It is always normal to equi-potential surfaces everywhere.
3. It lies in the direction of maximum increase of potential.
4. Negative potential gradient gives the electric field, that is,

$$\mathbf{E} = -\nabla V$$

5. The electric field and potential gradient are in opposite directions.
6. The potential gradient in different coordinate systems is given by

$$(\nabla V)_{\text{Cartesian}} = \frac{\partial V}{\partial x}\mathbf{a}_x + \frac{\partial V}{\partial y}\mathbf{a}_y + \frac{\partial V}{\partial z}\mathbf{a}_z$$

$$(\nabla V)_{\text{cylindrical}} = \frac{\partial V}{\partial \rho}\mathbf{a}_\rho + \frac{1}{\rho}\frac{\partial V}{\partial \phi}\mathbf{a}_\phi + \frac{\partial V}{\partial z}\mathbf{a}_z$$

$$(\nabla V)_{\text{spherical}} = \frac{\partial V}{\partial r}\mathbf{a}_r + \frac{1}{r}\frac{\partial V}{\partial \theta}\mathbf{a}_\theta + \frac{\partial V}{r\sin\theta\,\partial\phi}\mathbf{a}_\phi$$

Problem 2.36 If the potential function, V is given by $V = x^3y - xy^2 + 3z$, find the potential gradient.

Solution $V = x^3y - xy^2 + 3z$

The potential gradient is given by

$$(\nabla V) = \frac{\partial V}{\partial x}\mathbf{a}_x + \frac{\partial V}{\partial y}\mathbf{a}_y + \frac{\partial V}{\partial z}\mathbf{a}_z$$

$$\frac{\partial V}{\partial x} = \frac{\partial}{\partial x}(x^3 y - xy^2 + 3z)$$

$$= 3x^2 y - y^2$$

$$\frac{\partial V}{\partial y} = x^3 - 2xy$$

$$\frac{\partial V}{\partial z} = 3$$

$$\boxed{\nabla V = (3x^2 y - y^2)\,\mathbf{a}_x + (x^3 - 2xy)\,\mathbf{a}_y + 3\mathbf{a}_z,\ \text{V/m}}$$

2.19 EQUIPOTENTIAL SURFACE

Equipotential surface is one on which the potential is the same on the entire surface.

Gradient of the potential and the equipotential surface are orthogonal to each other.

2.20 POTENTIAL DUE TO ELECTRIC DIPOLE

An electric dipole is defined as a pair of opposite polarity with identical magnitude and with a small distance between them.

Potential due to a dipole

$$V = \frac{Qd\cos\theta}{4\pi\epsilon_0 r^2} = \frac{p\cos\theta}{4\pi\epsilon_0 r^2}\ \text{volts}$$

Here, Q = charge

d = distance between two charges

r = distance of the point from the centre of the pair of charges

p = dipole moment = Qd (C-m)

Electric field at a point due to a dipole

$$\mathbf{E} = \frac{Qd}{4\pi\epsilon_0 r^3}(2\cos\theta\,\mathbf{a}_r + \sin\theta\,\mathbf{a}_\theta)\ \text{V/m}$$

Proof Fig. 2.13 shows a dipole (A pair of charges)

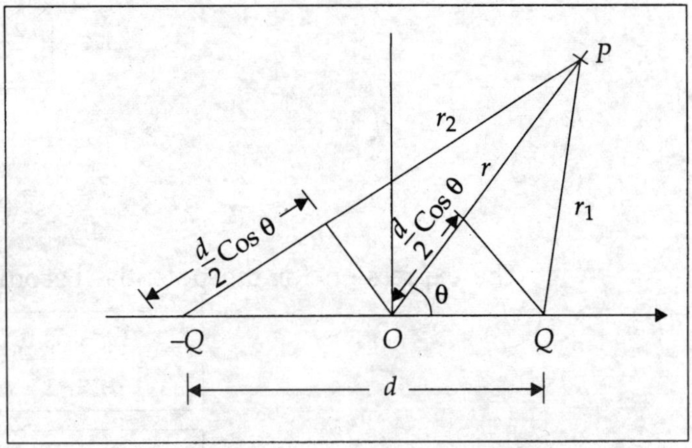

Fig. 2.13 An electric dipole

Let P be a point at which the potential and electric field are to be determined. Let r_1 be the distance of P from (–) ve charge and r_2 be the distance of P from (+) ve charge and r be the distance from the centre of the dipole.

From Fig. 2.13, we have

$$r_1 \approx r + \frac{d}{2}\cos\theta$$

$$r_2 \approx r - \frac{d}{2}\cos\theta$$

Potential at P due to Q and $-Q$ is

$$V = \frac{Q}{4\pi\epsilon_0 r_2} + \frac{-Q}{4\pi\epsilon_0 r_1}$$

$$V = \frac{Q}{4\pi\epsilon_0\left(r - \frac{d}{2}\cos\theta\right)} + \frac{(-Q)}{4\pi\epsilon_0\left(r + \frac{d}{2}\cos\theta\right)}$$

$$= \frac{Q}{4\pi\epsilon_0}\left[\frac{r + \frac{d}{2}\cos\theta - r + \frac{d}{2}\cos\theta}{\left(r^2 - \frac{d^4}{4}\cos^2\theta\right)}\right]$$

$$= \frac{Qd\cos\theta}{4\pi\epsilon_0\left(r^2 - \frac{d^2}{4}\cos^2\theta\right)}$$

if $$r >> d$$

$$r^2 >> \frac{d^2}{4}\cos^2\theta$$

The expression for the potential becomes

$$\boxed{V \approx \frac{Qd\cos\theta}{4\pi\epsilon_0 r^2} \text{ volts}}$$

or, $$\boxed{V \approx \frac{p\cos\theta}{4\pi\epsilon_0 r^2} \text{ volts}}$$

where $p = Qd =$ electric dipole moment.

If $d\cos\theta$ is written as

$$d\cos\theta = \mathbf{d}.\mathbf{a}_r$$

$$\mathbf{d} = d\mathbf{a}_x$$

Potential V becomes

$$V = \frac{Q\mathbf{d}\cdot\mathbf{a}_r}{4\pi\epsilon_0 r^2}$$

$$= \frac{\mathbf{p}\cdot\mathbf{a}_r}{4\pi\epsilon_0 r^2}$$

where $\mathbf{p} \equiv Q\mathbf{d}$ is known as dipole moment. It is a vector and has the direction of $\mathbf{d}$. Here $\mathbf{d}$ is the distance vector from $-Q$ to Q. If the centre of the dipole is at $\mathbf{r}'$ instead of at the origin, potential is given by

$$\boxed{V = \frac{\mathbf{p}\cdot(\mathbf{r}-\mathbf{r}')}{4\pi\epsilon_0 (r-r')^3}}$$

Problem 2.37 An electric dipole represented by $0.1a_y$ nC-m is at the origin. Find the potential at a point P (0, 10, 0).

Solution We have

$$V = \frac{\mathbf{p}.\mathbf{a}_r}{4\pi\epsilon_0 r^2}$$

$$\mathbf{p} = 0.1\mathbf{a}_y \text{ nC-m}$$
$$= 0.1\times10^{-9}\, \mathbf{a}_y \text{ C-m}$$
$$= 10^{-10}\, a_y \text{ C-m}$$

$$\frac{1}{4\pi\epsilon_0} = 9\times10^9$$

$$\mathbf{r} = (0,10,0)-(0,0,0)$$
$$= 10\mathbf{a}_y$$
$$|\mathbf{r}| = r = 10$$
$$\mathbf{a}_r = \frac{\mathbf{r}}{|\mathbf{r}|} = \frac{10\mathbf{a}_y}{10} = \mathbf{a}_y$$

$$V = \frac{9\times10^9\times10^{-10}}{10^2}$$

$$\boxed{V = 9\times10^{-3} = 9 \text{ mV}}$$

Problem 2.38 An electric dipole, $1.0a_y$ nC-m is located at (0, 0, 0). Find the potential at $\left(1, \frac{\pi}{4}, \frac{\pi}{2}\right)$.

Solution The potential due to an electric dipole is given by

$$V = \frac{p\cos\theta}{4\pi\epsilon_0 r^2}$$

where $p = 1.0\times10^{-9}$ C-m

$$\frac{1}{4\pi\epsilon_0} = 9\times10^9$$

$$r = 1.0,\ \theta = \frac{\pi}{4}$$

$$V = \frac{9\times10^9\times1.0\times10^{-9}\times0.707}{1} = 6.3 \text{ V}$$

Hence $\boxed{V = 6.3 \text{ volt}}$

2.21 ELECTRIC FIELD DUE TO DIPOLE

In spherical coordinates, the gradient of the potential is given by

$$\mathbf{E} = -\nabla V = -\left[\frac{\partial V}{\partial r}\mathbf{a}_r + \frac{1}{r}\frac{\partial V}{\partial \theta}\mathbf{a}_\theta + \frac{1}{r\sin\theta}\frac{\partial V}{\partial \phi}\mathbf{a}_\phi\right]$$

where $$V = \frac{Qd\cos\theta}{4\pi\epsilon_0 r^2}, \text{ volt}$$

It is evident from the above expression that V is not a function of ϕ. Therefore,

$$E_\phi = \frac{1}{r\sin\theta}\frac{\partial V}{\partial \phi} = 0$$

E is only a function of r and θ.

$$\mathbf{E} = -\left[\frac{\partial}{\partial r}\left(\frac{Qd\cos\theta}{4\pi\epsilon_0 r^2}\right)\mathbf{a}_r + \frac{1}{r}\frac{\partial}{\partial\theta}\left(\frac{Qd\cos\theta}{4\pi\epsilon_0 r^2}\right)\mathbf{a}_\theta\right]$$

$$= -\left[-\frac{2Qd\cos\theta}{4\pi\epsilon_0 r^3}\mathbf{a}_r - \frac{Qd\sin\theta}{4\pi\epsilon_0 r^3}\mathbf{a}_\theta\right]$$

$$\boxed{\mathbf{E} = \frac{Qd}{4\pi\epsilon_0 r^3}(2\cos\theta\,\mathbf{a}_r + \sin\theta\,\mathbf{a}_\theta), \text{V/m}}$$

or in terms of electric dipole moment, **E** is expressed as

$$\boxed{\mathbf{E} = \frac{p}{4\pi\epsilon_0 r^3}(2\cos\theta\,\mathbf{a}_r + \sin\theta\,\mathbf{a}_\theta), \text{V/m}}$$

Problem 2.39 An electric dipole, $1.0\mathbf{a}_z$ nC-m is at (0, 0, 0). Find the electric field at (0, 0, 1).

Solution We have

$$\mathbf{E} = \frac{\mathbf{p}.\mathbf{a}_r}{2\pi\epsilon_0 r^3}\mathbf{a}_r$$

$$= \frac{Qd\,\mathbf{a}_z.\mathbf{a}_z}{2\pi\epsilon_0 r^3}\mathbf{a}_r$$

$$Qd = p = 1.0\times 10^{-9}\text{ C-m}$$

$$\mathbf{r} = (0,0,1) - (0,0,0) = \mathbf{a}_z$$

$$r = 1$$

$$\mathbf{E} = \frac{18\times10^{9}\times10^{-9}}{1^{3}}\,\mathbf{a}_r$$

$$\boxed{\mathbf{E} = 18\mathbf{a}_r, \text{V/m}}$$

Problem 2.40 If an electric dipole located at the origin is represented by $0.1a_z$ nC-m, find **E** at $\left(1, \frac{\pi}{2}, \frac{\pi}{2}\right)$.

Solution We have

$$\mathbf{E} = \frac{Qd\cos\theta}{2\pi\epsilon_0 r^3}\mathbf{a}_r + \frac{Qd\sin\theta}{4\pi\epsilon_0 r^3}\mathbf{a}_\theta$$

$$Qd = p = 0.1\times10^{-9}\text{ C-m}$$

$$\theta = \frac{\pi}{4}, \phi = \frac{\pi}{2}$$

$$r = 1$$

$$\mathbf{E} = \frac{18\times10^{9}\times0.1\times10^{-9}}{1^{3}}\times\cos\frac{\pi}{4}\mathbf{a}_r + \frac{9\times10^{9}\times0.1\times10^{-9}}{1^{3}}\times\sin\frac{\pi}{4}\mathbf{a}_\theta$$

$$= 1.8\times0.707\mathbf{a}_r + 0.9\times0.707\mathbf{a}_\theta$$

$$\boxed{\mathbf{E} = 1.272\mathbf{a}_r + 0.636\mathbf{a}_\theta, \text{V/m}}$$

2.22 ELECTRIC FLUX

Electric flux is also known as **Electric Displacement Flux.**

Definition 1 Electric flux is defined as the displaced charge, that is,

Electric flux, $\Psi \equiv Q$, Coulomb.

Definition 2 Electric flux is defined as the surface integral of electric flux density, that is,

Electric flux, $\Psi \equiv \int_S \mathbf{D}.d\mathbf{S}$

2.23 SALIENT FEATURES OF ELECTRIC FLUX

1. It is independent of the medium.
2. The electric field creates a force on a charge and hence the charge moves along a certain path. This path is called the flux line.
3. The force between two charges acts along a certain path. This path is also called the flux line.
4. Magnitude of flux depends only on the charge from which it originates.
5. The flux lines are equal to the charge in Coulombs.
6. Flux line is only an imaginary line.
7. Its direction is the same as that of the electric field.
8. The flux lines from a point charge are shown in Fig. 2.14.

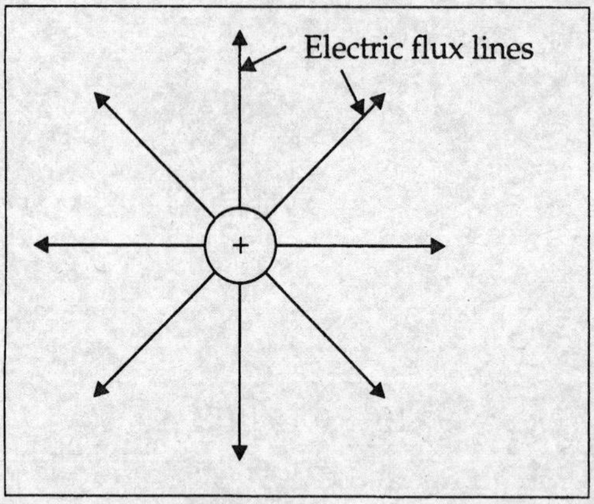

Fig. 2.14 Field from an isolated charge

9. The flux lines between a (+)ve and a (–)ve point charges are shown in Fig. 2.15.

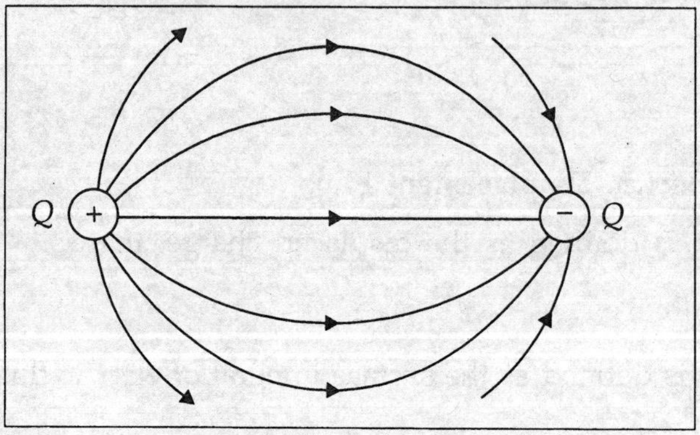

Fig. 2.15 The field in a system of equal but opposite charges

10. It is a scalar quantity.
11. The flux lines from a pair of (+)ve charges are shown in Fig. 2.16.

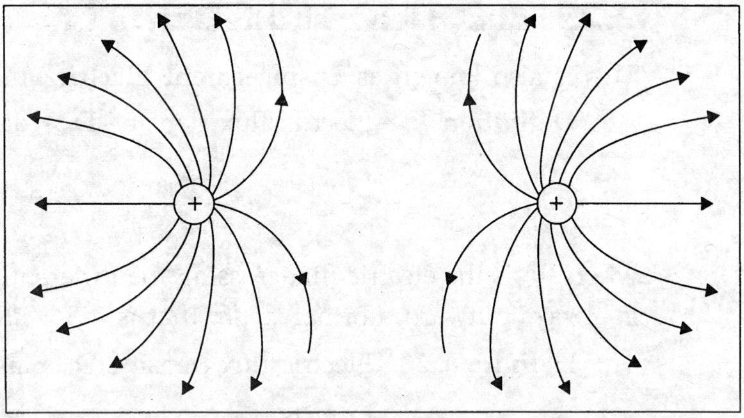

Fig. 2.16 Flux lines from two (+)ve charges

2.24 FARADAY'S EXPERIMENT TO DEFINE FLUX

Apparatus used One small metallic sphere, two hemispheres forming a full sphere larger than the small sphere, if held together over a dielectric material.

Procedure See the steps given below:

1. Inner sphere is charged to Q Coulombs.
2. Dielectric material is pasted radially over the sphere with uniform thickness.
3. Two hemispheres are held together on the dielectric strip so that they form an outer sphere.
4. Outer sphere is momentarily grounded to discharge its charge.
5. The two hemispheres are removed with insulated tools without disturbing the induced charge.
6. The charge on each sphere is measured.
7. The total induced charge on the outer sphere is found to be negative.
8. The magnitude of the induced charge is equal to that of the inner sphere.
9. The displacement took place from inner to outer sphere through the dielectric material.
10. This displacement is known as displacement flux.
11. The displacement flux is also called electric flux.
12. This flux is independent of the type of dielectric material.
13. It is independent of the separation between inner and outer spheres.
14. The final result of Faraday's experiment is,

Electric flux, $\psi = Q$, Coulombs.

2.25 ELECTRIC FLUX DENSITY

This is also known as **Displacement Electric Flux Density**.

Definition 1 Electric flux density, **D** is defined as

$$\mathbf{D} \equiv \frac{d\psi}{dS}\mathbf{a}_n, \mathrm{C/m^2}$$

where ψ is the electric flux crossing the differential area, $d\mathbf{S}$. The direction of $d\mathbf{S}$ is always outward, normal to dS, that is, $d\mathbf{S} = dS\ \mathbf{a}_n$.

Definition 2 Electric flux density, **D** is also defined as

$$\mathbf{D} \equiv \epsilon\,\mathbf{E}, \mathrm{C/m^2}$$

where ϵ_0 = permittivity of free space, F/m

E = electric field strength, V/m

2.26 SALIENT FEATURES OF ELECTRIC FLUX DENSITY, *D*

1. The unit of electric flux density is $\mathrm{C/m^2}$.
2. It is a vector.
3. It is inversely proportional to r^2, r being the radius of the sphere.
4. In free space, **D** is in the direction of **E**.
5. **D** in a Gaussian surface is determined from Gauss's law.
6. **D** is independent of the medium.
7. **D** is given by $\mathbf{D} = \dfrac{Q}{4\pi r^2}\mathbf{a}_r$
8. **D** in a general medium is given by

$$\mathbf{D} = \epsilon\,\mathbf{E}$$

9. **D** in a dielectric medium is given by

$$\mathbf{D} = \epsilon_0\mathbf{E} + \mathbf{P}$$

P = polarisation of medium

Problem 2.41 If an electric field in free space is given by

$$\mathbf{E} = \mathbf{a}_x + 2\mathbf{a}_y + 5\mathbf{a}_z \ \mathrm{V/m},$$

find the electric flux density.

Solution Electric field, $\mathbf{E} = \mathbf{a}_x + 2\mathbf{a}_y + 5\mathbf{a}_z$

$$\epsilon_0 = 8.854 \times 10^{-12} \text{ F/m}$$

$$\mathbf{D} = \epsilon_0 \mathbf{E}$$

$$= 8.854 \times 10^{-12} (\mathbf{a}_x + 2\mathbf{a}_y + 5\mathbf{a}_z)$$

$$\boxed{\mathbf{D} = (8.854\mathbf{a}_x + 17.7\mathbf{a}_y + 44.27\mathbf{a}_z), \text{PC/m}^2}$$

Problem 2.42 A point charge, Q = 10 nC is at the origin in free space. Find the electric field at P (1, 0, 1). Also find the electric flux density at P.

Solution

$$Q = 10 \text{ nC} = 10 \times 10^{-9} \text{ C}$$

$$P = (1, 0, 1)$$

$$\mathbf{r} = (1, 0, 1) - (0, 0, 0)$$

$$= \mathbf{a}_x + \mathbf{a}_z$$

$$r = |\mathbf{r}| = \sqrt{1+1} = \sqrt{2}$$

$$\mathbf{a}_r = \frac{1}{\sqrt{2}}(\mathbf{a}_x + \mathbf{a}_z)$$

$$\frac{1}{4\pi\epsilon_0} = 9 \times 10^9$$

Electric field,

$$\mathbf{E} = \frac{Q}{4\pi\epsilon_0 r^2}\mathbf{a}_r$$

$$= \frac{10 \times 10^{-9} \times 9 \times 10^9}{(\sqrt{2})^2}\mathbf{a}_r$$

$$= \frac{90}{2}\mathbf{a}_r$$

$$= \frac{90}{2 \times \sqrt{2}}(\mathbf{a}_x + \mathbf{a}_z)$$

$$\boxed{\mathbf{E} = 31.8(\mathbf{a}_x + \mathbf{a}_z), \text{V/m}}$$

The electric flux density,

$$\mathbf{D} = \epsilon_0 \mathbf{E}$$

$$= 8.854 \times 10^{-12} \times 31.8 \, (\mathbf{a}_x + \mathbf{a}_z)$$

$$\boxed{\mathbf{D} = 281.55 \, (\mathbf{a}_x + \mathbf{a}_z), \text{PC/m}^2}$$

Problem 2.43 What is the electric flux, ψ that passes the surface shown in Fig. 2.17, if the displacement flux density is $\mathbf{D} = y\mathbf{a}_x + x\mathbf{a}_y$ mC/m^2.

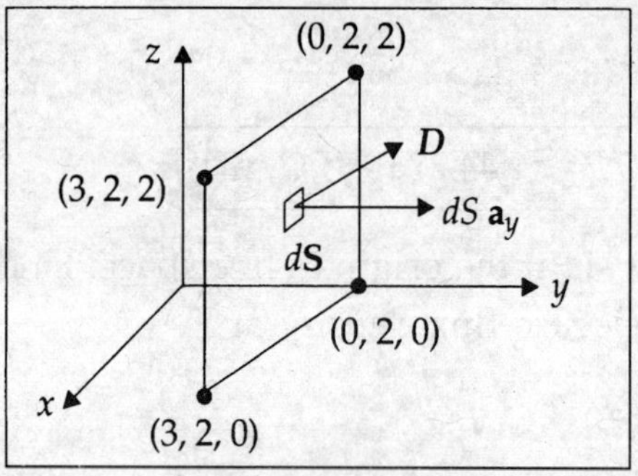

Fig. 2.17 Electric flux

Solution Electric flux density is given by

$$\mathbf{D} = y\mathbf{a}_x + x\mathbf{a}_y \text{ (mC/m}^2\text{)}$$

The differential area, d**S** is given by

$$d\mathbf{S} = dx\, dz\, \mathbf{a}_y$$

The differential flux, $d\psi$ passing through $d\mathbf{S}$ is

$$d\psi = \mathbf{D}.d\mathbf{S}$$

$$= (y\mathbf{a}_x + x\mathbf{a}_y) \times 10^{-3} .dx\, dz\, \mathbf{a}_y$$

$$= x\, dx\, dz \times 10^{-3} \text{ C}$$

$$\psi = \int_0^2 \int_0^3 x\, dx\, dz \text{ mC}$$

$$= \int_0^2 \left(\frac{x^2}{2} \right)_0^3 dz$$

$$\int_0^2 \frac{9}{2} \text{dz}$$

$$\boxed{\psi = 9 \text{ mC}}$$

2.27 GAUSS'S LAW AND APPLICATIONS

Generalised Faraday's law is Gauss's law.

Gauss's law It states that the net flux passing through any closed surface is equal to the charge enclosed by that surface, that is,

$$\oint_s \mathbf{D}.d\mathbf{s} = Q_{enc}$$

This is known as Gauss's law in integral form. Gauss's law is applicable only on Gaussian surfaces.

Proof Consider a spherical surface which encloses a charge Q at its centre (Fig. 2.18).

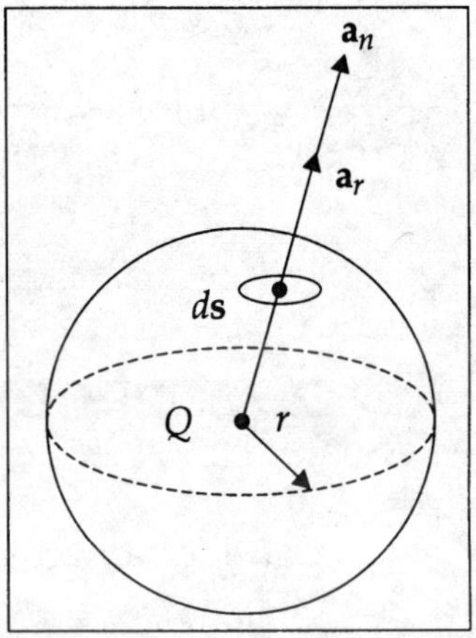

Fig. 2.18 Spherical surface enclosing a charge, Q

The differential area, ds is on the surface of the sphere whose direction is $\mathbf{a}_n$. Let r be the radius of the sphere.

The electric field, $\mathbf{E}$ at the spherical surface is given by

$$\mathbf{E} = \frac{Q}{4\pi\epsilon_0 r^2}\mathbf{a}_r$$

But

$$\mathbf{D} = \epsilon_0 \mathbf{E} = \frac{Q}{4\pi r^2}\mathbf{a}_r$$

Taking dot product with ds on both sides, we get

$$\mathbf{D}.ds = \frac{Q}{4\pi r^2}\mathbf{a}_r \,.\, ds\, \mathbf{a}_n$$

For the spherical surface under consideration, $\mathbf{a}_r$ and $\mathbf{a}_n$ are in the same direction.

$$\mathbf{D}.ds = \frac{Q}{4\pi r^2} ds$$

Taking surface integral on both sides, we get

$$\oint_s \mathbf{D}.ds = \int_s \frac{Q}{4\pi r^2} ds$$

$$= \frac{Q}{4\pi r^2} \oint_s d\mathbf{s}$$

$$= \frac{Q}{4\pi r^2} S$$

But $S = 4\pi r^2$ for a sphere. So,

$$\boxed{\oint_s \mathbf{D}.d\mathbf{s} = Q} \quad \text{Hence proved.}$$

2.28 PROOF OF GAUSS'S LAW (ON ARBITRARY SURFACE)

Consider an arbitrary surface of Fig. 2.19.

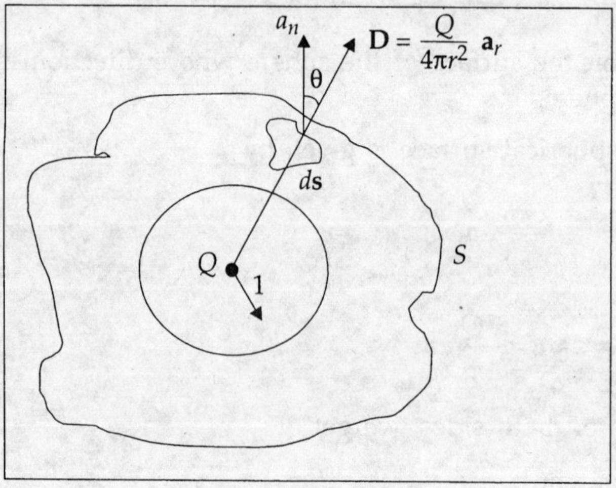

Fig. 2.19 Arbitrary surface to prove Gauss's law

Consider a sphere of radius one metre within the surface, S which encloses a charge, Q at its centre.

Electric flux density, **D** at $d\mathbf{s}$ on the arbitrary surface is

$$\mathbf{D} = \frac{Q}{4\pi r^2}\mathbf{a}_r$$

Taking dot product with $d\mathbf{s}$ on both sides

$$\mathbf{D}.d\mathbf{s} = \frac{Q}{4\pi r^2}\mathbf{a}_r \cdot \mathbf{a}_n \; ds$$

Here, if $\mathbf{a}_r$ and $\mathbf{a}_n$ make an angle of θ, then

$$\mathbf{D}.ds = \frac{Q}{4\pi r^2} ds\cos\theta$$

Now taking surface integral on both sides,

$$\oint_s \mathbf{D}.ds = \oint_s \frac{Q}{4\pi r^2} ds\cos\theta$$

$$= \frac{Q}{4\pi} \oint_s \frac{ds\cos\theta}{r^2}$$

On the right hand side, the integrand consists of $\frac{ds\cos\theta}{r^2}$. The numerator $ds\cos\theta$ represents the projection of area ds on the spherical surface whose radius is r. Hence $\frac{ds\cos\theta}{r^2}$ will be the projection of ds on the spherical surface of radius equal to unity. This is the solid angle subtended by an area ds at the location of the point charge. Therefore, $\oint_s \frac{ds\cos\theta}{r^2}$ is the total solid angle subtended by s at the point charge. It is the sum of the projections of all ds on the spherical surface of radius unity and centered at Q. This is equal to the area of the spherical surface of unity radius. Hence

$$\oint_s \mathbf{D}.ds = \frac{Q}{4\pi} 4\pi$$

$$\oint_s \mathbf{D}.ds = Q \qquad \text{Hence proved.}$$

2.29 GAUSS'S LAW IN POINT FORM

Gauss's law in point form states that the divergence of electric flux density is equal to the volume charge density, that is,

$$\nabla.\mathbf{D} = \rho_v$$

Proof Consider a differential parallelepiped as shown in Fig. 2.20.

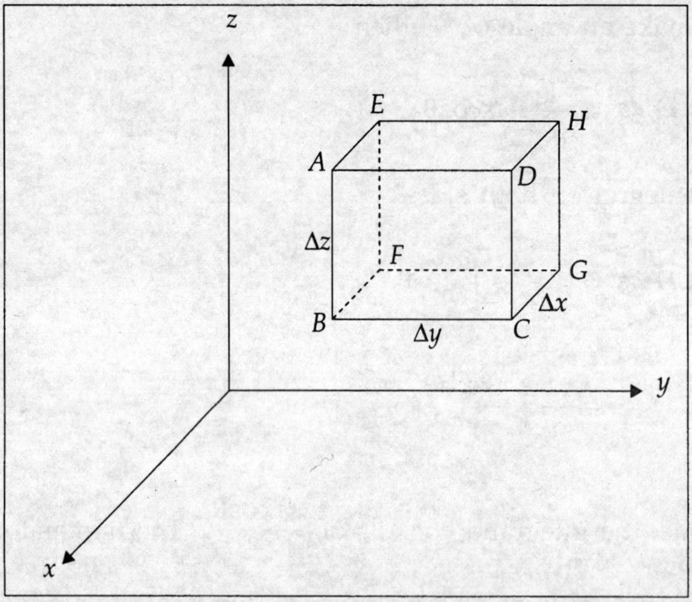

Fig. 2.20 Differential parallelepiped

Assumptions

1. The differential parallelepiped has the dimensions Δx, Δy and Δz.
2. The point, P is in the centre of the element.
3. The flux density $\mathbf{D}$ at the centre is given by

$$\mathbf{D} = \mathbf{D}_c = D_{cx}\ \mathbf{a}_x + D_{cy}\ \mathbf{a}_y + D_{cz}\ \mathbf{a}_z$$

The integral form of Gauss's law is

$$\oint_s \mathbf{D}.d\mathbf{S} = Q$$

$$\oint_s \mathbf{D}.d\mathbf{S} = \int_1 \mathbf{D}_1.d\mathbf{S}_1 + \int_2 \mathbf{D}_2.d\mathbf{S}_2 + \int_3 \mathbf{D}_3.d\mathbf{S}_3$$

$$+ \int_4 \mathbf{D}_4.d\mathbf{S}_4 + \int_5 \mathbf{D}_5.d\mathbf{S}_5 + \int_6 \mathbf{D}_6.d\mathbf{S}_6$$

where

Face 1 represents the face *ABCD*
Face 2 represents the face *EFGH*
Face 3 represents the face *ABFE*
Face 4 represents the face *DCGH*
Face 5 represents the face *ADHE*
Face 6 represents the face *BCGF*

$\mathbf{D}_1, \mathbf{D}_2, \mathbf{D}_3, \mathbf{D}_4, \mathbf{D}_5$ and $\mathbf{D}_6$ are the flux densities on the faces 1, 2, 3, 4, 5 and 6 respectively.

$d\mathbf{S}_1, d\mathbf{S}_2, d\mathbf{S}_3, d\mathbf{S}_4, d\mathbf{S}_5$ and $d\mathbf{S}_6$ are differential areas of the faces 1, 2, 3, 4, 5 and 6, respectively.

As the flux density at the centre is known, it is found on each face of the parallelepiped by considering the first two terms of Taylor's theorem.

Taylor's theorem states that if $f(x)$ has continuous derivatives in the neighbourhood of a point $x = a$, then

$$f(x) = f(a) + \frac{f'(a)}{1!}(x-a) + \frac{f''(a)}{2!}(x-a)^2 + \ldots$$

If $(x-a)$ is very small, we have

$$f(x) \approx f(a) + \frac{f'(a)}{1!}(x-a)$$

Accordingly, we can simplify $\mathbf{D}_1, \mathbf{D}_2, \mathbf{D}_3, \mathbf{D}_4, \mathbf{D}_5$ and $\mathbf{D}_6$

$$\begin{aligned} \mathbf{D}_1 . d\mathbf{S}_1 &= \mathbf{D}_1 . dS_1 \, \mathbf{a}_x \\ &= (D_{1x}\,\mathbf{a}_x + D_{1y}\,\mathbf{a}_y + D_{1z}\,\mathbf{a}_z) . dS_1\,\mathbf{a}_x \\ &= D_{1x}\, dS_1 \\ &\approx \left(D_{cx} + \frac{\partial D_x}{\partial x}\frac{\Delta x}{2} \right) dS_1 \end{aligned}$$

$$\begin{aligned} \int_1 \mathbf{D}_1 . d\mathbf{S}_1 &= D_{1x} \int dS_1 \\ &= D_{1x}\, \Delta S_1 \\ &\approx \left(D_{cx} + \frac{\partial D_x}{\partial x}\frac{\Delta x}{2} \right) \Delta y\, \Delta z \end{aligned}$$

Similarly $\quad \int_2 \mathbf{D}_2 . d\mathbf{S}_2 = -D_{2x}\,\Delta S_2 \qquad$ [As $d\mathbf{S}_2 = dS_2(-\mathbf{a}_x)$]

$$\approx \left[D_{cx} + \frac{\partial D_x}{\partial x}\left(-\frac{\Delta x}{2} \right) \right] \Delta y\, \Delta z$$

$$\int_1 \mathbf{D}_1 . d\mathbf{S}_1 + \int_2 \mathbf{D}_2 . d\mathbf{S}_2 \approx \frac{\partial D_x}{\partial x} \Delta_x \, \Delta y \, \Delta z$$

and $$\int_3 \mathbf{D}_3 . d\mathbf{S}_3 + \int_4 \mathbf{D}_4 . d\mathbf{S}_4 \approx \frac{\partial D_y}{\partial y} \Delta x \, \Delta y \, \Delta z$$

$$\int_5 \mathbf{D}_5 . d\mathbf{S}_5 + \int_6 \mathbf{D}_6 . d\mathbf{S}_6 \approx \frac{\partial D_z}{\partial z} \Delta x \, \Delta y \, \Delta z$$

$$\oint_s \mathbf{D} . d\mathbf{S} \approx \left(\frac{\partial D_x}{\partial x} + \frac{\partial D_y}{\partial y} + \frac{\partial D_z}{\partial z} \right) \Delta x \, \Delta y \, \Delta z$$

$$= Q$$

If $$\Delta \upsilon = \Delta x \; \Delta y \; \Delta z$$

$$\oint_s \frac{D . dS}{\Delta \upsilon} \approx \frac{\partial D_x}{\partial x} + \frac{\partial D_y}{\partial y} + \frac{\partial D_z}{\partial z} = \frac{Q}{\Delta \upsilon}$$

This becomes exact if $\Delta \upsilon \to 0$.

$$\underset{\Delta \upsilon \to 0}{\text{Lt}} \oint_s \frac{\mathbf{D} . d\mathbf{S}}{\Delta \upsilon} = \frac{\partial D_x}{\partial x} + \frac{\partial D_y}{\partial y} + \frac{\partial D_z}{\partial z} = \rho_\upsilon \qquad \left[\rho_\upsilon = \frac{Q}{\Delta \upsilon} \right]$$

But $$\frac{\partial D_x}{\partial x} + \frac{\partial D_y}{\partial y} + \frac{\partial D_z}{\partial z} = \nabla . \mathbf{D}$$

$$\boxed{\nabla . \mathbf{D} = \rho_\upsilon}$$ Hence proved.

2.30 DIVERGENCE OF A VECTOR, ELECTRIC FLUX DENSITY

The divergence of electric flux density is defined as

$$\nabla . \mathbf{D} \equiv \underset{\Delta \upsilon \to 0}{\text{Lt}} \oint_s \frac{\mathbf{D} . d\mathbf{S}}{\Delta \upsilon}$$

This means that the divergence of flux density is the outflow of electric flux from a closed surface per unit volume as the volume shrinks to zero.

The point form of Gauss's law in different coordinate systems is

$$(\nabla . \mathbf{D})_{\text{Cart.}} = \frac{\partial D_x}{\partial x} + \frac{\partial D_y}{\partial y} + \frac{\partial D_z}{\partial z} = \rho_\upsilon$$

$$(\nabla . \mathbf{D})_{cy} = \frac{1}{\rho}\frac{\partial}{\partial \rho}(\rho D_\rho) + \frac{1}{\rho}\frac{\partial D_\phi}{\partial \phi} + \frac{\partial D_z}{\partial z} = \rho_\upsilon$$

$$(\nabla . \mathbf{D})_{sph} = \frac{1}{r^2}\frac{\partial}{\partial r}(r^2 D_r) + \frac{1}{r\sin\theta}\frac{\partial}{\partial \theta}(\sin\theta\, D_\theta) + \frac{1}{r\sin\theta}\frac{\partial D_\phi}{\partial \phi} = \rho_\upsilon$$

2.31 APPLICATIONS OF GAUSS'S LAW

1. It is useful to find flux, ψ or flux density, **D** from the knowledge of enclosed charge and surface.
2. It is useful to find the electric field from the knowledge of enclosed charge and surface.
3. It is useful to find the enclosed charge from the knowledge of either **D** or **E**.

2.32 LIMITATIONS OF GAUSS'S LAW

1. It cannot be applied on Non-Gaussian surfaces.
2. It can be applied only if the surface encloses the volume completely.

2.33 SALIENT FEATURES OF GAUSS'S LAW

1. It relates volume integral to surface integral.
2. It is applicable only if the surface completely encloses the volume.
3. It does not specify any particular shape for the closed surface.
4. It does not require knowledge of the distance from the points on the surface.
5. It cannot be applied if the charges are outside the surface.
6. It is useful to find **D** or **E** from the knowledge of the charge enclosed and the surface.
7. It is applicable only on Gaussian surfaces.
8. It is useful to find the outward flow of flux from a closed surface from the enclosed charge.

Problem 2.44 If electric flux density, **D** is given by,

$$\mathbf{D} = [(2y^2 + z)\,\mathbf{a}_x + 4xy\,\mathbf{a}_y + x\mathbf{a}_z)]\ \mu\text{C/m}^2$$

find the volume charge density at (0, 0, 0) and (–1, 0, 4).

Solution $\mathbf{D} = [(2y^2 + z)\,\mathbf{a}_x + 4xy\,\mathbf{a}_y + x\mathbf{a}_z)]\ \mu\text{C/m}^2$

$$\nabla.\mathbf{D} = \frac{\partial D_x}{\partial x} + \frac{\partial D_y}{\partial y} + \frac{\partial D_z}{\partial z}$$

$$= 0 + 4x + 0 = 4x$$

But ρ_υ = volume charge density $= \nabla.\mathbf{D}$

$$\boxed{\rho_\upsilon/(0,0,0) = 0 \text{ and } \rho_\upsilon/(-1,0,4) = -4.0\mu\text{C/m}^3}$$

2.34 POISSON'S AND LAPLACE'S EQUATIONS

Poisson's equation $\nabla^2 V = -\rho_\upsilon/\epsilon$

Laplace's equation $\nabla^2 V = 0$

Proof The point form of Gauss's law is

$$\nabla.\mathbf{D} = \rho_\upsilon$$

But $\mathbf{D} = \epsilon\,\mathbf{E}$

and $\mathbf{E} = -\nabla V$

$$\nabla.\mathbf{D} = \nabla.\epsilon\,\mathbf{E} = \nabla.\epsilon(-\nabla V) = \rho_\upsilon$$

or, $\boxed{\nabla^2 V = -\rho_\upsilon/\epsilon}$ [as $\nabla.\nabla = \nabla^2$]

where ∇^2 is a scalar operator $\left(\frac{1}{m^2}\right)$ and is called Laplace's operator.

$$\nabla^2 = \frac{\partial^2}{\partial x^2} + \frac{\partial^2}{\partial y^2} + \frac{\partial^2}{\partial z^2}$$

In the regions where $\rho_\upsilon = 0$, Poisson's equation becomes

$$\boxed{\nabla^2 V = 0}$$

Laplace's equation in one dimensional form is given by

$$\nabla^2 V = \frac{\partial^2 V}{\partial x^2} = 0$$

Its solution is in the form of

$$V = mx + a$$

Laplace's equation in two dimensional form is given by

$$\nabla^2 V = \frac{\partial^2 V}{\partial x^2} + \frac{\partial^2 V}{\partial y^2} = 0$$

Its solution is given by

$$V(x, y) = \frac{1}{2\pi r} \oint_{\text{circle}} V \, dL$$

where r is the radius of a circle about a point (x, y).

Laplace's equation in three dimensional form is given by

$$\nabla^2 V = \frac{\partial^2 V}{\partial x^2} + \frac{\partial^2 V}{\partial y^2} + \frac{\partial^2 V}{\partial z^2} = 0$$

The value of V at point p is the average value of V over a spherical surface of radius r centered at p and it is given by

$$V(p) = \frac{1}{4\pi r^2} \oint_{s} V ds$$

Poisson's equation in different coordinates

$$(\nabla^2 V)_{\text{Cart.}} = \frac{\partial^2 V}{\partial x^2} + \frac{\partial^2 V}{\partial y^2} + \frac{\partial^2 V}{\partial z^2} = -\frac{\rho_v}{\epsilon}$$

$$(\nabla^2 V)_{cy} = \frac{1}{\rho} \frac{\partial}{\partial \rho}\left(\rho \frac{\partial V}{\partial \rho}\right) + \frac{1}{\rho^2} \frac{\partial^2 V}{\partial \phi^2} + \frac{\partial^2 V}{\partial z^2} = -\frac{\rho_v}{\epsilon}$$

$$(\nabla^2 V)_{sph} = \frac{1}{r^2} \frac{\partial}{\partial r}\left(r^2 \frac{\partial V}{\partial \rho}\right) + \frac{1}{r^2 \sin\theta} \frac{\partial V}{\partial \theta}\left(\sin\theta \frac{\partial V}{\partial \theta}\right) + \frac{1}{r^2 \sin^2\theta} \frac{\partial^2 V}{\partial \phi^2} = -\frac{\rho_v}{\epsilon}$$

2.35 APPLICATIONS OF POISSON'S AND LAPLACE'S EQUATIONS

1. They are useful in determining surface charge densities and electric field in the regions of interest.
2. They can be used to find the capacitance of different structures by appropriate application of boundary conditions.

2.36 UNIQUENESS THEOREM

It states that either Poisson's or Laplace's equation has only one solution.

Proof Assume that V_1 and V_2 are the solutions of Laplace's equation and V_{1b} and V_{2b} are the potentials on the boundaries. By hypothesis,

$$\nabla^2 V_1 = 0$$

and
$$\nabla^2 V_2 = 0$$

$$\nabla^2 (V_1 - V_2) = 0$$

As V_{1b} and V_{2b} are the values on the boundary we have

$$V_{1b} = V_{2b} = V_b$$

or,
$$V_{1b} - V_{2b} = 0$$

Consider a vector identity, namely

$$\nabla.(\psi \mathbf{A}) = \psi(\nabla.\mathbf{A}) + \mathbf{A}.\nabla\psi$$

where ψ is a scalar and **A** is a vector quantity.

This is valid for any scalar, ψ and any vector, **A**.

In the present case, let $\psi = V_1 - V_2$, $\mathbf{A} = \nabla(V_1 - V_2)$. Then, the above identities become

$$\nabla.(V_1 - V_2)\nabla.(V_1 - V_2) = (V_1 - V_2)\nabla.\nabla(V_1 - V_2) + \nabla(V_1 - V_2).\nabla(V_1 - V_2)$$

Take volume integral on both sides

$$\int_v \nabla.(V_1 - V_2)\nabla(V_1 - V_2)d\upsilon$$

$$= \int_v (V_1 - V_2)\nabla.\nabla(V_1 - V_2)\,d\upsilon + \int_v \nabla(V_1 - V_2).\nabla(V_1 - V_2)\,d\upsilon$$

Applying divergence theorem to the left hand side, volume integral is replaced by surface integral,

that is, $\int_v \nabla.[(V_1 - V_2)\nabla(V_1 - V_2)]\,d\upsilon = \oint_s (V_1 - V_2)\nabla(V_1 - V_2).ds$

But on the boundary, this becomes

$$\int_v \nabla.[(V_1 - V_2)\nabla(V_1 - V_2)]\,d\upsilon = \oint_s (V_{1b} - V_{2b})\nabla(V_{1b} - V_{2b}).ds$$

Right hand side is zero because $V_{1b} = V_{2b}$. By hypothesis

$$\int_v (V_1 - V_2)\nabla.[\nabla(V_1 - V_2)]\,d\upsilon = \int_v (V_1 - V_2)\nabla^2[(V_1 - V_2)]\,d\upsilon = 0$$

So the remaining integral is

$$\int_v [\nabla(V_1 - V_2)]^2 d\upsilon = 0$$

This is possible if

1. Integrand is zero.
2. Integrand is positive in some region and negative in some other region.

The second condition cannot be true as it is a square term.

So the integrand is zero,

that is, $$[\nabla(V_1 - V_2)]^2 = 0 \text{ or } \nabla(V_1 - V_2) = 0$$

If the gradient of a scalar is zero everywhere, then $(V_1 - V_2)$ cannot change with any coordinate. Hence $(V_1 - V_2)$ = constant.

The constant is easily evaluated by considering a point on the boundary. Here $V_1 - V_2 = V_1 b - V_2 b = 0$,

that is, $$\boxed{V_1 = V_2}$$ Hence proved.

Problem 2.45 Consider concentric spherical shells in free space in which $V = 0$ volts at r = 10 cm and V = 10 volts at r = 20 cm. Find **E** and **D**.

Solution Here V is a function of only r and not θ, and ϕ. Then Laplace's equation

$$\nabla^2 V = \frac{1}{r^2}\frac{\partial}{\partial r}\left(r^2 \frac{\partial V}{\partial r}\right) = 0$$

Integrating twice, we get

$$V = -\frac{A}{r} + B$$

The boundary conditions are

$V = 0\,\text{V at } r = 10\,\text{cm} = 0.1\,\text{m}$ and

$V = 10\,\text{V at } r = 20\,\text{cm} = 0.2\,\text{m}$,

that is, $$\frac{-A}{0.1} + B = 0 \text{ and}$$

$$\frac{-A}{0.2} + B = 10$$

$$A = 2.0 \text{ volt-m}, \ B = 20 \text{ V}$$

So,
$$\boxed{V = \frac{-2}{r} + 20 \text{ volt}}$$

But
$$\mathbf{E} = -\nabla V = \frac{-\partial V}{\partial r} \mathbf{a}_r$$

or,
$$\boxed{\mathbf{E} = \frac{-2.0}{r^2} \mathbf{a}_r, \text{V/m}}$$

$$\mathbf{D} = \epsilon_0 \mathbf{E} = \frac{-2.0}{r^2} \times 8.854 \times 10^{-12} \mathbf{a}_r$$

or,
$$\boxed{\mathbf{D} = \frac{-17.708}{r^2} \mathbf{a}_r, \text{PC/m}^2}$$

Problem 2.46 There exists a potential of $V = -2.5$ V on a conductor at 0.02 m and V = 15.0 V at r = 0.35 m. A dielectric material whose $\epsilon_r = 3.0$ exists between the conductors. Determine the surface charge densities on the conductors.

Solution As V is a function of only r, Laplace's equation is given by

$$\frac{1}{r^2} \frac{\partial}{\partial r} \left(r^2 \frac{\partial V}{\partial r} \right) = 0$$

Integrating twice, we get

$$V = -\frac{A}{r} + B$$

A and B can be found by the use of boundary conditions,

that is,
$$\frac{-A}{0.02} + B = -2.5$$

$$\frac{-A}{0.35} + B = 15.0$$

Solving, we get

$$A = 37.12 \times 10^{-2}, \text{ V-m}$$

$$B = 16.06$$

and
$$V = \frac{-0.3712}{r} + 16.06$$

But
$$\mathbf{E} = -\nabla V$$

$$= \frac{-d}{dr}\left(\frac{-0.3712}{r} + 16.06\right)\mathbf{a}_r$$

$$\boxed{\mathbf{E} = \frac{-0.3712}{r^2}\,\mathbf{a}_r,\ \text{V/m}}$$

and
$$\mathbf{D} = \in \mathbf{E} = \in_0 \in_r \mathbf{E}$$

$$= 8.854 \times 10^{-12} \times 3.0 \times \left(\frac{-0.3712}{r^2}\right)\mathbf{a}_r$$

or,
$$\boxed{\mathbf{D} = \frac{-9.8598}{r^2}\,\mathbf{a}_r\ \text{PC/m}^2}$$

On the conductor surfaces,

$$D_n = \rho_s = \text{surface density}$$

At
$$r = 0.02\,\text{m}$$

$$\rho_s = -2.4649 \times 10^{-8}\ \text{C/m}^2$$

or,
$$\boxed{\rho_s = -24.6\,\text{nC/m}^2}$$

At
$$r = 0.35\,\text{m}$$

$$\rho_s = \frac{-9.8598}{(35)^2} \times 10^{-12} \times 10^4$$

or,
$$\boxed{\rho_s = -\ 0.080\,\text{nC/m}^2}$$

Problem 2.47 In what manner does permittivity vary to satisfy Laplace's equation in a non-homogeneous, charge-free space?

Solution For charge-free space,

$$\rho_v = 0$$

$$\nabla.\mathbf{D} = 0$$

$$\nabla.\in \mathbf{E} = 0 \qquad [\text{as } \mathbf{D} = \in \mathbf{E}]$$

$$\nabla.(-\in \nabla V) = 0 \qquad [\text{as } \mathbf{E} = -\nabla V]$$

$$\nabla.(\in \nabla V) = 0$$

If $\in$ varies spatially, then

$$\nabla.(\in \nabla V) = \nabla V.\nabla \in + \nabla^2 V = 0$$

As
$$\nabla^2 V = 0$$

$$(\nabla V).(\nabla \in) = 0$$

This is true only when ∇V and $\nabla \in$ are perpendicular to each other. Hence permittivity should vary so that its gradient is perpendicular to the electric field.

Problem 2.48 If a potential $V = x^2 yz + Ay^3 z$, (a) find A so that Laplace's equation is satisfied (b) with the value of A, determine electric field at (2, 1, –1).

Solution (a) Laplace's equation is

$$\nabla^2 V = 0$$

or,
$$\frac{\partial^2 V}{\partial x^2} + \frac{\partial^2 V}{\partial y^2} + \frac{\partial^2 V}{\partial z^2} = 0$$

As
$$V = x^2 yz + Ay^3 z$$

The above equation becomes

$$2yz + 6Ayz = 0$$

So,
$$A = -\frac{1}{3}$$

(b) With
$$A = -\frac{1}{3}$$

$$V = x^2 yz - \frac{1}{3} y^3 z$$

But
$$\mathbf{E} = -\nabla V$$

$$= -\left[\frac{\partial V}{\partial x}\mathbf{a}_x + \frac{\partial V}{\partial y}\mathbf{a}_y + \frac{\partial v}{\partial z}\mathbf{a}_z\right]$$

$$= -[(2yzx)]\mathbf{a}_x + \left[x^2 z - \frac{1}{3}3y^2 z\right]\mathbf{a}_y + \left[x^2 y - \frac{1}{3}y^3\right]\mathbf{a}_z$$

$$= -2xyz\,\mathbf{a}_x - (x^2 z - y^2 z)\,\mathbf{a}_y - \left(x^2 y - \frac{1}{3}y^3\right)\mathbf{a}_z$$

$$\mathbf{E}/(2,1,-1) = -2\,[2\times 1\times(-1)]\,\mathbf{a}_x - [4\times(-1) - 1\times(-1)]\,\mathbf{a}_y - \left(4 - \frac{1}{3}\times 1\right)\mathbf{a}_z$$

$$\boxed{\mathbf{E} = 4\mathbf{a}_x + 3\mathbf{a}_y - 3.66\mathbf{a}_z \text{ V/m}}$$

2.37 BOUNDARY CONDITIONS ON *E* AND *D*

1. The tangential component of **E** is continuous across any boundary, that is, $\boxed{\mathbf{E}_{\tan 1} = \mathbf{E}_{\tan 2}}$

   or

   The tangential component of **E** in medium 1 is the same as that of **E** in medium 2 at any boundary.
2. The normal component of **D** is continuous across any boundary except at the surface of the conductor. In general,

$$\boxed{D_{n1} - D_{n2} = \rho_s}$$

ρ_s = surface charge density, (C/m^2). For any point other than the conductor boundary, $D_{n1} = D_{n2}$.

2.38 PROOF OF BOUNDARY CONDITIONS

Consider the rectangular loop on the boundary of two media (Fig. 2.21).

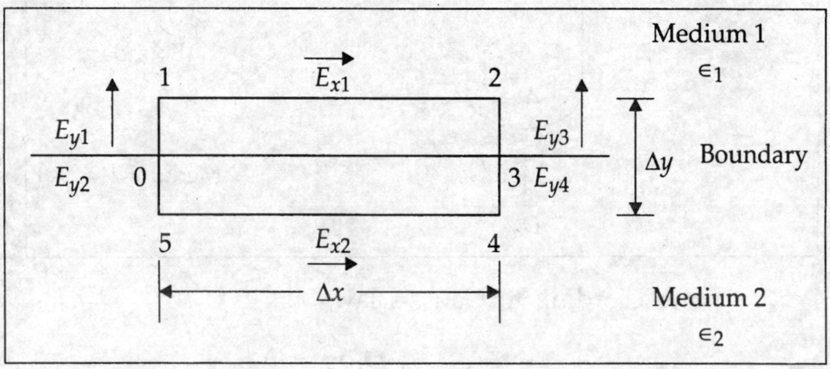

Fig. 2.21 Rectangular loop on boundary

It is well known that electric field is conservative and hence the line integral of $\mathbf{E}.\mathbf{dL}$ is zero around a closed path,

that is, $$\oint \mathbf{E}.d\mathbf{L} = 0$$

From the figure shown above, LHS is written as

$$\oint \mathbf{E}.d\mathbf{L} = \int_{01} + \int_{12} + \int_{23} + \int_{34} + \int_{45} + \int_{50}$$

$$= E_{y1}\frac{\Delta y}{2} + E_{x1}\,\Delta x - E_{y3}\frac{\Delta y}{2} - E_{y4}\frac{\Delta y}{2} - E_{x2}\,\Delta x + E_{y2}\frac{\Delta y}{2}$$

As $\Delta y \to 0$, we get

$$\oint \mathbf{E}.d\mathbf{L} = E_{x1}\,\Delta x - E_{x2}\,\Delta x = 0$$

Thus, $$E_{x1} = E_{x2}$$

It is obvious that E_{x1} and E_{x2} are the tangential components of **E** in medium 1 and 2 respectively.

So, $$\mathbf{E}_{\tan 1} = \mathbf{E}_{\tan 2}$$

Now consider a cylinder across the media 1 and 2 (Fig. 2.22).

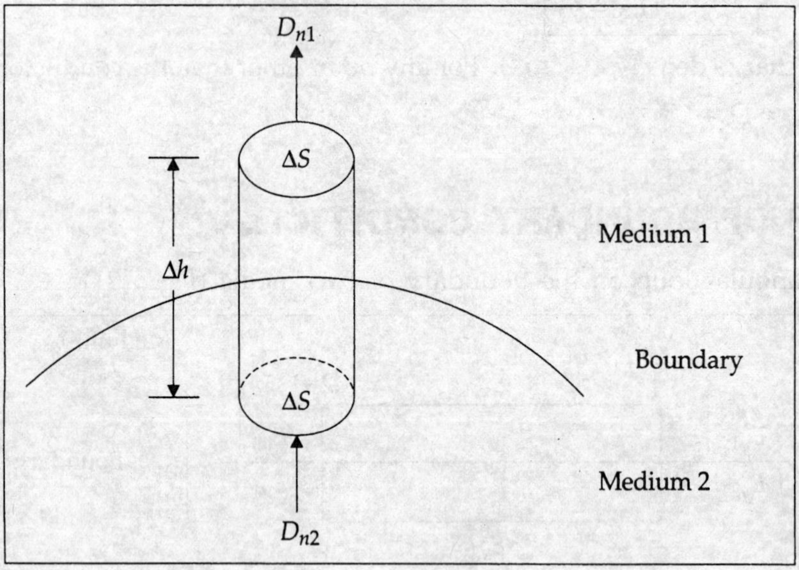

Fig. 2.22 Cylindrical surfaces on boundary

According to Gauss's law,

$$\oint \mathbf{D}.\mathbf{ds} = Q$$

Applying this to the cylindrical surface on the boundary spreading over medium 1 and medium 2, we get $\Delta h \to 0$

$$D_{n1}\,\Delta s - D_{n2}\,\Delta s = Q$$

or, $$D_{n1} - D_{n2} = \frac{Q}{\Delta S} = \rho_S$$

$$\boxed{D_{n1} - D_{n2} = \rho_s}$$ Hence proved.

Problem 2.49 The region $y < 0$ contains a dielectric material for which $\epsilon_{r1} = 2.0$ and the region $y > 0$ contains a dielectric material for which $\epsilon_{r2} = 4.0$. If $\mathbf{E}_1 = -3.0\mathbf{a}_x + 5.0\mathbf{a}_y + 7.0\mathbf{a}_z$ V/m, find the electric field, $\mathbf{E}_2$ and $\mathbf{D}_2$ in medium 2.

Solution As $y<0$ belongs to medium 1 and $y>0$ belongs to medium 2

$$\mathbf{E}_{\tan 1} = -3.0\mathbf{a}_x + 7.0\mathbf{a}_z \text{ V/m}$$
$$\mathbf{E}_{n1} = 5.0\mathbf{a}_y \text{ V/m}$$
$$\epsilon_{r1} = 2$$
$$\epsilon_{r2} = 4$$

The boundary condition on tangential component of **E** is

$$\mathbf{E}_{\tan 1} = \mathbf{E}_{\tan 2}$$
$$\mathbf{E}_{\tan 2} = -3.0\mathbf{a}_x + 7.0\mathbf{a}_z \text{ V/m}$$

and
$$\mathbf{E}_{n2} = \frac{\epsilon_{r1}}{\epsilon_{r2}}\mathbf{E}_{n1} = \frac{2}{4}\times 5\mathbf{a}_y$$
$$= 2.5\mathbf{a}_y$$
$$\mathbf{E}_2 = \mathbf{E}_{\tan 2} + \mathbf{E}_{n2}$$
$$\boxed{\mathbf{E}_2 = -3\mathbf{a}_x + 2.5\mathbf{a}_y + 7.0\mathbf{a}_z \text{ V/m}}$$
$$\mathbf{D}_2 = \epsilon_2 (\mathbf{E}_2)$$
$$= 4\epsilon_0 (\mathbf{E}_2)$$
$$\boxed{\mathbf{D}_2 = \epsilon_0 (-12\mathbf{a}_x + 10\mathbf{a}_y + 28\mathbf{a}_z) \text{ C/m}^2}$$

Problem 2.50 An electric field in medium 1 of $\epsilon_{r1}=7$ passes into a medium 2 of $\epsilon_{r2}=2$. When the field, **E** makes an angle of 60° as shown in Fig. 2.23 with the axis normal to the boundary line, find the angle made by the field with the normal in medium 2.

Solution From Fig. 2.23

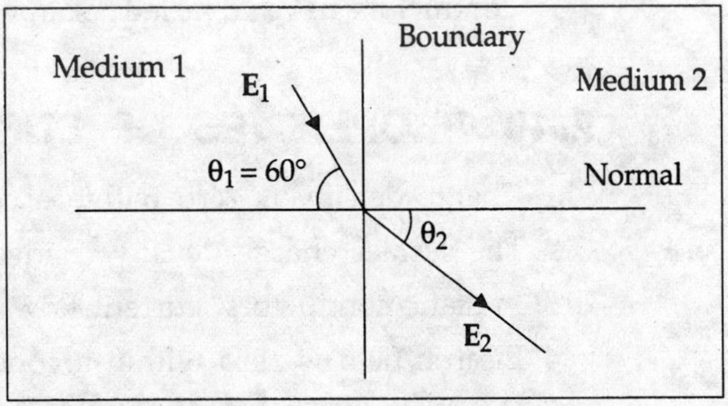

Fig. 2.23

$$\frac{E_{\tan 1}}{E_{n1}} = \tan\theta_1 = \tan 60°$$

and
$$\frac{E_{\tan 2}}{E_{n2}} = \tan\theta_2$$

According to the boundary conditions,

$$E_{\tan 1} = E_{\tan 2}$$

$$D_{n1} = D_{n2} = \epsilon_1 E_{n1} = \epsilon_2 E_{\tan 2}$$

$$E_{n2} = \frac{\epsilon_1}{\epsilon_2} E_{n1} = \frac{7}{2} E_{n1} = 3.5E_{n1}$$

and

$$E_{\tan 2} = E_{n2} \tan \theta_2$$

$$= 3.5E_{n1} \tan \theta_2$$

$$= E_{\tan 1} = E_{n1} \tan 60°$$

$$\tan \theta_2 = \frac{\tan 60°}{3.5} = 0.495$$

$$\boxed{\theta_2 = 26.40°}$$

2.39 CONDUCTORS IN ELECTRIC FIELD

Conductors

Definition 1 Conductors are materials which have very low resistance. Examples: Copper, Silver and Aluminium.

Definition 2 Conductors are materials for which no forbidden gap exists between valance band and conduction band.

Definition 3 A material is defined as a conductor if $\frac{\sigma}{\omega\epsilon} >> 1$.

Conductors have a variety of applications in all fields of life.

2.40 PROPERTIES OF CONDUCTORS

1. Charge density is zero within a conductor.
2. The surface charge density resides on the exterior surface of a conductor.
3. In static conductors, current flow is zero.
4. Electric field is zero within a conductor.
5. Conductivity is very large.
6. Resistivity is small.
7. Magnetic field is zero inside a conductor.
8. Good conductors reflect electric and magnetic fields completely.

9. A conductor consists of a large number of free electrons which constitute conduction current with the application of an electric field.
10. A conductor is an equipotential body.
11. The potential is same everywhere in the conductor.
12. $\mathbf{E} = -\nabla V = 0$ in a conductor.
13. In a perfect conductor, conductivity is infinity.
14. When an external field is applied to a conductor, the positive charges move in the direction **E** and the negative charges move in the opposite direction. This happens very quickly.
15. Free charges are confined to the surface of the conductor and hence surface charge density, J_s is induced. These charges create internal induced electric field. This field cancels the external field.

    It is interesting to note that copper and silver are not super conductors but aluminium is a superconductor for temperature below 1.14 K.

2.41 ELECTRIC CURRENT

The current through a given medium is defined as charge passing through the medium per unit time. It is a scalar, that is,

$$I \equiv \frac{dQ}{dt}, \text{Ampere}$$

Current is of three types.

1. Convection current
2. Conduction current
3. Displacement current

1. **Convection current** It is defined as the current produced by a beam of electrons flowing through an insulating medium. This does not obey Ohm's law. For example, current through a vacuum, liquid and so on is convection current.
2. **Conduction current** It is defined as the current produced due to flow of electrons in a conductor. This obeys Ohm's law. For example, current in a conductor like copper is conduction current.
3. **Displacement current** It is defined as the current which flows as a result of time-varying electric field in a dielectric material. For example, current through a capacitor when a time-varying voltage is applied is displacement current.

2.42 CURRENT DENSITIES

In electromagnetic field theory, it is of interest to describe the events at a point instead of in a large region. This is the reason why current densities are considered. Current densities are vector quantities.

Current Density is defined as the current at a given point through a unit normal area at that point. It is a vector and it has the unit of Ampere/ m^2. It is represented by **J**.

Current densities are of three types:

1. Convection current density
2. Conduction current density
3. Displacement current density

1. **Convection current density (A/m²)** It is defined as the convection current at a given point through a unit normal area at that point, that is,
Convection current density

$$\equiv \frac{dI}{d\mathbf{S}}$$

$$\equiv \frac{dI}{dS}\mathbf{a}_n$$

where dI = differential convection current

$d\mathbf{S}$ = differential area

$= dS\mathbf{a}_n$

$\mathbf{a}_n$ = outward unit normal to dS

As convection current density is confined to specific media, it is not of much interest in this book.

2. **Conduction current density, J_c (A/m²)** It is defined as the conduction current at a given point through a unit normal area at that point,

that is,
$$\mathbf{J}_c \equiv \sigma\mathbf{E}$$

and
$$\mathbf{J}_c \equiv \frac{dI}{dS}\mathbf{a}_n$$

Conduction current density exists in the case of conductors when an electric field is applied.

3. **Displacement current density, J_d (A/m^2)** It is defined as the rate of displacement electric flux density with time, that is,

$$\mathbf{J}_d \equiv \frac{\partial \mathbf{D}}{\partial t}$$

If I_d is the displacement current in a dielectric due to applied electric field, displacement current density is defined as

$$\mathbf{J}_d = \frac{dI_d}{dS}\mathbf{a}_n$$

As

$$\mathbf{D} = \in \mathbf{E}$$

$$\mathbf{J}_d = \frac{\partial \mathbf{D}}{\partial t} = \in \frac{\partial \mathbf{E}}{\partial t}$$

In fact, displacement current density exists due to displacement of bound charges in a dielectric by the applied electric field.

2.43 EQUATION OF CONTINUITY

Equation of continuity in integral form is $I = \int \mathbf{J}.d\mathbf{s}$

I = outward flow of current (A)

J = conduction current density (A/m^2)

$$\nabla.\mathbf{J} = -\dot{\rho}_\upsilon$$

where

$$\dot{\rho}_\upsilon = \frac{\partial \rho_\upsilon}{\partial t}$$

Proof If Q_i is the charge inside a closed surface, the rate of decrease of charge due to the outward flow of current is given by $\left(-\frac{dQ_i}{dt}\right)$.

From the principle of conservation of charge, we have

$$I = -\frac{dQ_i}{dt} = \oint_s \mathbf{J}.d\mathbf{s}$$

From divergence theorem, we have

$$\oint_s \mathbf{J}.d\mathbf{s} = \int_\upsilon \nabla.\mathbf{J}\,d\upsilon$$

So, $$\int_{\upsilon} \nabla . \mathbf{J} d\upsilon = -\frac{dQ_i}{dt}$$

$$= -\frac{d}{dt}\int_{\upsilon} \rho_\upsilon d_\upsilon = -\int_{\upsilon} \dot{\rho}_\upsilon d_\upsilon$$

Two volume integrals are equal if the integrands are equal. So,

$$\boxed{\nabla . \mathbf{J} = -\frac{\partial \rho_\upsilon}{\partial t} = -\dot{\rho}_\upsilon} \qquad \text{Hence proved.}$$

In the above equation the derivative became a partial derivative as the surface is kept constant.

2.44 RELAXATION TIME (T_r)

It is also called **rearrangement time.**

Relaxation time is defined as the time taken by a charge placed in a material to reach 36.8 per cent of its initial value. It is given by

$$T_r \equiv \frac{\epsilon}{\sigma}, \text{ sec}$$

where $\epsilon =$ permittivity (F/m), $\sigma =$ conductivity (mho/m).

Problem 2.51 Find the relaxation time of sea water whose $\epsilon_r = 81$ and $\sigma = 5$ mho/m.

Solution Relaxation time of sea water

$$T_r = \frac{\epsilon}{\sigma} = \frac{81 \times 8.854 \times 10^{-12}}{5}$$

$$= 143.37 \text{ picosecond}$$

Problem 2.52 Find the relaxation time of porcelain whose $\sigma = 10^{-10}$ mho/m, $\epsilon_r = 6$.

Solution Relaxation time of porcelain

$$T_r = \frac{\epsilon}{\sigma} = \frac{6 \times 8.854 \times 10^{-12}}{10^{-10}}$$

$$= 53.12 \times 10^{-2}$$

$$\boxed{T_r = 531.2 \text{ m sec}}$$

2.45 RELATION BETWEEN CURRENT DENSITY AND VOLUME CHARGE DENSITY

$$\boxed{\mathbf{J} = \rho_\upsilon \mathbf{V}}$$

$\mathbf{J}$ = conduction current density, A/m^2

$\mathbf{V}$ = velocity of the charge (m/s)

Proof We know that $I = \int \mathbf{J}.d\mathbf{s} = \dfrac{dQ}{dt}$

Consider an element charge (Fig. 2.24)

$$\Delta Q = \rho_\upsilon \Delta \upsilon = \rho \Delta s \, \Delta x$$

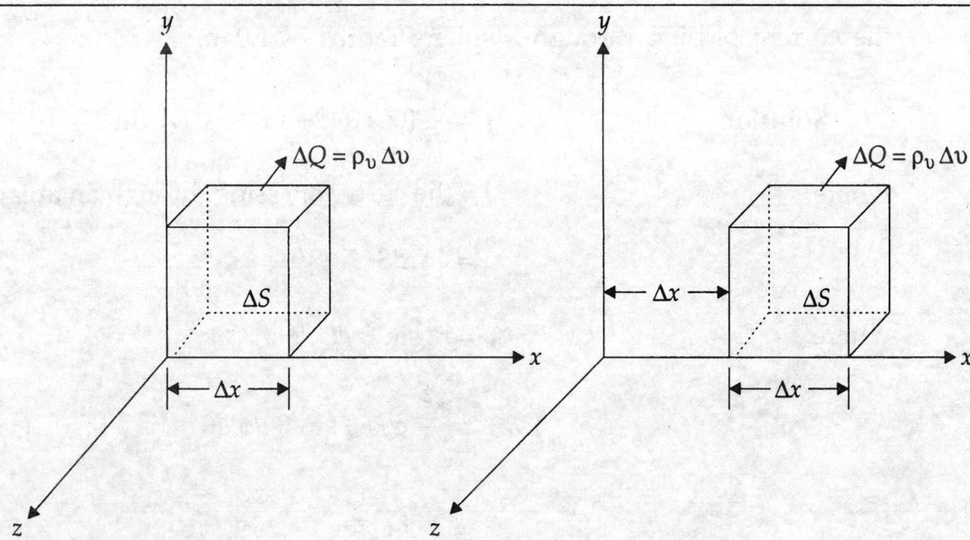

Fig. 2.24 Volume charge in motion

Assume that the charge element is oriented parallel to the coordinate axes. Let there be only an x-component of velocity. It moves a distance of Δx in a time Δt as in the figure. Therefore,

$$\Delta Q = \rho_\upsilon \, \Delta s \, \Delta x$$

The resultant current is

$$\Delta I = \frac{\Delta Q}{\Delta t} = \rho_\upsilon \Delta s \frac{\Delta x}{\Delta t} = \rho_\upsilon \Delta s \, V_x \qquad \left[\text{as } V_x = \frac{\Delta x}{\Delta t}\right]$$

where $V_x = x$-component of velocity of the charge.

$$\frac{\Delta I}{\Delta s} = \rho_\upsilon V_x$$

$$J_x = \rho_v V_x \, (\text{A/m}^2)$$

Similarly, if the charge moves in y and z-directions, we get

$$J_y = \rho_v V_y$$

$$J_z = \rho_v V_z$$

$$\mathbf{J} = \rho_v (V_x \mathbf{a}_x + V_y \mathbf{a}_y + V_z \mathbf{a}_z)$$

As $\quad \mathbf{V} = V_x \mathbf{a}_x + V_y \mathbf{a}_y + V_z \mathbf{a}_z$

$$\boxed{\mathbf{J} = \rho_v \mathbf{V}} \quad \text{Hence proved.}$$

Problem 2.53 If the current density, $\mathbf{J} = \frac{1}{r^2}(\cos\theta a_r + \sin\theta a_\theta)$, A/m^2, find the current passing through a sphere radius of 1.0 m.

Solution $\quad \mathbf{J} = \frac{1}{r^2}(\cos\theta a_r + \sin\theta a_\theta), \text{A/m}^2$

where $\quad I$ = the current passing through an area

$$= \int \mathbf{J}.d\mathbf{S}$$

where $\quad d\mathbf{S} = r^2 \sin\theta \, d\theta \, d\phi \, \mathbf{a}_r$

$$\mathbf{J}.d\mathbf{S} = \frac{r^2}{r^2} \cos\theta \sin\theta \, d\phi \, d\theta \qquad [\text{as } \mathbf{a}_\theta . \mathbf{a}_r = 0]$$

$$I = \int_0^{\pi} \int_0^{2\pi} \mathbf{J}.d\mathbf{S}$$

$$I = \int_0^{\pi} \int_0^{2\pi} \cos\theta \sin\theta \, d\phi \, d\theta$$

that is, $\quad I = 2\pi \int_0^{\pi} \sin\theta \, d(\sin\theta)$

$$= 2\pi \left(\frac{\sin^2\theta}{2} \right)_0^{\pi/2} = \pi$$

Problem 2.54 Find the electric flux density in free space if the electric field, $\mathbf{E} = 6\mathbf{a}_x - 2\mathbf{a}_y + 3\mathbf{a}_z$, V/m.

Solution Electric field in free space,

$$\mathbf{E} = 6\mathbf{a}_x - 2\mathbf{a}_y + 3\mathbf{a}_z, \text{V/m}$$

$$\epsilon_0 = 8.854 \times 10^{-12}$$

The electric flux density,

$$\mathbf{D} = \epsilon_0 \mathbf{E}$$

$$= 8.854 \times 10^{-12} (6\mathbf{a}_x - 2\mathbf{a}_y + 3\mathbf{a}_z)$$

$$\boxed{\mathbf{D} = (53.12a_x - 17.7a_y + 26.55a_z)\,\text{PC/m}^2}$$

2.46 DIELECTRIC MATERIALS IN ELECTRIC FIELD

Definition 1 An ideal dielectric material is one which does not contain free electrons.

Definition 2 An ideal dielectric material is one in which the charges are well bounded and cannot be set in motion easily.

Definition 3 An ideal dielectric material is one for which there exists a large forbidden gap between valance band and conduction band.

Definition 4 A material is defined as dielectric material if $\dfrac{\sigma}{\omega \epsilon} << 1$.

Definition 5 A material is defined as dielectric material if it does not conduct electric current and opposes the flow of current.

2.47 PROPERTIES OF DIELECTRIC MATERIALS

1. Conductivity is zero.
2. Volume charge density, $\rho_\upsilon = 0$.
3. Electric and magnetic fields exist in a dielectric material.
4. Resistivity is ∞.
5. Electric and magnetic fields penetrate the dielectric material freely.
6. There exists no free electrons.

Dielectrics in Electric Field

An atom of a dielectric consists of a nucleus and a bunch of electrons. Similarly, a molecule of a dielectric consists of nuclei and a set of electron bunches. The charge of the nucleus is positive and the charge of the electron bunch is negative.

The atoms and molecules are electrically neutral as they contain an equal number of negative and positive charges.

Dielectrics are classified into polar and non-polar type of materials.

Polar type of dielectrics

The centres of positive and negative charges of a molecule of polar type of dielectric material are separated by a small distance. Each pair acts as a dipole and there exists a dipole moment. However, such pairs are randomly distributed in a dielectric material. Hence the overall dipole moment is zero.

If such a material is kept in an electric field, all the positive charges move in the direction of the electric field and all the negative charges move in the opposite direction. As a result, dipole moment is induced by the electric field. Under these conditions, the material is said to be under a state of polarisation.

A polar type of molecule is shown in Fig. 2.25.

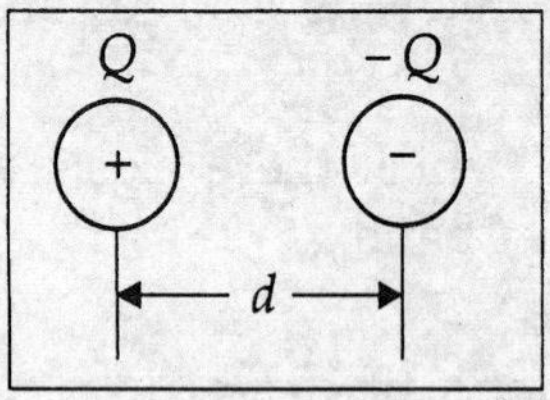

Fig. 2.25 Polar type of molecule

Examples of polar dielectrics are water, hydrochloric acid, sulphur dioxide and others.

Non-polar type of dielectrics

The centres of positive and negative charges of a molecule of non-polar type of dielectric material coincide as in Fig. 2.26, that is, there is no separation between them and hence dipole moment is zero.

However, when such a material is placed in an electric field, the centres of positive and negative charges are displaced and there exists a distance between them, that is, dipole moment is induced. Under these conditions, the material is said to be under a state of polarisation.

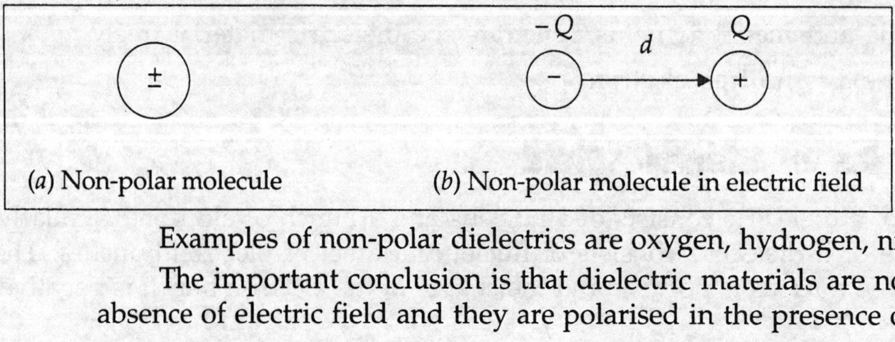

Fig. 2.26

Examples of non-polar dielectrics are oxygen, hydrogen, nitrogen and so on.

The important conclusion is that dielectric materials are not polarised in the absence of electric field and they are polarised in the presence of an electric field.

As a result, the electric flux density is greater than that in free space conditions with the same field intensity. The intensity of polarisation is described in terms of dipole moment and polarisation.

2.48 DIPOLE MOMENT, p

It is defined as the product of charge and distance between the centres of (+)ve and (–)ve charges of a molecule,

that is, $\mathbf{p} \equiv Q\mathbf{d}$, (C-m)

where $\mathbf{p}$ = dipole moment

Q = charge magnitude

$\mathbf{d}$ = distance vector from (–)ve to (+)ve charges of the dipole

If there are N dipoles in a dielectric material of volume $\Delta \upsilon$, the total dipole moment with the application of electric field is

$$\mathbf{p}_t \equiv \sum_{i=1}^{N} Q_i \, \mathbf{d}_i$$

2.49 POLARISATION, P

Polarisation is defined as the dipole moment per unit volume of the dielectric,

that is,

$$\mathbf{P} = \underset{\Delta \upsilon \to 0}{\text{Lt}} \sum_{i=1}^{N} \frac{Q_i \, \mathbf{d}_i}{\Delta \upsilon}$$

In some dielectrics, the polarisation, **P** is defined as

$$\mathbf{P} \equiv \chi_e \epsilon_0 \mathbf{E}$$

where χ_e = electric susceptibility of the dielectric

The charges are bound in a dielectric and the total positive bound charge on a surface, S enclosing the dielectric is given by

$$Q_b = -\int \mathbf{P}.d\mathbf{S}$$

On the other hand, the charge that remains inside the surface, S is $-Q_b$ and it is given by

$$Q_b = -\oint_{\upsilon} \nabla.\mathbf{P}\, d\upsilon = \oint_{\upsilon} \rho_b\, d\upsilon$$

If there is some free charge in the dielectric, the free charge volume charge density is ρ_υ. If ρ_b is the bound volume charge density, then the total volume charge density is given by

$$\rho_t = \rho_\upsilon + \rho_b$$
$$= \nabla.\in \mathbf{E}$$

or,

$$\rho_\upsilon = \rho_t - \rho_b$$
$$= \nabla.\in_0 \mathbf{E} + \nabla.\mathbf{P}$$
$$= \nabla.\mathbf{D}$$

$$\boxed{\mathbf{D} = \in_0 \mathbf{E} + \mathbf{P}}$$

As $\mathbf{P} \equiv \chi_e \in_0 \mathbf{E}$, we have

$$\mathbf{D} = \in_0 \mathbf{E} + \chi_e \in_0 \mathbf{E}$$
$$= \in_0 \mathbf{E}\,(1 + \chi_e)$$
$$\mathbf{D} = \in_0 \in_r \mathbf{E} = \in \mathbf{E}$$

where

$$\in_r = 1 + \chi_e = \frac{\in}{\in_0}$$

or,

$$\boxed{\chi_e = \in_r - 1}$$

In summary, we have

$$Q_b = \int_{\upsilon} \rho_b\, d\upsilon$$

$$Q = \int_{\upsilon} \rho_\upsilon\, d\upsilon$$

$$Q_t = \int_{\upsilon} \rho_t\, d\upsilon$$

$$\nabla.\mathbf{P} = -\rho_b$$
$$\nabla.\in_0 \mathbf{E} = \rho_t$$
$$\nabla.\mathbf{D} = \rho_\upsilon$$

Problem 2.55 A pair of negative and positive charges of $10\mu\text{C}$ each are separated by a distance of 0.1 m along the x-axis. Find the dipole moment.

Solution $$Q_1 = -10\mu C$$
$$Q_2 = 10\mu C$$
$$\mathbf{d} = 0.1\mathbf{a}_x$$
$$Q = 10\mu C$$

The dipole moment is

$$\mathbf{p} = Q\mathbf{d}$$
$$= 10\times 10^{-6} \times 0.1\mathbf{a}_x$$
$$\boxed{\mathbf{p} = 1.0\mathbf{a}_x \,\mu\text{C-m}}$$

Problem 2.56 If a dielectric material of $\in_r = 4.0$ is kept in an electric field $\mathbf{E} = 3.0\mathbf{a}_x + 2.0\mathbf{a}_y + \mathbf{a}_z$, V/m, find the polarisation.

Solution $$\in_r = 4.0$$
$$\mathbf{E} = 3.0\mathbf{a}_x + 2.0\mathbf{a}_y + \mathbf{a}_z$$

Polarisation in the dielectric,

$$\mathbf{P} = \chi_e \in_0 \mathbf{E}$$
$$\chi_e = \in_r - 1$$
$$= 4 - 1 = 3$$
$$\mathbf{P} = 3 \times 8.854 \times 10^{-12} \times (3\mathbf{a}_x + 2\mathbf{a}_y + \mathbf{a}_z)$$
$$= (79.68\mathbf{a}_x + 53.12\mathbf{a}_y + 26.56\mathbf{a}_z)\,\text{PC/m}^2$$

Problem 2.57 What are the magnitudes of electric flux densities and polarisation for a dielectric material in which **E** = 150 kV/m? Electric susceptibility of the dielectric material is 4.75.

Solution We have $$E = 150\,\text{kV/m}$$
$$\in_r = \chi_e + 1$$
$$\boxed{\in_r = 4.75 + 1 = 5.75}$$
$$D = \in E = \in_0 \in_r E$$
$$= 5.75 \times 8.854 \times 10^{-12} \times 150 \times 10^3$$
$$\boxed{D = 7.623\,\mu\text{C/m}^2}$$

Polarisation is

$$P = \chi_e \in_0 E$$
$$= 4.75 \times 8.854 \times 10^{-12} \times 150 \times 10^3$$

or,

$$\boxed{P = 6.308\,\mu\text{C/m}^2}$$

Problem 2.58 Find the polarisation, **P** in a homogeneous and isotropic dielectric material whose $\in_r = 3.0$ when $\mathbf{D} = 3.0\mathbf{a}_r\,\mu\text{C/m}^2$.

Solution Polarisation,

$$\mathbf{P} = \chi_e \epsilon_0 \mathbf{E}$$

and

$$\mathbf{D} = \epsilon_0 \epsilon_r \mathbf{E}, \ \chi_e = \epsilon_r - 1$$

$$\mathbf{D} = \epsilon_0 \mathbf{E} + \mathbf{P}$$

or,

$$\mathbf{P} = \mathbf{D} - \epsilon_0 \mathbf{E}$$

$$= \mathbf{D}\left(1 - \frac{1}{\epsilon_r}\right)$$

$$= \mathbf{D}\left(\frac{\epsilon_r - 1}{\epsilon_r}\right)$$

$$= \left(\frac{3-1}{3}\right) \times 3.0 \times 10^{-6} \, \mathbf{a}_r$$

$$\boxed{\mathbf{P} = 2\mathbf{a}_r \, \mu\text{C/m}^2}$$

Problem 2.59 If the polarisation $\mathbf{P} = 3\mathbf{a}_x \text{ nC/m}^2$ in a homogeneous and isotropic dielectric material whose $\chi_e = 4.5$, find $\mathbf{E}$ in the material.

Solution We have

$$\mathbf{P} = 3\mathbf{a}_x \text{ nC/m}^2$$

$$= \chi_e \epsilon_0 \mathbf{E}$$

or,

$$\mathbf{E} = \frac{\mathbf{P}}{\chi_e \epsilon_0}$$

$$= \frac{3 \times 10^{-9} \mathbf{a}_x}{4.5 \times 8.854 \times 10^{-12}}$$

$$\boxed{\mathbf{E} = 0.0753 \times 10^3 \, \mathbf{a}_x \text{ V/m}}$$

Problem 2.60 A dielectric slab $(\epsilon_r = 2)$ is placed under the influence of electric flux density $= 10\mathbf{a}_x \text{ C/m}^2$. The slab has a volume of 0.1 cm^3. Determine the polarisation in the slab and total dipole moment.

Solution The electric flux density,

$$\mathbf{D} = \epsilon_0 \mathbf{E} + \mathbf{P}$$

or,

$$\mathbf{P} = \mathbf{D} - \epsilon_0 \mathbf{E}$$

$$= \mathbf{D} - \frac{\epsilon_0 \epsilon_r}{\epsilon_r} \mathbf{E}$$

$$= \left(\mathbf{D} - \frac{\mathbf{D}}{\epsilon_r} \right)$$

$$= \mathbf{D} \left(1 - \frac{1}{\epsilon_r} \right)$$

$$\mathbf{P} = 10 \left(1 - \frac{1}{2} \right) \mathbf{a}_x$$

$$\mathbf{P} = 5\mathbf{a}_x \text{ C/m}^2$$

Dipole moment, $\mathbf{p}$ = polarisation × volume

Slab volume $= 0.1 \text{ cm}^3 = 0.1 \times (10^{-2})^3 = 10^{-7} \text{ m}^3$

Dipole moment, $\boxed{\mathbf{p} = 0.5\mathbf{a}_x \, \mu \text{ C-m}}$

Problem 2.61 What are the magnitudes of **P** and **D** for a dielectric material in which $E = 1.0 \text{ V/m}$ and $\chi_e = 5.0$?

Solution $\epsilon_r = \chi_e + 1 = 5.0 + 1 = 6.0$

$$|\mathbf{D}| = D = \epsilon_0 \epsilon_r E$$

$$= 6.0 \times 8.854 \times 10^{-12} \times 1.0$$

$$= 53.124 \text{ PC/m}^2$$

$$|\mathbf{P}| = P = \chi_e \epsilon_0 E$$

$$= 5 \times 8.854 \times 1.0^{-12} \times 1.0$$

$$\boxed{P = 44.272 \text{ PC/m}^2}$$

2.50 CAPACITANCE OF DIFFERENT CONFIGURATIONS

Any two conducting bodies separated by free space or a dielectric material exhibit a capacitance between them.

Capacitance is defined as the ratio of the absolute value of charge to the absolute value of the voltage difference,

that is, $C \equiv \left| \frac{Q}{V} \right|$ (Farad)

Capacitance depends only on the geometry of the system and properties of the dielectrics involved. It does not depend on Q and V.

The capacitance of a capacitor is the ability to store electric charge. It opposes sudden changes in voltage. Its reactance is $\frac{1}{2\pi f C}$ ohms. Here f is the frequency and C is capacitance in Farads.

Capacitance of a parallel plate capacitor It is given by

$$C = \frac{\in A}{d}$$

where $\in$ = permittivity of the dielectric between the conductors (F/m)

A = area of the conductor

d = distance between the conductors

Proof Consider Fig. 2.27.

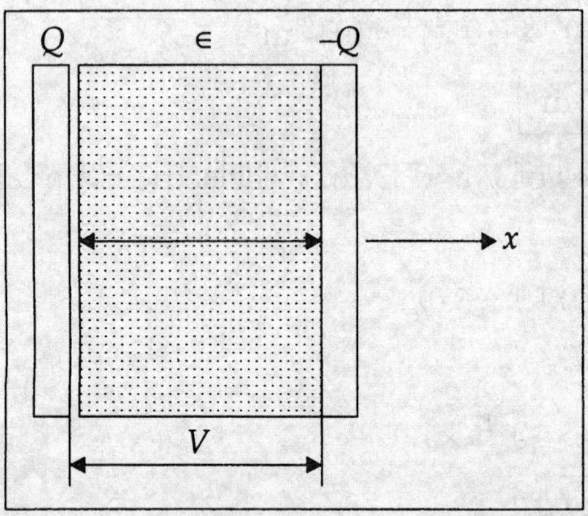

Fig. 2.27 Parallel plate capacitor

Let Q be the absolute charge on one of the plates, V be the potential difference between them, d be the distance between the plates, A be the area of the plate and $\in$ be the permittivity of the medium between the plates.

Then the electric flux density is

$$\mathbf{D} = \frac{Q}{A}\mathbf{a}_x \text{ or } \mathbf{E} = \frac{Q}{\in A}\mathbf{a}_x$$

Potential difference between the plates is

$$V = \mathbf{E}.d\mathbf{a}_x = \frac{Q}{\in A}\mathbf{a}_x . d\mathbf{a}_x = \frac{Q}{\in A}d$$

$$C = \frac{Q}{V} = \frac{\in A}{d}$$ Hence proved.

Capacitance of parallel plate capacitor of *n* dielectric slabs

It is given by

$$C = \frac{A}{\sum_{i=1}^{n} \frac{d_i}{\in_i}}$$

where d_i = width of the ith slab

ϵ_i = permittivity of the ith slab

Proof Let $\epsilon_1, \epsilon_2, \epsilon_3, \ldots, \epsilon_n$ be the permittivity of dielectric materials between the plates; $d_1, d_2, d_3, \ldots, d_n$ respectively be their thickness. (Fig. 2.28).

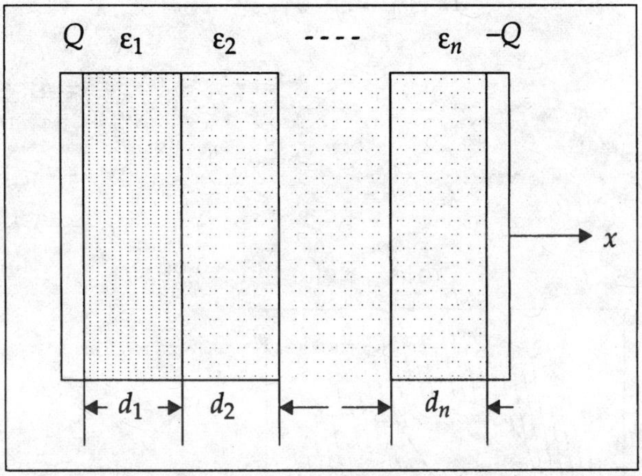

Fig. 2.28 Parallel plate capacitor with multi dielectrics

The potential difference across the capacitor is

$$V = V_1 + V_2 + \ldots + V_n$$

$$\text{But} \quad \left.\begin{aligned} V_1 &= \mathbf{E}_1 . d_1\, \mathbf{a}_x = E_1\, \mathbf{a}_x . d_1\, \mathbf{a}_x \\ V_2 &= \mathbf{E}_2 . d_2\, \mathbf{a}_x = E_2\, \mathbf{a}_x . d_2\, \mathbf{a}_x \\ V_n &= \mathbf{E}_n . d_n\, \mathbf{a}_x = E_n\, \mathbf{a}_x . d_n\, \mathbf{a}_x \end{aligned}\right\}$$

If Q is the charge on the plates, then the electric flux density, **D** is

$$\mathbf{D} = \frac{Q}{A}\mathbf{a}_x$$

Then

$$\mathbf{E}_1 = \frac{\mathbf{D}}{\epsilon_1}, \mathbf{E}_2 = \frac{\mathbf{D}}{\epsilon_2}, \mathbf{E}_3 = \frac{\mathbf{D}}{\epsilon_3}, \ldots, \mathbf{E}_n = \frac{\mathbf{D}}{\epsilon_n}$$

$$V = E_1\, d_1 + E_2\, d_2 + \ldots + E_n\, d_n$$

$$= \frac{D}{\epsilon_1} d_1 + \frac{D}{\epsilon_2} d_2 + \ldots + \frac{D}{\epsilon_n} d_n$$

or,

$$V = \frac{Q}{A}\left[\frac{d_1}{\epsilon_1} + \frac{d_2}{\epsilon_2} + \ldots + \frac{d_n}{\epsilon_n}\right]$$

Hence $$C = \frac{Q}{V} \frac{A}{\sum_{i=1}^{n} \left(\frac{d_i}{\epsilon_i}\right)} \text{ Farads}$$ Hence proved.

Capacitance between two concentric spheres
It is given by

$$C = 4\pi\epsilon \left(\frac{r_1 r_2}{r_2 - r_1}\right) \text{Farads}$$

where r_1 = radius of the inner sphere

r_2 = radius of the outer sphere

Proof Refer Fig. 2.29.

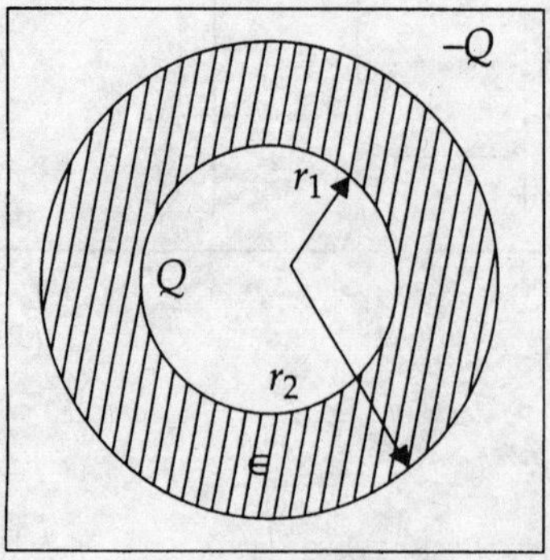

Fig. 2.29 Concentric spheres

The radial electric field is given

$$\mathbf{E} = \frac{Q}{4\pi\epsilon r^2}\mathbf{a}_r \qquad (r_1 \le r \le r_2)$$

Potential difference between the two spheres is given by

$$V = -\int_{r_2}^{r_1} \mathbf{E}.\mathbf{d}r\,\mathbf{a}_r$$

$$= -\int_{r_2}^{r_1} \frac{Q}{4\pi\epsilon\, r^2}\,\mathbf{a}_r\,.dr\,\mathbf{a}_r$$

$$= -\frac{Q}{4\pi\epsilon}\int_{r_2}^{r_1} \frac{1}{r^2}dr$$

$$= \frac{Q}{4\pi\epsilon}\left[\frac{1}{r}\right]_{r_2}^{r_1} = \frac{Q}{4\pi\epsilon}\left(\frac{1}{r_1} - \frac{1}{r_2}\right)$$

$$C = \frac{Q}{V} = \frac{4\pi\epsilon\, r_1 r_2}{(r_2 - r_1)} \text{ Farads}$$ Hence proved.

If r_2 is ∞, the structure will result in an isolated sphere.

The capacitance of an isolated sphere of radius $r_1 = r$ is

$$C = 4\pi\epsilon\, r, \text{ Farads.}$$

Capacitance of a coaxial cable

It is given by

$$C = \frac{2\pi\epsilon\, l}{ln\left(\frac{\rho_2}{\rho_1}\right)} \text{ Farads}$$

where l = length of the cable

ρ_1, ρ_2 = radius of inner and outer conductors

ϵ = permittivity of the dielectric between the conductors

Proof Refer Fig. 2.30.

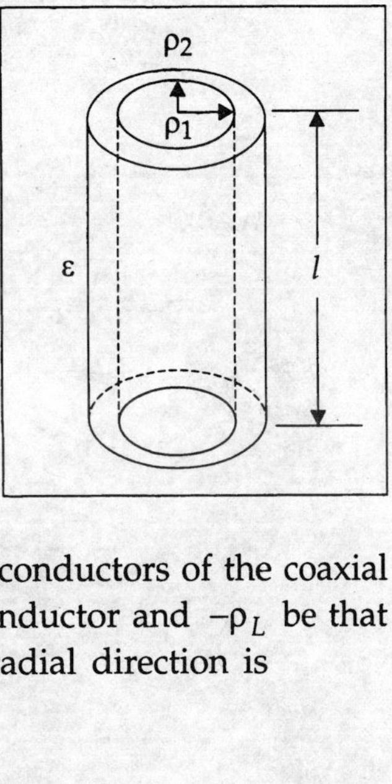

Fig. 2.30 Coaxial cable

Let ρ_1 and ρ_2 be the radii of the inner and outer conductors of the coaxial cable. Let ρ_L be the line charge density on the inner conductor and $-\rho_L$ be that on the outer conductor. Then the electric field in the radial direction is

$$\mathbf{E} = \frac{\rho_L}{2\pi\epsilon\,\rho}\mathbf{a}_\rho$$

The potential difference between the cylinders is

$$V = -\int_{\rho_2}^{\rho_1} \mathbf{E}.d\rho\,\mathbf{a}_\rho$$

$$= -\int_{\rho_2}^{\rho_1} \frac{\rho_L}{2\pi\in\rho} \mathbf{a}_\rho . d\rho\, \mathbf{a}_\rho$$

or,
$$V = -\frac{\rho_L}{2\pi\in} ln\left(\frac{\rho_2}{\rho_1}\right)$$

$$C = \frac{\rho_L}{V} = \frac{2\pi\in}{ln\left(\frac{\rho_2}{\rho_1}\right)} \text{ (Farads/m)}$$

The capacitance of a cable of length l metres is

$$C = \frac{2\pi\in l}{ln\left(\frac{\rho_2}{\rho_1}\right)}, \text{ (Farad)}$$

Hence proved.

Capacitance of parallel wires

It is given by

$$C = \frac{\pi\in}{ln\left(\frac{d}{r}\right)} \text{ (F/m)}$$

where d = distance between the wires
r = radius of each conducting wire

Proof Refer Fig. 2.31.

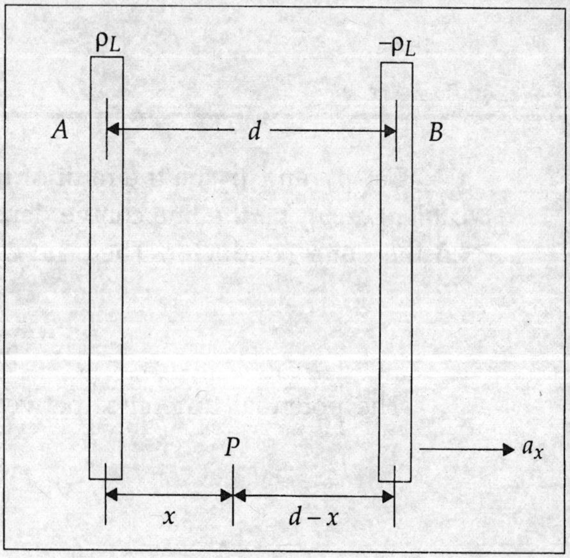

Fig. 2.31 Parallel wires like power transmission lines

Let ρ_L and $-\rho_L$ be the line charge densities of the lines A and B respectively. Let d be the distance between the wires and r be the radius of each wire.

Electric field between the wires at the point P due to ρ_L is

$$\mathbf{E}_2 = \frac{\rho_L}{2\pi\epsilon\, x}\mathbf{a}_x$$

Electric field at P due to $-\rho_L$ is

$$\mathbf{E}_2 = +\frac{\rho_L}{2\pi\epsilon\,(d-x)}\mathbf{a}_x$$

The potential difference, V

$$V = -\int_B^A \mathbf{E}.\, dx\mathbf{a}_x = \int_{(d-r)}^{r} \mathbf{E}.dx\mathbf{a}_x$$

$$= -\int_{(d-r)}^{r} \frac{\rho_L}{2\pi\epsilon\, x}\mathbf{a}_x.dx\mathbf{a}_x - \int_{(d-r)}^{r} \frac{\rho_L}{2\pi\epsilon\,(d-x)}\mathbf{a}_x.dx\mathbf{a}_x$$

$$= -\int_{(d-r)}^{r} \frac{\rho_L}{2\pi\epsilon\, x}dx - \int_{(d-r)}^{r} \frac{\rho_L}{2\pi\epsilon\,(d-x)}dx$$

$$= -\frac{\rho_L}{2\pi\epsilon}\left[\int_{(d-r)}^{r} \frac{1}{x}dx + \int_{(d-r)}^{r} \frac{1}{(d-x)}dx\right]$$

$$= -\frac{\rho_L}{2\pi\epsilon}\left[ln\left(\frac{x}{d-r}\right) + ln\left(\frac{r}{d-r}\right)\right]$$

$$= \frac{\rho_L}{2\pi\epsilon} 2ln\left(\frac{d-r}{r}\right)$$

that is, $$V = \frac{\rho_L}{\pi\epsilon}\, ln\,\frac{(d-r)}{r}$$

In all practical cases, $d >> r$. As a result $\frac{(d-r)}{r} \approx \frac{d}{r}$.

So, $$V = \frac{\rho_L}{\pi \epsilon} \ln\left(\frac{d}{r}\right)$$

The capacitance, $$C \equiv \frac{\rho_L}{V} \text{ F/m}$$

$$C \equiv \frac{\pi \epsilon}{\ln\left(\frac{d}{r}\right)} \text{ Farad/m}$$

or capacitance of a pair of wires of length l metres is, therefore, given by

$$C \equiv \frac{\pi \epsilon\, l}{\ln\left(\frac{d}{r}\right)} \text{ Farad}$$

Problem 2.62 Find the capacitance of an isolated sphere of radius 1 cm.

Solution The expression for the capacitance of an isolated sphere is

$$C = 4\pi\epsilon_0 r$$

Here $$r = 1\text{ cm} = 0.01\text{ m}$$

$$C = 4\pi\epsilon_0 \times 0.01\text{m}$$

$$= \frac{0.01}{9\times 10^9}$$

$$= 1.11\times 10^{-12}$$

or, $$\boxed{C = 1.11\text{ pF}}$$

Problem 2.63 A parallel plate capacitor has conducting plates of area equal to $0.04\,\text{m}^2$. The plates are separated by a dielectric material whose $\epsilon_r = 2$ with the plate separation of 1 cm. Find (a) its capacitance value (b) the charge on the plates when a potential difference of 10 V is applied (c) the energy stored.

Solution (a) The capacitance of parallel plate capacitor, C

$$C = \frac{\epsilon A}{d}$$

where $$\epsilon = \epsilon_0 \epsilon_r = 2\epsilon_0$$

$$A = 0.04\text{ m}^2$$

$$d = 1\text{ cm} = 0.01\text{ m}$$

$$C = \frac{8.854 \times 10^{-12} \times 2 \times 0.04}{0.01}$$

or,

$$\boxed{C = 70.832\,\text{pF}}$$

(b) We have

$$Q = CV,\; V = 10\,\text{volt}$$

$$= 70.832 \times 10^{-12} \times 10$$

$\therefore$

$$\boxed{Q = 708.32\ \text{PC}}$$

(c) Energy stored

$$W_E = \frac{1}{2}CV^2$$

$$= \frac{1}{2} \times 70.832 \times 10^{-12} \times 100$$

$$= 35.416 \times 10^{-10}$$

$$\boxed{W_E = 3.5416\,\text{nJ}}$$

Problem 2.64 Find the capacitance for a 10 km long coaxial cable shown in Fig. 2.32.

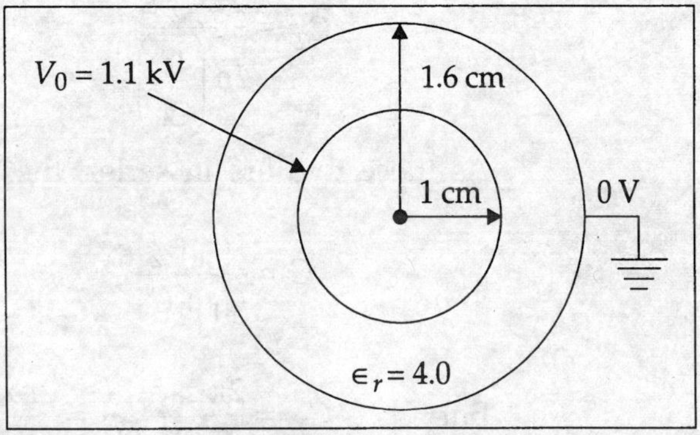

Fig. 2.32 Coaxial cable

Solution For the coaxial cable shown, the inner radius, a = 1 cm, and outer radius, b = 1.6 cm.

The capacitance of the cable is,

$$C = \frac{2\pi \epsilon L}{ln(b/a)}$$

$$= \frac{2\pi \times 4\epsilon_0 \times 10 \times 10^3}{ln\left(\frac{1.6}{1}\right)}$$

$$\boxed{C = 4.734\,\mu\text{C}}$$

Problem 2.65 The cable shown in Fig. 2.33 is 10 km long. If $r_1 = 10\,\text{mm}$, $r_2 = 15\,\text{mm}$, $r_3 = 20\,\text{mm}$, $\in_{r1} = 2.0$, $\in_{r2} = 4.0$, find the capacitance of the cable.

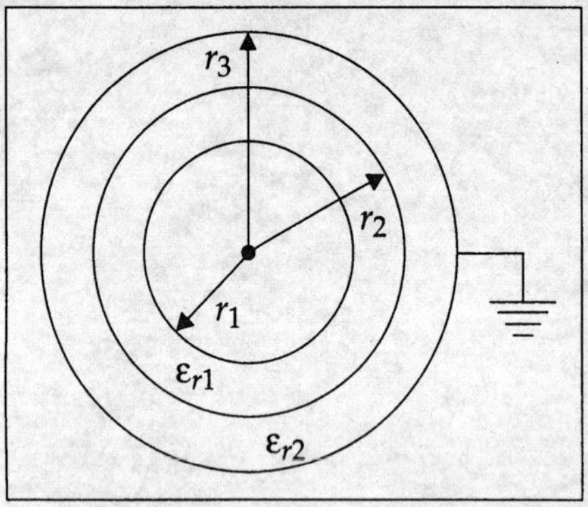

Fig. 2.33 Coaxial cable

Solution The capacitance of the two inner conductors

$$C_1 = \frac{2\pi \in_1 L}{ln\frac{r_2}{r_1}}$$

The capacitance of the two outer conductors

$$C_2 = \frac{2\pi \in_2 L}{ln\left(\frac{r_3}{r_2}\right)}$$

As these two are in series, the resultant capacitance, C

$$C = \frac{C_1 C_2}{C_1 + C_2}$$

But
$$C_1 = \frac{2\pi \in_0 \in_{r1} L}{ln\left(\frac{r_2}{r_1}\right)} = \frac{2\pi \times 8.854 \times 10^{-12} \times 2 \times 10^4}{ln\left(\frac{15}{10}\right)}$$

$$C_1 = 2.74\,\mu\text{F}$$

$$C_2 = \frac{2\pi \in_0 \in_{r2} L}{ln\left(\frac{r_3}{r_2}\right)} = \frac{2\pi \times 8.854 \times 10^{-12} \times 4 \times 10^4}{ln\left(\frac{20}{15}\right)}$$

$$C_2 = 7.73\,\mu\text{F}$$

$$C = \frac{C_1 C_2}{C_1 + C_2} = \frac{2.74 \times 7.73 \times 10^{-12}}{10.47 \times 10^{-6}}$$

$$\boxed{C = 2.023\ \mu\text{F}}$$

2.51 ENERGY STORED IN AN ELECTROSTATIC FIELD

Energy density in an electrostatic field is given by

$$W_E = \frac{1}{2} \epsilon_0 E^2 \text{ Joules/m}^3$$

When a positive charge is brought from a distance of infinity to a point in a field of another positive charge, work is done by an external source. Energy spent in doing so represents the potential energy.

If the external source is removed, the charge that is brought moves back, acquiring kinetic energy of its own and it is capable of doing some work.

If V is the potential at a point due to some fixed charge,

Work done $=$ potential energy

that is, $W_E = QV$

where Q is the charge brought by an external source,

V is the potential at the point due to a fixed charge.

Let us consider two charges Q_1 and Q_2 separated by a distance of infinity. If Q_1 is fixed, work done on bringing Q_2 towards Q_1 is given by

$$W_2 = Q_2 V_2^1$$

where $V_2^1 = \text{potential of } Q_1 \text{ at } Q_2$

Similarly, consider another charge, Q_3 which is at infinity from Q_1 and Q_2. Now work done in bringing Q_3 towards Q_1 and Q_2

$$W_3 = Q_3 V_3^1 + Q_3 V_3^2$$

This is because there exists force due to Q_1 and Q_2 after Q_2 is brought to Q_1.

In the above equation, V_3^1 and V_3^2 are the potentials at Q_3 due to Q_1 and Q_2 respectively. Therefore, total work done in bringing Q_2 and Q_3 is

$$W_e = W_2 + W_3$$

In a similar fashion, consider n charges. Then we have

$$W_t = Q_1 V_2^1 + (Q_3 V_3^1 + Q_3 V_3^2) + (Q_4 V_4^1 + Q_4 V_4^2 + Q_4 V_4^3) + (Q_n V_n^1 + Q_n V_n^2 + \ldots + Q_n V_n^{n-1})$$

that is, $$W_t = \sum_{i=2}^{n}\sum_{j=1}^{i-1} Q_i V_i^j$$

where V_i^j is the potential of Q_j at the location of Q_i. Note that

$$Q_i V_i^j = Q_i \frac{Q_j}{4\pi\epsilon_0 R_{ij}}$$

$$= Q_j \frac{Q_i}{4\pi\epsilon_0 R_{ji}} = Q_j V_j^i$$

W_t may be written as

$$W_t = Q_1 V_1^2 + (Q_1 V_1^3 + Q_2 V_2^3) + (Q_1 V_1^4 + Q_2 V_2^4 + Q_3 V_3^4) + \ldots$$
$$+ (Q_1 V_1^n + Q_2 V_2^n + \ldots + Q_{n-1} V_{n-1}^n)$$

Adding the above two equations and simplifying, we get

$$2W_t = Q_1 (V_1^2 + V_1^3 + V_1^4 + \ldots) + Q_2 (V_2^1 + V_2^3 + V_2^4 + \ldots) + Q_3 (V_3^1 + V_3^2 + V_3^4 + \ldots) + \ldots$$

$$= Q_1 \times (\text{potential at } Q_1 \text{ due to all other charges})$$
$$+ Q_2 \times (\text{potential at } Q_2 \text{ due to all other charges})$$
$$+ Q_n \times (\text{potential at } Q_n \text{ due to all other charges})$$
$$= Q_1 V_1 + Q_2 V_2 + \ldots Q_n V_n$$

$$2W_t = \sum_{i=1}^{n} Q_i V_i$$

or, $$W_t = \frac{1}{2}\sum_{i=1}^{n} Q_i V_i$$

This equation represents the potential energy stored in a system of n point charges.

If $$Q_i = \int_{\upsilon} \rho_\upsilon \, d\upsilon$$

$$W_E = \frac{1}{2}\int_{\upsilon} \rho_\upsilon V \, d\upsilon$$

But $\nabla . \mathbf{D} = \rho_\upsilon$ or $\nabla . \mathbf{E} = \rho_\upsilon / \epsilon_0$

$$W_t = \frac{1}{2}\int_{\upsilon} \epsilon_0 (\nabla.\mathbf{E})\, V\, d\upsilon$$

From standard vector identity,

$$(\nabla.\mathbf{E})V = \nabla.V\mathbf{E} - \mathbf{E}.\nabla V$$

this becomes

$$W_t = \frac{1}{2}\epsilon_0 \int_{\upsilon} (\nabla.V\mathbf{E} - \mathbf{E}.\nabla V)\, d\upsilon$$

$$= \frac{1}{2}\epsilon_0 \int_{\upsilon} \nabla.V\mathbf{E}\, d\upsilon + \frac{1}{2}\epsilon_0 \int_{\upsilon} \mathbf{E}.\mathbf{E}\, d\upsilon$$

Applying divergence theorem, first term on the right hand side can be written as

$$\underset{\text{complete space}}{\int_{\upsilon}} \nabla.V\mathbf{E}\, d\upsilon = \underset{\substack{\text{surface}\\\text{bounding}\\\text{the space}}}{\int} V\, \mathbf{E}.d\mathbf{S}$$

However, viewing from a surface bounding complete space, the charge distribution of finite volume appears as a point charge, say Q. We know that

$$\mathbf{E} = \frac{Q}{4\pi\epsilon_0 r^2}\mathbf{a}_r$$

and

$$V = \frac{Q}{4\pi\epsilon_0 r}$$

From the expressions of $\mathbf{E}$ and V, we get

$$\underset{\substack{\text{surface}\\\text{bounding}\\\text{the space}}}{\int} V\, \mathbf{E}.dS\mathbf{a}_n = \underset{r\to\infty}{\text{Lt}} \int_{\theta=0}^{\pi}\int_{\phi=0}^{2\pi} \frac{Q}{4\pi\epsilon_0 r} \times \frac{Q}{4\pi\epsilon_0 r^2}\mathbf{a}_r.r^2 \sin\theta\, d\theta\, d\phi\, \mathbf{a}_n$$

[as $\mathbf{E}$ is $\mathbf{a}_r$-directed, $d\mathbf{S}$ is $\mathbf{a}_n$-directed and $dS = r^2 \sin\theta\, d\theta\, d\phi$]

$$= \underset{r\to\infty}{\text{Lt}} \int_{\theta=0}^{\pi}\int_{\phi=0}^{2\pi} \frac{Q^2}{4\pi\epsilon_0 r} \sin\theta\, d\theta\, d\phi = 0$$

Hence W_t becomes

$$W_t = \frac{1}{2} \in_0 \int_{\substack{\text{complete}\\ \text{space}}} \mathbf{E}.\mathbf{E} d\upsilon$$

or,

$$W_t = \int_{\substack{\text{complete}\\ \text{space}}} \frac{1}{2} \in_0 E^2 \, d\upsilon \text{ Joules}$$

This expression is the total energy stored in an electrostatic field. Therefore, the integrand represents energy density, that is,

The energy density,

$$W_E = \frac{1}{2} \in_0 E^2, \text{ Joules/m}^3$$

Hence proved.

2.52 ENERGY IN A CAPACITOR

Energy stored in a capacitor is given by

$$W_c = \frac{1}{2} CV^2 \text{ Joules}$$

Proof Method 1 We know that energy stored in the electric field of a capacitor is given by

$$W_c = \frac{1}{2} \int_{\substack{\text{space between}\\ \text{the conductors}}} \mathbf{D}.\mathbf{E} \, d\upsilon$$

If the space between the conductors is occupied by a dielectric material whose relative permittivity is $\in_r$, then

$$\mathbf{D} = \in \mathbf{E} = \in_0 \in_r \mathbf{E}$$

$$W_c = \frac{1}{2} \int_{\upsilon} \in_0 \in_r E^2 \, d\upsilon$$

As the plates are assumed to be separated in x-direction,

$$\mathbf{D} = \frac{Q}{A} \mathbf{a}_x$$

and $$\mathbf{E} = \frac{V}{x}\,\mathbf{a}_x$$

$$W_c = \frac{1}{2}\int_{\upsilon} \frac{Q}{A}\mathbf{a}_x \cdot \frac{V}{x}\mathbf{a}_x\, d\upsilon$$

$$= \int_{\upsilon} \frac{1}{2}\frac{CV^2}{Ax}\, d\upsilon$$

$$= \frac{1}{2}CV^2 \frac{\upsilon}{\upsilon} \qquad [\text{as } Ax = \upsilon]$$

$$W_c = \frac{1}{2}CV^2 \text{ Joules} \qquad \text{Hence proved.}$$

Method 2 When a capacitor is charged, energy is stored in the electrostatic field which is set up in the dielectric medium.

Assume that a capacitor C is charged to a voltage, V. If the potential difference across the plates at any instant of charging is V, this is equal to the work done in shifting one coulomb of charge from one plate to another. If dQ is the charge transferred, work done is

$$dW_c = VdQ$$

As per the definition of C,

$$Q = CV$$

$$dQ = C\,dV$$

or, $$dW_c = CV\,dV$$

Total work done in producing a potential of V is

$$W_c = \int_0^V CV\,dV = C\frac{V^2}{2}\Big]_0^V = \frac{1}{2}CV^2$$

Energy stored in a capacitor

$$W_c = \frac{1}{2}CV^2 \text{ Joules} = \frac{1}{2}QV = \frac{Q^2}{2C} \text{ Joules} \qquad \text{Hence proved.}$$

POINTS/FORMULAE TO REMEMBER

- Line charge density is $\rho_L = \frac{Q}{L}$, C/m.
- Surface charge density is $\rho_s = \frac{Q}{s}$, C/m^2.
- Volume charge density, $\rho_s = \frac{Q}{\upsilon}$, C/m^3.
- Coulomb's law is $\mathbf{F} = \frac{1}{4\pi\epsilon_0} \frac{Q_1 Q_2}{r^2} \mathbf{a}_r$, Newton.
- $\epsilon_0 = 8.854 \times 10^{-12}$, F/m
- $\frac{1}{4\pi\epsilon_0} = 9 \times 10^9$ m/F
- $\mathbf{F}_{12} = -\mathbf{F}_{21}$
- Displacement electric flux = electric flux = Q.
- Electric field intensity = electric field, $\mathbf{E} = \frac{\mathbf{F}}{Q}$.
- $\mathbf{E} = -\nabla V$, V/m
- Direction of Coulomb's force is the same as electric field due to a point charge.
- Electric field due to a line charge is $\mathbf{E} = \int \frac{\rho_L \, dL}{4\pi\epsilon_0 r^2} \mathbf{a}_r$.
- Electric field due to an infinite line charge is $\mathbf{E} = \frac{\rho_L}{2\pi\epsilon_0 \rho} \mathbf{a}_\rho$.
- Electric field due to surface charge density, $\mathbf{E} = \frac{\rho_s}{2\epsilon_0} \mathbf{a}_n$.
- Electric field due to volume charge density, $\mathbf{E} = \int_\upsilon \frac{\rho_\upsilon}{4\pi\epsilon_0 r^2} d\upsilon \, \mathbf{a}_r$.

- Potential at a point is $V = \dfrac{Q}{4\pi\epsilon_0 r}$.
- $V_{AB} = -\int_A^B \mathbf{E}.d\mathbf{L}$
- Potential gradient, $\nabla V = -\mathbf{E}$.
- Potential due to electric dipole is $V = \dfrac{Qd\cos\theta}{4\pi\epsilon_0 r^2}$.
- Electric field due to a dipole is $\mathbf{E} = \dfrac{Qd}{4\pi\epsilon_0 r^3}(2\cos\theta\,\mathbf{a}_r + \sin\theta\,\mathbf{a}_\theta)$.
- Electric flux, $\psi = Q$.
- Gauss's law is $\int_s \mathbf{D}.d\mathbf{S} = Q$.
- Electric flux density, $\mathbf{D} = \dfrac{d\psi}{ds}\,\mathbf{a}_n\ \text{C/m}^2$
  $= \epsilon\,\mathbf{E}$.
- Point form of Gauss's law is $\nabla.\mathbf{D} = \rho_\upsilon$.
- Laplace's equation is $\nabla^2 V = 0$.
- Either Poisson's or Laplace's equation has unique solution.
- $\mathbf{E}_{\tan 1} = \mathbf{E}_{\tan 2}$
- $D_{n1} - D_{n2} = \rho_s$
- Conduction current density, $\mathbf{J}_c = \sigma\mathbf{E}$.
- Displacement current density, $\mathbf{J}_d = \dfrac{\partial \mathbf{D}}{\partial t}$.
- Equation of continuity is $\nabla.\mathbf{J} = -\dot{\rho}_\upsilon$.

- Relaxation time, $T_r = \dfrac{\epsilon}{\sigma}$.
- $\mathbf{J} = \rho_v \mathbf{V}$
- Electric dipole moment is $\mathbf{p} = Q\mathbf{d}$.
- Polarisation, $\mathbf{P} = \chi_e \epsilon_0 \mathbf{E} = \dfrac{\text{dipole moment}}{\text{volume}}$.
- Electric susceptibility, $\chi_e = \epsilon_r - 1$.
- Displacement flux density in dielectrics is $\mathbf{D} = \epsilon_0 \mathbf{E} + \mathbf{P}$.
- Capacitance, $C = \left|\dfrac{Q}{V}\right|$, Farad.
- Capacitance of parallel capacitor, $C = \dfrac{\epsilon A}{d}$.
- Capacitance of spherical condenser is $C = 4\pi\epsilon \left(\dfrac{r_1 r_2}{r_2 - r_1}\right)$.
- Capacitance of a coaxial cable is $C = \dfrac{2\pi\epsilon l}{ln\left(\dfrac{\rho_2}{\rho_1}\right)}$.
- Capacitance of parallel wires, $C = \dfrac{\pi\epsilon l}{ln\left(\dfrac{d}{r}\right)}$.
- Energy stored in electrostatic field is $W_E = \dfrac{1}{2}\epsilon_0 E^2$ Joules/m^3.
- Energy stored in a capacitor, $W_c = \dfrac{1}{2} CV^2$ Joules.

OBJECTIVE QUESTIONS

1. Coulomb's force depends on the medium in which the charges are placed. (Yes/No)
2. Constant of proportionality in Coulomb's law has units. (Yes/No)
3. The directions of electric field and Coulomb's force are the same. (Yes/No)
4. For a thin filament extending from $-\infty$ to ∞ along the z-axis, field varies with z and ϕ coordinates. (Yes/No)
5. For a line charge extending from $-\infty$ to ∞ along the z-axis, field at a point varies with ρ only. (Yes/No)
6. For N point charges, the electric field at a point is the vectorial sum of the fields due to each of N charges. (Yes/No)
7. Coulomb's law can be applied to find electric field at a point. (Yes/No)
8. Charge distributions produce electric field. (Yes/No)
9. Electric field at a point due to a point charge is inversely proportional to r. (Yes/No)
10. Electric field at a point due to a point charge is proportional to $1/r^2$. (Yes/No)
11. A surface of a conductor is an equipotential surface. (Yes/No)
12. Gradient of the potential and equipotential surface are orthogonal to each other. (Yes/No)
13. Electric dipole is a pair of two positive point charges. (Yes/No)
14. Electric dipole is a pair of a positive charge and a negative charge. (Yes/No)
15. Displacement electric flux is equal to the charge enclosed. (Yes/No)
16. Faraday's experiment says $\psi = Q$. (Yes/No)
17. Gauss's law is applicable on all surfaces. (Yes/No)
18. Gauss's law is applicable only on Gaussian surfaces. (Yes/No)
19. Gauss's law is applicable even if the charge is outside the closed surface. (Yes/No)
20. Displacement flux depends on the dielectric material. (Yes/No)

21. Laplace's equation has several solutions. (Yes/No)

22. $E_{tan1} = E_{tan2}$ (Yes/No)

23. $D_{n1} = D_{n2}$ (Yes/No)

24. Aluminium is a better conductor than silver at very low temperatures. (Yes/No)

25. For a good conductor, $\frac{\sigma}{\omega \epsilon} << 1$. (Yes/No)

26. Convection and conduction current densities are identical. (Yes/No)

27. Convection current obeys Ohm's law. (Yes/No)

28. Conduction current obeys Ohm's law. (Yes/No)

29. Displacement current in a conductor is greater than conduction current. (Yes/No)

30. Displacement current in dielectrics is greater than conduction current. (Yes/No)

31. The current density, $\mathbf{J} = \rho_\upsilon \mathbf{V}$. (Yes/No)

32. $\nabla . \mathbf{J} = \dot{\rho}_\upsilon$ (Yes/No)

33. $\nabla . \mathbf{J} = -\dot{\rho}_\upsilon$ (Yes/No)

34. Electric dipole moment is a vector. (Yes/No)

35. The polarisation, $\mathbf{P} = \chi_e \epsilon_0 \mathbf{E}$. (Yes/No)

36. Electric susceptibility has the unit of permittivity. (Yes/No)

37. Electric susceptibility has no units. (Yes/No)

38. Polarisation is a result of free electrons in dielectrics. (Yes/No)

39. Polarisation is a result of bound electrons in dielectrics. (Yes/No)

40. Electric flux density in dielectrics is greater than that in free space for the same field. (Yes/No)

41. Capacitance depends on dielectric material between the conductors. (Yes/No)

42. Gauss's law can be applied on arbitrary Gaussian surfaces. (Yes/No)

43. Potential is inversely proportional to r. (Yes/No)

44. The unit of potential is Joule/Coulomb. (Yes/No)

45. Potential obeys the superposition principle. (Yes/No)

46. Electric field obeys the superposition principle. (Yes/No)

47. Electrostatic energy is quadratic in the fields. (Yes/No)

48. Electrostatic energy does not obey superposition principle. (Yes/No)

49. $\oint \mathbf{E}.d\mathbf{L}=0$ is the same as $\nabla\times\mathbf{E}=0$. (Yes/No)

50. **E** is perpendicular to the surface just outside a conductor. (Yes/No)

51. $\rho_v = 0$ inside a conductor. (Yes/No)

52. Charge resides on the surface of a conductor. (Yes/No)

53. Direction of dipole moment is in the direction of applied electric field. (Yes/No)

54. Coulomb's force has the unit of ________.

55. The unit of constant of proportionality in Coulomb's law is ________.

56. The unit of line charge density is ________.

57. The unit of surface charge density is ________.

58. The unit of volume charge density is ________.

59. Electric field is defined as ________.

60. The unit of electric field is ________.

61. The unit of electric flux is ________.

62. Laplace's equation is ________.

63. Relaxation time in dielectrics is ________.

64. The unit of electric dipole moment is ________.

65. The unit of polarisation of dielectric is ________.

66. If $\epsilon_r = 3$, electric susceptibility is ________.

67. Application of electric field to dielectric material produces ________.

68. Energy density in electric field is ________.

69. Energy stored in a capacitor is ________.

70. If the charge is doubled everywhere, the total energy is ________.

71. Atomic polarisability has unit of ________.

Answers

1. Yes	2. Yes	3. Yes	4. No	5. Yes
6. Yes	7. Yes	8. Yes	9. No	10. Yes
11. Yes	12. Yes	13. No	14. Yes	15. Yes
16. Yes	17. No	18. Yes	19. No	20. No
21. No	22. Yes	23. No	24. Yes	25. No
26. No	27. No	28. Yes	29. No	30. Yes
31. Yes	32. No	33. Yes	34. Yes	35. Yes
36. No	37. Yes	38. No	39. Yes	40. Yes
41. Yes	42. Yes	43. Yes	44. Yes	45. Yes
46. Yes	47. Yes	48. Yes	49. Yes	50. Yes
51. Yes	52. Yes	53. Yes	54. Newton	55. metre/Farad
56. Coulombs/m		57. C/m^2	58. C/m^3	59. $\mathbf{E} = -\nabla V$
60. V/m	61. Coulomb	62. $\nabla^2 V = 0$	63. ϵ/σ	64. C-m
65. C/m^2	66. $\chi_e = 2$	67. Induced dipole moment		68. $0.5\epsilon E^2$
69. $0.5\ CV^2$	70. Quadrupled		71. Farad-m^2	

MULTIPLE CHOICE QUESTIONS

1. Coulomb's force is proportional to

(a) r (b) r^2 (c) $\frac{1}{r}$ (d) $\frac{1}{r^2}$

2. The proportionality constant in Coulomb's law has unit of

(a) Farads (b) Farads/metre

(c) Newton (d) metre/Farad

3. The value of proportionality constant in Coulomb's law is

(a) 9×10^9 (b) 9×10^{-9}

(c) 8.854×10^{-12} (d) $\frac{1}{36\pi}\times10^9$

4. The unit of electric field is

(a) Newton (b) Coulomb/Newton

(c) Newton/Coulomb (d) Coulomb/metre

5. If the direction of Coulomb's force on a unit charge is $\mathbf{a}_x$, the direction of electric field is

(a) $-\mathbf{a}_x$ (b) $\mathbf{a}_y$ (c) $\mathbf{a}_x$ (d) $\mathbf{a}_z$

6. The unit of electric flux is

(a) Coulomb (b) Coulomb/metre

(c) Weber (d) Weber/m^2

7. The electric field on x-axis due to a line charge extending from $-\infty$ to ∞ is

(a) $\frac{\rho_l}{2\pi\epsilon_0\rho}$ (b) $\frac{\rho_l}{2\epsilon_0\rho}$ (c) $\frac{\rho_l}{2\rho}$ (d) $\frac{\rho_l}{\epsilon_0\rho}$

8. Electrostatic field due to a dipole consists of

(a) $\frac{1}{r}$ term (b) $\frac{1}{r^2}$ term (c) $\frac{1}{r^3}$ term (d) r terms

9. Potential at all the points on the surface of a conductor is

(a) the same (b) not the same

(c) zero (d) infinity

10. Gradient of the potential and an equipotential surface
    (a) have the same direction (b) have opposite directions
    (c) are orthogonal to each other (d) have no directional relation

11. The potential at a point due to electric dipole consists of
    (a) r terms (b) $\frac{1}{r^2}$ term (c) $\frac{1}{r^3}$ term (d) $\frac{1}{r}$ term

12. The unit of electric dipole moment is
    (a) C/m (b) C-m (c) C/m^2 (d) C-m^2

13. The unit of polarisation in dielectric is
    (a) C/m^2 (b) C/m (c) C/m^3 (d) C-m^2

14. Point form of Gauss's law is
    (a) $\nabla.\mathbf{D}=\rho_v$ (b) $\nabla.\mathbf{D}=\rho_s$ (c) $\nabla.\mathbf{D}=\rho_v/\epsilon_0$ (d) $\nabla.\mathbf{D}=Q$

15. The Laplacian operator, ∇^2
    (a) has unit of m^2 (b) is a vector operator
    (c) has unit of 1/m^2 (d) has no unit

16. Laplace's equation has
    (a) two solutions (b) infinite solutions
    (c) no solution (d) only one solution

17. The surface charge density in a good dielectric is
    (a) zero (b) ρ_s (c) infinity (d) $-\rho_s$

18. Relaxation time of a medium with $\epsilon_r = 3.0$ and $\sigma = 3.0$ Mho/m is
    (a) 8.854 picosecond (b) 9 picosecond
    (c) 7.9686 picosecond (d) 1 second

19. The force magnitude between $Q_1 = 1$C and $Q_2 = 1$C when they are separated by 1 m in free space is
    (a) 9×10^9 N (b) 8.854×10^{-12} N
    (c) $\frac{1}{36\pi} \times 10^{-9}$ N (d) 9×1^{-9} N

20. When the force on 2 C due to fixed charge of 4 C is 2 N, the electric field at the charge of 2 C is

(a) 1 N/C (b) 4 N/C (c) 8 N/C (d) 16 N/C

21. If $\in_r = 2$ for a dielectric medium, its electric susceptibility is
(a) 1 (b) 2 (c) 3 (d) $2\in_0$

22. If a pair of (+)ve and (–)ve charges of 1 C each are separated by a distance of 1.0 μm, the magnitude of dipole moment is
(a) 2 C-μm (b) 1 C-μm (c) 0 C-μm (d) 1 C-μm

23. If dipole moment of 1 C-m in a dielectric material of volume 0.1 m^3 exists, the polarisation is
(a) 10 C/m^2 (b) 0.1 C/m^2 (c) 10 C/m (d) 0.1 C/m

24. If a charge element, whose volume charge density is 2.0 C/m^2, is moving with a velocity of $3\mathbf{a}_x$ m/s, the current density is
(a) $6\mathbf{a}_x$ A/m^2 (b) $6\mathbf{a}_x$ A/m (c) $1.5\mathbf{a}_x$ A/m^2 (d) $1.5\mathbf{a}_x$ A/m

25. If $\mathbf{E}_{\tan 1} = \mathbf{a}_x$ and $\mathbf{E}_{n1} = 0$, the electric field $\mathbf{E}_2$ in a dielectric medium 2 is
(a) $\mathbf{a}_x$ (b) $2\mathbf{a}_x$ (c) $\mathbf{a}_y$ (d) $\mathbf{a}_z$

26. Two point charges $Q_1 = 1\text{C}$ and $Q_2 = 3\text{C}$ are separated by 1.0 m. The force on Q_1 is
(a) zero (b) repulsive
(c) attractive (d) increasing linearly

27. A charge density of 10 nC/m^2 is distributed on a plane z = 10 m, the electric field intensity at the origin is
(a) $-180\pi\mathbf{a}_z$ V/m (b) $-10\pi\mathbf{a}_z$ V/m
(c) $-360\pi\mathbf{a}_z$ V/m (d) $-18\pi\mathbf{a}_z$ V/m

28. If a force, $\mathbf{F} = 4\mathbf{a}_x + \mathbf{a}_y + 2\mathbf{a}_z$ moves 1μC charge through a displacement of $4\mathbf{a}_x + 2\mathbf{a}_y - 6\mathbf{a}_z$, the resultant work done is
(a) 6 μJ (b) 12 μJ (c) 18 μJ (d) 24 μJ

29. If a potential of 1 V is applied across a capacitor of 10 PF, the energy stored is
(a) 5 PJ (b) 10 PJ (c) 100 PJ (d) 0.01 PJ

30. Example of non-polar type of dielectric is
(a) water (b) hydrochloric acid
(c) sulphur dioxide (d) oxygen

31. Example of polar type of dielectric is
(a) oxygen (b) water (c) hydrogen (d) nitrogen

32. If the voltage applied across a capacitor is increased, the capacitance value
(a) increases (b) decreases
(c) remains constant (d) becomes infinity

33. If the electric field intensity is 1 V/m in free space, the energy density is
(a) 4.427 PJ/m^3 (b) 8.854 PJ/m^3
(c) 4.427 PJ (d) 8.854 PJ

34. If electric susceptibility of a dielectric is 4, its relative permittivity is
(a) 5 (b) 4 (c) 3 (d) 2

35. The unit of electric flux is
(a) Coulomb (b) Coulomb/m (c) Weber (d) Tesla

36. Gauss's law is
(a) $\int_s \mathbf{D}.d\mathbf{S} = Q$ (b) $\oint_s \mathbf{D}.d\mathbf{S} = Q$
(c) $\oint \mathbf{D}\, d\mathrm{S} = Q$ (d) $\int_s \mathbf{D}\, d\mathrm{S} = Q$

37. Gauss's law in point form is
(a) $\nabla.\mathbf{D} = \rho_v$ (b) $\nabla.\mathbf{D} = \rho_s$
(c) $\nabla.\mathbf{D} = Q$ (d) $\nabla\mathbf{D} = \rho_v$

38. Equation of continuity is
(a) $\int_s \mathbf{J}.d\mathbf{S} = I$ (b) $\oint_s \mathbf{J}.d\mathbf{S} = I$
(c) $\int_s \mathbf{J}\, d\mathrm{S} = I$ (d) $\oint_s \mathbf{J}\, d\mathrm{S} = I$

39. Relaxation time is
(a) $\frac{\epsilon}{\sigma}$ (b) $\frac{\sigma}{\epsilon}$
(c) $\frac{\sigma}{\omega\epsilon}$ (d) $\frac{\omega\epsilon}{\sigma}$

40. In dielectrics,

(a) $\int_{\upsilon} \rho_{\upsilon}\, d\upsilon = Q_b$ (b) $\int_{\upsilon} \rho_b\, d\upsilon = -Q_b$

(c) $\int_{\upsilon} \rho\, d\upsilon = Q_b$ (d) $\int_{\upsilon} \rho_b\, d\upsilon = Q_b$

41. Potential has the unit of

(a) Joules/Coulomb (b) Joules

(c) Joules/m^3 (d) Joules/m^2

42. If a total charge of 10 coulombs is uniformly distributed along a filament of length 10 m, the line charge density is

(a) 1 C/m (b) 100 C-m (c) 100 C/m (d) 1 C-m

43. If a charge of 10 coulombs is uniformly distributed on the surface of a conductor of area 10 m^2, the surface charge density is

(a) 1 C/m^2 (b) 100 C/m^2 (c) 1 C-m^2 (d) 100 C-m^2

44. If a total charge of 1 C is contained in a tiny sphere of volume 0.1 m^3, the volume charge density is

(a) 10 C/m^3 (b) 0.1 C/m^3 (c) 10 C-m^3 (d) 10 C/m^2

45. Electric flux density is

(a) $\frac{Q}{4\pi\epsilon_0 r^2}$ (b) $\frac{Q}{4\pi r^2}$ (c) $\frac{Q}{4\pi r^2}\mathbf{a}_r$ (d) $\frac{Q}{4\pi\epsilon_0 r^2}\mathbf{a}_r$

46. Poisson's equation is

(a) $\nabla^2 V = \rho_\upsilon/\epsilon$ (b) $\nabla^2 V = -\rho_\upsilon/\epsilon$

(c) $\nabla^2 V = -\rho_\upsilon$ (d) $\nabla^2 V = -\dot{\rho}_\upsilon/\epsilon$

47. Boundary condition for the normal component of **E** on the boundary of a dielectric is

(a) $E_{n1} = E_{n2}$ (b) $E_{n1} - E_{n2} = \rho_s$

(c) $E_{n1} = \frac{\epsilon_2}{\epsilon_1} E_{n2}$ (d) $E_{n1} = 0$

48. Electric flux lines

(a) originate at (+)ve charge

(b) originate at (–)ve charge

(c) are closed loops

(d) originate at (+)ve charge and also terminate at (+)ve charge

49. Unit of electric flux is

(a) Coulomb (b) Weber

(c) Tesla (d) Weber/m

50. Potential due to a charge at a point situated at ∞ is

(a) zero (b) ∞ (c) $-\infty$ (d) 1

Answers

1. (d)	2. (d)	3. (a)	4. (c)	5. (c)
6. (a)	7. (a)	8. (c)	9. (a)	10. (c)
11. (b)	12. (b)	13. (a)	14. (a)	15. (c)
16. (d)	17. (a)	18. (a)	19. (a)	20. (a)
21. (a)	22. (d)	23. (a)	24. (a)	25. (a)
26. (b)	27. (a)	28. (a)	29. (a)	30. (d)
31. (b)	32. (c)	33. (a)	34. (a)	35. (a)
36. (b)	37. (a)	38. (b)	39. (a)	40. (d)
41. (a)	42. (a)	43. (a)	44. (a)	45. (a)
46. (b)	47. (c)	48. (a)	49. (a)	50. (a)

EXERCISE PROBLEMS

1. An infinite line charge, $\rho_L = 10\,\text{nC/m}$ parallel to z-axis is at $x = 3$, $y = 4$ in free space. Find **E** at
   (a) (0, 0, 0)
   (b) (0, 1, 2)
   (c) (1, 1, 1)

2. Determine the force on a point charge of 5 nC at (0, 0, 5) m due to uniformly distributed charge of 5 mC over a circular disc of radius $r \le 1\,\text{m}$ in z = 0 plane.

3. Find the total charge in the volume specified by $0 \le x \le 1\,\text{m}$, $0 \le y \le 1\,\text{m}$ and $0 \le z \le 1\,\text{m}$ when $\rho_v = 30x^2 y\,(\text{nC/m}^3)$.

4. Three infinitely long lines charged uniformly are parallel to the z-axis. They are separated by a distance of b m. The charge density of each is ρ_L = 2.0 PC/m. Find the electric field **E** at a point P on the y-axis at $y = a\,\text{m}$. If $a = b = 1\,\text{m}$, what is the electric field, **E**?

5. A dipole consists of two charges Q and – Q separated by $2b$ m as shown in Fig. 2.34. Determine the electric field at point P.

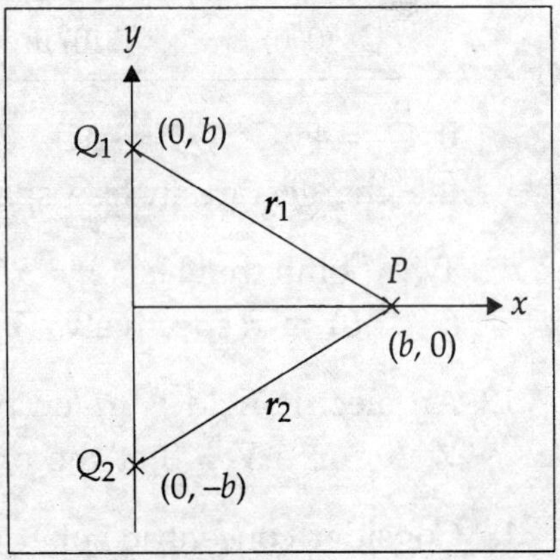

Fig. 2.34

6. An infinite length of uniform line charge has $\rho_L = 10\,\text{PC/m}$ and it lies along the z-axis. Determine electric field **E** at (4, 3, 3) m.

7. When six equal charges Q = 10 PC are located at (2, 0, 0), (3, 0, 0), (4, 0, 0), (5, 0, 0), (6, 0, 0) and (7, 0, 0), find the potential at the origin.

8. If the electric potential is as shown in Fig. 2.35, sketch the respective electric field distribution in a specified region.

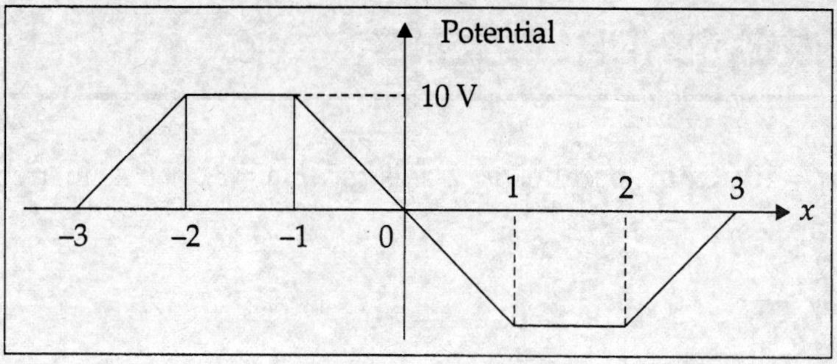

Fig. 2.35

9. (a) What is the force magnitude between two charges $Q_1 = 4\,\text{nC}$ and $Q_2 = 6\,\text{nC}$ which are separated by 0.1 m in free space.

(b) What will happen if they are kept in ice whose $\epsilon_r = 4.2$?

10. Consider the Fig. 2.36.

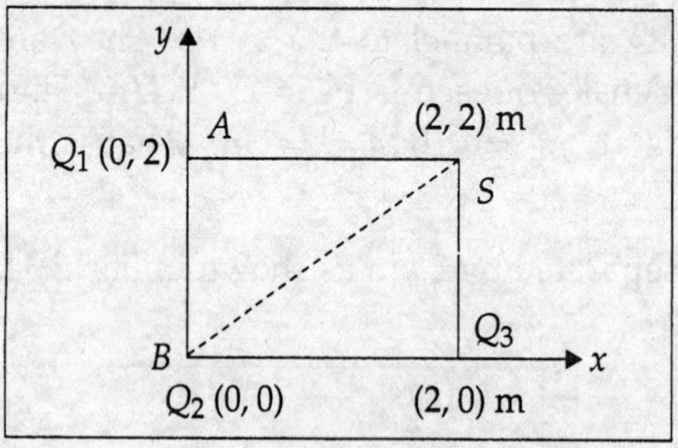

Fig. 2.36

If $Q_1 = 4\,\text{nC}$, $Q_2 = 4\,\text{nC}$, $Q_3 = 4\,\text{nC}$, find the electric field at point S. Assume that the charges are in free space.

11. Two point charges $Q_1 = 5.0\,\text{nC}$ and $Q_2 = 1.0\,\text{nC}$ are located at (− 1, 1, − 3) m and (3, 1, 0) m respectively. Determine the electric field at Q_1 and Q_2.

12. An electric field is given by $\mathbf{E} = 10y\,\mathbf{a}_x + 10x\,\mathbf{a}_y$, V/m. Find the potential function, V. Assume $V = 0$ at the origin.

13. Consider concentric spherical shells in free space in which $V = 0$ at $r = 10$ cm and $V = 10$ volts at $r = 20$ cm. Find **E** and **D**.

14. If electric flux density, **D** is given by $\mathbf{D} = [(2y^2 + z)\,\mathbf{a}_x + 4xy\,\mathbf{a}_y + x\mathbf{a}_z)]\,\mu\text{C/m}^2$, find the volume charge density at (0, 0, 0) and (− 1, 0, 4).

15. A sphere whose radius is 0.5 m contains a charge density of $\rho_v = (5 - 2r)\,\mu\text{C/m}^3$. Find **D** at a distance 10 m away from the centre of the sphere.

CHAPTER 3

STEADY MAGNETIC FIELDS

Steady magnetic fields are produced by steady currents.

The main objective of this chapter is to provide detailed concepts of magnetostatics. They include:

- applications of magnetostatic fields
- Faraday's induction law, Biot-Savart law, Ampere's circuit law, Ampere's force laws for current elements
- force on current element and between current elements
- Lorentz force equation
- Boundary conditions, scalar and vector magnetic potentials
- fields in magnetic materials
- inductances
- energy stored in magnetic field and inductors
- comparison between electric and magnetic parameters/circuits
- points/formulae to remember, objective and multiple choice questions and exercise problems.

Do you know?

Some birds, bees and certain animals are blessed with magnetic sense but man can sense magnetic fields only with a compass.

3.1 INTRODUCTION

Steady currents produce **steady magnetic fields**. Steady magnetic fields are magnetic fields which are constant with time. These fields are also called static magnetic fields or magnetostatic fields.

Magnetic fields have several applications. Some of them are listed below.

These fields are described by the magnetic field intensity, **H** and the magnetic flux density, **B**. These are related by $\mathbf{B} = \mu\mathbf{H}$. In fact, most of the relations in magnetostatics are derived from the knowledge of relations in electrostatics.

Steady magnetic fields are governed by Biot-Savart law and Ampere's circuit law.

3.2 APPLICATIONS OF MAGNETOSTATIC FIELDS

Magnetostatic fields are used in

1. magnetic separators
2. particle accelerators like cyclotrons
3. development of motors
4. compasses
5. microphones
6. telephone ringers
7. advertising displays
8. velocity selector
9. mass spectrometer
10. transformers
11. television focus controls
12. high speed velocity devices
13. magnetohydrodynamic generator
14. electromagnetic pump and so on

3.3 FUNDAMENTALS OF STEADY MAGNETIC FIELDS

Steady magnetic fields are also called static magnetic fields or magnetostatic fields. These are produced by a magnet or by a current element. It is well known that loadstone is a natural magnet. But it is a fairly weak magnet. Strong magnets are made of iron, nickel, cobalt, or steel. The two opposite ends of a magnet are called its poles (Fig. 3.1).

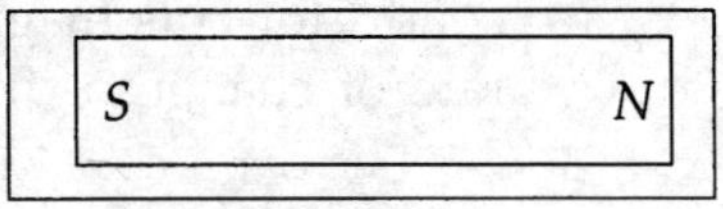

Fig. 3.1 Poles of a magnet

If a magnet is floated freely, one pole will point towards the north pole and is called the north pole of the magnet. The other pole is the south pole.

If a bar magnet is placed under a sheet of paper on which there are some iron fillings and then the paper is gently shaken, the fillings will tend to arrange themselves into a pattern like the one shown in Fig. 3.2.

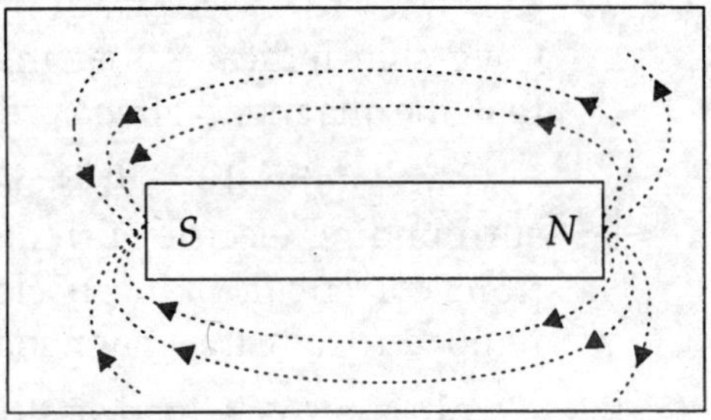

Fig. 3.2 Magnetic flux lines

It is observed that the iron fillings arrange themselves in a set of parallel lines going from one pole to another. These lines never cross or unite. These are called the magnetic lines of force or flux. The area covered by them is called the magnetic field. Flux lines move from N to S.

Electricity produces magnetism and magnets can produce electricity (Fig. 3.3).

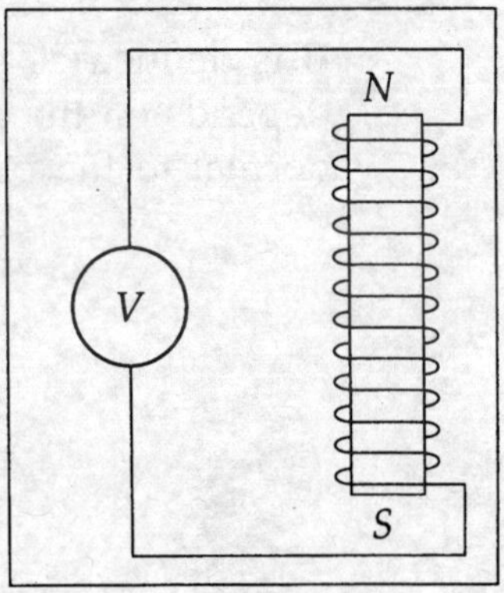

Fig. 3.3 Generation of electricity with a magnet

AC current can be produced by moving the magnet up and down in the coil or by rotating a conductor coil around a magnet. The latter method is preferred to produce large quantities of electric power.

3.4 FARADAY'S LAW OF INDUCTION

Faraday's law of induction states that emf is induced whenever the magnetic flux linkage changes. In other words, an emf is induced in a conductor when the conductor cuts magnetic flux. The magnitude of the induced emf is equal to the rate of change of flux linkage, that is, mathematically,

$$\text{Induced emf} = V = -\frac{Nd\phi}{dt} \text{ volt}$$

where V = induced emf (volt)

N = number of turns

ϕ = magnetic flux (wb)

(–)ve sign indicates that the induced emf sets up a current in such a direction that the magnetic effect produced by it opposes the very cause producing it.

Magnetic flux It is defined as the lines of force produced in the medium surrounding electric currents or magnets. The flux through an area is measured from current flow in an electric circuit bounding the area when this circuit is removed from the magnetic field.

It is also defined as the surface integral of the magnetic flux density, that is,

$$\phi \equiv -\int_s \mathbf{B}.d\mathbf{S}, \text{ weber}$$

3.5 MAGNETIC FLUX DENSITY, B (wb/m²)

B is defined as magnetic flux per unit area through a loop of small area. As ϕ depends on the orientation of the loop and its area, **S** is a vector. Its direction is normal to the plane of the loop.

$$\phi = \int_S \mathbf{B}.d\mathbf{S}$$

or,

$$\mathbf{B} \equiv \frac{d\phi}{d\mathbf{S}}\mathbf{a}_n$$

B is also defined as $\mathbf{B} = \mu\mathbf{H}$

where $\mathbf{H}$ = magnetic field (A/m)

μ = permeability of the medium (H/m)

$= \mu_0 \mu_r$

μ_0 = permeability of free space $= 4\pi\times10^{-7}$ H/m

μ_r = relative permeability of the medium

In this book, we are more interested to learn about the magnetic fields which are produced from current elements.

Definition of current element A current element is a conductor carrying current. It is represented by **IL**. Here I is the current and **L** is the length of the conductor.

Problem 3.1 If the magnetic flux density in a medium is given by $\mathbf{B} = \frac{1}{\rho}\cos\phi\,\mathbf{a}_\rho$, what is the flux crossing the surface defined by $-\frac{\pi}{4} < \phi \leq \frac{\pi}{4}$, $0 \leq z \leq 2\,\text{m}$.

Solution $$\mathbf{B} = \frac{1}{\rho}\cos\phi\,\mathbf{a}_\rho$$

By definition $$\phi = \int_s \mathbf{B}.d\mathbf{s}$$

where $$d\mathbf{s} = \rho\, d\phi\, dz\, \mathbf{a}_\rho$$

$$\phi = \int_s \frac{1}{\rho}\cos\phi\,\mathbf{a}_\rho\,.\rho\, d\phi\, dz\,\mathbf{a}_\rho$$

$$= \int_0^2 \int_{-\pi/4}^{\pi/4} \cos\phi\,\mathbf{a}_\rho\,.\rho\, d\phi\, dz\,\mathbf{a}_\rho$$

$$\boxed{\phi = 2.83\text{ wb}}$$

Problem 3.2 If a magnetic field, $\mathbf{H} = 3\mathbf{a}_x + 2\mathbf{a}_y$, A/m exists at a point in free space, what is the magnetic flux density at the point?

Solution $$\mathbf{H} = 3\mathbf{a}_x + 2\mathbf{a}_y$$

$$\mu_0 = \text{permeability of free space}$$
$$= 4\pi\times 10^{-7}\,\text{H/m}$$

$$\mathbf{B} = \mu_0\mathbf{H}$$

$$= 4\pi\times 10^{-7}(3\mathbf{a}_x + 2\mathbf{a}_y)$$

$$= (3.767\,\mathbf{a}_x + 2.513\,\mathbf{a}_y)\times 10^{-6}$$

$$\boxed{\mathbf{B} = (3.767\,\mathbf{a}_x + 2.513\,\mathbf{a}_y)\ \mu\text{wb/m}^2}$$

Problem 3.3 If the magnetic field, $\mathbf{H} = r \sin\phi\, \mathbf{a}_r + 2.5r \sin\theta \cos\phi\, \mathbf{a}_\phi$, A/m exists in a medium whose $\mu_r = 3.0$, find the magnetic flux density.

Solution $\mathbf{H} = r \sin\phi\, \mathbf{a}_r + 2.5\, r \sin\theta \cos\phi\, \mathbf{a}_\phi$, A/m

Relative permeability of the medium,

$$\mu_r = 2.5$$

The magnetic flux density,

$$\mathbf{B} = \mu\mathbf{H}$$

$$= \mu_0\, \mu_r\, \mathbf{H},\ \mu_0 = 4\pi \times 10^{-7}\ \text{H/m}$$

$$= 4\pi \times 10^{-7} \times 2.5\mathbf{H}$$

$$\boxed{\mathbf{B} = 3.14\ \text{r} \sin\phi\, \mathbf{a}_r + 7.85\, r \sin\theta \cos\phi\, \mathbf{a}_\phi\ \mu\text{wb/m}^2}$$

Problem 3.4 A circular coil of radius 2.0 cm is in a magnetic flux density of 10 wb/m^2. If the plane of the coil is perpendicular to the field, determine the total flux around the coil.

Solution The flux density = 10 wb/m^2

Area of the coil, $S = \pi r^2$

$$= \pi \times (2.0 \times 10^{-2})^2$$

$$= 4\pi \times 10^{-4}$$

Total flux, $\phi = \mathbf{B}\,.\,\mathbf{S}$

As **B** and **S** are in the same direction,

$$\mathbf{B}\,.\,\mathbf{S} = \text{BS}$$

$$\phi = 10 \times 4\pi \times 10^{-4}$$

$$\boxed{\phi = 12.56\ \text{mwb}}$$

Problem 3.5 Given magnetic flux density, $\mathbf{B} = \rho\, \mathbf{a}_\phi$, find the total flux crossing the surface $\phi = \pi/2$, $1 \le \rho \le 2$ m and $0 \le z \le 5$ m.

Solution $\phi = \iint \mathbf{B}.ds$

$$\mathbf{B} = \rho\, \mathbf{a}_\phi$$

$$ds = d\rho\ dz\ \mathbf{a}_\phi$$

$$\phi = \int_{z=0}^{5} \int_{\rho=1}^{2} \rho\, d\rho\, dz$$

$$\boxed{\phi = 7.5\ \text{wb}}$$

3.6 AMPERE'S LAW FOR CURRENT ELEMENT OR BIOT-SAVART LAW

Mathematically, Biot-Savart law is given by

$$d\mathbf{H} = \frac{Id\mathbf{L} \times \mathbf{a}_r}{4\pi r^2} \text{ A/m}$$

Here $d\mathbf{H}$ = magnetic field at a point, P

$Id\mathbf{L}$ = differential current element (A-m)

$\mathbf{a}_r$ = unit vector along the line joining the point P and the $Id\mathbf{L}$

r = distance of P from the current element (m)

Statement of Biot-Savart law When a differential current element produces a differential magnetic field, d**H**, the field magnitude at a point is proportional to the product of Id**L** and sine of the angle between the conductor and the line of the point to the conductor. It is also inversely proportional to the square of the distance from the element to the point.

The direction of d**H** is normal to the plane containing Id**L** and the unit vector along the line from the element to the point. This normal is in the direction of progress of a right-handed screw turned from d**L** through a small angle to the line from the element to the point. The constant of proportionality is $\frac{1}{4\pi}$.

The direction of the magnetic field is easily known using the right hand thumb rule. The magnetic field is in closed loops around the current element.

If a current element is held in the right hand with the thumb pointing upwards indicating the direction of current, then the direction of the remaining fingers indicate the direction of the magnetic field (Fig. 3.4), that is, if the current is upwards, the direction of magnetic field is anti-clockwise and if the current is downwards, the direction of magnetic field is clockwise.

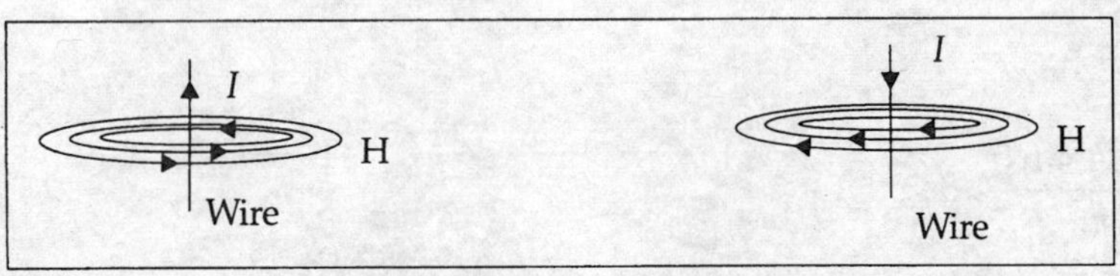

Fig. 3.4 Direction of **H**

3.7 FIELD DUE TO INFINITELY LONG CURRENT ELEMENT

The field produced by an infinitely long current element at a point is given by

$$\mathbf{H} = \frac{I}{2\pi\rho}\mathbf{a}_\phi$$

Here I = current in the element

ρ = distance of the point from the element

Proof Assume an infinitely long current element to be vertical with the current passing upwards. Then, the direction of **H** is along $\mathbf{a}_\phi$ (Fig. 3.5).

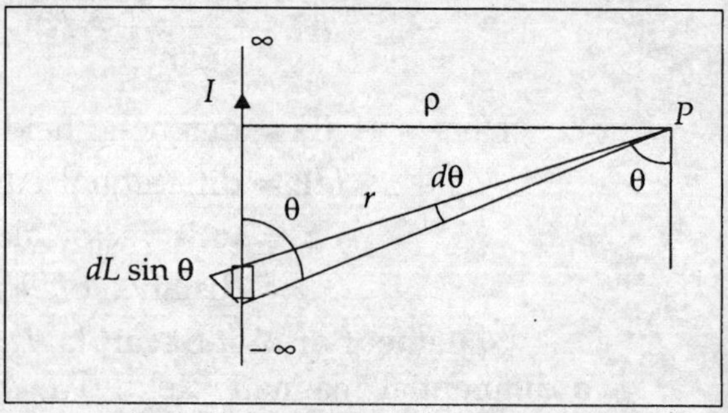

Fig. 3.5 Infinitely long current element

Let ρ be the distance of P from the element. Consider a differential element dL at a distance of r from P. Let the line joining P and IdL make an angle θ with the element. Now, by Biot-Savart law we have

dH due to I**dL** at P,

$$d\mathbf{H} = \frac{Id\mathbf{L}\times\mathbf{a}_r}{4\pi r^2}$$

$$= \frac{IdL\sin\theta}{4\pi r^2}\mathbf{a}_\phi$$

But $dL\sin\theta = r\,d\theta$

$$d\mathbf{H} = \frac{I}{4\pi}\frac{d\theta}{r}\mathbf{a}_\phi$$

As $\sin\theta = \dfrac{\rho}{r}$

$$d\mathbf{H} = \frac{I}{4\pi}\frac{\sin\theta}{\rho}d\theta\mathbf{a}_\phi$$

So **H** due to infinitely long current element is given by

$$\mathbf{H} = \frac{I}{4\pi\rho}\int_0^\pi \sin d\theta\ \mathbf{a}_\phi$$

$$\boxed{\mathbf{H} = \frac{I}{2\pi\rho}\mathbf{a}_\phi\ (\text{A/m})}$$

3.8 FIELD DUE TO A FINITE CURRENT ELEMENT

The magnetic field, **H** at a point, P due to a finite current element is

$$\mathbf{H} = \frac{I}{4\pi R}[\cos\alpha_2 - \cos\alpha_1]\,\mathbf{a}_\phi$$

where I = current in the element

R = distance of the point from the element axis

α_1 = angle made by the line joining the point and one end of the element with the axis of the element

α_2 = angle made by the line joining the point and the other end of the element with the axis

Proof Consider Fig. 3.6.

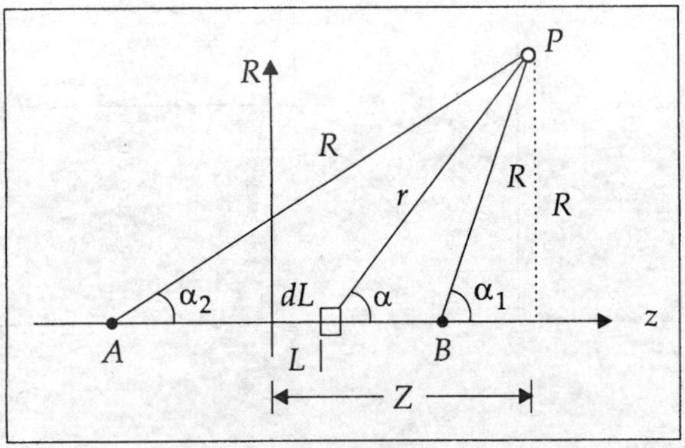

Fig. 3.6 Finite current element

The differential magnetic field, d**H** at the point P due to Id**L** (Fig. 3.6) is

$$d\mathbf{H} = \frac{IdL\sin\alpha}{4\pi r^2}\,\mathbf{a}_\phi$$

$$\mathbf{H} = \int \frac{IdL\sin\alpha}{4\pi r^2}\,\mathbf{a}_\phi$$

We have $$r^2 = (Z-L)^2 + R^2$$

and $$Z - L = R\cot\alpha$$

Thus, we get $$-dL = -R\,\text{cosec}^2\,\alpha\, d\alpha$$

$$dL = R\left[\frac{R^2 + (Z-L)^2}{R^2}\right] d\alpha$$

$$\mathbf{H} = \frac{I}{4\pi R}\int_{\alpha_2}^{\alpha_1} \sin\alpha \, d\alpha \, \mathbf{a}_\phi = \frac{1}{4\pi R}[\cos\alpha_2 - \cos\alpha_1]\mathbf{a}_\phi, \text{ A/m}$$

Problem 3.6 A thin conductor of finite length is along z-axis lying between $z = z_1$ and $z = z_2$. Find **H** at a point *P* in the *xy*-plane. What is **H** if $z_1 = \infty$ and $z_2 = -\infty$?

Solution Consider Fig. 3.7 in which a differential current element *Id***L** is shown on the current element.

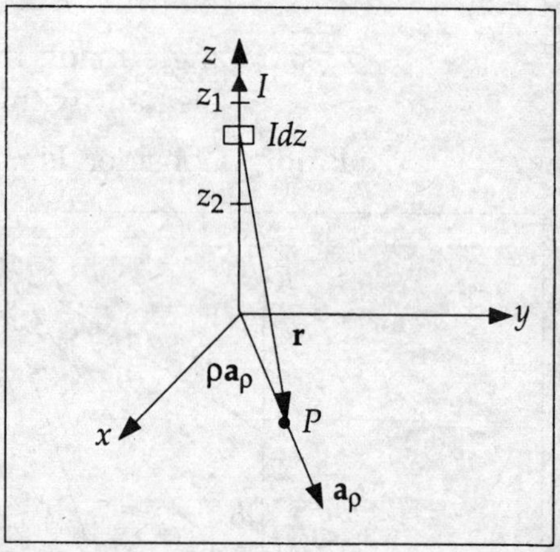

Fig. 3.7

Then, **r**

$$Id\mathbf{z}\times\mathbf{r} = Idz\mathbf{a}_z \times (\rho\mathbf{a}_\rho - z\mathbf{a}_z) = Idz\,\rho\,\mathbf{a}_\phi$$

$$\mathbf{H} = \frac{I\rho}{4\pi}\int_{z_1}^{z_2} \frac{dz}{(\rho^2 + z^2)^{3/2}}\mathbf{a}_\phi = \frac{I\rho}{4\pi\rho^2}\left[\frac{z}{(\rho^2+z^2)^{1/2}}\right]_{z_1}^{z_2}\mathbf{a}_\phi$$

$$\mathbf{H} = \frac{I}{4\pi\rho}\left[\frac{z_2}{\sqrt{\rho^2 + z_2^2}} - \frac{z_1}{\sqrt{\rho^2 + z_1^2}}\right]\mathbf{a}_\phi$$

If $z_1 \to \infty, z_2 \to -\infty$, **H** will be

$$\mathbf{H} = -\frac{I}{4\pi\rho}[1+1]\,\mathbf{a}_\phi$$

$$\mathbf{H} = -\frac{I}{2\pi\rho}\mathbf{a}_\phi$$

Problem 3.7 Determine the magnetic field intensity, **H** at the centre of a square current element. The length of each side is 2 m and the current, I = 1.0 Amp.

Solution Consider Fig. 3.8 and let each side be of length $2l$.

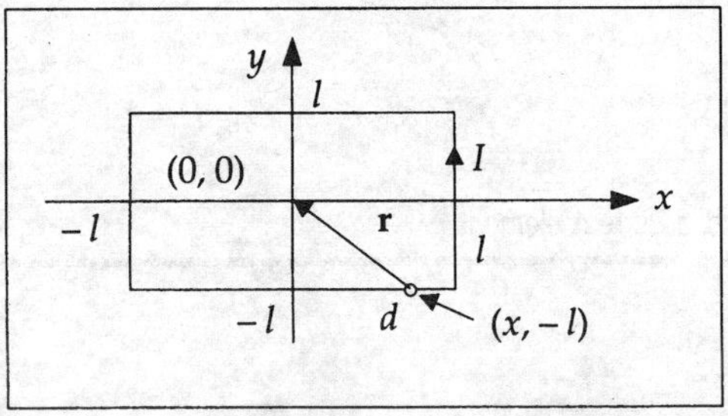

Fig. 3.8 Square current element

For the differential length, dx, we have

$$d\mathbf{H} = \frac{I dx \mathbf{a}_x \times \mathbf{r}}{4\pi r^2}$$

$$\mathbf{r} = (0, 0) - (x, -l) = -x\,\mathbf{a}_x + l\mathbf{a}_y$$

$$r = \sqrt{x^2 + l^2}$$

$$d\mathbf{H} = \frac{I dx \mathbf{a}_x \times (-x\mathbf{a}_x + l\mathbf{a}_y)}{4\pi(x^2 + l^2)^{3/2}} = \frac{I dx l \mathbf{a}_z}{4\pi\,(x^2 + l^2)^{3/2}}$$

The total field at the origin is

$$\mathbf{H} = \int_0^1 Il \frac{dx}{4\pi\,(x^2 + l^2)^{3/2}}\,\mathbf{a}_z = \frac{\sqrt{2}I}{\pi l}\,\mathbf{a}_z$$

$$\boxed{\mathbf{H} = \frac{I\sqrt{2}}{\pi l}\,\mathbf{a}_z}$$

If $2l$ = 2 m or l = 1 m and I = 1 Amp,

$$\boxed{\mathbf{H} = \frac{\sqrt{2}}{\pi}\,\mathbf{a}_z, \text{A/m}}$$

3.9 AMPERE'S WORK LAW OR AMPERE'S CIRCUIT LAW

Statement: Ampere's circuit law states that the line integral of the magnetic field **H** about any closed loop is equal to the current enclosed by the path. Mathematically,

$$\oint \mathbf{H}.d\mathbf{L} = I_{enc}$$

Proof Consider a circular loop as in Fig. 3.9 which encloses a current element. Let the current be in the upward direction. Then, the field is anti-clockwise ($\mathbf{a}_\phi$).

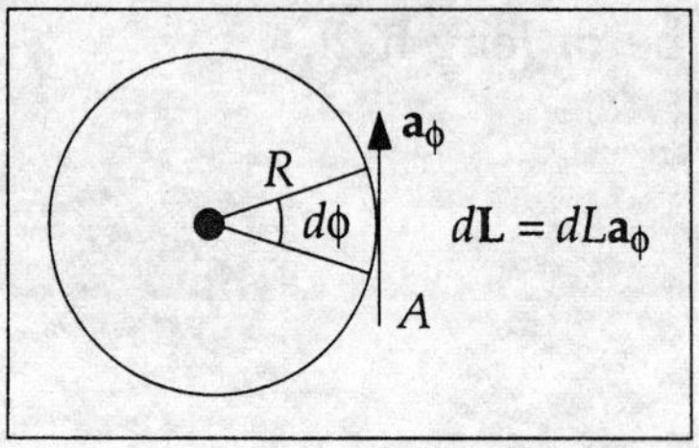

Fig. 3.9 Magnetic field around a current element

H at the point A is given by

$$\mathbf{H} = \frac{I_{enc}}{2\pi R}\mathbf{a}_\phi$$

Taking dot product with $d\mathbf{L}$ on both sides, we get

$$\mathbf{H}.d\mathbf{L} = \frac{I_{enc}}{2\pi R}\mathbf{a}_\phi.dL\mathbf{a}_\phi = \frac{I_{enc}}{2\pi R}dL$$

But $$dL = Rd\phi$$

$$\mathbf{H}.d\mathbf{L} = \frac{I_{enc}}{2\pi}d\phi$$

Taking line integral around the closed loop, we get

$$\oint \mathbf{H}.d\mathbf{L} = \int_0^{2\pi}\frac{I_{enc}}{2\pi}d\phi = I_{enc}$$

$$\oint \mathbf{H}.d\mathbf{L} = I_{enc}$$

This is called the integral form of Ampere's circuit law.

Problem 3.8 Determine the magnetic flux between the conductors of a coaxial cable of length 10 m. The radius of the inner conductor is a = 1 cm and that of the outer conductor is 2 cm. The current enclosed is 2A.

Solution By Ampere's circuit law

$$\oint \mathbf{H}.d\mathbf{L} = I$$

that is, $$HL = I$$

Here, $dL = 2\pi r$, r lies between a and b.

$$\mathbf{H} = \frac{I}{2\pi r}\mathbf{a}_\phi$$

or,

$$\mathbf{B} = \frac{\mu_0 I}{2\pi r}\mathbf{a}_\phi$$

$$\phi = \iint \mathbf{B}\,.\,d\mathbf{S} = \int_0^l \int_a^b \frac{\mu_0 I \mathbf{a}_\phi}{2\pi r}\,.\,\mathbf{a}_\phi\, dr\, dz = \int_0^{10} \int_{0.01}^{0.02} \frac{\mu_0 I}{2\pi r}\, dr\, dz$$

$$\boxed{\phi = 277.0\mu \text{ wb}}$$

3.10 DIFFERENTIAL FORM OF AMPERE'S CIRCUIT LAW

The differential form of Ampere's circuit law is given by

$$\nabla \times \mathbf{H} = \mathbf{J}$$

$\mathbf{J}$ = conduction current density (A/m^2)

Proof Consider a closed rectangular path in x-y plane which encloses a current in z-direction (Fig. 3.10).

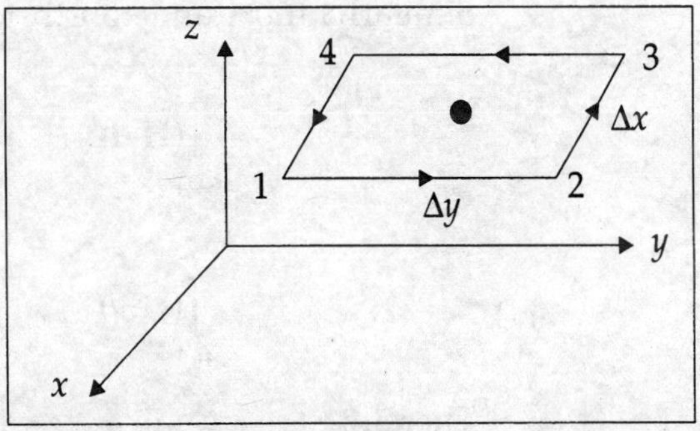

Fig. 3.10 Closed path in x-y plane enclosing a current element

Assume that $\mathbf{H}$ at the centre of the loop is known, that is,

$$\mathbf{H} = \mathbf{H}_0 = H_{x0}\,\mathbf{a}_x + H_{y0}\,\mathbf{a}_y + H_{z0}\,\mathbf{a}_z$$

Line integral of $\mathbf{H}$ around the loop is the sum of $\mathbf{H}.d\mathbf{L}$ on each side. Let us choose the direction of the loop to be 1-2-3-4. This corresponds to a current in the centre along z-direction, that is, we have

$$\oint \mathbf{H}.d\mathbf{L} = \int_{1-2} (\mathbf{H}.\Delta\mathbf{L})_{1-2} + \int_{2-3} (\mathbf{H}.\Delta\mathbf{L})_{2-3} + \int_{3-4} (\mathbf{H}.\Delta\mathbf{L})_{3-4} + \int_{4-1} (\mathbf{H}.\Delta\mathbf{L})_{4-1}$$

First term on the right side

$$= H_{x,1-2}\,\mathbf{a}_x + H_{y,1-2}\,\mathbf{a}_y + H_{z,1-2}\,\mathbf{a}_z$$

$$\int_{1-2} (\mathbf{H}.\Delta\mathbf{L})_{1-2} = H_{y,1-2}\,\Delta y$$

But by Taylor's theorem, $H_{y,1-2}$ can be obtained approximately by considering the first two terms, that is,

$$H_{y,1-2} \approx H_{yo} + \frac{\partial H_y}{\partial x}\left(\frac{\Delta x}{2}\right)$$

Thus $$\int_{1-2} (\mathbf{H}.\Delta\mathbf{L})_{1-2} \approx \left(H_{yo} + \frac{1}{2}\frac{\partial H_y}{\partial x}\Delta x\right)\Delta y$$

Second term on the right side

$$(\mathbf{H}.\Delta\mathbf{L})_{2-3} = (H_{x,2-3}\,\mathbf{a}_x + H_{y,2-3}\,\mathbf{a}_y + H_{z,2-3}\,\mathbf{a}_z).\Delta x(-\mathbf{a}_x)$$

$$= -H_{x,2-3}(\Delta x) = -\left(H_{x0} + \frac{1}{2}\frac{\partial H_x}{\partial y}\Delta y\right)\Delta x$$

Similarly, third and fourth terms are given by

$$\int_{3-4} (\mathbf{H}.d\mathbf{L})_{3-4} \approx \left(-H_{y0} + \frac{1}{2}\frac{\partial H_y}{\partial x}\Delta x\right)\Delta y$$

and $$\int_{4-1} (\mathbf{H}.d\mathbf{L})_{4-1} \approx \left(H_{x0} - \frac{1}{2}\frac{\partial H_x}{\partial y}\Delta y\right)\Delta x$$

Therefore,

$$\oint \mathbf{H}.d\mathbf{L} \approx \frac{1}{2}\frac{\partial H_y}{\partial x}\Delta x\,\Delta y - \frac{1}{2}\frac{\partial H_x}{\partial y}\Delta x\,\Delta y + \frac{1}{2}\frac{\partial H_y}{\partial x}\Delta x\,\Delta y - \frac{1}{2}\frac{\partial H_x}{\partial y}\Delta x\,\Delta y$$

$$\oint \mathbf{H}.d\mathbf{L} \approx \left(\frac{\partial H_y}{\partial x} - \frac{\partial H_x}{\partial y}\right)\Delta x\,\Delta y = I_z = J_z\,\Delta x\,\Delta y$$

or, $$\frac{\oint \mathbf{H}.d\mathbf{L}}{\Delta x\,\Delta y} \approx \frac{\partial H_y}{\partial x} - \frac{\partial H_x}{\partial y} \approx J_z$$

If the closed path is made to shrink, this expression becomes exact, that is,

$$\underset{\Delta x\,\Delta y \to 0}{\text{Lt}} \int \frac{\mathbf{H}.d\mathbf{L}}{\Delta x\,\Delta y} = \left(\frac{\partial H_y}{\partial x} - \frac{\partial H_x}{\partial y}\right) = J_z$$

or,
$$\left(\frac{\partial H_y}{\partial x} - \frac{\partial H_x}{\partial y}\right)\mathbf{a}_z = J_z\,\mathbf{a}_z \tag{3.1}$$

Similarly, if we choose closed paths oriented perpendicular to the remaining two coordinate axes, and by following exactly the above procedure, we get the expressions for the x and y components of the current density. These are

$$\underset{\Delta y\,\Delta z \to 0}{\text{Lt}} \int \frac{\mathbf{H}.d\mathbf{L}}{\Delta y\,\Delta z} = \frac{\partial H_z}{\partial y} - \frac{\partial H_y}{\partial z} = J_x$$

or,
$$\left(\frac{\partial H_z}{\partial y} - \frac{\partial H_y}{\partial z}\right)\mathbf{a}_x = J_x\,\mathbf{a}_x \tag{3.2}$$

and
$$\underset{\Delta z\,\Delta x \to 0}{\text{Lt}} \int \frac{\mathbf{H}.d\mathbf{L}}{\Delta z\,\Delta x} = \frac{\partial H_x}{\partial z} - \frac{\partial H_z}{\partial x} = J_y$$

or,
$$\left(\frac{\partial H_x}{\partial z} - \frac{\partial H_z}{\partial x}\right)\mathbf{a}_y = J_y\,\mathbf{a}_y \tag{3.3}$$

3.1 + 3.2 + 3.3 gives

$$\left(\frac{\partial H_y}{\partial x} - \frac{\partial H_x}{\partial y}\right)\mathbf{a}_x + \left(\frac{\partial H_z}{\partial y} - \frac{\partial H_y}{\partial z}\right)\mathbf{a}_y + \left(\frac{\partial H_x}{\partial z} - \frac{\partial H_z}{\partial x}\right)\mathbf{a}_z$$
$$= J_z\,\mathbf{a}_z + J_x\,\mathbf{a}_x + J_y\,\mathbf{a}_y = \mathbf{J}$$

where
$$\mathbf{J} = J_x\,\mathbf{a}_x + J_y\,\mathbf{a}_y + J_z\,\mathbf{a}_z \tag{3.4}$$

Left side of Equation (3.4) is nothing but curl of vector **H**,

that is,
$$\nabla \times \mathbf{H} = \underset{\Delta s_n \to 0}{\text{Lt}} \oint \frac{\mathbf{H}.d\mathbf{L}}{\Delta s_n} = \mathbf{J}$$

$$\nabla \times \mathbf{H} = \mathbf{J}$$

This is the differential form of Ampere's circuit law.

It is important to note the following.

1. $\int (\nabla \times \mathbf{H}).d\mathbf{S}$ depends only on the boundary line and it does not depend on the particular surface used.
2. $\oint_s (\nabla \times \mathbf{H}).d\mathbf{S} = 0$ for any closed surface. This is because the boundary line shrinks to a point like the mouth of a balloon.

Ampere's circuit law in different coordinate systems is

$$\text{Curl } \mathbf{H} = (\nabla \times \mathbf{H})_{cart}$$

$$= \left(\frac{\partial H_z}{\partial y} - \frac{\partial H_y}{\partial z}\right)\mathbf{a}_x + \left(\frac{\partial H_x}{\partial z} - \frac{\partial H_z}{\partial x}\right)\mathbf{a}_y + \left(\frac{\partial H_y}{\partial x} - \frac{\partial H_x}{\partial y}\right)\mathbf{a}_z$$

$$(\nabla \times \mathbf{H})_{cy} = \frac{1}{\rho}\left(\frac{\partial H_z}{\partial \phi} - \frac{\partial H_\phi}{\partial z}\right)\mathbf{a}_\rho + \left(\frac{\partial H_\rho}{\partial z} - \frac{\partial H_z}{\partial \rho}\right)\mathbf{a}_\phi + \frac{1}{\rho}\left(\frac{1}{\rho}\frac{\partial(\rho H_\phi)}{\partial \rho} - \frac{1}{\rho}\frac{\partial H_\rho}{\partial \phi}\right)\mathbf{a}_z$$

$$= J_\rho \, \mathbf{a}_p + J_\phi \, \mathbf{a}_\phi + J_{za_z}$$

$$(\nabla \times \mathbf{H})_{sp} = \frac{1}{r\sin\theta}\left[\frac{\partial(H_\phi \sin\theta)}{\partial \theta} - \frac{\partial H_\theta}{\partial \phi}\right]\mathbf{a}_r + \frac{1}{r}\left[\frac{1}{\sin\theta}\frac{\partial H_r}{\partial \phi} - \frac{\partial(rH_\phi)}{\partial r}\right]\mathbf{a}_\theta$$

$$+ \frac{1}{r}\left[\frac{\partial(rH_\theta)}{\partial r} - \frac{\partial H_r}{\partial \theta}\right]\mathbf{a}_\phi$$

$$= J_r \, \mathbf{a_r} + J_\theta \, \mathbf{a}_\theta + J_\phi \, \mathbf{a}_\phi$$

Problem 3.9 If **H** is given by $\mathbf{H} = y\cos 2x \, \mathbf{a}_x + (y + e^x)\mathbf{a}_z$, determine **J** at the origin.

Solution The differential form of Ampere's circuit law is

$$\nabla \times \mathbf{H} = \mathbf{J}$$

$$= \begin{bmatrix} \mathbf{a}_x & \mathbf{a}_y & \mathbf{a}_z \\ \dfrac{\partial}{\partial x} & \dfrac{\partial}{\partial y} & \dfrac{\partial}{\partial z} \\ y\cos 2x & 0 & (y + e^x) \end{bmatrix}$$

$$= \mathbf{a}_x\left[\frac{\partial}{\partial y}(y + e^x)\right] - \mathbf{a}_y\left[\frac{\partial}{\partial x}(y + e^x) - \frac{\partial}{\partial z}(y\cos 2x)\right] - \mathbf{a}_z\left[\frac{\partial}{\partial y}(y\cos 2x)\right]$$

$$\mathbf{J} = \mathbf{a}_x - e^x\,\mathbf{a}_y - \cos 2x\,\mathbf{a}_z$$

At $$(0,0,0)\,\mathbf{J} = (\mathbf{a}_x - \mathbf{a}_y - \mathbf{a}_z)\ \text{A/m}^2$$

Problem 3.10 What is the magnetic field, **H** in Cartesian coordinates due to z-directed current element? Find **J** if $I = 2A$.

Solution We have

$$\mathbf{H} = \frac{I}{2\pi\rho}\mathbf{a}_\phi,\ \text{A/m}$$

But $$\rho = \sqrt{x^2+y^2}$$

$$\mathbf{a}_\phi = \frac{-y\mathbf{a}_x + x\mathbf{a}_y}{\sqrt{x^2+y^2}}$$

$$\mathbf{H} = \frac{2}{2\pi}\left(\frac{-y\mathbf{a}_x + x\mathbf{a}_y}{x^2+y^2}\right)$$

$$\nabla\times\mathbf{H} = \frac{1}{\pi}\begin{vmatrix} \mathbf{a}_x & \mathbf{a}_y & \mathbf{a}_z \\ \dfrac{\partial}{\partial x} & \dfrac{\partial}{\partial y} & \dfrac{\partial}{\partial z} \\ \dfrac{-y}{x^2+y^2} & \dfrac{x}{x^2+y^2} & 0 \end{vmatrix}$$

$$= \left[\frac{\partial}{\partial x}\left(\frac{x}{x^2+y^2}\right) - \frac{\partial}{\partial y}\left(\frac{-y}{x^2+y^2}\right)\right]\mathbf{a}_z = 0$$

$$\boxed{\mathbf{J} = 0}$$

Problem 3.11 Determine **J** at $(2, \pi, 0)$ in cylindrical coordinates if the magnetic field, **H** is given by $\mathbf{H} = 5\rho\sin\phi\,\mathbf{a}_z$, mA/m^2.

Solution In cylindrical coordinates, we have

$$(\nabla\times\mathbf{H})_{cy} = \frac{1}{p}\left(\frac{\partial H_z}{\partial\phi} - \frac{\partial H_\phi}{\partial z}\right)\mathbf{a}_\rho + \left(\frac{\partial H_\rho}{\partial z} - \frac{\partial H_z}{\partial\rho}\right)\mathbf{a}_\phi$$

$$+\left[\frac{1}{\rho}\frac{\partial}{\partial\rho}(\rho H_\phi) - \frac{1}{\rho}\frac{\partial H_\rho}{\partial\phi}\right]\mathbf{a}_z$$

Here $H_\rho = 0$ and $H_\phi = 0$.

So
$$(\nabla \times \mathbf{H})_{cy} = 10^{-3} \frac{1}{\rho}\frac{\partial H_z}{\partial \phi}\,\mathbf{a}_\rho - \frac{\partial H_z}{\partial \rho}\,\mathbf{a}_\phi$$

$$= 5\times 10^{-3} \frac{1}{\rho}\frac{\partial}{\partial \phi}(5\rho \sin\phi)\,\mathbf{a}_\rho - \frac{\partial}{\partial \rho}(5\rho \sin\phi)\,\mathbf{a}_\phi$$

$$(\nabla \times \mathbf{H}) = 5\times 10^{-3}\,[\cos\phi\,\mathbf{a}_\rho - \sin\phi\,\mathbf{a}_\phi] = \mathbf{J}$$

$$\boxed{\mathbf{J}/(2, \pi, 0) = -5\mathbf{a}_\rho \text{ mA/m}^2}$$

Problem 3.12 If the magnetic field, $\mathbf{H} = 100 \sin\theta\,\mathbf{a}_\theta$ A/m in spherical coordinates, determine $\mathbf{J}$ at $\left(10, \frac{\pi}{2}, 0\right)$.

Solution

$$(\nabla \times \mathbf{H})_{\text{sph}} = \frac{1}{r\sin\theta}\left[\frac{\partial}{\partial \theta}(H_\phi \sin\theta) - \frac{\partial H_\theta}{\partial \phi}\right]\mathbf{a}_r + \frac{1}{r}\left[\frac{1}{\sin\theta}\frac{\partial H_r}{\partial \phi} - \frac{\partial (rH_\phi)}{\partial r}\right]\mathbf{a}_\theta$$

$$+ \frac{1}{r}\left[\frac{\partial}{\partial r}(rH_\theta) - \frac{\partial H_r}{\partial \theta}\right]\mathbf{a}_\phi$$

But in the present problem,

$$H_r = 0 \text{ and } H_\phi = 0$$

$$\nabla \times \mathbf{H} = \frac{1}{r\sin\theta}\left[-\frac{\partial}{\partial \phi}(100\sin\theta)\right]\mathbf{a}_r + \frac{1}{r}\frac{\partial}{\partial r}(100 r \sin\theta)\,\mathbf{a}_\phi$$

that is,
$$\nabla \times \mathbf{H} = \frac{100\sin\theta}{r}\,\mathbf{a}_\phi$$

$$\mathbf{J} = (\nabla \times \mathbf{H}) \quad \left(10, \frac{\pi}{2}, 0\right) = 5.0\,\mathbf{a}_\phi$$

$$\boxed{\mathbf{J} = 5.0\mathbf{a}_\phi \text{ A/m}^2}$$

Problem 3.13 An infinitely long current element on x-axis carries a current of 1.0 mA in $\mathbf{a}_x$ direction. Determine $\mathbf{H}$ at the point P (5, 2, 1).

Solution We have

$$\mathbf{H} = \frac{1}{2\pi r}[\mathbf{a}_L \times \mathbf{a}_r]$$

$$\mathbf{r} = 5\mathbf{a}_x + 2\mathbf{a}_y + \mathbf{a}_z - 5\mathbf{a}_x = 2\mathbf{a}_y + \mathbf{a}_z$$

$$r = \sqrt{4+1} = \sqrt{5}$$

$$\mathbf{H} = \frac{1.0\times10^{-3}}{2\pi\sqrt{5}}\left[\frac{\mathbf{a}_x \times (2\mathbf{a}_x + \mathbf{a}_z)}{\sqrt{5}}\right]$$

$$\mathbf{H} = \frac{10^{-3}}{2\pi\times5}(-\mathbf{a}_y + 2\mathbf{a}_z)$$

$$= \frac{10^{-3}}{31.415}(-\mathbf{a}_y + 2\mathbf{a}_z)$$

$$\boxed{\mathbf{H} = (-31.83\mathbf{a}_y + 63.66\mathbf{a}_z)\ \mu\text{A/m}}$$

Problem 3.14 What is the current density which produces a magnetic field of $\mathbf{H} = 28\sin x\,\mathbf{a}_y$?

Solution By Ampere's circuit law, the current density, **J** is

$$\mathbf{J} = \nabla\times\mathbf{H} = \begin{vmatrix} \mathbf{a}_x & \mathbf{a}_y & \mathbf{a}_z \\ \frac{\partial}{\partial x} & \frac{\partial}{\partial y} & \frac{\partial}{\partial z} \\ 0 & 28\sin x & 0 \end{vmatrix}$$

$$\boxed{\mathbf{J} = 28\cos x\,\mathbf{a}_z}$$

3.11 STOKE'S THEOREM

Stoke's theorem relates a line integral to the surface integral and vice-versa, that is,

$$\oint_C \mathbf{H}.d\mathbf{L} = \int_S (\nabla\times\mathbf{H}).d\mathbf{S}$$

3.12 FORCE ON A MOVING CHARGE DUE TO ELECTRIC AND MAGNETIC FIELDS

If there is a charge or a moving charge, Q in an electric field, **E**, there exists a force on the charge. This force is given by

$$\mathbf{F}_E = Q\mathbf{E}$$

If a charge, Q moving with a velocity, **V** is placed in a magnetic field, $\mathbf{B}(=\mu\mathbf{H})$, then there exists a force on the charge (Fig. 3.11). This force is given by

$$\mathbf{F}_H = Q(\mathbf{V}\times\mathbf{B})$$

$\mathbf{B}$ = magnetic flux density, (wb/m^2)

$\mathbf{V}$ = velocity of the charge, m/s

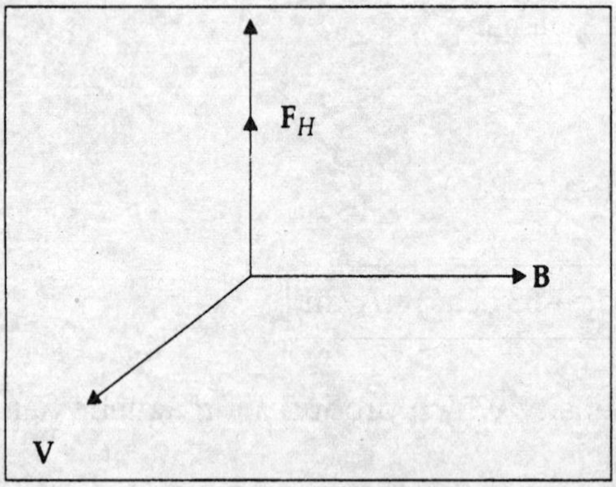

Fig. 3.11 Direction of field, velocity and force

If the charge, Q is placed in both electric and magnetic fields, then the force on the charge is

$$\mathbf{F} = Q(\mathbf{E} + \mathbf{V}\times\mathbf{B})$$

This equation is known as **Lorentz force equation.**

Problem 3.15 A charge of 12 C has velocity of $5\mathbf{a}_x + 2\mathbf{a}_y - 3\mathbf{a}_z$ m/s. Determine **F** on the charge in the field of (a) $\mathbf{E} = 18\mathbf{a}_x + 5\mathbf{a}_y + 10\mathbf{a}_z$ V/m (b) $\mathbf{B} = 4\mathbf{a}_x + 4\mathbf{a}_y + 3\mathbf{a}_z$ wb/m^2.

Solution (a) The force, **F** on the charge, Q due to **E** is

$$\mathbf{F} = Q\mathbf{E} = 12(18\mathbf{a}_x + 5\mathbf{a}_y + 10\mathbf{a}_z)$$

$$= 216\mathbf{a}_x + 60\mathbf{a}_y + 120\mathbf{a}_z$$

or,
$$F = Q|\mathbf{E}| = 12\sqrt{18^2 + 5^2 + 10^2}$$

$$F = 254.27\text{ N}$$

(b) The force **F** on the charge due to **B** is

$$\mathbf{F} = Q\,[\mathbf{V}\times\mathbf{B}]$$

Here,
$$\mathbf{V} = 5\mathbf{a}_x + 2\mathbf{a}_y - 3\mathbf{a}_z \text{ m/s}$$

$$\mathbf{B} = 4\mathbf{a}_x + 4\mathbf{a}_y + 3\mathbf{a}_z \text{ wb/m}^2$$

$$\boxed{\mathbf{F} = 12\,[18\mathbf{a}_x - 27\mathbf{a}_y + 12\mathbf{a}_z]}$$

$$F = 12\sqrt{(324+729+144)}$$

$$\boxed{F = 415.17 \text{ N}}$$

Problem 3.16 An electron has a velocity of 1 km/s along $\mathbf{a}_x$ in a magnetic field whose magnetic flux density is $\mathbf{B} = 0.2\mathbf{a}_x - 0.3\mathbf{a}_y + 0.5\mathbf{a}_z$ wb/m^2.

(a) Determine the electric field intensity if no force is applied to the electron.

(b) Also find the force on the electron under the influence of both **E** and **B** when $\mathbf{E} = (\mathbf{a}_x + \mathbf{a}_y + \mathbf{a}_z)$ KV/m.

Solution The expression for force **F** due to **E** and **B** is

$$\mathbf{F} = Q\,[\mathbf{E} + \mathbf{V}\times\mathbf{B}]$$

(a) If $\mathbf{F} = 0$,

$$Q\,[\mathbf{E} + \mathbf{V}\times\mathbf{B}] = 0$$

or,
$$\mathbf{E} = -\mathbf{V}\times\mathbf{B} = -\begin{vmatrix} \mathbf{a}_x & \mathbf{a}_y & \mathbf{a}_z \\ 10^3 & 0 & 0 \\ 0.2 & -0.3 & 0.5 \end{vmatrix}$$

$$\mathbf{E} = -\,[-\,0.5\times 10^3\,\mathbf{a}_y - 0.3\times 10^3\,\mathbf{a}_y]$$

or,
$$\boxed{\mathbf{E} = (0.5\mathbf{a}_y + 0.3\mathbf{a}_z) \text{ KV/m}}$$

(b) The force on the electron due to **E** and **B** is

$$\mathbf{F} = Q\,[\mathbf{E} + \mathbf{V}\times\mathbf{B}]$$

The charge of the electron,

$$Q = -1.6\times 10^{-19} \text{ C}$$

$$\mathbf{F} = -1.6\times 10^{-19}\,[\mathbf{a}_x + \mathbf{a}_y + \mathbf{a}_z + 0.5\mathbf{a}_y + 0.3\mathbf{a}_z]\times 10^3$$

$$\boxed{\mathbf{F} = (-1.6\mathbf{a}_x - 2.4\mathbf{a}_y - 2.08\mathbf{a}_z)\times 10^{-16} \text{ N}}$$

3.13 APPLICATIONS OF LORENTZ FORCE EQUATION

The solution of Lorentz force equation is extremely useful to determine:

(i) Electron orbits in magnetrons.
(ii) Proton paths in cyclotrons.
(iii) Plasma characteristics in Magneto Hydro Dynamic Generator (MHDG).
(iv) Motional characteristics of a charged body in combined electric and magnetic fields.

3.14 FORCE ON A CURRENT ELEMENT IN A MAGNETIC FIELD

The force on a current element when placed in a magnetic field, **B** is

$$\mathbf{F} = I\mathbf{L} \times \mathbf{B}$$

or,
$$\boxed{F = ILB \sin\theta, \text{ Newton}}$$

where θ is the angle between the direction of the current element and the direction of magnetic flux density

$\mathbf{B}$ = magnetic flux density, wb/m^2

$I\mathbf{L}$ = currentelement, Amp-m

Proof Consider a differential charge, dQ to be moving with a velocity, **V** in a magnetic field, $\mathbf{H} = \left(\frac{\mathbf{B}}{\mu}\right)$. Then the differential force on the charge is given by

$$d\mathbf{F} = dQ\,(\mathbf{V} \times \mathbf{B})$$

But
$$dQ = \rho_\upsilon\, d\upsilon$$

$$d\mathbf{F} = \rho_\upsilon\, d\upsilon\,(V \times B)$$

$$= (\rho_\upsilon\, \mathbf{V} \times \mathbf{B})\, d\upsilon$$

But
$$\rho_\upsilon \mathbf{V} = \mathbf{J}$$

$$d\mathbf{F} = \mathbf{J}\, d\upsilon \times \mathbf{B}$$

$\mathbf{J}\, d\upsilon$ is nothing but $Id\mathbf{L}$,

$$d\mathbf{F} = Id\mathbf{L} \times \mathbf{B}$$

or,
$$\mathbf{F} = I\mathbf{L} \times \mathbf{B}, \text{ Newton}$$

Problem 3.17 A current element 4 cm long is along y-axis with a current of 10 mA flowing in y-direction. Determine the force on the current element due to the magnetic field if the magnetic field $\mathbf{H} = \dfrac{5\mathbf{a}_x}{\mu}$ A/m.

Solution The force on a current element under the influence of magnetic field is

$$\mathbf{F} = I\mathbf{L} \times \mathbf{B}$$

Here,

$$I\mathbf{L} = 10 \times 10^{-3} \times 0.04\mathbf{a}_y$$
$$= 4 \times 10^{-4}\,\mathbf{a}_y$$
$$\mathbf{H} = \frac{5\mathbf{a}_x}{\mu}, \text{ A/m}$$
$$\mathbf{B} = 5\mathbf{a}_x \text{ wb/m}^2$$
$$\mathbf{F} = 4 \times 10^{-4}\,\mathbf{a}_y \times 5\mathbf{a}_x$$

or

$$\mathbf{F} = (0.4\mathbf{a}_y \times 5\mathbf{a}_x) \times 10^{-3}$$

$$\boxed{\mathbf{F} = -2.0\mathbf{a}_z \text{ mN}}$$

Problem 3.18 In a magnetic flux density of $\mathbf{B} = 1.0\mathbf{a}_x + 3.0\mathbf{a}_y$ wb/m^2, a current element, $10\mathbf{a}_z$ mA-m is placed. Find the force on the current element.

Solution The expression for force on a current element due to a magnetic field is given by

$$\mathbf{F} = I\mathbf{L} \times \mathbf{B}$$

Here,

$$I\mathbf{L} = 10\mathbf{a}_z \text{ mA-m}$$
$$= 10 \times 10^{-3}\,\mathbf{a}_z, \text{ A-m}$$
$$\mathbf{B} = (1.0\mathbf{a}_x + 3.0\mathbf{a}_y) \text{ wb/m}^2$$
$$\mathbf{F} = 10\mathbf{a}_z \times (1.0\mathbf{a}_x \times 3.0\mathbf{a}_y) \times 10^{-3}$$

$$\boxed{\mathbf{F} = -30\mathbf{a}_x + 10\mathbf{a}_y, \text{ mN}}$$

3.15 AMPERE'S FORCE LAW

Ampere's force law states that there exists a force between two current elements $I_1\,d\mathbf{L}_1$ and $I_2\,d\mathbf{L}_2$ and it is given by

$$\mathbf{F} = \frac{\mu I_1 I_2}{4\pi} \oint\oint \frac{d\mathbf{L}_2 \times (d\mathbf{L}_1 \times \mathbf{a}_r)}{r^2}$$

Here μ = permeability of the medium in which the current elements are placed

$\mathbf{a}_r$ = distance between the elements

$\mathbf{a}_r$ = unit vector along the line joining the two current elements

Proof Consider Fig. 3.12. Here, two differential current elements $I_1\,d\mathbf{L}_1$ and $I_2\,d\mathbf{L}_2$ are separated by a distance, r.

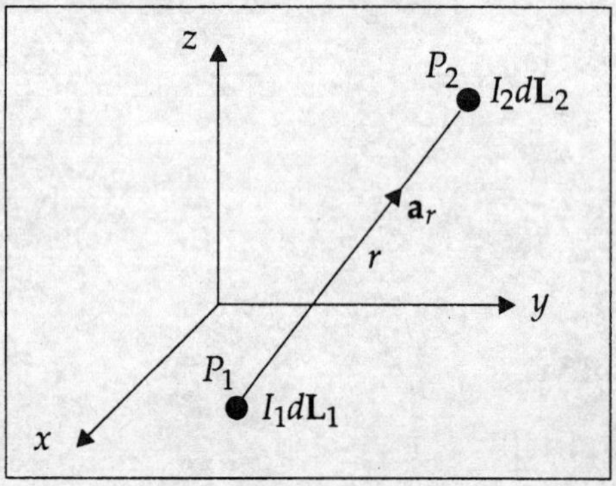

Fig. 3.12 Locations of differential current elements

Let the differential current element $I_1\,d\mathbf{L}_1$ be at point P_1 and $I_2\,d\mathbf{L}_2$ be at point P_2. The magnetic field at P_2 due to $I_1\,d\mathbf{L}_1$ is given by

$$d\mathbf{H} = \frac{I d\mathbf{L}_1 \times \mathbf{a}_r}{4\pi r^2}$$

or,

$$d\mathbf{B} = \frac{\mu I d\mathbf{L}_1 \times \mathbf{a}_r}{4\pi r^2}$$

This field exerts a force on the current element $I_2\,d\mathbf{L}_2$ at point P_2 and it is given by

$$d\,(d\mathbf{F}) = I_2\,d\mathbf{L}_2 \times d\mathbf{B}$$

This is the differential of a differential force on a differential current element due to a differential field, $d\mathbf{B}$. From the above two expressions we get,

$$d\,(d\mathbf{F}) = \frac{\mu I_2\,d\mathbf{L}_2 \times (I_1\,d\mathbf{L}_1 \times \mathbf{a}_r)}{4\pi r^2}$$

$$= \frac{\mu I_1 I_2}{4\pi}\,\frac{d\mathbf{L}_2 \times (d\mathbf{L}_1 \times \mathbf{a}_r)}{r^2}$$

$$\mathbf{F} = \frac{\mu I_1 I_2}{4\pi} \oint\oint \frac{d\mathbf{L}_2 \times (d\mathbf{L}_1 \times \mathbf{a}_r)}{r^2}$$ Hence proved.

3.16 BOUNDARY CONDITIONS ON H AND B

1. The tangential component of magnetic field, **H** is continuous across any boundary except at the surface of a perfect conductor, that is,

$$\mathbf{H}_{\tan 1} - \mathbf{H}_{\tan 2} = \mathbf{J}_s$$

At non-conducting boundaries, $J_s = 0$.

2. The normal component of magnetic flux density, **B** is continuous across any discontinuity, that is,

$$\mathbf{B}_{n1} = \mathbf{B}_{n2}$$

Proof Consider Fig. 3.13 in which a differential rectangular loop across a boundary separating medium 1 and medium 2 are shown.

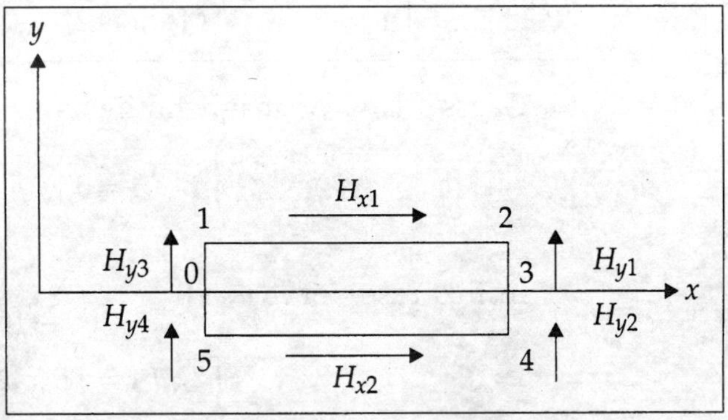

Fig. 3.13 A rectangular loop across a boundary

From Ampere's circuit law, we have

$$\oint \mathbf{H}.d\mathbf{L} = \int_{50} + \int_{01} + \int_{12} + \int_{23} + \int_{34} + \int_{45}$$

$$= H_{y4}\frac{\Delta y}{2} + H_{y3}\frac{\Delta y}{2} + H_{x1}\,\Delta x - H_{y1}\frac{\Delta y}{2} - H_{y2}\frac{\Delta y}{2} - H_{x2}\,\Delta x = I$$

As $\Delta y \to 0$, we get

$$\int \mathbf{H}.d\mathbf{L} = H_{x1}\,\Delta x - H_{x2}\,\Delta x = I$$

or, $$H_{x1} - H_{x2} = \frac{I}{\Delta x} = J_s$$

Here, H_{x1} and H_{x2} are tangential components in medium 1 and 2, respectively.

So, $$\mathbf{H}_{\tan 1} - \mathbf{H}_{\tan 2} = \mathbf{J}_s$$

Now consider a cylinder shown in Fig. 3.14.

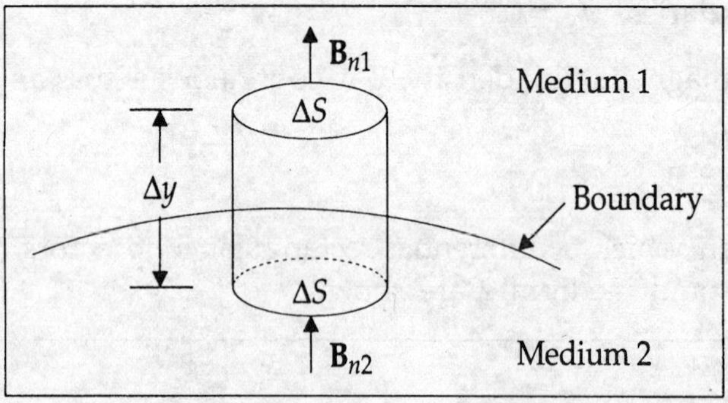

Fig. 3.14 A differential cylinder across the boundary

Gauss's law for magnetic fields is

$$\oint_s \mathbf{B}.d\mathbf{S} = 0$$

In this case, for $\Delta y \to 0$

$$\int_s \mathbf{B}.d\mathbf{S} = \int_s B_{n1}\,\mathbf{a}_y.dS\,\mathbf{a}_y + \int_s B_{n2}\,\mathbf{a}_y.dS\,(-\mathbf{a}_y)$$

that is, $$B_{n1}\,\Delta S - B_{n2}\,\Delta S = 0$$

Therefore, $$\boxed{\mathbf{B}_{n1} = \mathbf{B}_{n2}}$$

Problem 3.19 Two homogeneous, linear and isotropic media have an interface at $x = 0$. $x < 0$ describes medium 1 and $x > 0$ describes medium 2. $\mu_{r_1} = 2$ and $\mu_{r_2} = 5$. The magnetic field in medium 1 is

$$150\mathbf{a}_x - 400\mathbf{a}_y + 250\mathbf{a}_z \text{ A/m}.$$

Determine:

(a) Magnetic field in medium 2

(b) Magnetic flux density in medium 1

(c) Magnetic flux density in medium 2.

Solution The magnetic field in medium 1 is

$$\mathbf{H}_1 = 150\mathbf{a}_x - 400\mathbf{a}_y + 250\mathbf{a}_z \text{ A/m}$$

Consider Fig. 3.15.

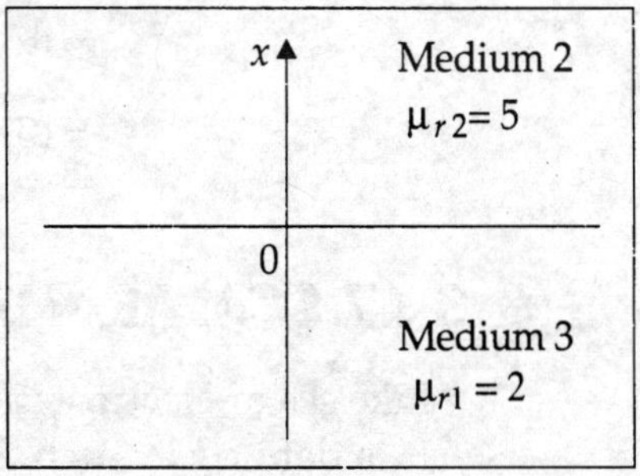

Fig. 3.15

(a)
$$\mathbf{H}_1 = \mathbf{H}_{\tan 1} + \mathbf{H}_{n_1}$$
$$\mathbf{H}_{\tan 1} = -400\mathbf{a}_y + 250\mathbf{a}_z \text{ A/m}$$
$$\mathbf{H}_{n1} = 150\mathbf{a}_x$$

The boundary condition is
$$\mathbf{H}_{\tan 1} = \mathbf{H}_{\tan 2}$$
$$\mathbf{H}_{\tan 2} = -400\mathbf{a}_y + 50\mathbf{a}_z \text{ A/m}$$

The boundary condition on **B** is $\mathbf{B}_{n_1} = \mathbf{B}_{n_2}$

that is,
$$\mu_1 \mathbf{H}_{n1} = \mu_2 \mathbf{H}_{n2}$$
$$\mathbf{H}_{n2} = \frac{\mu_1}{\mu_2}\mathbf{H}_{n1}$$
$$= \frac{2}{5} \times 150\mathbf{a}_x$$
$$= 60\,\mathbf{a}_x$$
$$\mathbf{H}_2 = \mathbf{H}_{\tan 2} + \mathbf{H}_{n2}$$
$$\boxed{\mathbf{H}_2 = 60\mathbf{a}_x - 400\mathbf{a}_y + 250\mathbf{a}_z}$$

(b)
$$\mathbf{B}_1 = \mu_1 \mathbf{H}_1$$
$$= \mu_0 \mu_r \mathbf{H}_1$$
$$= 4\pi \times 10^{-7} \times 2\,(150\mathbf{a}_x - 400\mathbf{a}_y + 250\mathbf{a}_z)$$
$$= (376.5\mathbf{a}_x - 1004\mathbf{a}_y + 627.5\mathbf{a}_z)\,\mu\text{wb/m}^2$$

(c) $$\mathbf{B}_2 = \mu_2 \mathbf{H}_2$$

$$= 4\pi \times 10^{-7} \times 5\,(60\mathbf{a}_x - 400\mathbf{a}_y + 250\mathbf{a}_z)$$

$$= (376.98\mathbf{a}_x - 2513.2\mathbf{a}_y + 1570.75\mathbf{a}_z)\ \mu\text{wb/m}^2$$

3.17 SCALAR MAGNETIC POTENTIAL

Like scalar electrostatic potential, it is possible to have scalar magnetic potential. It is defined in such a way that its negative gradient gives the magnetic field, that is,

$$\mathbf{H} = -\nabla V_m$$

V_m = scalar magnetic potential (Amp)

Taking curl on both sides, we get

$$\nabla \times \mathbf{H} = -\nabla \times \nabla V_m$$

But curl of the gradient of any scalar is always zero.

So, $$\nabla \times \mathbf{H} = 0$$

But by Ampere's circuit law $\nabla \times \mathbf{H} = \mathbf{J}$

or, $$\mathbf{J} = 0$$

In other words, scalar magnetic potential exists in a region where $\mathbf{J} = 0$.

$$\mathbf{H} = -\nabla V_m \quad (\mathbf{J} = 0)$$

The scalar potential satisfies Laplace's equation, that is, we have

$$\nabla.\mathbf{B} = \mu_0\, \nabla.\mathbf{H} = 0 = \mu \nabla\,(-\nabla V_m) = 0$$

or, $$\nabla^2 V_m = 0\ (\mathbf{J} = 0)$$

Characteristics of Scalar Magnetic Potential (V_m)

1. The negative gradient of V_m gives $\mathbf{H}$, or $\mathbf{H} = -\nabla V_m$
2. It exists where $\mathbf{J} = 0$
3. It satisfies Laplace's equation.

4. It is directly defined as

$$V_m = -\int_A^B \mathbf{H}.d\mathbf{L}$$

5. It has the unit of Ampere.

Problem 3.20 The magnetic field in a current free region is $\mathbf{H} = \frac{1}{\rho}\mathbf{a}_\phi$. The region is defined by $1 \le \rho \le 2$ m, $0 \le \phi \le 2\pi$ and $0 \le z \le 2$ m. Find the scalar magnetic potential at (4, 50°, 2).

Solution We have

$$V_m = -\int \mathbf{H}.d\mathbf{L}$$

Here
$$\mathbf{H} = \frac{1}{\rho}\mathbf{a}_\phi \text{ A/m}$$

$$d\mathbf{L} = d\rho\, \mathbf{a}_\rho + \rho\, d\phi\, \mathbf{a}_\phi + dz\, \mathbf{a}_z$$

So,
$$V_m = -\int \frac{1}{\rho}\mathbf{a}_\phi.(d\rho\, \mathbf{a}_\phi + \rho\, d\phi\, \mathbf{a}_\phi + dz\, \mathbf{a}_z)$$

$$= -\int \frac{1}{\rho}\cdot \rho\, d\phi$$

$$= -\phi$$

$$V_m/(4, 50°, 2) = -50 \times \frac{\pi}{180}$$

$$\boxed{V_m = -0.8726 \text{ Amp}}$$

3.18 VECTOR MAGNETIC POTENTIAL

Vector magnetic potential exists in regions where **J** is present. It is defined in such a way that its curl gives the magnetic flux density, that is,

$$\mathbf{B} \equiv \nabla \times \mathbf{A}$$

where **A** = vector magnetic potential (wb/m).

It is also defined as

$$\mathbf{A} \equiv \oint \frac{\mu_0 I d\mathbf{L}}{4\pi R} \left(\frac{\text{Henry-Amp}}{\text{m}}\right)$$

or,

$$\mathbf{A} \equiv \int_s \frac{\mu_0 \mathbf{K} ds}{4\pi R}, \qquad (\mathbf{K} = \text{current sheet})$$

or,

$$\mathbf{A} \equiv \int_v \frac{\mu_0 \mathbf{J} dv}{4\pi R}$$

Characteristics of Vector Magnetic Potential

1. It exists even when **J** is present.
2. It is defined in two ways

$$\mathbf{B} \equiv \nabla \times \mathbf{A} \qquad \text{and}$$

$$\mathbf{A} \equiv \int_v \frac{\mu_0 \mathbf{J} dv}{4\pi R}$$

3. $\nabla^2 \mathbf{A} = -\mu_0 \mathbf{J}$
4. $\nabla^2 \mathbf{A} = 0$ if $\mathbf{J} = 0$
5. Vector magnetic potential, **A** has applications to obtain radiation characteristics of antennas, apertures and also to obtain radiation leakage from transmission lines, waveguides and microwave ovens.
6. **A** is used to find near and far-fields of antennas.

Problem 3.21 The vector magnetic potential, **A** due to a direct current in a conductor in free space is given by $\mathbf{A} = (x^2 + y^2)\,\mathbf{a}_z$ μwb/m². Determine the magnetic field produced by the current element at (1, 2, 3).

Solution $\mathbf{A} = (x^2 + y^2)\,\mathbf{a}_z$ μwb/m²

We have $\mathbf{B} = \nabla \times \mathbf{A}$

$$= 10^{-6} \begin{vmatrix} \mathbf{a}_x & \mathbf{a}_y & \mathbf{a}_z \\ \frac{\partial}{\partial x} & \frac{\partial}{\partial y} & \frac{\partial}{\partial z} \\ 0 & 0 & (x^2 + y^2) \end{vmatrix}$$

$$= \left[\frac{\partial}{\partial y}(x^2 + y^2)\,\mathbf{a}_x + \left(-\frac{\partial}{\partial x}(x^2 + y^2)\right)\mathbf{a}_y\right] \times 10^{-6}$$

$$= [(2y)\,\mathbf{a}_x - (2x)\,\mathbf{a}_y] \times 10^{-6}$$

$$\mathbf{B}/\text{at } (1, 2, 3) = 4\mathbf{a}_x - (2)\,\mathbf{a}_y \times 10^{-6}$$

$$= (4\mathbf{a}_x - 2\mathbf{a}_y) \times 10^{-6}$$

$$\mathbf{H} = \frac{1}{\mu_0}[4\mathbf{a}_x - 2\mathbf{a}_y] \times 10^{-6}$$

$$= \frac{1}{4\pi \times 10^{-7}}(4\mathbf{a}_x - 2\mathbf{a}_y) \times 10^{-6}$$

$$\boxed{\mathbf{H} = (3.978\mathbf{a}_x - 4.774\mathbf{a}_y), \text{A/m}}$$

Problem 3.22 When vector magnetic potential is given by

$$\mathbf{A} = \frac{1}{r^3}(2.0\cos\theta\,\mathbf{a}_r + \sin\theta\,\mathbf{a}_\theta),$$

find the magnetic flux density.

Solution We have

$$\mathbf{B} = \nabla \times \mathbf{A} = \frac{1}{r\sin\theta}\left[\frac{\partial}{\partial\theta}(A_\phi \sin\theta) + -\frac{\partial A_\theta}{\partial\phi}\right]\mathbf{a}_r + \frac{1}{r}\left[\frac{1}{\sin\theta}\frac{\partial A_r}{\partial\phi} - \frac{\partial(r A_\phi)}{\partial r}\right]\mathbf{a}_\theta$$

$$+ \frac{1}{r}\left[\frac{\partial}{\partial r}(r A_\theta) - \frac{\partial A_r}{\partial\theta}\right]\mathbf{a}_\phi$$

Here $\quad A_r = \dfrac{2.0}{r^3}\cos\theta$

$$A_\theta = \frac{1}{r^3}\sin\theta$$

$$\boxed{\mathbf{B} = \nabla \times \mathbf{A} = 0}$$

3.19 FORCE AND TORQUE ON A LOOP OR COIL

Consider Fig. 3.16 in which a rectangular loop is placed under a uniform magnetic flux density, **B**.

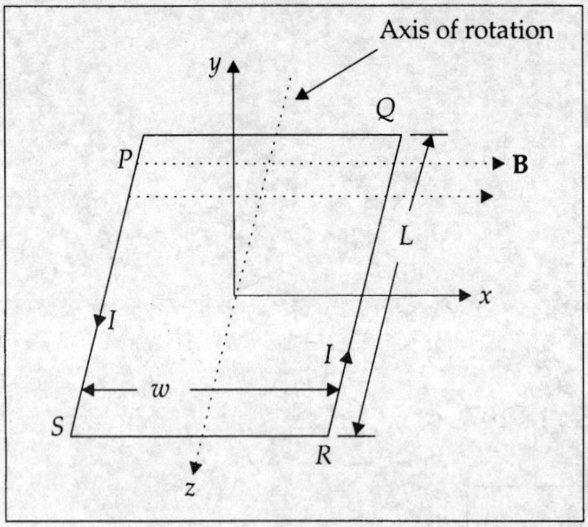

Fig. 3.16 Rectangular conductor loop in *x-z* plane

From Fig. 3.16, the force on QR due to $\mathbf{B}$ is

$$\mathbf{F}_1 = I\mathbf{L}\times\mathbf{B} = -IL\mathbf{a}_z \times B\mathbf{a}_x$$

$$\mathbf{F}_1 = -ILB\mathbf{a}_y$$

that is, the force, $\mathbf{F}_1$ on QR moves it downwards. Now the force on PS is

$$\mathbf{F}_2 = I\mathbf{L}\times\mathbf{B} = -IL\mathbf{a}_z \times \mathbf{B}\mathbf{a}_x$$

$$\mathbf{F}_2 = IL\mathbf{B}\mathbf{a}_y$$

Force, $\mathbf{F}_2$ on PS moves it upwards. It may be noted that the sides PQ and SR will not experience force as they are parallel to the field, $\mathbf{B}$.

The forces on QR and PS exert a torque. This torque tends to rotate the coil about its axis.

The torque, $\mathbf{T}$ is nothing but a mechanical moment of force. The torque on the loop is defined as the vector product of moment arm and force,

that is,

$$\mathbf{T} \equiv \mathbf{r}\times\mathbf{F}, \text{N-m}$$

where $\mathbf{r}$ = moment arm

$\mathbf{F}$ = force

Applying this definition to the loop considered above, the expression for torque is given by

$$\mathbf{T} = \mathbf{r}_1 \times \mathbf{F}_1 + \mathbf{r}_2 \times \mathbf{F}_2$$

$$= \frac{w}{2}\mathbf{a}_x \times (-IL\mathbf{B}\mathbf{a}_y) + \left(-\frac{w}{2}\mathbf{a}_x\right) \times (IL\mathbf{B}\mathbf{a}_y)$$

$$= -BILw\mathbf{a}_z$$

or,

$$\mathbf{T} = -BIS\mathbf{a}_z$$

where $S = wL$ = area of the loop

The torque in terms of magnetic dipole moment, **m** is

$$\mathbf{T} = \mathbf{m} \times \mathbf{B}, \text{ N-m}$$

where $\mathbf{m} = I1w\,\mathbf{a}_y$

$= IS\,\mathbf{a}_y$

Problem 3.23 A rectangular coil is placed in a field of $\mathbf{B} = (2\mathbf{a}_x + \mathbf{a}_y)$ wb/m^2. The coil is in y-z plane and has dimensions of 2 m × 2 m. It carries a current of 1 A. Find the torque about the z-axis.

Solution $\mathbf{m} = IS\,\mathbf{a}_n = 1 \times 4\mathbf{a}_x$

$\mathbf{T} = \mathbf{m} \times \mathbf{B} = 4\mathbf{a}_x \times (2\mathbf{a}_x + \mathbf{a}_y)$

$\mathbf{T} = 4\mathbf{a}_z$, N-m

Problem 3.24 Determine the torque on a square coil of 0.2 m × 0.2 m carrying a current of 3.0 A in a field of 10 wb/m^2.

Solution $\mathbf{T} = BIS = 10 \times 3 \times 0.04$

$\mathbf{T} = 1.2$ N-m.

3.20 MATERIALS IN MAGNETIC FIELDS

A material is said to be magnetic if $\chi_m \neq 0, \mu_r \neq 1$.

A material is said to be non-magnetic if $\chi_m = 0, \mu_r = 1$.

The term 'Magnetism' is commonly discussed in terms of magnets with basic examples like north pole, compass needle, horse shoe magnets and so on.

Magnetic properties are described in terms of magnetic susceptibility and relative permeability of the materials.

Magnetic materials are classified into

1. Diamagnetic materials
2. Paramagnetic materials
3. Ferromagnetic materials

Diamagnetic Materials

A material is said to diamagnetic if its electric susceptibility, $\chi_m < 0$ and $\mu_r \leq 1.0$.

Examples are copper, lead, silicon, diamond and bismuth.

Characteristics of diamagnetic materials

1. Magnetic fields due to the motion of orbiting electrons and spinning electrons cancel each other.
2. Permanent magnetic moment of each atom is zero.
3. These materials are widely affected by magnetic field.
4. Magnetic susceptibility χ_m is (–)ve.
5. $\mu_r = 1$
6. $\mathbf{B} = 0$
7. Most of the materials exhibit diamagnetism.
8. They are linear magnetic materials.
9. Diamagnetism is not temperature dependent.
10. These materials acquire magnetisation opposite to **H** and hence they are called diamagnetic materials.

Paramagnetic Materials

A material for which $\chi_m > 0$ and $\mu_r \geq 1$ is said to be paramagnetic.

Examples are air, tungsten, potassium and platinum.

Characteristics of paramagnetic materials

1. They have non-zero permanent magnetic moment.
2. Magnetic fields due to orbiting and spinning electrons do not cancel each other.
3. Paramagnetism is temperature dependent.
4. χ_m lies between 10^{-5} and 10^{-3}.
5. These are used in MASERS.
6. $\chi_m > 0$
7. $\mu_r \geq 1$
8. They are linear magnetic materials.

These materials acquire magnetisation parallel to **H** and hence they are called paramagnetic materials.

Ferromagnetic Materials

A material for which $\chi_m >> 0, \mu_r >> 1$ is said to be ferromagnetic.

Examples are iron, nickel, cobalt and their alloys.

Characteristics of ferromagnetic materials

1. They exhibit large permanent dipole moment.
2. $\chi_m >> 0$
3. $\mu_r >> 1$
4. They are strongly magnetised by magnetic field.
5. They retain magnetism even if the magnetic field is removed.
6. They lose their ferromagnetic properties when the temperature is raised.
7. If a permanent magnet made of iron is heated above its curie temperature, 770°C, it loses its magnetisation completely.
8. They are non-linear magnetic materials.
9. $\mathbf{B} = \mu\mathbf{H}$ does not hold good as μ depends on **B**.
10. In these materials, magnetisation is not determined by the field present. It depends on the magnetic history of the object.

3.21 MAGNETISATION IN MATERIALS

A material consists of atoms. The atoms in turn consist of electrons and nucleus. The electrons revolve in orbits around the nucleus and also rotate about their axes. The rotation about their axes is called spinning.

The phenomenon of orbiting and spinning of electrons produce an internal magnetic field. This field is similar to the field produced by a current loop.

The equivalent current loop exhibits a magnetic moment and the current loop is equivalent to a magnetic dipole. The magnetic moment is given by

$$\mathbf{m} \equiv I_b\, d\mathbf{S} \equiv I_b\, dS\, \mathbf{a}_n$$

where $\mathbf{a}_n$ = unit normal to the plane of the loop

I_b = bound current and is called as amperian produced by bound charges

The dipole moments are arbitrarily oriented in a material and hence the net magnetic dipole moment is zero.

However, when such a material is kept in an external magnetic field, randomness disappears and dipole moments align themselves in the direction of the applied field. As a result, the net magnetic dipole moment is zero.

Now consider a bar magnet consisting of north and south poles. It is important to note that these poles, known as magnetic charges, cannot be isolated. They always appear in pairs. In a given material, they are randomly arranged.

Dipole moment is defined as

$$\mathbf{m} \equiv Q_m \mathbf{d}$$

where Q_m = pole strength of magnet, A-m

$\mathbf{d}$ = pole separation of magnet, m

In fact, **m** is equivalent to the current loop given by

$$\mathbf{m} = I_b\, d\mathbf{S}$$

So,

$$\mathbf{m} \equiv Q_m \mathbf{d} = I_b\, d\mathbf{S},\ \text{A-m}^2$$

A tiny bar magnet and a current loop are shown in Fig. 3.17.

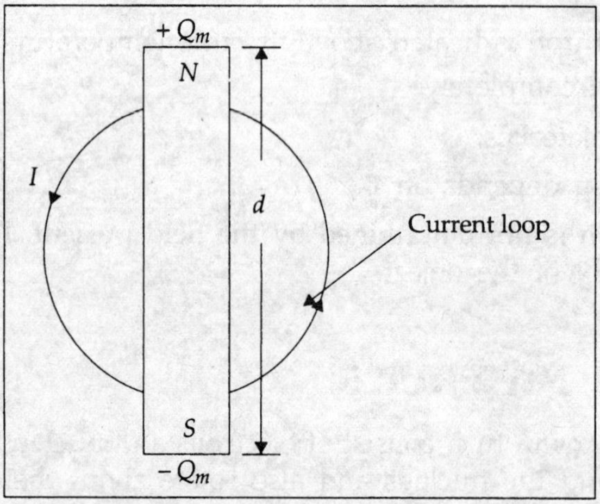

Fig. 3.17 Bar magnet and current loop

Magnetic Dipole Moment, m

It is defined as

$$\mathbf{m} = I_b\, d\mathbf{S},\ \text{Amp-m}^2$$

Here, the bound current, *Ib* is in a bound path enclosing a differential area, $d\mathbf{S}$.

If there are *n* magnetic dipoles per unit volume and if we consider a volume, Δv, then total magnetic dipole moment is given by

$$\mathbf{m}_{\text{total}} = \sum_{i=1}^{n\Delta v} \mathbf{m}_i$$

Magnetisation, M Magnetisation, **M** is defined as the magnetic dipole per unit volume, that is, mathematically,

$$\mathbf{M} \equiv \underset{\Delta v \to 0}{\text{Lt}} \frac{1}{\Delta v} \sum_{i=1}^{n\Delta v} \mathbf{m}_i$$

M is also defined as

$$\mathbf{M} \equiv \chi_m \mathbf{H}$$

χ_m = magnetic susceptibility $= (\mu_r - 1)$.
Relation between the currents and the fields are

$$I_b = \oint \mathbf{M} . d\mathbf{L}$$

$$I = \oint \mathbf{H} . d\mathbf{L}$$

$$I_T = \oint \frac{\mathbf{B}}{\mu_0} . d\mathbf{L}$$

where $$I_T = I + I_b$$

or, $$I = I_T - I_b$$

that is, $$\oint \mathbf{H} . d\mathbf{L} = \oint \left(\frac{\mathbf{B}}{\mu_0} - \mathbf{M} \right) . d\mathbf{L}$$

or, $$\mathbf{H} = \frac{\mathbf{B}}{\mu_0} - \mathbf{M}$$

$$\mathbf{B} = \mu_0 (\mathbf{H} + \mathbf{M})$$

The currents, in terms of current densities, are given by

$$I_b = \oint_S \mathbf{J}_b . d\mathbf{S}$$

$$I = \oint_S \mathbf{J} . d\mathbf{S}$$

$$I_T = \oint_S \mathbf{J}_T . d\mathbf{S}$$

The relations in differential form are given by

$$\nabla \times \mathbf{M} = \mathbf{J}_b$$

$$\nabla \times \mathbf{H} = J$$

$$\nabla \times \frac{\mathbf{B}}{\mu_0} = \mathbf{J}_T$$

Problem 3.25 An isotropic material has a magnetic susceptibility of 3 and the magnetic flux density, $\mathbf{B} = 10y\,\mathbf{a}_x$ mwb/m^2. Determine μ_r, μ, $\mathbf{J}_b$, $\mathbf{J}$, $\mathbf{M}$ and $\mathbf{H}$.

Solution The relation between relative permeability and magnetic susceptibility is

$$\mu_r = 1 + \chi_m = 1 + 3 = 4$$

$$\mu = \mu_0 \mu_r = 4 \times 4\pi \times 10^{-7}$$

$$\boxed{\mu = 16\pi \times 10^{-7} \text{ H/m}}$$

We have

$$\begin{aligned}
\mathbf{J}_b &= \nabla \times \mathbf{M} \\
&= \nabla \times \chi_m \mathbf{H} \\
&= \frac{\chi_m}{\mu} \nabla \times \mathbf{B} \\
&= \frac{3}{16\pi \times 10^{-7}} \nabla \times \mathbf{B} \\
&= \frac{3}{16\pi \times 10^{-7}} \nabla \times (10y\,\mathbf{a}_x) \times 10^{-3} \\
&= \frac{3}{16\pi \times 10^{-7}} \times 10\mathbf{a}_z \times 10^{-3} \\
&= -0.597 \times 10^4\,\mathbf{a}_z
\end{aligned}$$

$$\boxed{\mathbf{J}_b = -5.97\mathbf{a}_z \text{ KA/m}^2}$$

Now

$$\begin{aligned}
\mathbf{J} &= \nabla \times \mathbf{H} \\
&= \frac{1}{\chi_m} \nabla \times \mathbf{M} \\
&= \frac{1}{\chi_m} \mathbf{J}_b \\
&= \frac{-5.97\,\mathbf{a}_z}{3} \text{ KA/m}^2
\end{aligned}$$

$$\boxed{\mathbf{J} = -1.99\mathbf{a}_z \text{ KA/m}^2}$$

$$\mathbf{M} = \frac{\chi_m}{\mu} \mathbf{B}$$

or,

$$\mathbf{M} = 5.97y\,\mathbf{a}_x \text{ KA/m}$$

and

$$\mathbf{H} = \frac{\mathbf{B}}{\mu} = \frac{10y\,\mathbf{a_x}}{16\pi \times 10^{-7}} \times 10^{-3}$$

$$\boxed{\mathbf{H} = 1990y\,\mathbf{a}_x \text{ A/m}}$$

Problem 3.26 A magnetic material has $\mu_r = 10/\pi$ and is in a magnetic field of strength, $\mathbf{H} = 5\rho^3\,\mathbf{a}_\phi$ A/m. Find the magnetisation.

Solution The magnetic flux density,

$$\mathbf{B} = \mu\mathbf{H} = \mu_r\mu_0\mathbf{H} = \frac{10}{\pi}\times 4\pi\times 10^{-7}\times 5\rho^3\,\mathbf{a}_\phi$$

$$\mathbf{B} = 20\times 10^{-6}\rho^3\mathbf{a}_\phi \text{ wb/m}^2$$

But
$$\mathbf{B} = \mu_0(\mathbf{H}+\mathbf{M})$$

or,
$$\mu_0\mu_r\mathbf{H} = \mu_0[\mathbf{H}+\mathbf{M}]$$

or,
$$\mathbf{M} = \mathbf{H}[\mu_r - 1]$$

$$= \mathbf{H}\left[\frac{10}{\pi} - 1\right]$$

$$= 2.18\times 5\rho^3\mathbf{a}_\phi$$

$$\boxed{\mathbf{M} = 10.90\rho^3\mathbf{a}_\phi}$$

3.22 INDUCTANCE

Inductor It is a coil of wire wound according to various designs with or without a core of magnetic material to concentrate the magnetic field.

Inductance, L In a conductor, device or circuit, an inductance is the inertial property caused by an induced reverse voltage that opposes the flow of current when a voltage is applied. It also opposes a sudden change in current that has been established.

Definition of Inductance, L (Henry):

The inductance, L of a conductor system is defined as the ratio of magnetic flux linkage to the current producing the flux, that is,

$$L \equiv \frac{N\phi}{I}\text{ (Henry)}$$

Here
N = number of turns

ϕ = flux produced

I = current in the coil

$$1 \text{ Henry} \equiv 1 \text{ wb/Amp}$$

L is also defined as $\frac{2W_H}{I^2}$, or

$$L \equiv \frac{2W_H}{I^2}$$

where W_H = energy in **H** produced by I.

In fact, a straight conductor carrying current has the property of inductance. Aircore coils are wound to provide a few pico henries to a few micro henries. These are used at *IF* and *RF* frequencies in tuning coils, interstage coupling coils and so on.

The requirements of such coils are:

1. Stability of inductance under all operating conditions
2. High ratio inductive reactance to effective loss resistance at the operating frequency
3. Low self capacitance
4. Small size and low cost
5. Low temperature coefficient

3.23 STANDARD INDUCTANCE CONFIGURATIONS

Toroid

It consists of a coil wound on annular core. One side of each turn of the coil is threaded through the ring to form a Toroid (Fig. 3.18).

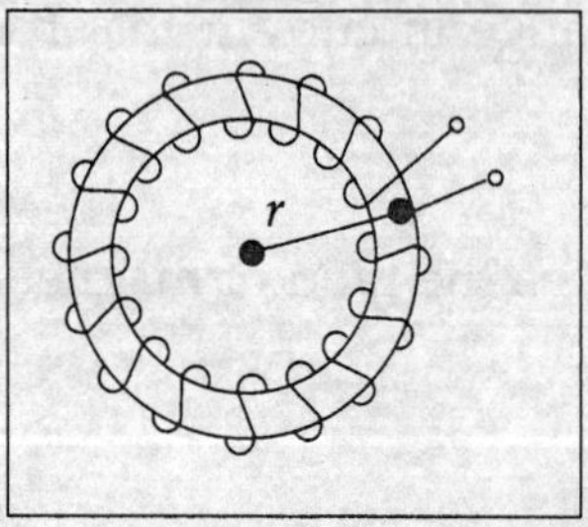

Fig. 3.18 Toroid

$$\text{Inductance of Toroid,} \quad L = \frac{\mu_0 N^2 S}{2\pi r}$$

Here N = number of turns

r = average radius

S = cross-sectional area

$$\text{Magnetic field in a Toroid, } H = \frac{NI}{2\pi r}$$

I is the current in the coil.

Quality factor of the coil is

$$Q = \frac{X_L}{R}$$

Specifications of inductors The specifications are:

1. Inductance value
2. Type of core
3. Type of winding, for example, single layer, multi-layer, standard and so on
4. Frequency
5. Q of the coil
6. Coupling factor
7. Self capacitance
8. Stability

Solenoid

It is a coil of wire which has a long axial length relative to its diameter. The coil is tubular in form. It is used to produce a known magnetic flux density along its axis.

A solenoid is also used to demonstrate electromagnetic induction. A bar of iron, which is free to move along the axis of the coil, is usually provided for this purpose. A typical solenoid is shown in Fig. 3.19.

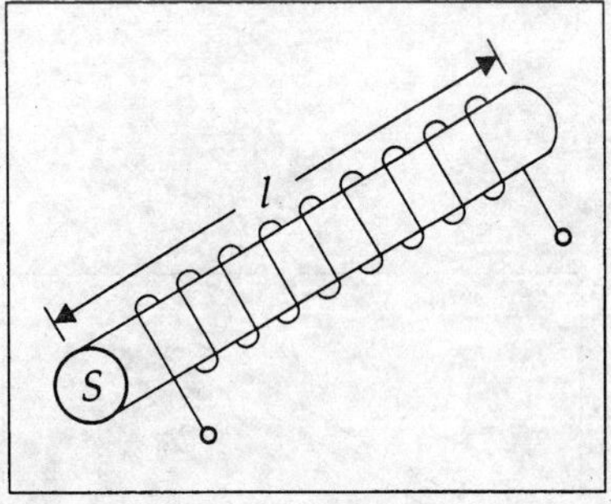

Fig. 3.19 Solenoid

The inductance, L of a solenoid is

$$L = \frac{\mu_0 N^2 S}{l}$$

l = length of solenoid

S = cross-sectional area

N = number of turns

The magnetic field in a solenoid is

$$H = \frac{NI}{l}$$

I is the current.

Coaxial Cable

It is a cable in which there are two concentric cylinders.

The inductance of a coaxial cable is

$$L = \frac{\mu_0 l}{2\pi} \ln \frac{b}{a}$$

Here l = length of the cable

b = outer radius

a = inner radius

The magnetic field, H in the cable is

$$H = \frac{Ir}{2\pi a^2}, \ (r \le a)$$

Parallel Conductors of Radius, a

The inductance of parallel conductors (Fig. 3.20) is

$$L = \frac{\mu_0 l}{\pi} \cosh^{-1} \frac{d}{2a}, \text{ Henry}$$

a = radius of the conductor

d = distance between the conductors

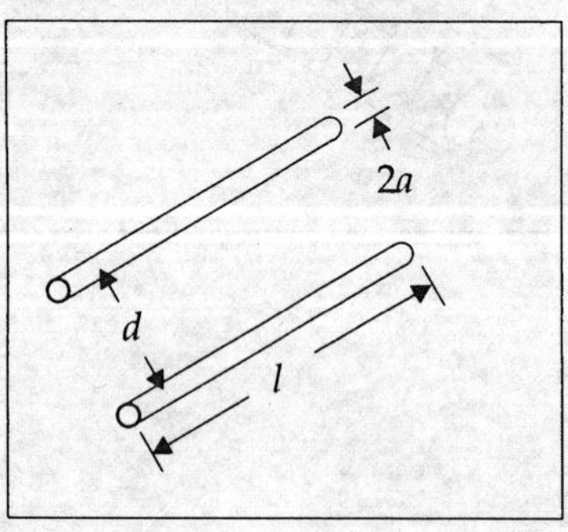

Fig. 3.20 Parallel conductors

It should be noted that the inductance, L depends on physical parameters but not on ϕ or I. L depends on permeability of the medium and a geometrical factor having the unit of length.

Problem 3.27 The radii of the inner and outer conductors of a coaxial cable are 2 mm and 6 mm respectively and $\mu = \mu_0$. Find the inductance of a 10 m long cable.

Solution The inductance of a coaxial cable is

$$L = \frac{\mu l}{2\pi} \ln \frac{r_2}{r_1}$$

where $l =$ length of the cable = 10 m

$r_1 =$ radius of inner conductor = 2 mm

$r_2 =$ radius of outer conductor = 6 mm

$$L = \frac{4\pi \times 10^{-7}}{2\pi} \ln\left(\frac{6.0}{2.0}\right) \times 10$$

$$L = 2.2\,\mu H$$

Problem 3.28 A Toroid has air core and has a cross-sectional area of 10 mm^2. It has 1000 turns and its mean radius is 10 mm. Find its inductance.

Solution The inductance of a Toroidal coil is given by

$$L = \frac{\mu N^2 A}{2\pi r}$$

where $\mu = \mu_0 = 4\pi \times 10^{-7}$ H/m

$N = 1000$

$A = 10\text{ mm}^2 = 10 \times 10^{-6}\text{ m}^2$

$r = 10\text{ mm} = 10 \times 10^{-3}\text{ m}$

$$L = \frac{4\pi \times 10^{-7} \times (10^3)^2 \times 10 \times 10^{-6}}{2\pi \times 10 \times 10^{-3}} = 2 \times 10^{-7} \times 10^3 = 2 \times 10^{-4}$$

$$\boxed{L = 0.2\text{ mH}}$$

Problem 3.29 A solenoid has 400 turns with a length of 2 m. It has a circular cross-section of 0.1 m^2. Find its inductance.

Solution The inductance of a solenoid is

$$L = \frac{\mu_0 N^2 S}{l}$$

Here $\mu_0 = 4\pi \times 10^{-7}$ H/m

$N = 400$

$S = 0.1\,\text{m}^2$

$l = 2\,\text{m}$

$$L = \frac{4\pi \times 10^{-7} \times 16 \times 10^4 \times 0.1}{2} = 32\pi \times 10^{-4}$$

or, $\boxed{L = 10.05\,\text{mH}}$

3.24 ENERGY DENSITY IN A MAGNETIC FIELD

Energy density in a magnetic field,

$$W_H = \frac{1}{2}\mu H^2 \text{ and } W_H = \frac{1}{2}\mathbf{B}.\mathbf{H}.$$

Proof **Method 1** (From fundamentals)

Faraday' law of induction is

$$V = \frac{-N\, d\phi(t)}{dt}$$

If the current is to be increased, the electric source must supply energy. The differential work done in a time, dt is

$$dW = -Vi\, dt = iN\, d\phi$$

The total work done $= N \int i d\phi$

For a linear magnetic circuit,

$$L = \frac{N d\phi}{di}$$

or, $$Nd\phi = L di$$

$$dW = Li\, di$$

If W_i is the initial energy in the coil corresponding to the initial current I_i and W_f is the final energy when the current is I_f, the increase of energy in the coil is

$$\int_{W_i}^{W_f} dW = \int_{I_i}^{I_f} Li\, di$$

or,
$$W = W_f - W_i = \frac{1}{2} LI_f^2 - \frac{1}{2} LI_i^2$$

If $I_i = 0$ and the current at any time t is I, the energy stored in the magnetic circuit is given by

$$W = \frac{1}{2} LI^2$$

For linear circuits,

$$L = \frac{N\phi}{I}$$

$$W = \frac{1}{2} N\phi I$$

We know that

$$\phi = \int_S \mathbf{B}.d\mathbf{S} = BS$$

and the total current enclosed by a path,

$$NI = \oint \mathbf{H}.d\mathbf{L} = Hl$$

Here S is the cross-sectional area of the coil and l is its length.

So,
$$W = \frac{1}{2} HBSl, \; Sl = \upsilon$$

This represents the total energy stored. Therefore, the energy density is given by

$$W_H = \frac{W}{\upsilon} = \frac{W}{Sl}$$

or,
$$W_H = \frac{1}{2} HB$$

$$W_H = \frac{1}{2}\mu H^2 \text{ Joules/m}^3 \qquad [\text{as } B = \mu H]$$

This is written in vector form as

$$W_H = \frac{1}{2}\mathbf{B.H} \text{ J/m}^3 \qquad \text{Hence proved.}$$

Method 2 The inductance of a solenoid is given by

$$L = \frac{\mu_0 N^2 S}{l}$$

where N = number of turns

S = area of cross-section of a solenoid

l = length of the solenoid

The energy stored in an inductor is given by

$$W_H = \frac{1}{2} LI^2$$

$$= \frac{1}{2}\mu_0 \frac{N^2 S}{l} I^2$$

$$= \frac{1}{2}\mu_0 \left(\frac{NI}{l}\right)^2 lS$$

Here lS = volume of space inside the coil.

But $H = \frac{NI}{l}$ in a solenoid

Energy stored $= \frac{1}{2}\mu_0 H^2 (lS)$ Joules

or, Energy density, $W_H = \frac{1}{2}\mu_0 H^2 \text{ Joules/m}^3$ Hence proved.

3.25 ENERGY STORED IN AN INDUCTOR

Energy stored in an inductor,

$$W_L = \frac{1}{2} LI^2$$

The energy density stored in a magnetic field is given as

$$W_H = \frac{1}{2}\,\mathbf{B}.\mathbf{H} = \frac{1}{2}\,\mu_0\,H^2 \qquad [\text{as } \mathbf{B} = \mu_0\,\mathbf{H}]$$

Total energy stored $= \int_{\upsilon} \frac{1}{2}\,\mu_0\,H^2\,d\upsilon = \frac{1}{2}\,\mu_0\,H^2\,\upsilon$

But υ = volume of space inside the coil = (lS), m^3

and $H = \dfrac{NI}{l}$

N = number of turns

I = current in the coil

l = length of the coil

So, energy stored $= \dfrac{1}{2}\,\mu_0\left(\dfrac{NI}{l}\right)^2 lS = \dfrac{1}{2}\left(\dfrac{\mu_0}{2}\,\dfrac{N^2\,I}{l}\right)I^2$

As the inductance of a solenoid is

$$L = \frac{\mu_0\,N^2\,S}{l} \text{ (Henry)}$$

The energy stored in an inductor is

$$W_L = \frac{1}{2}\,LI^2$$

Hence proved.

3.26 EXPRESSION FOR INDUCTANCE, *L*, IN TERMS OF FUNDAMENTAL PARAMETERS

The inductance, L, depends on physical parameters and medium but not on ϕ and I.

It is given by

$$L = \frac{\mu}{4\pi}\left[\oint\left(\oint \frac{d\mathbf{l}}{R}\right)\right].d\mathbf{l}$$

Proof **Self-inductance** is defined as

$$L \equiv \frac{2W_H}{I^2}$$

$$W_H = \frac{1}{2}\int \mathbf{B}.\mathbf{H}\, d\upsilon$$

$$L = \int_{\upsilon} \frac{\mathbf{B}.\mathbf{H}}{I^2}\, d\upsilon$$

But $\mathbf{B} = \nabla \times \mathbf{A}$

$\mathbf{A}$ = vector magnetic potential

$$L = \frac{1}{I^2}\int_{\upsilon} \mathbf{H}.(\nabla \times \mathbf{A})\, d\upsilon$$

Standard vector identity is

$$\nabla.(\mathbf{A}\times\mathbf{H}) = \mathbf{H}.(\nabla\times\mathbf{A}) - \mathbf{A}.(\nabla\times\mathbf{H})$$

or,

$$\mathbf{H}.(\nabla\times\mathbf{A}) = \nabla.(\mathbf{A}\times\mathbf{H}) + \mathbf{A}.(\nabla\times\mathbf{H})$$

$$L = \frac{1}{I^2}\left[\int_{\upsilon} \nabla.(\mathbf{A}\times\mathbf{H})\, d\upsilon + \int_{\upsilon} \mathbf{A}.(\nabla\times\mathbf{H})\, d\upsilon\right]$$

By divergence theorem,

$$\int_{\upsilon} \nabla.(\mathbf{A}\times\mathbf{H})\, d\upsilon + \int_{s} (\mathbf{A}\times\mathbf{H}).d\mathbf{S}$$

and $\nabla\times\mathbf{H} = \mathbf{J}$

Now the expression for L becomes

$$L = \frac{1}{I^2}\left[\oint_{S} (\mathbf{A}\times\mathbf{H}).d\mathbf{S} + \int_{\upsilon} \mathbf{A}.\mathbf{J}\, d\upsilon\right]$$

But the surface integral is zero as the surface encloses the volume containing all the magnetic energy and this requires **A** and **H** to be zero on the boundary surface, that is, the first term on the right = 0.

$$L = \frac{1}{I^2}\oint_{S} \mathbf{A}.\mathbf{J}\, d\upsilon$$

But

$$\mathbf{A} = \int_{\upsilon} \frac{\mu\mathbf{J}}{4\pi R}\, d\upsilon$$

$$L = \frac{1}{I^2}\left[\int_{\upsilon}\left(\int_{\upsilon} \frac{\mu\mathbf{J}}{4\pi R}\, d\upsilon\right).\mathbf{J}\, d\upsilon\right]$$

As $\mathbf{J}dv$ is nothing but $I\,d\mathbf{l}$, L becomes

$$L = \frac{\mu}{4\pi}\left[\oint\left(\oint\frac{d\mathbf{l}}{R}\right).\,d\mathbf{l}\right]$$ Hence proved.

3.27 MUTUAL INDUCTANCE

When the current in an inductor changes, flux varies and it cuts any other inductor nearby, producing induced voltage in both inductors.

The coil L_1 of Fig. 3.21 is connected to a generator which produces varying current in the turns. In fact, the coil L_2 is not connected physically to L_1. But the turns are linked by the magnetic field. Hence a varying current in L_1 induces voltage across L_1 and also across L_2.

When the induced voltage produces a current in L_2, its varying magnetic field induces voltage in L_1. Then the two coils L_1 and L_2 are said to have mutual inductance as the current in one produces voltage in the other.

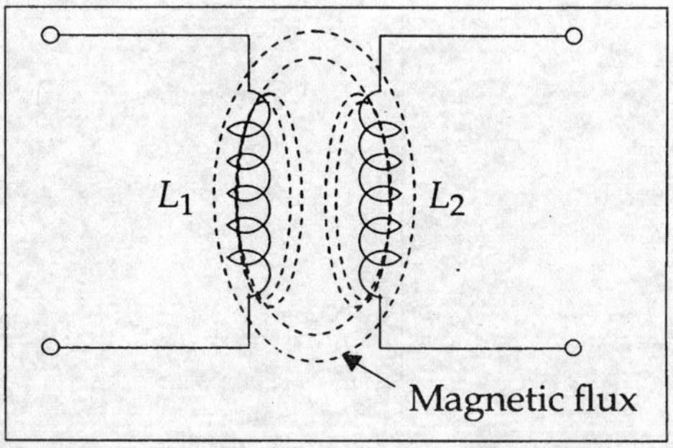

Fig. 3.21 Mutual inductance (M) between L_1 and L_2 due to flux linkage

Definition of Mutual Inductance

Mutual inductance, M between two coils is defined as the total flux linkage in one coil per unit change in current in the other coil. Mathematically, it is given by

$$M_{21} = \frac{N_2\,\phi_{21}}{I_1} \text{ or } M_{12} = \frac{N_1\,\phi_{12}}{I_2} \text{ Henry}$$

It may be noted that

$$M_{12} = M_{21}$$

Coefficient of Coupling

Coefficient of coupling between two coils is defined as the fraction of total flux from one coil linking another coil, that is,

$$k = \text{coefficient of coupling} \equiv \frac{\text{flux linkages between } L_1 \text{ and } L_2}{\text{flux produced by } L_1}$$

Calculation of Mutual Inductance, M

Mutual inductance increases with higher values of primary and secondary inductances and high coefficient of coupling.

$$M = k\sqrt{L_1 \times L_2}\text{, Henry}$$

3.28 COMPARISON BETWEEN ELECTRIC AND MAGNETIC FIELDS/CIRCUITS/PARAMETERS

S. no.	*Law/ parameter*	*Electric field/ electric circuit*	*Magnetic field/ magnetic circuit*
1.	Differential source	dQ	$Id\mathbf{L}$
2.	Source	Q	$I\mathbf{L}$
3.	Fundamental laws	Coulomb's law $\mathbf{F} = \frac{Q_1 Q_2}{4\pi\epsilon_0 r^2}\mathbf{a}_r$ Gauss's law $\int_S \mathbf{D}\,.\,d\mathbf{S} = Q_{enc}$	Biot-Savart law $d\mathbf{H} = \frac{Id\mathbf{L} \times \mathbf{a}_r}{4\pi r^2}$ Ampere's circuit law $\oint_L \mathbf{H}\,.\,d\mathbf{L} = I_{enc}$
4.	Force	Electric force $\mathbf{F}_E = Q\mathbf{E}$	Magnetic force $\mathbf{F}_H = Q(\mathbf{V} \times \mathbf{B})$
5.	Force property	Electric force does some work	Magnetic force does not do work
6.	Field	Electric field intensity, $\mathbf{E} = -\nabla V$	Magnetic field intensity $\mathbf{H} = -\nabla V_m$
7.	Unit of field	**E** is in volt/metre	**H** is in Ampere/metre
8.	Field direction	Electric field diverges from (+)ve charge	Magnetic field curls around current element
9.	Divergence of the field	$\nabla\,.\,\mathbf{E} = \rho_v/\epsilon_0$	$\nabla\,.\,\mathbf{H} = 0$
10.	Curl of the field	$\nabla \times \mathbf{E} = 0$	$\nabla \times \mathbf{H} = \mathbf{J}$
11.	Source property	Charges of opposite polarity attract each other Charges of same polarity repel each other	Current elements of opposite current direction repel each other Current elements of same current direction attract each other

S. no.	Law/ parameter	Electric field/ electric circuit	Magnetic field/ magnetic circuit
12.	Flux density	$\mathbf{D}=\frac{\Psi}{S}\mathbf{a}_n, \text{C/m}^2$	$\mathbf{B}=\frac{\phi}{S}\mathbf{a}_n, \text{wb/m}^2$
13.	Relation between field and flux density	$\mathbf{D}=\in\mathbf{E}$	$\mathbf{B}=\mu\mathbf{H}$
14.	Potentials	$V=-\int\mathbf{E}.d\mathbf{L}$ (volt) and $V=\frac{1}{4\pi\in}\int_{\upsilon}\frac{\rho_{\upsilon}d\upsilon}{r}$	$V_m=-\int\mathbf{H}.d\mathbf{L}\,(\mathbf{J}=0)$ $A=\frac{\mu}{4\pi}\int_{\upsilon}\frac{\mathbf{J}.d\upsilon}{r}$ (wb/m)
15.	Flux	$\Psi=\int\mathbf{D}.d\mathbf{S}$ $\Psi=Q=CV$	$\phi=\int\mathbf{B}.d\mathbf{S}$ $\phi=LI$
16.	Energy density	$W_E=\frac{1}{2}\in E^2$ $=\frac{1}{2}\mathbf{D}.\mathbf{E}$	$W_H=\frac{1}{2}\mu H^2$ $=\frac{1}{2}\mathbf{B}.\mathbf{H}$
17.	Current density	Conduction current density, $\mathbf{J}=\sigma\mathbf{E}$ Displacement current density, $\frac{\partial\mathbf{D}}{\partial t}$	Magnetic current density $\frac{\partial\mathbf{B}}{\partial t}$
18.	Units of current densities	$\mathbf{J}_c\to\text{A/m}^2$ $\mathbf{J}_d\to\text{A/m}^2$	$\mathbf{J}_m\to\text{V/m}^2$
19.	Response	emf (volt)	mmf (amp)
20.	Scalar potentials	$V=IR$ R = resistance $=\frac{l}{\sigma s}$	$V_m=\phi\Re$ $\Re$ = reluctance $=\frac{l}{\mu s}$
21.	Field property	$\oint\mathbf{E}.d\mathbf{L}=0$	$\oint\mathbf{H}.d\mathbf{L}=I_{enc}$
22.	Passive parameters	Conductivity, $\sigma=\frac{1}{\text{resistivity}}$ (mho/m)	Permeability, $\mu=\frac{1}{\text{reluctivity}}$ (H/A)
23.	Passive parameter	Conductance $=\frac{1}{\text{resistance}}$	Permeance $=\frac{1}{\text{reluctance}}$

S. no.	*Law/ parameter*	*Electric field/ electric circuit*	*Magnetic field/ magnetic circuit*
24.	Material property	Electric susceptibility, χ_e	Magnetic susceptibility, χ_m
25.	Field parameters	Polarisation, $\mathbf{P} = \chi_e \epsilon_0 \mathbf{E}$	Magnetisation, $\mathbf{M} = \chi_m \mathbf{H}$
26.	Gauss's law in integral form	$\oint \mathbf{D}.d\mathbf{S} = Q$	$\oint \mathbf{B}.d\mathbf{S} = 0$
27.	Passive parameters	Capacitance, $C = \left\lvert\frac{Q}{V}\right\rvert$	Inductance, $L = \frac{N\phi}{I}$
28.	Gauss's law in point form	$\nabla . \mathbf{D} = \rho_v$	$\nabla . \mathbf{B} = O$
29.	Boundary condition	$\mathbf{E}_{\tan 1} = \mathbf{E}_{\tan 2}$	$H_{\tan 1} - H_{\tan 2} = J_s$
30.	Boundary condition	$D_{n1} - D_{n2} = \rho_s$	$\mathbf{B}_{n1} = \mathbf{B}_{n2}$
31.	Circuit	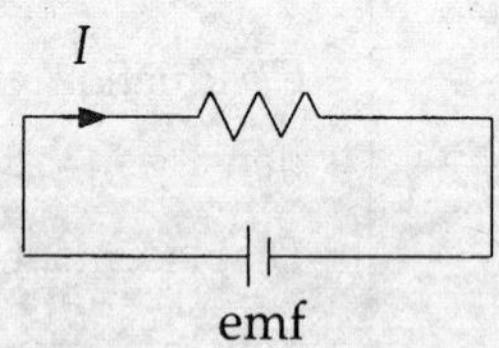	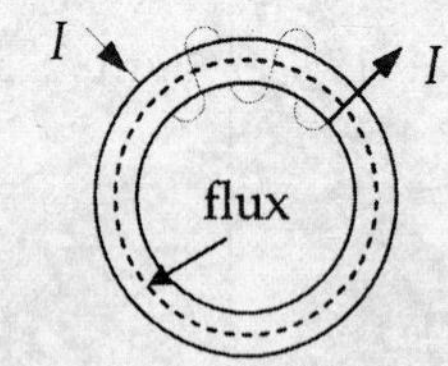

POINTS/FORMULAE TO REMEMBER

- Static magnetic fields are produced from current elements.
- There exists no isolated magnetic poles.
- Magnetic field lines are closed loops.
- Magnetic field and magnetic flux density are related by $\mathbf{B} = \mu\mathbf{H}$.
- Biot-Savart law is $d\mathbf{H} = \dfrac{I\,d\mathbf{L} \times \mathbf{a}_r}{4\pi r^2}$, A/m.
- Magnetic field due to a vertically unpolarised infinitely long current element is $\mathbf{H} = \dfrac{I}{2\pi\rho}\,\mathbf{a}_\phi$.
- Magnetic field due to a finite element is $\mathbf{H} = \dfrac{I}{4\pi R}(\cos\alpha_2 - \cos\alpha_1)\,a_\phi$.
- Ampere's circuit law in differential form is $\nabla \times \mathbf{H} = \mathbf{J}$.
- Ampere's circuit law in integral form is $\oint \mathbf{H}.d\mathbf{L} = I_{enc}$.
- Stoke's theorem is $\oint_c \mathbf{H}.d\mathbf{L} = \int_s (\nabla \times \mathbf{H}).d\mathbf{S}$.
- Force on a charge due to electric field is $\mathbf{F}_E = Q\mathbf{E}$.
- Force on a moving charge due to magnetic field is $\mathbf{F}_H = Q\,(\mathbf{V} \times \mathbf{B})$.
- Lorentz force equation is $\mathbf{F} = Q\,(\mathbf{E} + \mathbf{V} \times \mathbf{B})$.
- Force on a current element in a magnetic field is $\mathbf{F} = I\mathbf{L} \times \mathbf{B}$.
- Ampere's force law is $\mathbf{F} = \dfrac{\mu I_1 I_2}{4\pi} \oint\oint \dfrac{d\mathbf{L}_2 \times (d\mathbf{L}_1 \times \mathbf{a}_r)}{r^2}$.
- $H_{\tan 1} - H_{\tan 2} = J_s$
- $\mathbf{B}_{n1} = \mathbf{B}_{n2}$
- Scalar magnetic potential is $V_m = -\int_A^B \mathbf{H}.d\mathbf{L}$.

- Scalar magnetic potential has the unit of Ampere.
- Vector magnetic potential is $\mathbf{A} = \int_{\upsilon} \frac{\mu_0 \mathbf{J} d\upsilon}{4\pi R}$.
- **A** has the unit of wb/m.
- The torque on the loop is $\mathbf{T} = \mathbf{r} \times \mathbf{F}$.
- For diamagnetic materials, $\chi_m < 0$ and $\mu_r \leq 1.0$.
- For paramagnetic materials, $\chi_m > 0$, $\mu_r \geq 1$.
- For ferromagnetic materials, $\chi_m >> 0$, $\mu_r >> 1$.
- Examples of diamagnetic materials are copper, lead, diamond, silicon, bismuth and so on.
- Examples of paramagnetic materials are air, tungsten, potassium, platinum and so on.
- Examples of ferromagnetic materials are iron, nickel, cobalt and their alloys.
- Magnetic dipole moment is $\mathbf{m} = I_b\, d\mathbf{S}$, Amp-m^2.
- Magnetisation is $\mathbf{M} = \chi_m \mathbf{H}$, A/m^2.
- Magnetic susceptibility is $\chi_m = \mu_r - 1$.
- Magnetic flux density in magnetic materials is $\mathbf{B} = \mu_0 [\mathbf{H} + \mathbf{M}]$.
- Self inductance, L is defined as $L = \frac{N\phi}{I}$, Henry.
- Inductance of a Toroid is $L = \frac{\mu_0 N^2 S}{2\pi r}$.
- Magnetic field in a Toroid is $H = \frac{NI}{2\pi r}$.
- Quality factor of a coil is $Q = \frac{X_L}{R}$.
- Inductance of a solenoid is $L = \frac{\mu_0 N^2 S}{l}$.
- Magnetic field in a solenoid is $H = \frac{NI}{l}$.

- Inductance of a coaxial cable is $L=\dfrac{\mu_0 l}{2\pi}\ln\dfrac{b}{a}$.
- The magnetic field in a coaxial cable is $H=\dfrac{Ir}{2\pi a^2}$.
- The inductance of parallel wires is $L=\dfrac{\mu_0 l}{\pi}\cosh^{-1}\dfrac{d}{2a}$.
- Energy stored in a static magnetic field is $W_H=\dfrac{1}{2}\mu H^2$.
- Energy stored in an inductor is $W_L=\dfrac{1}{2}LI^2$.
- Inductance in terms of fundamental parameters is $L=\dfrac{\mu}{4\pi}\left[\oiint\dfrac{d\mathbf{l}}{R}\right].d\mathbf{l}$.
- Mutual inductance between two coils is

$$M_{12}=\frac{N_1\phi_{12}}{I_2}$$

$$M_{21}=\frac{N_2\phi_{21}}{I_1}$$

$$M_{12}=M_{21}$$

- Coefficient of coupling is $k=\dfrac{\text{flux linkages between } L_1 \text{ and } L_2}{\text{flux produced by } L_1}$.
- Mutual inductance is calculated from $M=k\sqrt{L_1\times L_2}$, Henry.

OBJECTIVE QUESTIONS

1. Static magnetic fields are produced from charges at rest. (Yes/No)
2. South and north poles of a magnet can be isolated. (Yes/No)
3. Magnetic field is not conservative. (Yes/No)
4. Electricity produces magnetism. (Yes/No)
5. Magnets produce electricity. (Yes/No)
6. A steady current, **J** flowing in a wire generates a magnetic induction field, **B**. (Yes/No)
7. A magnetic induction field, **B** does not produce a current, **J**. (Yes/No)
8. Magnetic flux is conservative. (Yes/No)
9. Magnetic charge means magnetic pole. (Yes/No)
10. Magnetic flux density is a vector. (Yes/No)
11. Scalar magnetic potential exists where **J** is present. (Yes/No)
12. Torque is a vector. (Yes/No)
13. Inductance depends on current and flux. (Yes/No)
14. The value of inductance depends on physical parameters only. (Yes/No)
15. $\int_{\text{surface}} (\nabla \times \mathbf{H}).d\mathbf{S}$ depends on the boundary line only. (Yes/No)
16. $\int_{\text{surface}} (\nabla \times \mathbf{H}).d\mathbf{S}$ does not depend on the particular surface. (Yes/No)
17. $\mathbf{H} = \nabla V_m$ (Yes/No)
18. $\nabla^2 V_m = \rho_\upsilon$ (Yes/No)
19. $\mathbf{A} = \nabla \times \mathbf{B}$ (Yes/No)
20. $\mathbf{H} = \nabla \times \mathbf{A}$ (Yes/No)
21. $\mathbf{B} = \nabla \times \mathbf{A}$ (Yes/No)
22. $M_{12} = M_{21}$ (Yes/No)
23. Inductance of a solenoid is proportional to N^2. (Yes/No)

24. Inductance of a solenoid is proportional to the area of cross-section. (Yes/No)

25. Inductance coil should have very low temperature coefficient. (Yes/No)

26. $\nabla \times \mathbf{M} = \mathbf{J}_b$ (Yes/No)

27. Magnetic field is measured by ________.

28. SQUID means ________.

29. The unit of magnetic flux is ________.

30. The unit of magnetic flux density is ________.

31. Ampere's circuit law is ________.

32. Biot-Savart law is ________.

33. Differential form of Ampere's circuit law is ________.

34. Lorentz force equation is ________.

35. The boundary condition on **H** is ________.

36. The boundary condition on **B** is ________.

37. The unit of scalar magnetic potential is ________.

38. The unit of vector magnetic potential is ________.

39. Magnetic dipole moment has the unit of ________.

40. Magnetisation has the unit of ________.

41. Energy stored in an inductor is ________.

42. Energy stored in a magnetostatic field is ________.

43. Torque is ________.

44. The unit of torque is ________.

45. Magnetisation is ________.

46. Bound current is called ________.

47. Quality factor of the coil is ________.

48. Two specifications of an inductance are ________.

49. Magnetic field in a solenoid is ________.

50. Magnetic field in a Toroid is ________.

Answers

1. No 2. No 3. Yes 4. Yes 5. Yes
6. Yes 7. Yes 8. Yes 9. Yes 10. Yes
11. No 12. Yes 13. No 14. Yes 15. Yes
16. Yes 17. No 18. No 19. No 20. No
21. Yes 22. Yes 23. Yes 24. Yes 25. Yes
26. Yes 27. SQUID
28. Super Conducting Quantum Interference device 29. Weber
30. Wb/m^2 31. $\oint \mathbf{H}.d\mathbf{L} = I_{enc}$ 32. $d\mathbf{H} = \frac{I\,d\mathbf{L} \times \mathbf{a}_r}{4\pi r^2}$ 33. $\nabla \times \mathbf{H} = \mathbf{J}$
34. $\mathbf{F} = Q\,[\mathbf{E} + \mathbf{V} \times \mathbf{B}]$ 35. $H_{\tan 1} - H_{\tan 2} = J_s$ 36. $B_{n1} = B_{n2}$
37. Ampere 38. Wb/m 39. $Amp\text{-}m^2$ 40. A/m 41. $\frac{1}{2}LI^2$
42. $\frac{1}{2}\mu H^2$ 43. $\mathbf{r} \times \mathbf{F}$ 44. Newton-metre 45. $\chi_m \mathbf{H}$
46. Amperion current 47. $\frac{X_L}{R}$
48. Inductance value, Q of the coil 49. $\frac{NI}{l}$ 50. $\frac{NI}{2\pi r}$

MULTIPLE CHOICE QUESTIONS

1. If flux density is 10 wb/m^2 and the area of the coil is 2 m^2, the flux is

(a) 10 wb (b) 20 wb (c) 5 wb (d) 40 wb

2. If the magnetic field, $\mathbf{H} = 4\mathbf{a}_x$, A/m, flux density in free space is

(a) $1.6\pi \mathbf{a}_x \,\mu\, wb/m^2$ (b) $16\pi \mathbf{a}_x \,\mu\, wb/m^2$

(c) $1.6\pi \,\mu\, wb/m$ (d) $160\pi \mathbf{a}_x$ wb/m

3. If the curl of the magnetic field is $2.0\mathbf{a}_x$ A/m^2, the current density is

(a) $2.0\mathbf{a}_x$ A/m^2 (b) $1.0\mathbf{a}_x$ A/m^2

(c) 2.0 A/m (d) 1.0 A/m

4. $\oint(\nabla \times \mathbf{H}).d\mathbf{S}$ is

(a) zero (b) I_{enc} (c) J (d) $\oint \mathbf{H}.d\mathbf{S}$

5. If a charge of 2.0 C is placed in an electric field of 2.0 V/m, the force on the charge is

(a) 4.0 N (b) 1.0 N (c) 2.0 N (d) zero

6. If a charge, 1.0 C is moving with a velocity $2.0\mathbf{a}_x$ in a magnetic field of $\mathbf{B} = 1.0\mathbf{a}_y$, the force on the charge is

(a) $2.0\mathbf{a}_z$ (b) $1.0\mathbf{a}_y$ (c) $2.0\mathbf{a}_x$ (d) $1.0\mathbf{a}_z$

7. The force produced by $B = 2.0\, wb/m^2$ on a current element of 2.0 A-m, is

(a) 4.0 N (b) 1.0 N (c) 2.0 N (d) 0.5 N

8. The magnetic field in an ideal conductor is

(a) zero (b) infinite

(c) finite (d) the same as its outside field

9. If the normal component of **B** in medium 1 is $1.0\mathbf{a}_x$ wb/m^2, the normal component in medium 2 is

(a) $0.5\mathbf{a}_x$ wb/m^2 (b) $1.0\mathbf{a}_x$ wb/m^2

(c) $2.0\mathbf{a}_x$ wb/m^2 (d) 1.0 wb/m^2

10. The unit of scalar magnetic potential is

(a) Ampere (b) Volt

(c) Amp/m (d) Volt/m

11. $\nabla \times \nabla V_m$ is

(a) zero (b) $\nabla^2 V_m$ (c) J (d) $\nabla . \nabla V_m$

12. $\nabla \times \mathbf{A}$ is

(a) **H** (b) **B** (c) **J** (d) 0

13. Torque has the unit of

(a) N-m (b) N/m (c) N-m^2 (d) N

14. Magnetisation, **M** is defined as

(a) $\chi_m \mathbf{H}$ (b) $\chi_m \mu_0 \mathbf{H}$ (c) $\chi_m \mathbf{B}$ (d) $\mathbf{B}/\mu_0$

15. The unit of magnetic dipole moment is

(a) A-m (b) A-m^2 (c) A/m (d) C-m

16. The dipole moment of a magnet is

(a) $Q_m \mathbf{d}$ (b) $Q_m \mathbf{d}$ (c) $Q_m I$ (d) $Q_m \mathbf{S}$

17. If $\mu = 1.0 \mu\text{H/m}$ for a medium, H = 2.0 A/m, the energy stored in the field is

(a) 0.5J/m^3 (b) $1.0 \mu\text{J/m}^3$

(c) $2.0 \mu\text{J/m}^3$ (d) 1.0J/m^3

18. If a current of 1.0 amp flowing in an inductor, L = 2 Henry, the energy stored in an inductance is

(a) 2.0 J (b) 1.0 J (c) 2.0 J/m^3 (d) 0.5 J

19. $\oint \mathbf{B} . d\mathbf{S}$ is

(a) zero (b) Q (c) H (d) J

20. The unit of magnetic susceptibility is

(a) Nil (b) Amp (c) H/m (d) Wb

Answers

1. (b)	2. (a)	3. (a)	4. (a)	5. (a)
6. (a)	7. (a)	8. (a)	9. (b)	10. (a)
11. (a)	12. (b)	13. (a)	14. (a)	15. (b)
16. (a)	17. (c)	18. (b)	19. (a)	20. (a)

EXERCISE PROBLEMS

1. For a circular current element of radius, b = 2 m, determine the magnetic field **H** on the axis of the loop if I = 2A, h = 2 m. Also find **H** at the centre of the loop.

2. Three thin conductors along x, y and z-axes carry a current of 1.0 mA. Determine the magnetic field, **H** at the point (2, 3, 4).

3. What is the current that produces a magnetic field inside a conductor of circular cross-section given by

$$\mathbf{H} = \frac{1}{r}\left[\frac{1}{K^2}\sin Kr - \frac{r}{K_1}\cos Kr\right]\mathbf{a}_\phi,\ \text{A/m}$$

The radius of the conductor is 0.01 m. Assume $K = 50.0\pi\,(\text{m}^{-1})$.

4. If the separation between two infinitely long current elements is 5 m and the elements carry 2 Ampere in opposite direction, find the magnetic field at 1 m from one current element.

5. The radius of a circular coil is 1.0 cm. The plane of the coil is perpendicular to a magnetic flux density of 10 mwb/m^2. Determine the total flux threading the coil.

6. A charge of 2.0 C moving with a velocity of $\mathbf{V} = (\mathbf{a}_x + \mathbf{a}_y)$ m/s experiences no force in electric and magnetic fields. If the magnetic field intensity is

$$\frac{1}{\mu_0}[2\mathbf{a}_x + 2\mathbf{a}_y + \mathbf{a}_z],\ \text{A/m},$$

find the electric field.

7. If the magnetic field is $\mathbf{H} = \dfrac{0.01}{\mu_0}\mathbf{a}_x$, A/m, what is the force on a charge of 1.0 pc moving with a velocity of $10^6\,\mathbf{a}_x$ m/s?

8. A uniform magnetic flux density, **B** = 1.0 wb/m^2 is present in a medium whose $\mu = 100\mu_0$, as shown in the Fig. 3.22.

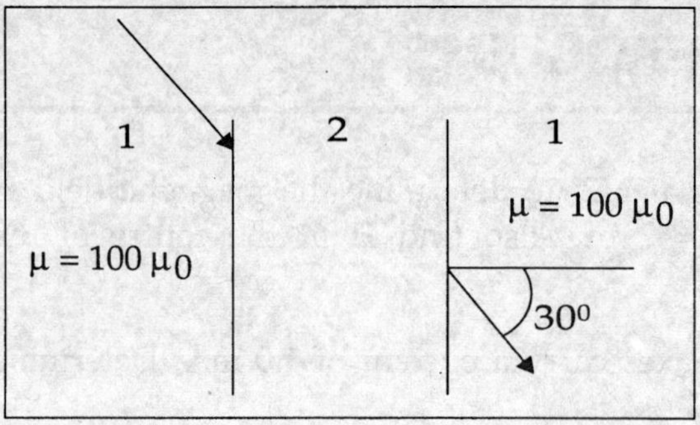

Fig. 3.22

If the air gap is cut as in the figure, determine **B** in the air gap.

9. What is the inductance of a coaxial cable of length 10 m if $r_1 = 2\,\text{mm}$, $r_2 = 6\,\text{mm}$? Assume $\mu_r = l$.

10. What is the inductance of a pair of transmission lines separated by 1.868 m, if the diameter of each wire is 0.01 m and the medium between the lines has $\mu = 2\mu_0$? Length of the line is 10 m.

CHAPTER

4

MAXWELL'S EQUATIONS

Maxwell's equations are very popular and they are known as Electromagnetic Field Equations.

The main aim of this chapter is to provide sufficient background and concepts on Maxwell's equations. They include:

- Maxwell's equations for static and time varying fields in free space and conductive media in differential and integral form
- meaning and proof of Maxwell's equations
- comprehensive boundary conditions in scalar and vector form
- time varying and retarded potentials
- Helmholtz theorem and Lorentz gauge condition
- points/formulae to remember, objective and multiple choice questions and exercise problems.

Do you know?

In 1884, Oliver Heaviside and Willard Gibbs used vector calculus to reformulate Maxwell's original (1865) system of equations to a simpler representation.

4.1 INTRODUCTION

Static electric field has applications in cathode ray oscilloscopes for deflecting charged particles, in ink-jet printers to increase the speed of printing and print quality, in sorting of materials in mining industry and in the development of electrostatic voltmeters.

Static magnetic field has applications in magnetic separators to separate magnetic materials from non-magnetic materials, in cyclotrons for imparting high energy to charged particles, in velocity selector and mass separators. It is also used in magnetohydrodynamic generators.

On the other hand, electromagnetic fields in their time varying form constitute electromagnetic waves. These fields/waves are useful in all communication and radar systems. In fact, the EM waves are carriers of information, mainly in free space, between the transmitter and the receiver. The electric and magnetic fields of the EM waves are related through Maxwell's equations.

Most of the problems related to antennas can be solved with the help of Maxwell's equations and boundary conditions. In view of this, in this chapter, Maxwell's equations are presented in detail. Maxwell's equations are very popular and important and hence they are also referred to as electromagnetic field equations. The fields **E**, **H**, **D** and **B** in static form are represented in Cartesian coordinates as

$$\mathbf{E}(x, y, z)$$

$$\mathbf{H}(x, y, z)$$

$$\mathbf{D}(x, y, z)$$

and $$\mathbf{B}(x, y, z)$$

In cylindrical coordinates, they are:

$$\mathbf{E}(\rho, \phi, z)$$

$$\mathbf{H}(\rho, \phi, z)$$

$$\mathbf{D}(\rho, \phi, z)$$

and $$\mathbf{B}(\rho, \phi, z)$$

and in spherical coordinates, they are:

$$\mathbf{E}(r, \theta, \phi)$$

$$\mathbf{H}(r, \theta, \phi)$$

$$\mathbf{D}(r, \theta, \phi)$$

and $$\mathbf{B}(r, \theta, \phi)$$

The time varying fields **E** (t, x, y, z), **H** (t, x, y, z) constitute electromagnetic waves which have wide applications in all communications, radars and also in bio-medical engineering.

In these applications, time varying fields are of more practical value than static electric and magnetic fields.

It may be noted that electrostatic fields are usually produced by static charges. Magnetostatic fields are produced from the motion of electric charges with uniform velocity (direct current) or static magnetic charges namely magnetic dipoles.

On the other hand, time varying fields constituting EM waves are produced by time varying currents, that is, any pulsating current produces radiation fields which are nothing but time varying fields. In brief, note that:

1. Static charges produce electrostatic fields.
2. Steady currents (DC currents) produce magnetostatic fields

   and

   static magnetic charges (magnetic dipoles) also produce magnetostatic fields.
3. Time varying currents produce EM waves or EM fields.

4.2 EQUATION OF CONTINUITY FOR TIME VARYING FIELDS

Equation of continuity in point form is

$$\nabla . \mathbf{J} = -\dot{\rho}_v$$

where

$$\mathbf{J} = \text{conduction current density} (\text{A/m}^2)$$

$$\rho_v = \text{volume charge density} (\text{c/m}^3),\ \dot{\rho}_v = \frac{\partial \rho_v}{\partial t}$$

$$\nabla = \text{vector differential operator} \left(\frac{1}{m}\right)$$

$$= \mathbf{a}_x \frac{\partial}{\partial x} + \mathbf{a}_y \frac{\partial}{\partial y} + \mathbf{a}_z \frac{\partial}{\partial z}$$

Proof Consider a closed surface enclosing a charge Q. There exists an outward flow of current given by

$$I = \oint_S \mathbf{J}.d\mathbf{S}$$

This is equation of continuity in **integral form**.

Here, I is the current flowing away through a closed surface, $d\mathbf{S}$ is the differential area on the surface whose direction is always outward normal to the surface. As there is outward flow of current, there will be a rate of decrease of charge given by $\frac{-dQ}{dt}$, where Q is the enclosed charge. From the principle of conservation of charge, we have

$$I = \oint_S \mathbf{J}.d\mathbf{S} = \frac{-dQ}{dt}$$

From divergence theorem, we have

$$\oint_S \mathbf{J}.d\mathbf{S} = \int_\upsilon (\nabla.\mathbf{J})\, d\upsilon$$

Thus, $$\int_\upsilon (\nabla.\mathbf{J})\, d\upsilon = \frac{-dQ}{dt}$$

By definition, $$Q = \int_\upsilon \rho_\upsilon\, d\upsilon$$

where $$\rho_\upsilon = \text{volume charge density } (\text{C/m}^3)$$

So, $$\int_\upsilon (\nabla.\mathbf{J})\, d\upsilon = \int \frac{-\partial\rho_\upsilon}{\partial t}\, d\upsilon$$

$$= \int -\dot{\rho}_\upsilon\, d\upsilon$$

where $$\dot{\rho}_\upsilon = \frac{\partial\rho_\upsilon}{\partial t}$$

Two volume integrals are equal only if their integrands are equal.

Thus, $$\boxed{\nabla.\mathbf{J} = -\dot{\rho}_\upsilon}$$ Hence proved.

4.3 MAXWELL'S EQUATIONS FOR TIME VARYING FIELDS

These are basically four in number.

Maxwell's equations in **differential form** are given by

$$\nabla\times\mathbf{H} = \dot{\mathbf{D}} + \mathbf{J}$$

$$\nabla \times \mathbf{E} = -\dot{\mathbf{B}}$$
$$\nabla . \mathbf{D} = \rho_{\upsilon}$$
$$\nabla . \mathbf{B} = 0$$

Here $\mathbf{H}$ = magnetic field strength (A/m)

$\mathbf{D}$ = electric flux density, (C/m^2)

$\dot{\mathbf{D}} = \dfrac{\partial \mathbf{D}}{\partial t}$ = displacement electric current density (A/m^2)

$\mathbf{J}$ = conduction current density (A/m^2)

$\mathbf{E}$ = electric field (V/m)

$\mathbf{B}$ = magnetic flux density wb/m^2 or Tesla

$\dot{\mathbf{B}} = \dfrac{\partial \mathbf{B}}{\partial t}$ = time-derivative of magnetic flux density (wb/m^2-sec)

$\dot{\mathbf{B}}$ is called magnetic current density (V/m^2) or Tesla/sec

ρ_{υ} = volume charge density (C/m^3)

Maxwell's equations for time varying fields in **integral form** are given by

$$\oint_L \mathbf{H}.d\mathbf{L} = \int_S (\dot{\mathbf{D}} + \mathbf{J}).d\mathbf{S}$$

$$\oint_L \mathbf{E}.d\mathbf{L} = -\int_S \dot{\mathbf{B}}.d\mathbf{S}$$

$$\oint_S \mathbf{D}.d\mathbf{S} = \int_{\upsilon} \rho_{\upsilon}\, d\upsilon$$

$$\oint_S \mathbf{B}.d\mathbf{S} = 0$$

Here, $d\mathbf{L}$ is the differential length and $d\mathbf{S}$ is the differential area whose direction is always outward normal to the surface.

4.4 MEANING OF MAXWELL'S EQUATIONS

It is easy to understand the meaning of Maxwell's equations from their integral form.

1. The first Maxwell's equation states that the magnetomotive force around a closed path is equal to the sum of electric displacement and conduction currents through any surface bounded by the path.

2. The second law states that the electromotive force around a closed path is equal to the minus of the time derivative of the magnetic flux flowing through any surface bounded by the path

or

it can also be stated that the electromotive force around a closed path is equal to the inflow of magnetic current through any surface bounded by the path.

3. The third law states that the total electric displacement flux passing through a closed surface (Gaussian surface) is equal to the total charge inside the surface.
4. The fourth law states that the total magnetic flux passing through any closed surface is zero.

4.5 CONVERSION OF DIFFERENTIAL FORM OF MAXWELL'S EQUATION TO INTEGRAL FORM

1. Consider the first Maxwell's equation

$$\nabla \times \mathbf{H} = \dot{\mathbf{D}} + \mathbf{J}$$

Take surface integral on both sides.

$$\int_s (\nabla \times \mathbf{H}) . d\mathbf{S} = \int_s (\dot{\mathbf{D}} + \mathbf{J}) . d\mathbf{S}$$

Applying Stoke's theorem to LHS, we can write

$$\int_s (\nabla \times \mathbf{H}) . d\mathbf{S} = \oint_L \mathbf{H} . d\mathbf{L}$$

Hence $$\oint_L \mathbf{H} . d\mathbf{L} = \int_s (\mathbf{D} + \mathbf{J}) . d\mathbf{S} \quad \text{first law}$$

2. Consider the second Maxwell's equation

$$\nabla \times \mathbf{E} = -\dot{\mathbf{B}}$$

Take surface integral on both sides.

$$\int_s (\nabla \times \mathbf{E}) . d\mathbf{S} = -\int_s \dot{\mathbf{B}} . d\mathbf{S}$$

Applying Stoke's theorem to LHS, we get

$$\int_s (\nabla \times \mathbf{E}) . d\mathbf{S} = \oint_L \mathbf{E} . d\mathbf{L}$$

Therefore $$\oint_L \mathbf{E}.d\mathbf{L} = -\int_s \dot{\mathbf{B}}.d\mathbf{S} \text{ second law}$$

3. Consider the third Maxwell's equation

$$\nabla.\mathbf{D} = \rho_\upsilon$$

Take volume integral on both sides.

$$\int_\upsilon (\nabla.\mathbf{J})\, d\upsilon = \int_\upsilon \rho_\upsilon\, d\upsilon$$

Applying divergence theorem to LHS, we get

$$\int_\upsilon (\nabla.\mathbf{J})\, d\upsilon = \oint_s \mathbf{D}.d\mathbf{S}$$

Therefore $$\oint_s \mathbf{D}.d\mathbf{S} = \int_\upsilon \rho_\upsilon\, d\upsilon \text{ third law}$$

4. Consider the fourth Maxwell's equation

$$\nabla.\mathbf{B} = 0$$

Take volume integral on both sides.

$$\int_\upsilon \nabla.\mathbf{B} d\upsilon = 0$$

Applying divergence theorem to LHS, we get

$$\int_\upsilon \nabla.\mathbf{B} d\upsilon = \oint_s \mathbf{B}.d\mathbf{S}$$

Therefore, $$\oint_s \mathbf{B}.d\mathbf{S} = 0 \text{ fourth law}$$

4.6 MAXWELL'S EQUATIONS FOR STATIC FIELDS

Maxwell's equations for static fields are:

$$\nabla\times\mathbf{H} = \mathbf{J} \leftrightarrow \oint_L \mathbf{H}.d\mathbf{L} = \int_s \mathbf{J}.d\mathbf{S}$$

$$\nabla\times\mathbf{E} = 0 \leftrightarrow \oint_L \mathbf{E}.d\mathbf{L} = 0$$

$$\nabla . \mathbf{D} = \rho_v \leftrightarrow \oint_S \mathbf{D}.d\mathbf{S} = \int_v \rho_v \, dv$$

$$\nabla . \mathbf{B} = 0 \leftrightarrow \oint_S \mathbf{B}.d\mathbf{S} = 0$$

As the fields are static, all the field terms which have time derivatives are zero, that is, $\dot{\mathbf{B}} = 0$, $\dot{\mathbf{D}} = 0$.

4.7 CHARACTERISTICS OF FREE SPACE

Free space is characterised by the following parameters:

Relative permittivity, $\epsilon_r = 1$

Relative permeability, $\mu_r = 1$

Conductivity, $\sigma = 0$

Conduction current density, $\mathbf{J} = 0$

Volume charge density, $\rho_v = 0$

Intrinsic impedance or characteristic impedance

$$\eta = 120\pi \text{ or } 377\Omega$$

4.8 MAXWELL'S EQUATIONS FOR FREE SPACE

$$\nabla \times \mathbf{H} = \dot{\mathbf{D}} \leftrightarrow \oint_L \mathbf{H}.d\mathbf{L} = \int_s \dot{\mathbf{D}}.d\mathbf{S}$$

$$\nabla \times \mathbf{E} = -\dot{\mathbf{B}} \leftrightarrow \oint_L \mathbf{E}.d\mathbf{L} = -\int_s \dot{\mathbf{B}}.d\mathbf{S}$$

$$\nabla . \mathbf{D} = 0 \leftrightarrow \oint_s \mathbf{D}.d\mathbf{S} = 0$$

$$\nabla . \mathbf{B} = 0 \leftrightarrow \oint_s \mathbf{B}.d\mathbf{S} = 0$$

4.9 MAXWELL'S EQUATIONS FOR STATIC FIELDS IN FREE SPACE

$$\nabla \times \mathbf{H} = 0 \leftrightarrow \oint_L \mathbf{H}.d\mathbf{L} = 0$$

$$\nabla \times \mathbf{E} = 0 \leftrightarrow \oint_L \mathbf{E}.d\mathbf{L} = 0$$

$$\nabla . \mathbf{D} = 0 \leftrightarrow \oint_s \mathbf{D}.d\mathbf{S} = 0$$

$$\nabla . \mathbf{B} = 0 \leftrightarrow \oint_s \mathbf{B}.d\mathbf{S} = 0$$

4.10 PROOF OF MAXWELL'S EQUATIONS

1. From Ampere's circuital law, we have

$$\nabla \times \mathbf{H} = \mathbf{J}$$

Take dot product on both sides

$$\nabla . \nabla \times \mathbf{H} = \nabla . \mathbf{J}$$

As the divergence of curl of a vector is zero,

$$\text{RHS} = \nabla . \mathbf{J} = 0$$

But the equation of continuity in point form is

$$\nabla . \mathbf{J} = \frac{-\partial \rho_\upsilon}{\partial t} = -\dot{\rho}_\upsilon$$

This means that if $\nabla \times \mathbf{H} = \mathbf{J}$ is true, it is resulting in $\nabla . \mathbf{J} = 0$.

As the equation of continuity is more fundamental, Ampere's circuital law should be modified. Hence we can write

$$\nabla \times \mathbf{H} = \mathbf{J} + \mathbf{F}$$

Take dot product on both sides

$$\nabla . \nabla \times \mathbf{H} = \nabla . \mathbf{J} + \nabla . \mathbf{F}$$

that is, $$\nabla . \nabla \times \mathbf{H} = 0 = \nabla . \mathbf{J} + \nabla . \mathbf{F}$$

Substituting the value of $\nabla . \mathbf{J}$ from the equation of continuity in the above expression, we get

$$\nabla . \mathbf{F} + (-\dot{\rho}_\upsilon) = 0$$

or, $$\nabla . \mathbf{F} = \dot{\rho}_\upsilon$$

The point form of Gauss's law is

$$\nabla . \mathbf{D} = \rho_\upsilon$$

or, $$\nabla . \dot{\mathbf{D}} = \dot{\rho}_\upsilon$$

From the above expressions, we get

$$\nabla . \mathbf{F} = \nabla . \dot{\mathbf{D}}$$

The divergence of two vectors are equal only if the vectors are identical,

that is, $\mathbf{F} = \dot{\mathbf{D}}$

So, $\boxed{\nabla \times \mathbf{H} = \dot{\mathbf{D}} + \mathbf{J}}$ Hence proved.

2. According to Faraday's law,

$$\text{emf} = \frac{-d\phi}{dt}$$

$$\phi = \text{magnetic flux, (wb)}$$

and by definition,

$$\text{emf} = \oint_L \mathbf{E}.d\mathbf{L}$$

$$\oint_L \mathbf{E}.d\mathbf{L} = \frac{-d\phi}{dt}$$

But
$$\phi = \int_s \mathbf{B}.d\mathbf{S}$$

$$\oint_L \mathbf{E}.d\mathbf{L} = -\int_s \frac{\partial \mathbf{B}}{\partial t} \cdot d\mathbf{S}$$

$$= -\int_s \dot{\mathbf{B}}.d\mathbf{S}, \; \dot{\mathbf{B}} = \frac{\partial \mathbf{B}}{\partial t}$$

Applying Stoke's theorem to LHS, we get

$$\int_L \mathbf{E}.d\mathbf{L} = \int_s (\nabla \times \mathbf{E}).d\mathbf{S}$$

$$\int_s (\nabla \times \mathbf{E}).d\mathbf{S} = \int_s -\dot{\mathbf{B}}.d\mathbf{S}$$

Two surface integrals are equal only if their integrands are equal,

that is, $\boxed{\nabla \times \mathbf{E} = -\dot{\mathbf{B}}}$ Hence proved.

3. From Gauss's law in electric field, we have

$$\oint_s \mathbf{D}.d\mathbf{S} = Q = \int_\upsilon \rho_\upsilon \, d\upsilon$$

Applying divergence theorem to LHS, we get

$$\oint_s \mathbf{D}.d\mathbf{S} = \int_\upsilon (\nabla.\mathbf{D})\, d\upsilon = \int_\upsilon \rho_\upsilon\, d\upsilon$$

Two volume integrals are equal if their integrands are equal,

that is, $\nabla.\mathbf{D} = \rho_\upsilon$

Hence proved.

4. We have Gauss's law for magnetic fields as

$$\oint_s \mathbf{B}.d\mathbf{S} = 0$$

RHS is zero as there are no isolated magnetic charges and the magnetic flux lines are closed loops.

Applying divergence theorem to LHS, we get

$$\int_\upsilon \nabla.\mathbf{B}\, d\upsilon = 0$$

or, $\nabla.\mathbf{B} = 0$ Hence proved.

4.11 SINUSOIDAL TIME VARYING FIELD

In practice, electric and magnetic fields vary sinusoidally. It is well-known that any periodic variation can be described in terms of sinusoidal variations with fundamental and harmonic frequencies.

The fields can be represented by

$$\tilde{\mathbf{E}} = E_m \cos \omega t$$

or,

$$\tilde{\mathbf{E}} = E_m \sin \omega t$$

where $\omega = 2\pi f$, $f =$ frequency variation of the field, E_m is the maximum field strength.

It means that a sinusoidal time factor is attached to the field. It is also possible to represent the fields using phasor notation. The time varying field $\tilde{\mathbf{E}}\,(r, t)$ is related to phasor field $E\,(r)$ as

$$\tilde{\mathbf{E}}\,(r, t) = \text{Re}\,\{\mathbf{E}(r)\, e^{j\omega t}\}$$

or,

$$\tilde{\mathbf{E}}\,(r, t) = \text{Im}\,\{\mathbf{E}\,(r)\, e^{j\omega t}\}$$

It may be noted that cosinusoidal variation is also considered to be sinusoidal in usage.

4.12 MAXWELL'S EQUATIONS IN PHASOR FORM

1. Consider the first Maxwell's equation

$$\nabla \times \tilde{\mathbf{H}} = \frac{\partial \tilde{\mathbf{D}}}{\partial t} + \tilde{\mathbf{J}}$$

If

$$\tilde{\mathbf{H}} = \text{Re}\,\{\mathbf{H}e^{j\omega t}\},$$

$$\tilde{\mathbf{D}} = \text{Re}\,\{\mathbf{D}e^{j\omega t}\},$$

$$\tilde{\mathbf{J}} = \text{Re}\,\{\mathbf{J}e^{j\omega t}\},$$

it becomes, $\nabla \times \text{Re}\,\{\mathbf{H}e^{j\omega t}\} = \frac{\partial}{\partial t}\text{Re}\,\{\mathbf{D}e^{j\omega t}\} + \text{Re}\,\{\mathbf{J}e^{j\omega t}\}$

or, $\text{Re}\,\{\nabla \times \mathbf{H} - j\omega\mathbf{D} - \mathbf{J})\,e^{j\omega t}\} = 0$

or,

$$\boxed{\nabla \times \mathbf{H} = j\omega\mathbf{D} + \mathbf{J}}$$

This is the first Maxwell's equation in phasor form.

2. Consider the second Maxwell's equation

$$\nabla \times \tilde{\mathbf{E}} = -\frac{\partial \tilde{\mathbf{B}}}{\partial t}$$

If

$$\tilde{\mathbf{E}} = \text{Re}\,\{\mathbf{E}e^{j\omega t}\}$$

and

$$\tilde{\mathbf{B}} = \text{Re}\,\{\mathbf{B}e^{j\omega t}\}$$

the second Maxwell's equation becomes,

$$\nabla \times \text{Re}\,\{\mathbf{E}e^{j\omega t}\} = -\frac{\partial}{\partial t}\text{Re}\,(\mathbf{B}e^{j\omega t})$$

or, $\text{Re}\,[(\nabla \times \mathbf{E} + j\omega\mathbf{B})\,e^{j\omega t}] = 0$

or,

$$\boxed{\nabla \times \mathbf{E} = -j\omega\mathbf{B}}$$

This is the second Maxwell's equation in phasor form.

3. Similarly, consider the third Maxwell's equation,

$$\nabla.\tilde{\mathbf{D}} = \rho_v$$

that is, $\nabla.\text{Re}\,(\mathbf{D}e^{j\omega t}) = \rho_v$

or,

$$\boxed{\nabla.\mathbf{D} = \rho_v}$$

4. Consider the fourth Maxwell's equation

$$\nabla.\tilde{\mathbf{B}} = 0$$

that is, $\nabla . \text{Re}\,(\mathbf{B}e^{j\omega t}) = 0$

or, $$\nabla . \mathbf{B} = 0$$

In summary, Maxwell's equations in phasor form are:

$$\nabla \times \mathbf{H} = j\omega\mathbf{D} + \mathbf{J}$$
$$\nabla \times \mathbf{E} = -j\omega\mathbf{B}$$
$$\nabla . \mathbf{D} = \rho_v$$
$$\nabla . \mathbf{B} = 0$$

4.13 INFLUENCE OF MEDIUM ON THE FIELDS

When the sources of electric and magnetic fields exist in a medium, the medium has influence on the characteristics of the fields. The **constitutive relations,** namely,

$$\mathbf{D} = \epsilon\,\mathbf{E}$$
$$\mathbf{B} = \mu\,\mathbf{H}$$
$$\mathbf{J} = \sigma\mathbf{E}$$

describe the characteristics.

Here ϵ = permittivity (F/m), μ = permeability (H/m) and σ = conductivity (mho/m) of the medium.

4.14 TYPES OF MEDIA

Medium can be divided into five types:

(i) Homogeneous medium
(ii) Non-homogeneous medium
(iii) Isotropic medium
(iv) Anistropic medium
(v) Source-free regions

(i) **Homogeneous medium** It is a medium for which ϵ, μ and σ are constant throughout the medium.

Example Free space.

(ii) **Non-homogeneous medium** It is a medium for which ϵ, μ and σ are not constants and are different from point to point in the medium.

Example Human body.

(iii) **Isotropic medium** It is a medium for which ϵ is a scalar constant.

Example Free space.

(iv) **Anistropic medium** It is a medium for which ϵ is not a scalar constant.

Example Human body.

(v) **Source-free region** It is a medium in which there are no field sources.

Problem 4.1 Given $\mathbf{E} = 10 \sin(\omega t - \beta z)\, \mathbf{a}_y$ V/m in free space, determine **D**, **B**, **H**.

Solution

$$\mathbf{E} = 10 \sin(\omega t - \beta z)\, \mathbf{a}_y \text{ V/m}$$

$$\mathbf{D} = \epsilon_0 \mathbf{E}, \epsilon_0 = 8.854 \times 10^{-12} \text{ F/m}$$

$$\boxed{\mathbf{D} = 10\epsilon_0 \sin(\omega t - \beta z)\, \mathbf{a}_y \text{ C/m}^2}$$

Second Maxwell's equation is

$$\nabla \times \mathbf{E} = -\dot{\mathbf{B}}$$

that is,

$$\nabla \times \mathbf{E} = \begin{vmatrix} \mathbf{a}_x & \mathbf{a}_y & \mathbf{a}_z \\ \dfrac{\partial}{\partial x} & \dfrac{\partial}{\partial y} & \dfrac{\partial}{\partial z} \\ 0 & E_y & 0 \end{vmatrix}$$

or,

$$\nabla \times \mathbf{E} = \mathbf{a}_x \left[-\frac{\partial}{\partial z} E_y \right] + 0 + \mathbf{a}_z \left[\frac{\partial}{\partial x} E_y \right]$$

As

$$E_y = 10 \sin(\omega t - \beta z) \text{ V/m}$$

$$\frac{\partial E_y}{\partial x} = 0$$

Now $\nabla \times \mathbf{E}$ becomes

$$\nabla \times \mathbf{E} = -\frac{\partial E_y}{\partial z} \mathbf{a}_x$$

$$= 10\beta \cos(\omega t - \beta z)\, \mathbf{a}_x$$

$$= -\frac{\partial \mathbf{B}}{\partial t}$$

$$\mathbf{B} = -\int 10\beta \cos(\omega t - \beta z)\, dt\, \mathbf{a}_x$$

or,

$$\boxed{\mathbf{B} = \frac{10\beta}{\omega} \sin(\omega t - \beta z)\, \mathbf{a}_x, \text{ wb/m}^2}$$

and

$$\boxed{\mathbf{H} = \frac{\mathbf{B}}{\mu_0} = \frac{10\beta}{\mu_0 \omega} \sin(\omega t - \beta z)\, \mathbf{a}_x, \text{ A/m}}$$

Problem 4.2 If the electric field strength, **E** of an electromagnetic wave in free space is given by $\mathbf{E} = 2\cos\omega\left(t - \frac{z}{\upsilon_0}\right)\mathbf{a}_y$ V/m, find the magnetic field, **H**.

Solution We have

$$\frac{\partial \mathbf{B}}{\partial t} = -\nabla \times \mathbf{E}$$

$$= -\begin{vmatrix} \mathbf{a}_x & \mathbf{a}_y & \mathbf{a}_z \\ \frac{\partial}{\partial x} & \frac{\partial}{\partial y} & \frac{\partial}{\partial z} \\ 0 & E_y & 0 \end{vmatrix}$$

$$= -\left[\mathbf{a}_x\left(-\frac{\partial}{\partial z}E_y\right) + \mathbf{a}_y(0) + \mathbf{a}_z\left(\frac{\partial}{\partial x}E_y\right)\right]$$

$$= \frac{\partial E_y}{\partial z}\mathbf{a}_x$$

$$= \frac{2\omega}{\upsilon_0}\sin\omega\left(t - \frac{z}{\upsilon_0}\right)\mathbf{a}_x$$

$$\mathbf{B} = \frac{2\omega}{\upsilon_0}\int \sin\omega\left(t - \frac{z}{\upsilon_0}\right)dt\,\mathbf{a}_x$$

or, $$\mathbf{B} = \frac{-2\omega}{\upsilon_0\omega}\cos\omega\left(t - \frac{z}{\upsilon_0}\right)\mathbf{a}_x$$

or, $$\mathbf{H} = \frac{\mathbf{B}}{\mu_0} = \frac{-2}{\upsilon_0\mu_0}\cos\omega\left(t - \frac{z}{\upsilon_0}\right)\mathbf{a}_x \qquad \eta_0 = \sqrt{\frac{\mu_0}{\epsilon_0}} = 120\pi\,\Omega$$

Thus, $$\mathbf{H} = \frac{-2}{\eta_0}\cos\omega\left(t - \frac{z}{\upsilon_0}\right)\mathbf{a}_x \qquad \left[\upsilon_0 = \frac{1}{\sqrt{\mu_0\epsilon_0}}\right]$$

$$\boxed{\mathbf{H} = -\frac{1}{60\pi}\cos\omega\left(t - \frac{z}{\upsilon_0}\right)\mathbf{a}_x \text{ A/m}}$$

Problem 4.3 The parallel plates in a capacitor have an area of 4×10^{-4} m^2 and are separated by 0.4 cm. A voltage of $10 \sin 10^3 t$ volts is applied to the capacitor. Find the displacement current when the dielectric material between the plates has a relative permittivity of 4.

Solution We have

$$\mathbf{D} = \in \mathbf{E}$$

or,

$$D = \in E$$

The displacement current density, J_d is

$$J_d = \frac{\partial D}{\partial t} = \frac{\partial (\in E)}{\partial t}$$

But

$$E = \frac{V}{d}$$

where d = plate separation = 0.4 cm = 0.004 m.

So,

$$J_d = \frac{d}{\partial t}\left(\in \frac{V}{d}\right)$$

$$= \frac{\in}{d} \frac{dV}{dt}$$

$$= \frac{\in_0 \in_r}{d} \frac{dV}{dt}$$

The displacement current, I_d is

$$I_d = J_d \times A = \frac{\in_0 \in_r A}{d} \frac{dV}{dt},$$

A = area of the plate

$$= C \frac{dV}{dt} \qquad \left[\text{as } C = \frac{\in_0 \in_r A}{d}\right]$$

$$= \frac{4 \times 8.854 \times 10^{-12} \times 4 \times 10^{-4}}{4 \times 10^{-3}} \frac{dV}{dt}$$

$$= 35.42 \times 10^{-13} \frac{dV}{dt}$$

But

$$V = 10 \sin 10^3 t \text{ volt}$$

$$\frac{dV}{dt} = 10 \times 10^3 \cos 10^3 t$$

So, $$\boxed{I_d = 35.42 \cos 10^3 t \text{ nA}}$$

Problem 4.4 In free space, the magnetic field of an EM wave is given by $\mathbf{H} = 0.4\omega\epsilon_0 \cos(\omega t - 50x)\,\mathbf{a}_z$ A/m. Find the electric field, **E** and displacement current density, $\dot{\mathbf{D}}$.

Solution $$\mathbf{H} = 0.4\omega\epsilon_0 \cos(\omega t - 50x)\,\mathbf{a}_z \text{ A/m}$$

We have $$\nabla \times \mathbf{H} = \dot{\mathbf{D}} + \mathbf{J}$$

But $\mathbf{J} = 0$ for free space.

So, $$\nabla \times \mathbf{H} = \frac{\partial \mathbf{D}}{\partial t} = \epsilon_0 \frac{\partial \mathbf{E}}{\partial t}$$

$$\epsilon_0 \frac{\partial \mathbf{E}}{\partial t} = \begin{vmatrix} \mathbf{a}_x & \mathbf{a}_y & \mathbf{a}_z \\ \frac{\partial}{\partial x} & \frac{\partial}{\partial y} & \frac{\partial}{\partial z} \\ 0 & 0 & H_z \end{vmatrix}$$

$$= \mathbf{a}_x \left[\frac{\partial}{\partial y} H_z\right] + \mathbf{a}_y \left[-\frac{\partial}{\partial x} H_z\right] + [0]\,\mathbf{a}_z$$

But $$\frac{\partial H_z}{\partial y} = 0$$ [**H** is not a function of y]

So, $$\epsilon_0 \frac{\partial \mathbf{E}}{\partial t} = -\frac{\partial H_z}{\partial x}\,\mathbf{a}_y$$

$$\frac{\partial H_z}{\partial x} = \frac{\partial}{\partial x}[0.4\omega\epsilon_0 \cos(\omega t - 50x)]$$

$$= +0.4\omega\epsilon_0\, 50 \times \sin(\omega t - 50x)$$

$$= 20\epsilon_0\, \omega \sin(\omega t - 50x)$$

that is, $$\epsilon_0 \frac{\partial \mathbf{E}}{\partial t} = -20\epsilon_0\, \omega \sin(\omega t - 50x)\,\mathbf{a}_y$$

or, $$\frac{\partial \mathbf{E}}{\partial t} = -20\omega \sin(\omega t - 50x)\,\mathbf{a}_y$$

$$\mathbf{E} = \int -20\omega \sin(\omega t - 50x)\, dt\, \mathbf{a}_y$$

$$= \frac{-20}{\omega} \omega (-1) \cos (\omega t - 50x) \mathbf{a}_y$$

$$\boxed{\mathbf{E} = 20 \cos (\omega t - 50x) \mathbf{a}_y \text{ V/m}}$$

The displacement current density, $\mathbf{J}_d$

$$\mathbf{J}_d = \frac{\partial \mathbf{D}}{\partial t}$$

$$= \epsilon_0 \frac{\partial \mathbf{E}}{\partial t}$$

$$= \epsilon_0 \frac{\partial}{\partial t} (20 \cos (\omega t - 50x) \mathbf{a}_y$$

$$\boxed{\mathbf{J}_d = -20 \epsilon_0 \omega \sin (\omega t - 50x) \mathbf{a}_y \text{ A/m}^2}$$

Problem 4.5 If there is a magnetic field represented by

$$\mathbf{B} = 2 \sin (\omega t - \beta x) \mathbf{a}_x + 2y \cos (\omega t - \beta x) \mathbf{a}_y$$

in a medium where $\rho_v = 0$, $\sigma = 0$ and $J = 0$, find the electric field. Assume $\epsilon_r = 1$, $\mu_r = 1$.

Solution We have $\nabla \times \mathbf{H} = \dot{\mathbf{D}} + \mathbf{J}$

But $\qquad \mathbf{J} = 0$

as $\qquad \mathbf{D} = \epsilon_0 \mathbf{E}, \mathbf{B} = \mu_0 \mathbf{H}$

We can write $\qquad \nabla \times \mathbf{H} = \frac{1}{\mu_0} \nabla \times \mathbf{B} = \epsilon_0 \frac{\partial \mathbf{E}}{\partial t}$

or, $\qquad \nabla \times \mathbf{B} = \mu_0 \epsilon_0 \frac{\partial \mathbf{E}}{\partial t}$

$$\text{LHS} = \nabla \times \mathbf{B} = \begin{vmatrix} \mathbf{a}_x & \mathbf{a}_y & \mathbf{a}_z \\ \frac{\partial}{\partial x} & \frac{\partial}{\partial y} & \frac{\partial}{\partial z} \\ B_x & B_y & 0 \end{vmatrix}$$

$$= \mathbf{a}_x \left[-\frac{\partial}{\partial z} B_y \right] + \mathbf{a}_y \left[\frac{\partial}{\partial z} B_x + 0 \right] + \mathbf{a}_z \left[\frac{\partial}{\partial x} B_y - \frac{\partial}{\partial y} B_x \right]$$

As B_x and B_y are independent of y and z,

$$\frac{\partial B_y}{\partial z} = 0, \frac{\partial B_x}{\partial y} = 0$$

$$\nabla \times \mathbf{B} = \mathbf{a}_z \left[\frac{\partial B_y}{\partial x} - \frac{\partial B_x}{\partial y} \right]$$

But
$$\frac{\partial B_y}{\partial x} = \frac{\partial\,(2y \cos\,(\omega t - \beta x)}{\partial x}$$

$$= +2y\beta \sin\,(\omega t - \beta x)$$

$$= 2\beta y \sin\,(\omega t - \beta x)$$

$$\nabla \times \mathbf{B} = 2\beta y \sin\,(\omega t - \beta x)\,\mathbf{a}_z$$

$$= \mu_0 \in_0 \frac{\partial \mathbf{E}}{\partial t}$$

or,
$$\frac{\partial \mathbf{E}}{\partial t} = \frac{\mathbf{a}_z}{\mu_0 \in_0} [2\beta y \sin\,(\omega t - \beta x)]$$

$$\mathbf{E} = \frac{\mathbf{a}_z 2\beta y}{\mu_0 \in_0} \int \sin\,(\omega t - \beta x)\,dt$$

$$\mathbf{E} = \frac{-2\beta y}{\omega \mu_0 \in_0} \cos\,(\omega t - \beta x)\,\mathbf{a}_z \text{ V/m}$$

Problem 4.6 An electric field in a medium which is source-free is given by $\mathbf{E} = 1.5 \cos\,(10^8 t - \beta z)\,\mathbf{a}_x$ V/m, where E_m is the amplitude of **E**, ω is the angular frequency and β is the phase constant. Obtain **B, H, D**. Assume $\in_r = 1$, $\mu_r = 1$, $\sigma = 0$.

Solution We have

$$\nabla \times \mathbf{E} = -\frac{\partial \mathbf{B}}{\partial t}$$

$$\nabla \times \mathbf{E} = \begin{vmatrix} \mathbf{a}_x & \mathbf{a}_y & \mathbf{a}_z \\ \frac{\partial}{\partial x} & \frac{\partial}{\partial y} & \frac{\partial}{\partial z} \\ E_x & 0 & 0 \end{vmatrix}$$

$$= \mathbf{a}_x [0] + \mathbf{a}_y \left[\frac{\partial E_x}{\partial z} \right] - \mathbf{a}_z \left[\frac{\partial E_x}{\partial y} \right]$$

As E_x is not a function of y,

$$\frac{\partial E_x}{\partial y} = 0$$

$$\nabla \times \mathbf{E} = \frac{\partial E_x}{\partial z} \mathbf{a}_y$$

or,
$$\frac{\partial \mathbf{B}}{\partial t} = -\frac{\partial E_x}{\partial z} \mathbf{a}_y$$

$$= -1.5\beta \sin(10^8 t - \beta z) \mathbf{a}_y$$

$$\mathbf{B} = -\int 1.5\beta \sin(10^8 t - \beta z)\, dt\, \mathbf{a}_y$$

$$\boxed{\mathbf{B} = \frac{1.5\beta}{10^8} \cos(10^8 t - \beta z) \mathbf{a}_y \text{ wb/m}^2}$$

As
$$\mathbf{B} = \mu \mathbf{H}, \mu = \mu_0 = 4\pi \times 10^{-7} \text{ H/m}$$

$$\boxed{\mathbf{H} = \frac{\mathbf{B}}{\mu} = 12\beta \cos(10^8 t - \beta z) \mathbf{a}_y, \text{ mA/m}}$$

But
$$\mathbf{D} = \in \mathbf{E} = \in_0 \mathbf{E}, \in_0 = 8.854 \times 10^{-12} \text{ F/m}$$

So,
$$\boxed{\mathbf{D} = 13.28 \cos(10^8 t - \beta z) \mathbf{a}_x, \text{ pC/m}^2}$$

Problem 4.7 Verify whether the following fields

$$\mathbf{E} = 2 \sin x \sin t \mathbf{a}_y \text{ and}$$

$$\mathbf{H} = \frac{2}{\mu_0} \cos x \cos t \mathbf{a}_z$$

satisfy Maxwell's equations in free space.

Solution
$$\nabla \times \mathbf{H} = \dot{\mathbf{D}} \qquad [\text{as } \mathbf{J} = 0]$$

that is,
$$\frac{-\partial H_z}{\partial x} \mathbf{a}_y = \in_0 \frac{\partial E_y}{\partial t} \mathbf{a}_y$$

or,
$$\frac{2}{\mu_0} \sin x \cos t = 2\in_0 \sin x \cos t$$

$$\frac{2}{\mu_0} = 2\epsilon_0$$

or, $$\mu_0 \epsilon_0 = 1$$

which cannot be satisfied. Therefore, the given fields do not satisfy Maxwell's equations.

Problem 4.8 In a medium of conduction current density given by

$$J = 3.0 \sin(\omega t - 10z)\, \mathbf{a}_y + \cos(\omega t - 10z)\, \mathbf{a}_z \text{ mA/m}^2,$$

find the volume charge density.

Solution $J = 3.0 \sin(\omega t - 10z)\, \mathbf{a}_y + \cos(\omega t - 10z)\, \mathbf{a}_z$

By the equation of continuity, we have

$$-\dot{\rho}_\upsilon = \frac{\partial \rho_\upsilon}{\partial t}$$

$$= \nabla . \mathbf{J}$$

$$= \frac{\partial J_x}{\partial x} + \frac{\partial J_y}{\partial y} + \frac{\partial J_z}{\partial z}$$

$$= 0 + 0 + 10 \sin(\omega t - 10z)$$

that is, $$\dot{\rho}_\upsilon = -10 \sin(\omega t - 10z)$$

or, $$\rho_\upsilon = \int -10 \sin(\omega t - 10z)\, dt$$

$$\boxed{\rho_\upsilon = +10\omega \cos(\omega t - 10z) \text{ C/m}^3}$$

Problem 4.9 If the electric field strength of a radio broadcast signal at a TV receiver is given by

$$\mathbf{E} = 5.0 \cos(\omega t - \beta y)\, \mathbf{a}_z, \text{ V/m},$$

determine the displacement current density. If the same field exists in a medium whose conductivity is given by 2.0×10^3 (mho)/cm, find the conduction current density.

Solution E at a TV receiver in free space

$$= 5 \cos(\omega t - \beta y)\, \mathbf{a}_z \text{ V/m}$$

Electric flux density,

$$\mathbf{D} = \epsilon_0 \mathbf{E} = 5\epsilon_0 \cos(\omega t - \beta z)\, \mathbf{a}_z \text{ C/m}^2$$

The displacement current density

$$\mathbf{J}_d = \dot{\mathbf{D}} = \frac{\partial \mathbf{D}}{\partial t}$$

$$= \frac{\partial}{\partial t}[5\epsilon_0 \cos(\omega t - \beta y)\, \mathbf{a}_z]$$

$$\boxed{\mathbf{J}_d = -5\epsilon_0\, \omega \sin(\omega t - \beta y)\, \mathbf{a}_z \text{ A/m}^2}$$

The conduction current density,

$$\mathbf{J}_c = \sigma \mathbf{E}$$

$$\sigma = 2.0 \times 10^3 \text{ (mho)/cm}$$

$$= 2 \times 10^5 \text{ mho/m}$$

$$\mathbf{J}_c = 2 \times 10^5 \times 5 \cos(\omega t - \beta y)\, \mathbf{a}_z$$

$$\boxed{\mathbf{J}_c = 10^6 \cos(\omega t - \beta y)\, \mathbf{a}_z, \text{ A/m}^2}$$

Problem 4.10 Find the electric flux density and volume charge density if the electric field,

$$\mathbf{E} = x^2\, \mathbf{a}_x + 2y^2\, \mathbf{a}_y + z^2\, \mathbf{a}_z \text{ V/m}$$

in a medium whose $\epsilon_r = 2$.

Solution $\mathbf{E} = x^2\, \mathbf{a}_x + 2y^2\, \mathbf{a}_y + z^2\, \mathbf{a}_z$ V/m

$$\mathbf{D} = \epsilon\, \mathbf{E} = \epsilon_0\, \epsilon_r\, \mathbf{E}$$

$$= 8.854 \times 10^{-12} \times 2 \times (x^2\, \mathbf{a}_x + 2y^2\, \mathbf{a}_y + z^2\, \mathbf{a}_z)$$

$$\boxed{\mathbf{D} = 17.708x^2\, \mathbf{a}_x + 35.416y^2\, \mathbf{a}_y + 17.708z^2\, \mathbf{a}_z, \text{ pc/m}^2}$$

From Maxwell's equation, we have

$$\nabla . \mathbf{D} = \text{Div}(\mathbf{D}) = \frac{\partial}{\partial x} D_x + \frac{\partial}{\partial y} D_y + \frac{\partial}{\partial z} D_z = \rho_v$$

$$\boxed{\rho_v = 35.416x + 70.832y + 35.416z \text{ C/m}^3}$$

4.15 SUMMARY OF MAXWELL'S EQUATIONS FOR DIFFERENT CASES

Maxwell's equations in differential form (General)	*Maxwell's equations in integral form (General)*	*Maxwell's equations in phasor form*	*Maxwell's equations in free space (differential form)*	*Maxwell's equations for static fields (differential form)*	*Maxwell's equations in free space (integral form)*	*Maxwell's equations in free space for static fields*
$\nabla\times\mathbf{H}=\dot{\mathbf{D}}+\mathbf{J}$	$\oint_L \mathbf{H}.d\mathbf{L}=\int_s(\dot{\mathbf{D}}+\mathbf{J}).d\mathbf{S}$	$\nabla\times\mathbf{H}=j\omega\mathbf{D}+\mathbf{J}$	$\nabla\times\mathbf{H}=\dot{\mathbf{D}}$	$\nabla\times\mathbf{H}=\mathbf{J}$	$\oint_L \mathbf{H}.d\mathbf{L}=\int_s \dot{\mathbf{D}}.d\mathbf{S}$	$\nabla\times\mathbf{H}=0$ or $\oint_L \mathbf{H}.d\mathbf{L}=0$
$\nabla\times\mathbf{E}=-\dot{\mathbf{B}}$	$\oint_L \mathbf{E}.d\mathbf{L}=-\int_s \dot{\mathbf{B}}.d\mathbf{S}$	$\nabla\times\mathbf{E}=-j\omega\mathbf{B}$	$\nabla\times\mathbf{E}=-\dot{\mathbf{B}}$	$\nabla\times\mathbf{E}=0$	$\oint_L \mathbf{E}.d\mathbf{L}=-\int_s \dot{\mathbf{B}}.d\mathbf{S}$	$\nabla\times\mathbf{E}=0$ or $\oint_L \mathbf{E}.d\mathbf{L}=0$
$\nabla.\mathbf{D}=\rho_\upsilon$	$\oint_s \mathbf{D}.d\mathbf{S}=\int_\upsilon \rho_\upsilon\, d\upsilon$	$\nabla.\mathbf{D}=\rho_\upsilon$	$\nabla.\mathbf{D}=0$	$\nabla.\mathbf{D}=\rho_\upsilon$	$\oint_s \mathbf{D}.d\mathbf{S}=\int_\upsilon \rho_\upsilon\, d\upsilon$	$\nabla.\mathbf{D}=0$ or $\oint_s \mathbf{D}.d\mathbf{S}=0$
$\nabla.\mathbf{B}=0$	$\oint_s \mathbf{B}.d\mathbf{S}=0$	$\nabla.\mathbf{B}=0$	$\nabla.\mathbf{B}=0$	$\nabla.\mathbf{B}=0$	$\oint_s \mathbf{B}.d\mathbf{S}=0$	$\nabla.\mathbf{B}=0$ or $\oint_s \mathbf{B}.d\mathbf{S}=0$

4.16 CONDITIONS AT A BOUNDARY SURFACE

Relations between the main fields **E**, **D**, **H**, **B** are expressed in terms of Maxwell's equations. These are valid at any point in a continuous medium.

As Maxwell's equations contain space derivatives, they cannot give information at points of discontinuity in the medium. However, Maxwell's equations in integral form can be used to get the information at the boundary surface between different media.

The electromagnetic fields that are solved using Maxwell's equations must satisfy boundary conditions at the interface between different media.

The **boundary conditions** on electric and magnetic fields at any surface of discontinuity are given by:

1. The tangential component of electric field, **E** is continuous across any discontinuity, that is,

$$\mathbf{E}_{\tan 1} = \mathbf{E}_{\tan 2}$$

Subscript tan 1 is the tangential component of the field at the boundary in medium 1. Subscript tan 2 represents it in medium 2.

2. The tangential component of magnetic field, **H** is continuous across any surface except at the surface of a perfect conductor. At the surface of a perfect conductor, the tangential component of **H** is discontinuous by quantity equal to the surface current density (A/m).

$$\mathbf{H}_{\tan 1} - \mathbf{H}_{\tan 2} = \mathbf{J}_s$$

3. The normal component of **B** is continuous across any discontinuity.

$$\mathbf{B}_{n1} = \mathbf{B}_{n2}$$

4. The normal component of **D** is continuous except at a surface which has surface charge density. At the surface where surface charge density exists, the normal component of **D** is discontinuous by a quantity equal to the surface charge density.

$$\mathbf{D}_{n1} - \mathbf{D}_{n2} = \rho_s$$

Subscript $n1$ represents normal component of the field in medium 1 and $n2$ represents it in medium 2.

4.17 PROOF OF BOUNDARY CONDITIONS ON E, D, H AND B

To prove $E_{\tan 1} = E_{\tan 2}$

Consider the rectangular loop on the boundary of two media (Fig. 4.1).

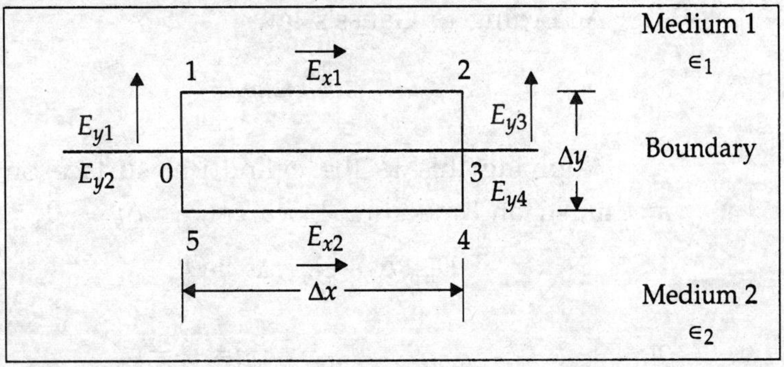

Fig. 4.1 Rectangular loop on boundary

It is well-known that electric field is conservative and hence the line integral of **E** . d**L** is zero around a closed path.

So, $\oint \mathbf{E}.d\mathbf{L} = 0$

From the figure shown above, LHS is written as

$$\text{LHS} = \int_{01} + \int_{12} + \int_{23} + \int_{34} + \int_{45} + \int_{50}$$

$$= E_{y1}\frac{\Delta y}{2} + E_{x1}\,\Delta x - E_{y3}\frac{\Delta y}{2} - E_{y4}\frac{\Delta y}{2} - E_{x2}\,\Delta x + E_{y2}\frac{\Delta y}{2}$$

As $\Delta y \to 0$, we get

$$\oint \mathbf{E}.d\mathbf{L} = E_{x1}\,\Delta x - E_{x2}\,\Delta x = 0$$

or, $$E_{x1} = E_{x2}$$

It is obvious that E_{x1} and E_{x2} are the tangential components of **E** in medium 1 and 2 respectively.

So, $\mathbf{E}_{\tan 1} = \mathbf{E}_{\tan 2}$

Now consider a cylinder across the media 1 and 2 (Fig. 4.2).

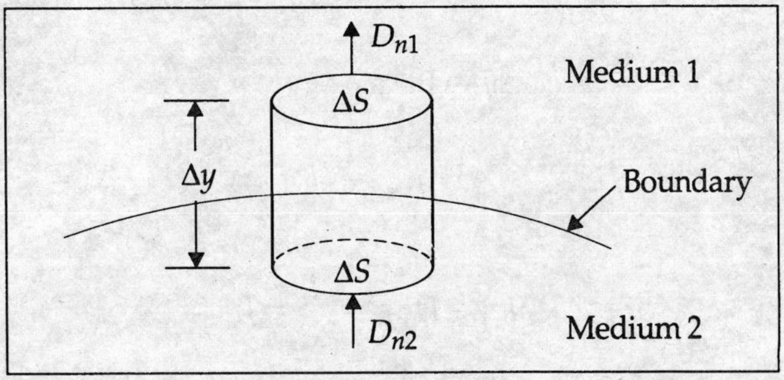

Fig. 4.2 Cylinder across boundary

According to Gauss's law,

$$\oint \mathbf{D}.d\mathbf{s} = Q$$

Applying this to the cylindrical surface on the boundary spreading over medium 1 and medium 2, we get, as $\Delta y \to 0$

$$D_{n1}\ \Delta s - D_{n2}\ \Delta s = Q$$

or, $$D_{n1} - D_{n2} = \frac{Q}{\Delta s} = \rho_s$$

$$\mathbf{D}_{n1} - \mathbf{D}_{n2} = \rho_s \qquad \text{Hence proved.}$$

To prove $\mathbf{H}_{\tan 1} - \mathbf{H}_{\tan 2} = \mathbf{J}_s$

Consider the rectangular loop on the boundary of two media (Fig. 4.3).

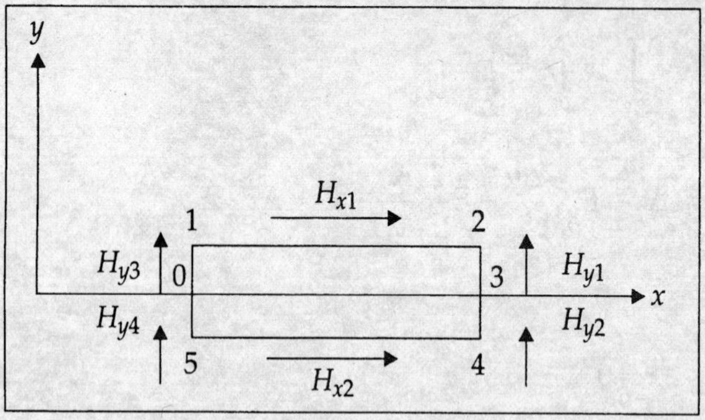

Fig. 4.3 A rectangular loop across a boundary

From Ampere's circuit law, we have

$$\oint \mathbf{H}.d\mathbf{L} = \int_{50} + \int_{01} + \int_{12} + \int_{23} + \int_{34} + \int_{45}$$

$$= H_{y4}\frac{\Delta y}{2} + H_{y3}\frac{\Delta y}{2} + H_{x1}\ \Delta x - H_{y1}\frac{\Delta y}{2} - H_{y2}\frac{\Delta y}{2} - H_{x2}\ \Delta x = I$$

As $\Delta y \to 0$, we get

$$\oint \mathbf{H}.d\mathbf{L} = H_{x1}\ \Delta x - H_{x2}\ \Delta x = I$$

or, $$H_{x1} - H_{x2} = \frac{I}{\Delta x} = J_s$$

Here, H_{x1} and H_{x2} are nothing but tangential components in medium 1 and 2 respectively.

$$\mathbf{H}_{\tan 1} - \mathbf{H}_{\tan 2} = \mathbf{J}_s \qquad \text{Hence proved.}$$

To prove $\mathbf{B}_{n1} = \mathbf{B}_{n2}$

Now consider a cylinder shown in Fig. 4.4.

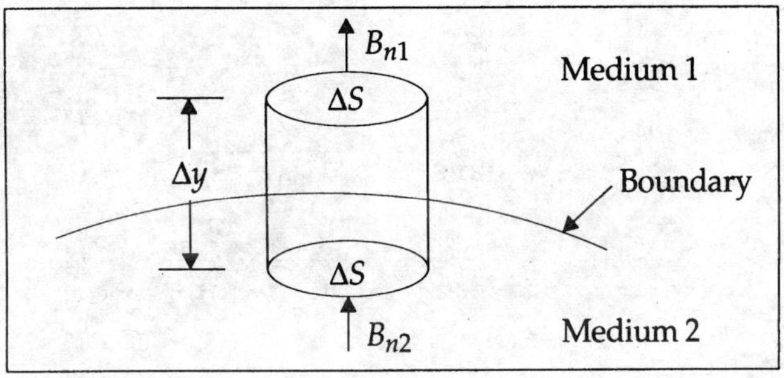

Fig. 4.4 A differential cylinder across the boundary

Gauss's law for magnetic fields is

$$\int_S \mathbf{B} . d\mathbf{S} = 0$$

$$\text{LHS} = \int_s B_{n1}\, \mathbf{a}_y . dS\, \mathbf{a}_y + \int_s B_{n2}\, \mathbf{a}_y . dS\, (-\mathbf{a}_y)$$

Applying LHS to the cylindrical surfaces shown, we get, for $\Delta y \to 0$

$$B_{n_1}\, \Delta S - B_{n_2}\, \Delta S = 0$$

$$\mathbf{B}_{n1} = \mathbf{B}_{n2} \qquad \text{Hence proved.}$$

Given the fields in one medium, it is usually required to determine fields in a second medium. This requires the knowledge of both tangential and normal components for each field. The above boundary conditions yield either tangential or normal component. The second unknown component is determined from the constitutive relations between **E** and **D**, **H** and **B**.

We have $\mathbf{E}_{\tan 1} = \mathbf{E}_{\tan 2}$

But $\mathbf{D} = \in \mathbf{E}$

$$\frac{\mathbf{D}_{\tan 1}}{\in_1} = \frac{\mathbf{D}_{\tan 2}}{\in_2}$$

$$\mathbf{D}_{\tan 2} = \frac{\in_2}{\in_1} \mathbf{D}_{\tan 1} = \frac{\in_{r2}}{\in_{r1}} \mathbf{D}_{\tan 1}$$

We have $$\mathbf{D}_{n1} - \mathbf{D}_{n2} = \rho_s$$

$$\epsilon_1 \mathbf{E}_{n1} - \epsilon_2 \mathbf{E}_{n2} = \rho_s$$

We have $$\mathbf{B}_{n1} = \mathbf{B}_{n2}$$

But $$\mathbf{B} = \mu \mathbf{H}$$

$$\mu_1 \mathbf{H}_{n1} = \mu_2 \mathbf{H}_{n2}$$

$$\mathbf{H}_{n1} = \frac{\mu_2}{\mu_1} \mathbf{H}_{n2} = \frac{\mu_{r2}}{\mu_{r1}} \mathbf{H}_{n2}$$

We have $$\mathbf{H}_{\tan 1} - \mathbf{H}_{\tan 2} = \mathbf{J}_s$$

$$\frac{\mathbf{B}_{\tan 1}}{\mu_1} - \frac{\mathbf{B}_{\tan 2}}{\mu_2} = \mathbf{J}_s$$

4.18 COMPLETE BOUNDARY CONDITIONS IN SCALAR FORM

$$E_{\tan 1} - E_{\tan 2} = 0$$

$$\epsilon_1 E_{n1} - \epsilon_2 E_{n2} = \rho_s$$

$$H_{\tan 1} - H_{\tan 2} = J_s$$

$$\mu_{r1} H_{n1} - \mu_{r2} H_{n2} = 0$$

$$B_{n1} - B_{n2} = 0$$

$$\frac{B_{\tan 1}}{\mu_1} - \frac{B_{\tan 2}}{\mu_2} = J_s$$

$$\frac{D_{\tan 1}}{\epsilon_{r1}} - \frac{D_{\tan 2}}{\epsilon_{r2}} = 0$$

$$D_{n1} - D_{n2} = \rho_s$$

$$J_{n_1} = J_{n_2}$$

$$J_{t_1} = \frac{\sigma_1}{\sigma_2} J_{t_2}$$

4.19 BOUNDARY CONDITIONS IN VECTOR FORM

$$\mathbf{a}_n \times (\mathbf{E}_1 - \mathbf{E}_2) = 0$$

$$\mathbf{a}_n \times (\mathbf{H}_1 - \mathbf{H}_2) = \mathbf{J}_s$$

$$\mathbf{a}_n \cdot (\mathbf{B}_1 - \mathbf{B}_2) = 0$$

$$\mathbf{a}_n \cdot (\mathbf{D}_1 - \mathbf{D}_2) = \rho_s$$

$$\mathbf{a}_n \cdot (\mathbf{J}_1 - \mathbf{J}_2) = 0$$

$$\mathbf{a}_n \times \left(\frac{\mathbf{J}_1}{\sigma_1} - \frac{\mathbf{J}_2}{\sigma_2} \right) = 0$$

$$\mathbf{a}_n \cdot (\epsilon_1 \mathbf{E}_1 - \epsilon_2 \mathbf{E}_2) = \rho_s$$

$$\mathbf{a}_n \cdot \left(\frac{\mathbf{H}_1}{\mu_1} - \frac{\mathbf{H}_2}{\mu_2} \right) = 0$$

$$\mathbf{a}_n \times \left(\frac{\mathbf{B}_1}{\mu_1} - \frac{\mathbf{B}_2}{\mu_2} \right) = \mathbf{J}_s$$

Problem 4.11 $x < 0$ defines region 1 and $x > 0$ defines region 2. Region 1 is characterised by $\mu_{r_1} = 3.0$ and region 2 is characterised by $\mu_{r_2} = 5.0$. If the magnetic field in region 1 is given by $\mathbf{H}_1 = 4.0\mathbf{a}_x + 1.5\mathbf{a}_y - 3.0\mathbf{a}_z$, A/m, find $\mathbf{H}_2$ and H_2.

Solution $\quad \mathbf{H}_1 = 4.0\mathbf{a}_x + 1.5\mathbf{a}_y - 3.0\mathbf{a}_z, \text{A/m}$

For the regions given,

$$\mathbf{H}_{t_1} = 1.5\mathbf{a}_y - 3.0\mathbf{a}_z, \text{A/m}$$

and $\quad \mathbf{H}_{n_1} = 4.0\mathbf{a}_x$

As $\quad \mathbf{H}_{t_1} = \mathbf{H}_{t_2}$

$$\mathbf{H}_{t_2} = 1.5\mathbf{a}_y - 3.0\mathbf{a}_z, \text{A/m}$$

and $\quad \mathbf{H}_{n_2} = \dfrac{\mu_1}{\mu_2} \mathbf{H}_{n_1}$

$$= \frac{3}{5} \times 4.0\mathbf{a}_x$$

or, $\quad \mathbf{H}_{n_2} = 2.4\mathbf{a}_x$

and $\quad \mathbf{H}_2 = \mathbf{H}_{t_2} + \mathbf{H}_{n_2}$

$$\boxed{\mathbf{H}_2 = 2.4\mathbf{a}_x + 1.5\mathbf{a}_y - 3.0\mathbf{a}_z, \text{A/m}}$$

The magnitude of $\mathbf{H}_2$ is

$$H_2 = \sqrt{2.4^2 + 1.5^2 + (-3.0)^2}$$

$$\boxed{H_2 = 4.1243\,\text{A/m}}$$

Problem 4.12 In a three-dimensional space, divided into region 1 ($x < 0$) and region 2 ($x > 0$), $\sigma_1 = \sigma_2 = 0$. $\mathbf{E}_1 = 1\mathbf{a}_x + 2\mathbf{a}_y + 3\mathbf{a}_z$. Find $\mathbf{E}_2$ and $\mathbf{D}_2 . \in_{r_1} = 1$ and $\in_{r_2} = 2$.

Solution

$$\mathbf{E}_{t1} = 2\mathbf{a}_y + 3\mathbf{a}_z, \text{V/m}$$

$$\mathbf{E}_{n1} = \mathbf{a}_x, \quad \in_{r1} = 1$$

$$\mathbf{D}_{t1} = \in_0 (2\mathbf{a}_y + 3\mathbf{a}_z), \text{C/m}^2$$

As

$$\mathbf{E}_{t1} = \mathbf{E}_{t2}$$

$$\mathbf{E}_{t2} = 2\mathbf{a}_y + 3\mathbf{a}_z, \text{V/m}$$

As

$$\mathbf{D}_{n1} = \mathbf{D}_{n2}$$

$$\in_1 \mathbf{E}_{n1} = \in_2 \mathbf{E}_{n2}$$

or,

$$\mathbf{E}_{n2} = \frac{\in_1}{\in_2} \mathbf{E}_{n1} = \frac{1}{2} \mathbf{a}_x$$

$$\mathbf{E}_2 = \mathbf{E}_{t2} + \mathbf{E}_{n2}$$

or,

$$\boxed{\mathbf{E}_2 = 0.5\mathbf{a}_x + 2\mathbf{a}_y + 3\mathbf{a}_z \text{ V/m}}$$

$$\mathbf{D}_2 = \in_2 \mathbf{E}_2$$

$$= 2\in_0 (0.5\mathbf{a}_x + 2\mathbf{a}_y + 3\mathbf{a}_z)$$

or,

$$\boxed{\mathbf{D}_2 = \in_0 (\mathbf{a}_x + 4\mathbf{a}_y + 6\mathbf{a}_z), \text{C/m}^2}$$

4.20 TIME VARYING POTENTIALS

It is always useful to relate electric and magnetic fields in terms of their sources. However, it is often more convenient to relate potentials in terms of sources and then the fields in terms of potentials.

In the first method potentials already developed for static fields are generalised to obtain time varying fields. Scalar electrostatic potential is given by

$$V(r) = \frac{1}{4\pi\in_0} \int_{\upsilon} \frac{\rho_\upsilon (r)\, d\upsilon}{r}$$

and vector magnetic potential is

$$\mathbf{A}(\mathbf{r}) = \frac{\mu_0}{4\pi} \int_{\upsilon} \frac{\mathbf{J}(\mathbf{r})\, d\upsilon}{r}$$

But the time varying potentials are expressed in the form of

$$V(r, t) = \frac{1}{4\pi\epsilon_0} \int_{\upsilon} \frac{\rho_\upsilon(r, t)}{r}\, d\upsilon$$

and

$$\mathbf{A}(r, t) = \frac{\mu_0}{4\pi} \int_{\upsilon} \frac{\mathbf{J}(r, t)}{r}\, d\upsilon$$

These time varying potentials are due to time varying charge and current distributions.

The above expressions do not take care of propagation delay. To obtain far-field expressions, this delay time must be taken into account.

4.21 RETARDED POTENTIALS

These are defined as the potentials in which a time delay or retarded time is taken into account. They are expressed as:

$$V(r, t) = \frac{1}{4\pi\epsilon} \int_{\upsilon} \frac{\rho_\upsilon(r, t - r/\upsilon_0)}{r}\, d\upsilon, \text{ (volt)}$$

$$\mathbf{A}(r, t) = \frac{\mu}{4\pi} \int_{\upsilon} \frac{\mathbf{J}(r, t - r/\upsilon_0)}{r}\, d\upsilon, \text{ (wb/m)}$$

where retarded time or delay time is

$$t_d = (t - r/\upsilon_0)$$

$$\upsilon_0 = \text{velocity of propagation of the EM wave}$$

From the above approach, there is no change in the expressions for **E** and **H**, that is,

$$\mathbf{E} = -\nabla V$$

and

$$\mathbf{H} = \frac{1}{\mu_0} \nabla \times \mathbf{A}$$

4.22 MAXWELL'S EQUATIONS APPROACH TO RELATE POTENTIALS, FIELDS AND THEIR SOURCES

By definition, vector magnetic potential, **A** is related to magnetic flux density as

$$\mathbf{B} \equiv \nabla \times \mathbf{A}$$

But the second Maxwell's equation is

$$\nabla \times \mathbf{E} = -\dot{\mathbf{B}}$$

From the above two expressions, we have

$$\nabla \times \mathbf{E} = -\frac{\partial}{\partial t}(\nabla \times \mathbf{A}) = -\nabla \times \dot{\mathbf{A}}$$

or, $$\nabla \times \mathbf{E} + \nabla \times \dot{\mathbf{A}} = 0$$

or, $$\nabla \times (\mathbf{E} + \dot{\mathbf{A}}) = 0$$

This is true only if $(\mathbf{E} + \dot{\mathbf{A}})$ is the gradient of a scalar. Therefore, setting $(\mathbf{E} + \dot{\mathbf{A}}) = -\nabla V$, we get **E** as

$$\mathbf{E} = -\nabla V - \dot{\mathbf{A}}$$

Note that **E** is not equal to $(-\nabla V)$ for time varying fields but it is given by the above expression.

So, $\mathbf{E} = -\nabla V$ is valid only for static fields

Now consider

$$\mathbf{B} = \nabla \times \mathbf{A}$$

Take curl on both sides,

$$\nabla \times \mathbf{B} = \nabla \times \nabla \times \mathbf{A}$$

or, $$\frac{1}{\mu_0} \nabla \times \nabla \times \mathbf{A} = \nabla \times \mathbf{H}$$

But $$\nabla \times \mathbf{H} = \dot{\mathbf{D}} + \mathbf{J} = \epsilon_0 \dot{\mathbf{E}} + \mathbf{J}$$

From the above expressions, we get

$$\epsilon_0 \dot{\mathbf{E}} + \mathbf{J} = -\epsilon_0 \nabla \dot{V} - \epsilon_0 \ddot{\mathbf{A}} + \mathbf{J}$$

$$\frac{1}{\mu_0} \nabla \times \nabla \times \mathbf{A} = -\epsilon_0 \nabla V - \epsilon_0 \ddot{\mathbf{A}} + \mathbf{J}$$

But vector identity gives

$$\nabla \times \nabla \times \mathbf{A} = \nabla \nabla . \mathbf{A} - \nabla^2 \mathbf{A}$$

that is, $$\nabla\nabla.\mathbf{A}-\nabla^2\mathbf{A}=-\mu_0\in_0\nabla\dot{V}-\mu_0\in_0\ddot{\mathbf{A}}+\mu_0\mathbf{J}$$

The third Maxwell's equation is

$$\nabla.\mathbf{E}=\rho_\upsilon/\in_0$$

This becomes $$\nabla.\mathbf{E}=-\nabla.\nabla V-\nabla.\dot{\mathbf{A}}=\rho_\upsilon/\in_0$$

or, $$\nabla^2V+\nabla.\dot{\mathbf{A}}=-\rho_\upsilon/\in_0$$

and $$\nabla^2\mathbf{A}-\mu_0\in_0\ddot{\mathbf{A}}=-\mu_0\mathbf{J}+\mu_0\in_0\nabla\dot{V}+\nabla\nabla.\mathbf{A}$$

Each of the equations contain both the potentials and one of the sources. It is difficult to solve such type of equations. We should be able to get equations containing one potential and its own source to solve them easily. Helmholtz theorem helps to do this.

4.23 HELMHOLTZ THEOREM

It states that any vector field like A due to a finite source is uniquely specified if and only if its curl and the divergence are specified.

To specify **A**, we know its curl, that is,

$$\nabla\times\mathbf{A}=\mathbf{B}$$

But the $\nabla.\mathbf{A}$ is specified as

$$\nabla.\mathbf{A}=-\mu_0\in_0\dot{V}$$

$$\nabla^2\mathbf{A}-\mu_0\in_0\ddot{\mathbf{A}}=-\mu_0\mathbf{J}$$

and $$\nabla^2V-\mu_0\in_0\ddot{V}=-\rho_\upsilon/\in_0$$

These are the expressions between time varying potentials and their sources.

The corresponding expressions for static fields are:

$$\nabla^2\mathbf{A}=-\mu_0\mathbf{J}$$

$$\nabla^2V=-\rho_\upsilon/\in_0$$

4.24 LORENTZ GAUGE CONDITION

It is given by

$$\nabla.\mathbf{A}=-\mu_0\in_0\dot{V}$$

Potential functions for sinusoidal fields are given by

$$\nabla^2 \mathbf{A} + \omega^2 \mu_0 \in_0 \mathbf{A} = -\mu_0 \mathbf{J}$$

$$\nabla^2 V + \omega^2 \mu_0 \in_0 V = -\rho_\upsilon / \in_0$$

Problem 4.13 If the retarded scalar electric potential, $V = x - \upsilon_0 t$ and vector magnetic potential, $\mathbf{A} = \left(\frac{x}{\upsilon_0} - t\right)\mathbf{a}_x$, where υ_0 is the velocity of propagation,

(a) find $\nabla . \mathbf{A}$

(b) find **B**, **H**, **E**, and **D**

(c) show that $\nabla . \mathbf{A} = -\mu_0 \in_0 \frac{\partial V}{\partial t}$

Solution (a) $$\nabla . \mathbf{A} = \frac{\partial A_x}{\partial x} + \frac{\partial A_y}{\partial y} + \frac{\partial A_z}{\partial z}$$

But $$A_y = 0,\ A_z = 0$$

$$\nabla . \mathbf{A} = \frac{\partial A_x}{\partial x} = \frac{\partial}{\partial x}\left(\frac{x}{\upsilon_0} - t\right) = \frac{1}{\upsilon_0}$$

$$\boxed{\nabla . \mathbf{A} = 1/\upsilon_0}$$

(b) We have $$\mathbf{B} = \nabla \times \mathbf{A}$$

Here $$\mathbf{A} = \left(\frac{x}{\upsilon_0} - t\right)\mathbf{a}_x$$

$$A_x = \left(\frac{x}{\upsilon_0} - t\right)$$

$$\nabla \times \mathbf{A} = \begin{vmatrix} \mathbf{a}_x & \mathbf{a}_y & \mathbf{a}_z \\ \frac{\partial}{\partial x} & \frac{\partial}{\partial y} & \frac{\partial}{\partial z} \\ A_x & 0 & 0 \end{vmatrix}$$

$$= \mathbf{a}_x[0] + \mathbf{a}_y\left[\frac{\partial A_x}{\partial z}\right] + \mathbf{a}_z\left[-\frac{\partial A_x}{\partial y}\right]$$

As $A_x \neq f(y)$, $\frac{\partial A_x}{\partial y} = 0$ and A_x is not function of z,

$$\frac{\partial A_x}{\partial z} = 0$$

Therefore, $\nabla \times \mathbf{A} = 0$

and $\boxed{\mathbf{B} = 0}$

As $\mathbf{H} = \dfrac{1}{\mu_0}\mathbf{B}, \ \mathbf{H} = 0$

$$\mathbf{E} = -\nabla V - \dot{\mathbf{A}}$$

$$= -\nabla V - \frac{\partial \mathbf{A}}{\partial t}$$

$$\nabla V = +\frac{\partial V}{\partial x}\mathbf{a}_x + \frac{\partial V}{\partial y}\mathbf{a}_y + \frac{\partial V}{\partial z}\mathbf{a}_z$$

Here $\dfrac{\partial V}{\partial y} = 0, \ \dfrac{\partial V}{\partial z} = 0$

$$\nabla V = \frac{\partial V}{\partial x}\mathbf{a}_x = \mathbf{a}_x$$

$$\frac{\partial \mathbf{A}}{\partial t} = \frac{\partial}{\partial t}(x/\upsilon_0 - t)\,\mathbf{a}_x$$

$$= -\mathbf{a}_x$$

or, $\mathbf{E} = -\nabla V - \dot{\mathbf{A}}$

$$\boxed{\mathbf{E} = -\mathbf{a}_x + \mathbf{a}_x = 0}$$

and $\boxed{\mathbf{D} = \epsilon_0 \mathbf{E} = 0}$

(c) $\dfrac{\partial V}{\partial t} = \dfrac{\partial}{\partial t}(x - \upsilon_0 t) = -\upsilon_0$

But $\nabla . \mathbf{A} = \dfrac{1}{\upsilon_0}$

$$\mu_0 \epsilon_0 \frac{\partial V}{\partial t} = -\upsilon_0 \times \frac{1}{\upsilon_0^2} = -\frac{1}{\upsilon_0}$$

or, $-\mu_0 \epsilon_0 \dfrac{\partial V}{\partial t} = \dfrac{1}{\upsilon_0}$

$\boxed{\nabla . \mathbf{A} = -\mu_0 \epsilon_0 \dfrac{\partial V}{\partial t}}$ Hence proved.

POINTS/FORMULAE TO REMEMBER

- Maxwell's equations are electromagnetic field equations.
- Maxwell's equations give relations between different fields and sources.
- Equation of continuity is $\nabla . \mathbf{J} = -\dot{\rho}_{\upsilon}$.
- Most general Maxwell's equations are

$$\nabla \times \mathbf{H} = \dot{\mathbf{D}} + \mathbf{J}, \quad \oint_L \mathbf{H}.d\mathbf{L} = \int_s \left(\dot{\mathbf{D}} + \mathbf{J} \right).d\mathbf{S}$$

$$\nabla \times \mathbf{E} = -\dot{\mathbf{B}}, \quad \int_L \mathbf{E}.d\mathbf{L} = -\int_s \mathbf{B}.d\mathbf{S}$$

$$\nabla . \mathbf{D} = \rho_{\upsilon}, \quad \int \mathbf{D}.d\mathbf{S} = \int_{\upsilon} \rho_{\upsilon}\, d\upsilon$$

$$\nabla . \mathbf{B} = 0, \quad \oint_s \mathbf{B}.d\mathbf{S} = 0$$

- For static fields, $\dot{\mathbf{B}} = 0$, $\dot{\mathbf{D}} = 0$.
- In good dielectrics, $\mathbf{J} = 0$.
- In good conductors, $\dot{\mathbf{D}} = 0$.
- For free space, $\rho_{\upsilon} = 0$, $J = 0$, $\sigma = 0$.
- For conducting media, $\rho_{\upsilon} = 0$.
- $\dot{\mathbf{B}} = \dfrac{\partial B}{\partial t}$ is known as the magnetic current density.
- Magnetic current density has the unit of volt/m^2.
- Maxwell's equations in phasor form are:

$$\nabla \times \mathbf{H} = j\omega \mathbf{D} + \mathbf{J}$$

$$\nabla \times \mathbf{E} = -j\omega \mathbf{B}$$

$$\nabla . \mathbf{D} = \rho_\upsilon$$

$$\nabla . \mathbf{B} = 0$$

- The constitutive relations are:

$$\mathbf{D} = \in \mathbf{E}$$

$$\mathbf{B} = \mu \mathbf{H}$$

$$\mathbf{J} = \sigma \mathbf{E}$$

- For a homogeneous medium, $\in$, μ, σ are constants throughout the medium.
- For isotropic medium, $\in$ is a scalar constant.
- Retarded potentials are:

$$V(r, t) = \frac{1}{4\pi \in} \int_\upsilon \frac{\rho_\upsilon \left(r, t - \frac{r}{\upsilon_0} \right)}{r} d\upsilon, \text{ volt}$$

$$\mathbf{A}(r, t) = \frac{\mu}{4\pi} \int_\upsilon \frac{\mathbf{J} \left(r, t - \frac{r}{\upsilon_0} \right)}{r} d\upsilon, \text{ wb/m}$$

- For time varying fields, $\mathbf{E} = -\nabla V - \dot{\mathbf{A}}$.
- According to Helmholtz theorem, if the curl and divergence of a vector are specified, it exhibits unique meaning.
- Lorentz gauge condition is $\nabla . \mathbf{A} = -\mu_0 \in_0 \dot{V}$.

OBJECTIVE QUESTIONS

1. $\mathbf{E} = -\nabla V$ is valid for all types of fields. (Yes/No)
2. $\frac{\partial \mathbf{B}}{\partial t}$ is magnetic current density. (Yes/No)
3. Theta polarisation is called linear polarisation. (Yes/No)
4. **B** and **D** can be related. (Yes/No)
5. The characteristic impedance of a medium is $\sqrt{\frac{\epsilon}{\mu}}$ (Yes/No)
6. $\oint \mathbf{B}.d\mathbf{S} = Q$ (Yes/No)
7. $\nabla.\mathbf{E} = -\dot{\mathbf{B}}$ (Yes/No)
8. $\nabla.\mathbf{J} = \dot{\rho}_\upsilon$ (Yes/No)
9. In free space, $\nabla.\mathbf{D} = 0$. (Yes/No)
10. In free space, $\nabla \times \mathbf{H} = \mathbf{J}$. (Yes/No)
11. For static fields, $\nabla \times \mathbf{H} = \mathbf{D}$. (Yes/No)
12. The unit of **D** is wb/m^2. (Yes/No)
13. The unit of $\dot{\mathbf{D}}$ is A/m^2. (Yes/No)
14. The unit of **J** is A/m. (Yes/No)
15. **E** and **D** have the same direction. (Yes/No)
16. **B** and **H** have the same direction. (Yes/No)
17. Ampere's circuit law and Maxwell's first equation are the same. (Yes/No)
18. $\mathbf{B}_{n1}$ is always equal to $\mathbf{B}_{n2}$. (Yes/No)
19. $\mathbf{D}_{n1}$ is always equal to $\mathbf{D}_{n2}$. (Yes/No)
20. $\mathbf{E}_{\tan 1}$ is always equal to $\mathbf{E}_{\tan 2}$. (Yes/No)

21. $\mathbf{D}_{t1}$ is always equal to $\mathbf{D}_{t2}$. (Yes/No)

22. Surface current density in dielectrics is zero. (Yes/No)

23. $\rho_\upsilon = 0$ for free space. (Yes/No)

24. In dielectrics, displacement current density is greater than conduction current density. (Yes/No)

25. In conductors, $J_c > J_d$. (Yes/No)

26. $\nabla . \mathbf{B} = 0$ because there exists no isolated magnetic poles. (Yes/No)

27. The unit of vector magnetic potential is _________.

28. The unit of magnetic current density is _________.

29. If $\mathbf{E} = \cos(6 \times 10^7 t - \beta z)\, \mathbf{a}_x$, β is _________.

30. For time varying fields, $\mathbf{E}$ = _________.

Answers

1. Yes	2. Yes	3. Yes	4. Yes	5. No
6. No	7. No	8. No	9. Yes	10. No
11. No	12. No	13. Yes	14. No	15. Yes
16. Yes	17. No	18. Yes	19. No	20. Yes
21. No	22. Yes	23. Yes	24. Yes	25. Yes
26. Yes	27. Wb/m	28. V/m	29. 0.2 rad/m	30. $-\nabla V - \dot{\mathbf{A}}$

MULTIPLE CHOICE QUESTIONS

1. The first Maxwell's equation in free space

(a) $\nabla \times \mathbf{H} = \dot{\mathbf{D}} + \mathbf{J}$ (b) $\nabla \times \mathbf{H} = \dot{\mathbf{D}}$

(c) $\nabla \times \mathbf{H} = 0$ (d) $\nabla \times \mathbf{H} = \mathbf{J}$

2. Absolute permeability of free space is

(a) $4\pi \times 10^{-7}$ A/m (b) $4\pi \times 10^{-7}$ H/m

(c) $4\pi \times 10^{-7}$ F/m (d) $4\pi \times 10^{-7}$ H/m^2

3. For static magnetic field,

(a) $\nabla \times \mathbf{B} = \rho$ (b) $\nabla \times \mathbf{B} = \mu \mathbf{J}$

(c) $\nabla . \mathbf{B} = \mu_0 J$ (d) $\nabla \times \mathbf{B} = 0$

4. The electric field in free space

(a) $\dfrac{\mathbf{D}}{\epsilon_0}$ (b) $\dfrac{\mathbf{D}}{\mu_0}$ (c) $\epsilon_0 \mathbf{D}$ (d) $\dfrac{\sigma}{\epsilon_0}$

5. Displacement current density is

(a) $\mathbf{D}$ (b) $\mathbf{J}$ (c) $\partial \mathbf{D}/\partial t$ (d) $\partial \mathbf{J}/\partial t$

6. The time varying electric field is

(a) $\mathbf{E} = -\nabla V$ (b) $\mathbf{E} = -\nabla V - \dot{\mathbf{A}}$

(c) $\mathbf{E} = -\nabla V - \mathbf{B}$ (d) $\mathbf{E} = -\nabla V - \mathbf{D}$

7. A field can exist if it satisfies

(a) Gauss's law (b) Faraday's law

(c) Coulomb's law (d) All Maxwell's equations

8. If $\sigma = 2.0$ mho/m, $E = 10.0$ V/m, the conduction current density is

(a) 5.0 A/m^2 (b) 20.0 A/m^2 (c) 40.0 A/m^2 (d) 20 A

9. Maxwell's equations give the relations between

(a) different fields (b) different sources

(c) different boundary conditions (d) none of these

10. The boundary condition on $\mathbf{E}$ is

(a) $\mathbf{a}_n \times (\mathbf{E}_1 - \mathbf{E}_2) = 0$ (b) $\mathbf{a}_n . (\mathbf{E}_1 - \mathbf{E}_2) = 0$

(c) $\mathbf{E}_1 = \mathbf{E}_2$ (d) none of these

11. The boundary conditions on **H** is

(a) $\mathbf{a}_n \times (\mathbf{H}_1 - \mathbf{H}_2) = \mathbf{J}_s$ (b) $\mathbf{a}_n . (\mathbf{H}_1 - \mathbf{H}_2) = 0$

(c) $\mathbf{a}_n \times (\mathbf{H}_1 - \mathbf{H}_2) = \mathbf{J}_s$ (d) $\mathbf{a}_n . (\mathbf{H}_1 - \mathbf{H}_2) = 0$

12. If E = 2 V/m, of a wave in free space, (H) is

(a) $\dfrac{1}{60\pi}$ A/m (b) 60π A/m

(c) 120π A/m (d) 240π A/m

13. The cosine of the angle between the two vectors is

(a) sum of the products of the directions of the two vectors

(b) difference of the products of the directions of the two vectors

(c) product of the products of the directions of the two vectors

(d) none of these

14. The electric field intensity, E at a point (1, 2, 2) due to (1/9) nc located at (0, 0, 0) is

(a) 33 V/m (b) 0.333 V/m

(c) 0.33 V/m (d) zero

15. If **E** is a vector, then $\nabla . \nabla \times \mathbf{E}$ is

(a) 0 (b) 1

(c) does not exist (d) none of these

16. The Maxwell's equation, $\nabla . \mathbf{B} = 0$ is due to

(a) $\mathbf{B} = \mu \mathbf{H}$ (b) $\mathbf{B} = \dfrac{H}{\mu}$

(c) non-existence of a mono pole (d) none of these

17. For free space,

(a) $\sigma = \infty$ (b) $\sigma = 0$

(c) $J \neq 0$ (d) none of these

18. The electric field for time varying potentials

(a) $\mathbf{E} = -\nabla V$ (b) $\mathbf{E} = -\nabla V - \mathbf{A}$

(c) $\mathbf{E} = \nabla V$ (d) $\mathbf{E} = -\nabla V + \mathbf{A}$

19. The intrinsic impedence of the medium whose $\sigma = 0$, $\epsilon_r = 9$, $\mu_r = 1$ is

(a) $40\pi\Omega$ (b) 9Ω

(c) $120\pi\Omega$ (d) $60\pi\Omega$

20. For time varying EM fields
    (a) $\nabla \times \mathbf{H} = \mathbf{J}$
    (b) $\nabla \times \mathbf{H} = \dot{\mathbf{D}} + \mathbf{J}$
    (c) $\nabla \times \mathbf{E} = 0$
    (d) none of these

21. The wavelength of a wave with a propagation constant $= 0.1\pi + j0.2\pi$ is
    (a) 10 m
    (b) 20 m
    (c) 30 m
    (d) 25 m

22. The electric field just above a conductor is always
    (a) normal to the surface
    (b) tangential to source
    (c) zero
    (d) ∞

23. The normal components of **D** are
    (a) continuous across a dielectric boundary
    (b) discontinuous across a dielectric boundary
    (c) zero
    (d) ∞

24. If $J_c = 1\,\text{mA/m}^2$ in a medium whose conductivity is $\sigma = 10\,\text{Mho/m}$, E is
    (a) 0.1 V/m
    (b) 10μ V/m
    (c) 1.0μ V/m
    (d) 10 V/m.

25. If $J_d = 2\,\text{mA/m}^2$ in a medium whose $\epsilon_r = 2$, $\sigma = 4.95\,\text{Mho/m}$ at a frequency of 1 GHz, J_c is
    (a) 8.9 mA/m^2
    (b) 89 mA/m^2
    (c) 0.89 mA/m^2
    (d) 89 A/m^2

Answers

1. (b)	2. (b)	3. (b)	4. (a)	5. (c)
6. (b)	7. (d)	8. (b)	9. (a)	10. (a)
11. (b)	12. (a)	13. (a)	14. (c)	15. (a)
16. (c)	17. (b)	18. (a)	19. (a)	20. (b)
21. (a)	22. (a)	23. (a)	24. (a)	25. (b)

EXERCISE PROBLEMS

1. If **A** is the vector magnetic potential, prove $\mathbf{E} = -\nabla V - \frac{\partial \mathbf{A}}{\partial t}$.
2. If $\mathbf{E} = 2.0 \sin kx \cos \omega t\, \mathbf{a}_x$ in free space, find the volume charge density.
3. If $\mathbf{E} = 10 \cos(\omega t - kz)\, \mathbf{a}_x$ V/m, find **D**, **H**, **B** in free space.
4. If the electric field strength of an EM wave in free space has an amplitude of 5.0 V/m, find the magnetic field strength.
5. When the amplitude of a magnetic field in free space is 10 mA/m, find the magnitude of the electric field.
6. The electric field of an EM wave is $\mathbf{E} = 15 \cos \omega \left(t - \frac{z}{\upsilon_0} \right) \mathbf{a}_y$. Find **H**.
7. If $\tilde{\mathbf{E}} = 2 \cos(10^8 t - 20x + 40°)\, \mathbf{a}_z$, what is the phasor form of **E**?

CHAPTER

5

ELECTROMAGNETIC FIELDS AND WAVES

EM waves can neither be seen nor sensed nor are they audible.

The main aim of this chapter is to provide overall concepts of EM waves and their characteristics. They include:

- applications of EM waves
- wave equations and solutions
- propagation characteristics
- waves in conductors and dielectrics
- polarisation
- normal and oblique incidence of EM waves
- reflection and transmission coefficients
- Brewster angle and total internal reflection
- Surface impedance and Poynting theorem
- solved problems, points/formulae to remember, objective and multiple choice questions and exercise problems.

Do you know?

The electricity generated from every power socket has associated low frequency electromagnetic fields.

5.1 INTRODUCTION

A wave means a recurring function of time at a point.

Definition of wave It is defined as a physical phenomenon. In its re-occurrence, there is a time delay which is proportional to the space separation between two adjacent locations.

In general, wave is a carrier of energy or information and is a function of time as well as space.

As far as we are concerned, a wave means Electromagnetic wave (or simply EM wave). Maxwell predicted the existence of EM waves and established it through his well-known Maxwell's equations. The same EM waves were investigated by Heinrich Hertz. Hertz conducted several experiments which could generate and detect EM waves. These radio waves are called Hertzian waves.

Examples radio, radar beams and TV signals.

5.2 APPLICATIONS OF EM WAVES

They have a wide range of applications in all types of communications like police radio, television, satellite, ionospheric, tropospheric, wireless, cellular, mobile communications and so on and in all types of radars like Doppler radar, MTI radar, speed trap radar, airport surveillance radar, weather forecasting radar, remote sensing radar, ground mapping radar, IFF radar, astronomy radar, fire control radar and so on. EM waves are also used in radiation therapy, microwave ovens and others.

Advantage in using EM waves for communication purposes is that the medium between the transmitter and receiver requires no maintenance. This is because free space is the best medium for EM wave propagation.

5.3 WAVE EQUATIONS IN FREE SPACE

Wave equations in free space are given by

$$\nabla^2 \mathbf{E} = \mu_0 \epsilon_0 \ddot{\mathbf{E}}$$

$$\nabla^2 \mathbf{H} = \mu_0 \epsilon_0 \ddot{\mathbf{H}}$$

where $$\ddot{\mathbf{E}} = \frac{\partial^2 \mathbf{E}}{\partial t^2}$$

and $$\ddot{\mathbf{H}} = \frac{\partial^2 \mathbf{H}}{\partial t^2}$$

Proof Free space is characterised by $\epsilon_r = 1, \mu_r = 1$ or $\epsilon = \epsilon_0$, $\mu = \mu_0$, and $\sigma = 0$, $\rho_\upsilon = 0$ and $\mathbf{J} = 0$. Due to these characteristics of free space, Maxwell's second equation becomes,

$$\nabla \times \mathbf{E} = -\dot{\mathbf{B}}$$

$$= -\mu_0 \dot{\mathbf{H}} \qquad [\text{as } \mathbf{B} = \mu_0 \mathbf{H}]$$

Taking curl on both sides, we get

$$\nabla \times \nabla \times \mathbf{E} = -\mu_0 \nabla \times \dot{\mathbf{H}}$$

Using standard vector identity, LHS is given by

$$\nabla \times \nabla \times \mathbf{E} = \nabla \nabla . \mathbf{E} - \nabla^2 \mathbf{E}$$

and for the first Maxwell's equation, RHS is

$$-\mu_0 \nabla \times \dot{\mathbf{H}} = -\mu_0 \ddot{\mathbf{D}} = -\mu_0 \epsilon_0 \ddot{\mathbf{E}}$$

$$\nabla \nabla . \mathbf{E} - \nabla^2 \mathbf{E} = -\mu_0 \epsilon_0 \ddot{\mathbf{E}}$$

But $$\nabla . \mathbf{D} = \nabla . \epsilon_0 \mathbf{E} = \epsilon_0 \nabla . \mathbf{E} = 0$$

$$\nabla^2 \mathbf{E} = \epsilon_0 \mu_0 \ddot{\mathbf{E}}$$ Hence proved.

Now consider the first Maxwell's equation

$$\nabla \times \mathbf{H} = \epsilon_0 \dot{\mathbf{E}}$$

Taking curl on both sides

$$\nabla \times \nabla \times \mathbf{H} = \epsilon_0 \nabla \times \dot{\mathbf{E}}$$

$$\nabla \nabla . \mathbf{H} - \nabla^2 \mathbf{H} = \epsilon_0 (-\ddot{\mathbf{B}}) = -\mu_0 \epsilon_0 \ddot{\mathbf{H}}$$

$$\nabla^2 \mathbf{H} = \mu_0 \epsilon_0 \ddot{\mathbf{H}} \qquad [\text{as } \nabla . \mathbf{H} = 0]$$

Hence proved.

5.4 WAVE EQUATIONS FOR A CONDUCTING MEDIUM

Wave equations for a conducting medium ($\rho_\upsilon = 0$, $\sigma \neq 0$ and $\mathbf{J} \neq 0$) are given by

$$\nabla^2 \mathbf{E} = \mu \epsilon \ddot{\mathbf{E}} + \mu \sigma \dot{\mathbf{E}}$$

and $$\nabla^2 \mathbf{H} = \mu \epsilon \ddot{\mathbf{H}} + \mu \sigma \dot{\mathbf{H}}$$

Proof Consider the second Maxwell's equation

$$\nabla \times \mathbf{E} = -\dot{\mathbf{B}} = -\mu \dot{\mathbf{H}}$$

Take curl on both sides

$$\nabla \times \nabla \times \mathbf{E} = -\mu \nabla \times \dot{\mathbf{H}}$$

$$\nabla \nabla . \mathbf{E} - \nabla^2 \mathbf{E} = -\mu\, [\in \ddot{\mathbf{E}} + \dot{\mathbf{J}}]$$

$$= -\mu \in \ddot{\mathbf{E}} - \mu\sigma \dot{\mathbf{E}}$$

$$\nabla^2 \mathbf{E} = \mu \in \ddot{\mathbf{E}} + \mu\sigma \dot{\mathbf{E}} \qquad [\text{as } \nabla . \mathbf{E} = 0]$$

Hence proved.

Similarly, consider the first Maxwell's equation

$$\nabla \times \mathbf{H} = \dot{\mathbf{D}} + \mathbf{J} = \in \dot{\mathbf{E}} + \sigma \mathbf{E}$$

Take curl on both sides

$$\nabla \times \nabla \times \mathbf{H} = \in \nabla \times \dot{\mathbf{E}} + \sigma \nabla \times \mathbf{E}$$

$$\nabla \nabla . \mathbf{H} - \nabla^2 \mathbf{H} = \in (-\mu \ddot{\mathbf{H}}) - \mu\sigma \dot{\mathbf{H}}$$

$$\nabla^2 \mathbf{H} = \mu \in \ddot{\mathbf{H}} + \mu\sigma \dot{\mathbf{H}} \qquad [\text{as } \nabla . \mathbf{H} = 0]$$

Hence proved.

5.5 UNIFORM PLANE WAVE EQUATION

Definition of uniform plane wave An EM wave propagating in *x*-direction is said to be a **uniform plane wave** if its fields **E** and **H** are independent of *y* and *z*-directions.

It is defined as a wave whose electric and magnetic fields have constant amplitude over the equiphase surfaces. These waves exist only in free space at an infinite distance from the source.

A uniform plane wave propagating in *x*-direction has no *x*-components of **E** and **H**, that is, $E_x = 0$, $H_x = 0$.

The electric and magnetic fields of an EM wave are always perpendicular to each other. A typical wave is shown in Fig. 5.1.

$E_x = 0$ and $H_x = 0$ for a uniform plane wave

Proof The plane wave equation in free space is given by

$$\nabla^2 \mathbf{E} = \mu_0 \in_0 \ddot{\mathbf{E}}$$

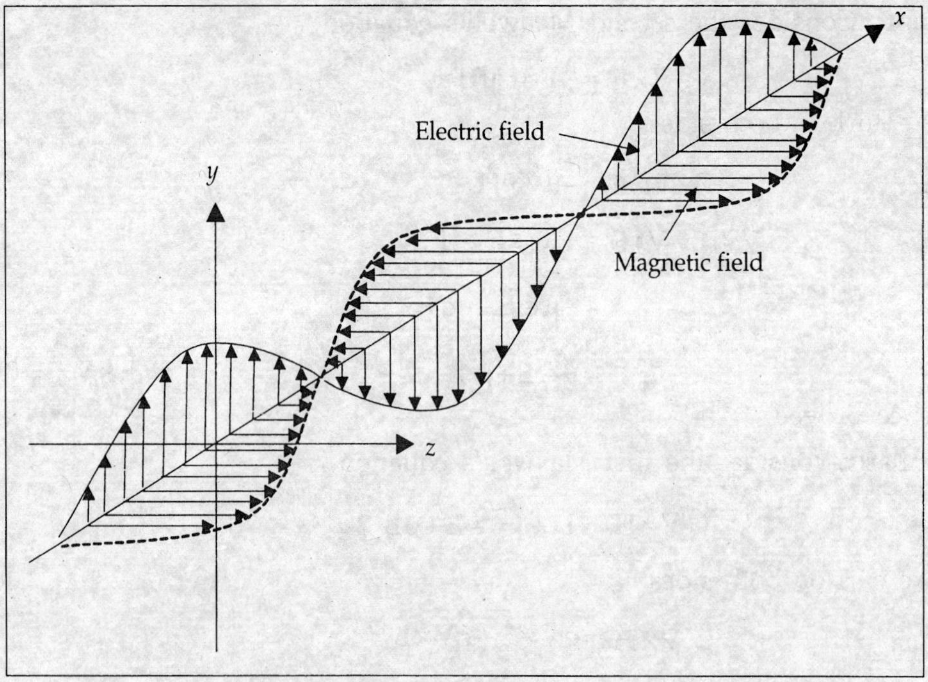

Fig. 5.1 Electromagnetic wave

that is,
$$\frac{\partial^2 \mathbf{E}}{\partial x^2} + \frac{\partial^2 \mathbf{E}}{\partial y^2} + \frac{\partial^2 \mathbf{E}}{\partial z^2} = \mu_0 \in_0 \frac{\partial^2 \mathbf{E}}{\partial t^2}$$

As per the definition of uniform plane wave,

$$\mathbf{E} \neq f\ (y)$$

and
$$\mathbf{E} \neq f\ (z)$$

$$\frac{\partial^2 \mathbf{E}}{\partial y^2} = 0, \frac{\partial^2 \mathbf{E}}{\partial z^2} = 0$$

Hence the wave equation becomes

$$\frac{\partial^2 \mathbf{E}}{\partial x^2} = \mu_0 \in_0 \frac{\partial^2 \mathbf{E}}{\partial t^2}$$

that is,
$$\frac{\partial^2 E_x}{\partial x^2}\mathbf{a}_x + \frac{\partial^2 E_y}{\partial x^2}\mathbf{a}_y + \frac{\partial^2 E_z}{\partial x^2}\mathbf{a}_z$$

$$= \mu_0 \in_0 \left[\frac{\partial^2 E_x}{\partial t^2}\mathbf{a}_x + \frac{\partial^2 E_y}{\partial t^2}\mathbf{a}_y + \frac{\partial^2 E_z}{\partial t^2}\mathbf{a}_z\right]$$

Equating the respective components on both sides, we get

$$\frac{\partial^2 E_x}{\partial x^2} = \mu_0 \epsilon_0 \frac{\partial^2 E_x}{\partial t^2} \tag{5.1}$$

$$\frac{\partial^2 E_y}{\partial x^2} = \mu_0 \epsilon_0 \frac{\partial^2 E_y}{\partial t^2}$$

$$\frac{\partial^2 E_z}{\partial x^2} = \mu_0 \epsilon_0 \frac{\partial^2 E_z}{\partial t^2}$$

Also we have

$$\nabla.\mathbf{D} = 0 \qquad [\text{as } \rho_\upsilon = 0]$$

or,

$$\nabla.\epsilon_0 \mathbf{E} = 0$$

that is,

$$\nabla.\mathbf{E} = 0$$

So,

$$\frac{\partial E_x}{\partial x} + \frac{\partial E_y}{\partial y} + \frac{\partial E_z}{\partial z} = 0$$

As

$$\frac{\partial E_y}{\partial y} = 0, \frac{\partial E_z}{\partial z} = 0, \text{ we have}$$

$$\frac{\partial E_x}{\partial x} = 0 \tag{5.2}$$

Substituting Equation (5.2) in (5.1), we get

$$\frac{\partial^2 E_x}{\partial t^2} = 0$$

This means that E_x should have one of the following solutions.

1. $E_x = 0$
2. $E_x = a$ constant with time
3. E_x increases uniformly with time, that is, $E_x = Kt$ where K is constant.

If $E_x = a$ constant and $E_x = Kt$, it will not be a part of wave motion. Therefore,

$$E_x = 0$$

Similarly, $H_x = 0$

This means that the components of electric and magnetic fields of a uniform plane wave in the direction of propagation are zero.

5.6 GENERAL SOLUTION OF UNIFORM PLANE WAVE EQUATION

The wave equation in free space is

$$\nabla^2 \mathbf{E} = \mu_0 \epsilon_0 \ddot{\mathbf{E}}$$

Applying the conditions of uniform plane wave equation, the above equation becomes

$$\nabla^2 \mathbf{E} = \frac{\partial^2 \mathbf{E}}{\partial x^2} \qquad [\text{as } \mathbf{E} \neq f(y), \mathbf{E} \neq f(z)]$$

$$\frac{\partial^2 \mathbf{E}}{\partial x^2} = \mu_0 \epsilon_0 \frac{\partial^2 \mathbf{E}}{\partial t^2} \tag{5.3}$$

Equating the respective components on either side and as $E_x = 0$, we have

$$\frac{\partial^2 E_y}{\partial x^2} = \mu_0 \epsilon_0 \frac{\partial^2 E_y}{\partial t^2}$$

$$\frac{\partial^2 E_z}{\partial x^2} = \mu_0 \epsilon_0 \frac{\partial^2 E_z}{\partial t^2}$$

Equation (5.3) has a general solution given by

$$\mathbf{E} = f_1(x - \upsilon_0 t) + f_2(x + \upsilon_0 t)$$

where f_1 and f_2 are functions of $(x - \upsilon_0 t)$ and $(x + \upsilon_0 t)$ respectively.

$$\upsilon_0 = \frac{1}{\sqrt{\mu_0 \epsilon_0}} = \text{velocity of propagation}$$

x is the direction of propagation of the wave.

$f_1(x - \upsilon_0 t)$ represents a forward wave and

$f_2(x + \upsilon_0 t)$ represents a reflected wave.

This reflected wave is present when there is a conductor which acts as a reflector. Otherwise, it is absent. As we are considering free space propagation, **E** will be $f_1(x - \upsilon_0 t)$ only, that is,

$$\mathbf{E} = f(x - \upsilon_0 t)$$

This is the solution of uniform plane wave equation in free space. The behaviour is represented typically in Fig. 5.2.

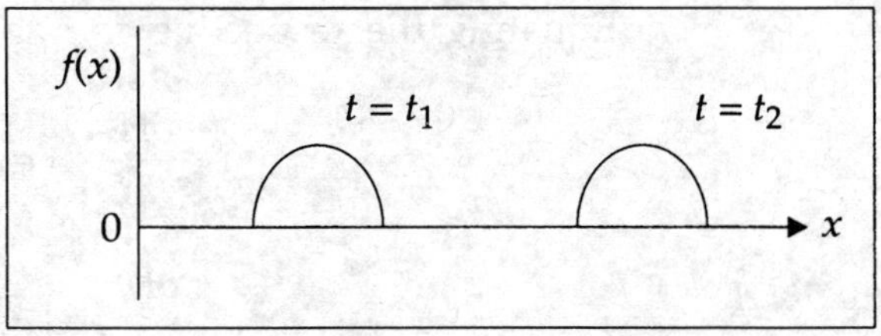

Fig. 5.2 A wave along *x*-direction

5.7 RELATION BETWEEN E AND H IN UNIFORM PLANE WAVE

The relation between **E** and **H** is

$$\left|\frac{\mathbf{E}}{\mathbf{H}}\right| = \frac{\mathrm{E}}{\mathrm{H}} = 120\pi\,\Omega \approx 377\,\Omega$$

Proof First Maxwell's equation is

$$\nabla\times\mathbf{H} = \begin{vmatrix} \mathbf{a}_x & \mathbf{a}_y & \mathbf{a}_z \\ \frac{\partial}{\partial x} & \frac{\partial}{\partial y} & \frac{\partial}{\partial z} \\ H_x & H_y & H_z \end{vmatrix} = \dot{\mathbf{D}} = \epsilon_0\,\dot{\mathbf{E}}$$

$$= \mathbf{a}_x\left[\frac{\partial H_z}{\partial y} - \frac{\partial H_y}{\partial z}\right] + \mathbf{a}_y\left[\frac{\partial H_x}{\partial z} - \frac{\partial H_z}{\partial x}\right] + \mathbf{a}_z\left[\frac{\partial H_y}{\partial x} - \frac{\partial H_x}{\partial y}\right] \tag{5.4}$$

But $\frac{\partial H_z}{\partial y} = 0,\ \frac{\partial H_y}{\partial z} = 0,\ H_x = 0$

Equation (5.4) becomes

$$\nabla\times\mathbf{H} = -\frac{\partial H_z}{\partial x}\mathbf{a}_y + \frac{\partial H_y}{\partial x}\mathbf{a}_z = \epsilon_0\,\dot{\mathbf{E}}$$

Similarly,

$$\nabla\times\mathbf{E} = -\frac{\partial E_z}{\partial x}\mathbf{a}_y + \frac{\partial E_y}{\partial x}\mathbf{a}_z = -\mu_0\,\dot{\mathbf{H}}$$

or,

$$-\frac{\partial H_z}{\partial x}\mathbf{a}_y + \frac{\partial H_y}{\partial x}\mathbf{a}_z = \epsilon_0\frac{\partial E_y}{\partial t}\mathbf{a}_y + \epsilon_0\frac{\partial E_z}{\partial t}\mathbf{a}_z$$

and $$-\frac{\partial E_z}{\partial x}\mathbf{a}_y + \frac{\partial E_y}{\partial x}\mathbf{a}_z = -\mu_0 \frac{\partial H_y}{\partial t}\mathbf{a}_y - \mu_0 \frac{\partial H_z}{\partial t}\mathbf{a}_z$$

Equating the respective components, we get

$$-\frac{\partial H_z}{\partial x} = \epsilon_0 \frac{\partial E_y}{\partial t} \tag{5.5}$$

$$\frac{\partial H_y}{\partial x} = \epsilon_0 \frac{\partial E_z}{\partial t} \tag{5.6}$$

$$\frac{\partial E_z}{\partial x} = \mu_0 \frac{\partial H_y}{\partial t}$$

$$\frac{\partial E_y}{\partial x} = -\mu_0 \frac{\partial H_z}{\partial t}$$

Writing E_y in the form of

$$E_y = f(x - \upsilon_0 t),\ \upsilon_0 = \frac{1}{\sqrt{\mu_0 \epsilon_0}}$$

that is, $$\frac{\partial E_y}{\partial t} = \frac{\partial f}{\partial (x - \upsilon_0 t)} \frac{\partial (x - \upsilon_0 t)}{\partial t} = -\upsilon_0 f' \tag{5.7}$$

where $$f' = \frac{\partial f}{\partial (x - \upsilon_0 t)}$$

From Equations (5.5) and (5.7), we have

$$\frac{\partial H_z}{\partial x} = -\epsilon_0 \frac{\partial E_y}{\partial t} = \epsilon_0 \upsilon_0 f'$$

or, $$H_z = \int \epsilon_0 \upsilon_0 f' dx + A$$

As the constant, A cannot be a part of wave motion, we can put $A = 0$.

So, $$H_z = \epsilon_0 \upsilon_0 \int f' dx$$

As $$\frac{\partial f}{\partial x} = f' \frac{\partial (x - \upsilon_0 t)}{\partial x} = f'$$

$$H_z = \sqrt{\frac{\epsilon_0}{\mu_0}} \int \frac{\partial f}{\partial x} dx$$

or, $$H_z = \sqrt{\frac{\epsilon_0}{\mu_0}} f = \sqrt{\frac{\epsilon_0}{\mu_0}} E_y$$

or, $$\frac{E_y}{H_z} = \sqrt{\frac{\mu_0}{\epsilon_0}}$$

Similarly, if we take

$$E_z = f(x - \upsilon_0 t)$$

$$\frac{\partial E_z}{\partial t} = -\upsilon_0 f' \tag{5.8}$$

From Equations (5.6) and (5.8), we have

$$\frac{\partial H_y}{\partial x} = \epsilon_0 \frac{\partial E_z}{\partial t} = -\epsilon_0 \upsilon_0 f'$$

$$H_y = -\epsilon_0 \upsilon_0 \int f' dx$$

$$= -\epsilon_0 \upsilon_0 f$$

$$= -\epsilon_0 \upsilon_0 E_z$$

or, $$\boxed{\frac{E_z}{H_y} = -\sqrt{\frac{\mu_0}{\epsilon_0}}}$$

$$\frac{E}{H} = \sqrt{\frac{E_y^2 + E_z^2}{H_y^2 + H_z^2}} = \sqrt{\frac{H_z^2\left(\frac{\mu_0}{\epsilon_0}\right) + H_y^2\left(\frac{\mu_0}{\epsilon_0}\right)}{H_y^2 + H_z^2}}$$

$$\frac{E}{H} = \sqrt{\frac{\mu_0}{\epsilon_0}} = \sqrt{\frac{4\pi \times 10^{-7}}{\frac{1}{36\pi \times 10^9}}} = 120\pi\Omega$$

$$\frac{E}{H} = 120\pi\Omega$$

Hence proved.

The intrinsic impedance or characteristic impedance, η_0 is defined as

$$\eta_0 = \frac{E}{H} = 120\pi\Omega$$

5.8 PROOF OF *E* AND *H* OF EM WAVE BEING PERPENDICULAR TO EACH OTHER

Consider

$$\mathbf{E}.\mathbf{H} = E_x H_x + E_y H_y + E_z H_z = E_y H_y + E_z H_z \quad [\text{as } E_x = 0,\ H_x = 0]$$

As

$$\frac{E_y}{H_z} = \sqrt{\frac{\mu_0}{\epsilon_0}},\ \frac{E_z}{H_y} = -\sqrt{\frac{\mu_0}{\epsilon_0}}$$

We get

$$\mathbf{E}.\mathbf{H} = E_x H_x + E_y H_y + E_z H_z$$

$$= \sqrt{\frac{\mu_0}{\epsilon_0}} H_y H_z - \sqrt{\frac{\mu_0}{\epsilon_0}} H_y H_z = 0$$

$$\mathbf{E}.\mathbf{H} = 0$$

The dot product of two vectors **E** and **H** is zero only when the two vectors are perpendicular to each other. Hence proved.

5.9 WAVE EQUATIONS IN PHASOR FORM

Wave equations in free space are

$$\nabla^2 \mathbf{H} = \mu_0 \epsilon_0 \ddot{\mathbf{H}} \quad \text{and}$$

$$\nabla^2 \mathbf{E} = \mu_0 \epsilon_0 \ddot{\mathbf{E}}$$

If

$$\tilde{\mathbf{E}} = \text{Re}\{\mathbf{E} e^{j\omega t}\}$$

$$\tilde{\mathbf{H}} = \text{Re}\{\mathbf{H} e^{j\omega t}\}$$

The wave equations become

$$\nabla^2 \mathbf{H} = -\omega^2 \mu_0 \epsilon_0 \mathbf{H}$$

$$\nabla^2 \mathbf{E} = -\omega^2 \mu_0 \epsilon_0 \mathbf{E}$$

These are wave equations in phasor form.

Note that a single time derivative of the field gives a factor of $j\omega$ and a double time derivative of the field gives a factor of $-\omega^2$.

Similarly, the wave equations in conductive medium are

$$\nabla^2 \mathbf{E} = (-\omega^2 \mu \in + j\omega\mu\sigma)\, \mathbf{E}$$
$$\nabla^2 \mathbf{H} = (-\omega^2 \mu \in + j\omega\mu\sigma)\mathbf{H}$$

These can be represented in the following form

$$\left.\begin{aligned} \nabla^2 \mathbf{E} &= \gamma^2 \mathbf{E} \\ \nabla^2 \mathbf{H} &= \gamma^2 \mathbf{H} \end{aligned}\right\}$$

where $\gamma =$ propagation constant $\left(\frac{1}{m}\right)$

$$\gamma = \sqrt{-\omega^2 \mu \in + j\omega\mu\sigma}$$

5.10 WAVE PROPAGATION IN LOSSLESS MEDIUM

The wave equation is

$$\nabla^2 \mathbf{E} = -\omega^2 \mu \in \mathbf{E}$$

that is,
$$\frac{\partial^2 \mathbf{E}}{\partial x^2} = -\omega^2 \mu \in \mathbf{E}$$

or,
$$\frac{\partial^2 \mathbf{E}}{\partial x^2} = -\beta^2 \mathbf{E}$$

where
$$\beta = \omega\sqrt{\mu \in}$$

The y-component of **E** may be written as

$$E_y = Ae^{-j\beta x} + Be^{j\beta x}$$

where A and B are arbitrary complex constants. Then

$$\tilde{E}_y(x, t) = \mathrm{Re}\{E_y(x)e^{j\omega t}\}$$
$$= \mathrm{Re}\{Ae^{j(\omega t - \beta x)} + Be^{j(\omega t + \beta x)}\}$$

If A and B are real, it becomes

$$\tilde{E}_y(x, t) = A\cos(\omega t - \beta x) + B\cos(\omega t + \beta x)$$

This is the sum of two waves. They travel in opposite directions. If $A = B$, the waves combine together and form a standing wave. Such waves do not progress.

The wave velocity (υ) It is defined as the velocity of propagation of the wave. It is also defined as

$$\upsilon \equiv \frac{\omega}{\beta}$$

where $\omega = 2\pi f =$ angular frequency

$\beta =$ phase shift constant, radians/m

Phase shift constant, β It is defined as a measure of the phase shift in radians per unit length.

Wavelength of the wave, λ It is defined as that distance through which the sinusoidal wave passes through a full cycle of 2π radians,

that is, $$\lambda \equiv \frac{2\pi}{\beta}$$

Phase velocity, (υ_p) It is defined as the velocity of some point in the sinusoidal waveform.

Intrinsic or **characteristic impedance** of a medium which has a finite value of conductivity is given by

$$\eta \equiv \sqrt{\frac{j\omega\mu}{\sigma + j\omega\epsilon}}$$

5.11 PROPAGATION CHARACTERISTICS OF EM WAVES IN FREE SPACE

The wave equation in free space is

$$\nabla^2 \mathbf{E} = -\omega^2 \mu_0 \epsilon_0 \mathbf{E}$$

$$= \gamma^2 \mathbf{E}$$

The propagation constant, $\gamma(\text{m}^{-1})$

$$\gamma = \sqrt{-\omega^2 \mu_0 \epsilon_0} = j\omega\sqrt{\mu_0 \epsilon_0}$$

$$= j\beta$$

The phase constant, $\beta\,(\text{rad/m})$

$$\beta = \omega\sqrt{\mu_0 \epsilon_0}$$

The propagation characteristics of EM wave in free space are:

1. Propagation constant, $\gamma = j\omega\sqrt{\mu_0 \epsilon_0}$, (m^{-1})

2. Phase shift constant, $\beta = \omega\sqrt{\mu_0 \epsilon_0}$, (rad/m)

3. Velocity of propagation, $\upsilon_0 = \frac{1}{\sqrt{\mu_0 \epsilon_0}}, \text{(m/s)}$

4. Velocity of propagation of an EM wave is the same as phase velocity,

$$\upsilon_p = \frac{\omega}{\beta} = \frac{\omega}{\omega\sqrt{\mu_0 \epsilon_0}} = \upsilon_0 \text{ (m/s)}$$

5. $\lambda = \frac{2\pi}{\beta} = \frac{2\pi}{\omega\sqrt{\mu_0 \epsilon_0}}, \text{(m)}$

6. $\frac{E}{H} = 120\pi\Omega = \sqrt{\frac{\mu_0}{\epsilon_0}}, \; (\Omega)$

7. Attenuation constant, $\alpha = 0$

Problem 5.1 If a wave with a frequency of 100 MHz propagates in free space, find the propagation constant.

Solution Propagation constant, γ in free space is

$$\gamma = j\omega\sqrt{\mu_0 \epsilon_0} = j\frac{\omega}{\upsilon_0} \qquad \left[\text{As } \upsilon_0 = \frac{1}{\sqrt{\mu_0 \epsilon_0}}\right]$$

$$= j\frac{2\pi f}{\upsilon_0} = j\frac{2\pi \times 100 \times 10^6}{3 \times 10^8}$$

$$= \frac{j2\pi \times 10^8}{3 \times 10^8} = j2.094$$

$$\boxed{\gamma = j2.0941 \text{m}^{-1}}$$

Problem 5.2 If H field is given by $\mathbf{H}(z, t) = 48 \cos(10^8 t + 40z)\mathbf{a}_y$, A/m, identify the amplitude, frequency and phase constant. Find the wavelength.

Solution Amplitude of the magnetic field

$$= 48 \text{ A/m}$$

$$\omega = 10^8$$

or,
$$f = \frac{10^8}{2\pi} = 15.915 \text{MHz}$$

$$\beta = 40 \text{ rad/m}$$

Now wavelength, $\lambda = \frac{2\pi}{\beta} = \frac{2\pi}{40} = 0.157 \text{ m}$

Problem 5.3 When the amplitude of the magnetic field in a plane wave is 2 A/m, (a) determine the magnitude of the electric field for the plane wave in free space (b) determine the magnitude of the electric field when the wave propagates in a medium which is characterised by $\sigma = 0, \mu = \mu_0$ and $\epsilon = 4\epsilon_0$.

Solution We have

(a) $$\frac{E}{H} = \eta_0 = \sqrt{\frac{\mu_0}{\epsilon_0}} = 120\pi\Omega \quad \text{for free space}$$

But $$H = 2 \text{ A/m}$$

$$E = \eta_0 H$$

$$= 120\pi \times 2 = 240\pi \text{ V/m}$$

$$\boxed{E = 240\pi \text{ V/m}}$$

(b) $\sigma = 0, \epsilon_r = 4, \mu_r = 1$

$$\eta_0 = \sqrt{\frac{\mu}{\epsilon}} = \sqrt{\frac{\mu_0}{4\epsilon_0}} = \frac{1}{2}\sqrt{\frac{\mu_0}{\epsilon_0}}$$

$$= \frac{1}{2} \times 120\pi = 60\pi\Omega$$

$$E = \eta_0 H = 60\pi \times 2 = 120\pi \text{ V/m}$$

$$\boxed{E = 120\pi \text{ V/m}}$$

Problem 5.4 If $\epsilon_r = 9, \mu = \mu_0$, for the medium in which a wave with a frequency, $f = 0.3$ GHz is propagating, determine the propagation constant and intrinsic impedance of the medium when $\sigma = 0$.

Solution The expression for propagation constant, γ is

$$\gamma = \sqrt{j\omega\mu(\sigma + j\omega\epsilon)} = j\omega\sqrt{\mu\epsilon} \text{ for } \sigma = 0$$

As $$\mu_r = 1, \epsilon_r = 9, \sigma = 0, f = 0.3 \text{ GHz}$$

$$\gamma = j\,2\pi \times 3 \times 10^8 \times \sqrt{4\pi \times 10^{-7} \times 9 \times 8.854 \times 10^{-12}}$$

$$= j\,6\pi\times10^{8}\times2\times3\sqrt{8.854\times10^{-19}\times\pi}$$

$$= j\,36\pi\times10^{8}\times10^{-9}\times1.66780$$

$$\boxed{\gamma = j\,18.8624\,(1/\text{m})}$$

Intrinsic impedance,

$$\eta = \sqrt{\frac{j\omega\mu}{\sigma + j\omega\epsilon}} = \sqrt{\frac{\mu}{\epsilon}}$$

$$\eta = \sqrt{\frac{\mu_0}{9\epsilon_0}} = \frac{1}{3}\sqrt{\frac{\mu_0}{\epsilon_0}} = \frac{1}{3}\times120\pi$$

$$\boxed{\eta = 40\pi\Omega}$$

Problem 5.5 The wavelength of an x-directed plane wave in a lossless medium is 0.25 m and the velocity of propagation is 1.5×10^{10} cm/s. The wave has z-directed electric field with an amplitude equal to 10 V/m. Find the frequency and permittivity of the medium. The medium has $\mu=\mu_0$.

Solution $\upsilon=1.5\times10^{10}\text{cm/sec}=1.5\times10^{8}\text{m/s}$

Frequency of the wave,

$$f=\frac{\upsilon}{\lambda}=\frac{1.5\times10^{8}}{25\times10^{-2}}=0.06\times10^{10}$$

$$\boxed{f=600\,\text{MHz}}$$

We have

$$\upsilon = \frac{1}{\sqrt{\mu\epsilon}} = \frac{1}{\sqrt{\mu_0\epsilon_0\epsilon_r}}$$

$$= \frac{\upsilon_0}{\sqrt{\epsilon_r}} = 1.5\times10^{8}$$

or,

$$\sqrt{\epsilon_r} = \frac{3\times10^{8}}{1.5\times10^{8}} = 2$$

$$\boxed{\epsilon_r = 4}$$

Problem 5.6 Identify frequency, phase constant when the electric field of an EM wave is given by $E=5.0\sin(10^{8}t-4.0x)\,\mathbf{a}_z$. Also find λ.

Solution $\mathbf{E} = 5.0 \sin(10^8 t - 4.0x)\mathbf{a}_z$

$$\omega = 10^8 \quad \text{or} \quad 2\pi f = 10^8$$

$$f = \frac{10^8}{2\pi} = 15.915 \text{ MHz}$$

$$\beta = 4.0 \text{ rad/m}$$

$$\lambda = \frac{v_0}{f} = \frac{3\times 10^8}{15.915\times 10^6} = 18.850 \text{ m}$$

5.12 PROPAGATION CHARACTERISTICS OF EM WAVES IN CONDUCTING MEDIUM

The wave equation in a conducting medium in phasor form is

$$\nabla^2 \mathbf{E} - \gamma^2 \mathbf{E} = 0 \tag{5.9}$$

where the propagation constant, γ is

$$\gamma = \sqrt{j\omega\mu(\sigma + j\omega\epsilon)}$$

$$= \sqrt{-\omega^2 \mu\epsilon + j\omega\mu\sigma}$$

or,

$$\gamma = \alpha + j\beta$$

where α is called the attenuation constant, dB/m

β is called phase constant, rad/m.

Phase constant, β is also called wave number and it is the imaginary part of propagation constant.

One solution of Equation (5.9) is

$$E(x) = E_0 e^{-\gamma x}$$

where x is the direction of propagation and in time varying form,

$$\tilde{\mathbf{E}}(x, \mathbf{t}) = \text{Re}\{E_0 e^{j\omega t}\}$$

$$= \text{Re}\{E_0 e^{-\gamma x + \omega t}\}$$

$$\tilde{\mathbf{E}}(x, \mathbf{t}) = \bar{e}^{\alpha x}\, \text{Re}\{E_0 e^{\omega t - \beta x}\}$$

This is the equation of EM wave propagating in x-direction and attenuated by a factor $e^{-\alpha x}$

Attenuation constant, α (dB/m) It is defined as a constant which indicates the rate at which the wave amplitude reduces as it propagates from one point to another. It is the real part of propagation constant.

Expressions for α and β in a conducting medium

$$\alpha = \omega\sqrt{\frac{\mu\epsilon}{2}\left(\sqrt{1+\frac{\sigma^2}{\omega^2\epsilon^2}}-1\right)}, \text{dB/m}$$

$$\beta = \omega\sqrt{\frac{\mu\epsilon}{2}\left(\sqrt{1+\frac{\sigma^2}{\omega^2\epsilon^2}}+1\right)}, \text{rad/m}$$

Proof From the wave equation, propagation constant, γ

$$\gamma = \sqrt{-\omega^2\mu\epsilon + j\omega\mu\sigma} = \alpha + j\beta$$

Squaring both sides, we get

$$\begin{aligned}\gamma^2 &= -\omega^2\mu\epsilon + j\omega\mu\sigma\\ &= (\alpha + j\beta)^2\\ &= \alpha^2 - \beta^2 + 2j\alpha\beta\end{aligned}$$

Equating real and imaginary parts,

$$\alpha^2 - \beta^2 = -\omega^2\mu\epsilon \tag{5.10}$$

$$2\alpha\beta = \omega\mu\sigma$$

or,

$$\alpha\beta = (\omega\mu\sigma)/2$$

Thus,

$$\alpha = \frac{\omega\mu\sigma}{2\beta} \quad \text{or} \quad \beta = \frac{\omega\mu\sigma}{2\alpha} \tag{5.11}$$

From Equations (5.10) and (5.11), we get

$$-\omega^2\mu\epsilon = \alpha^2 - \frac{\omega^2\mu^2\sigma^2}{4\alpha^2}$$

or,

$$-4\alpha^2\omega^2\mu\epsilon = 4\alpha^4 - \omega^2\mu^2\sigma^2$$

or, $4\alpha^4 + 4\alpha^2\omega^2\mu\epsilon - \omega^2\mu^2\sigma^2 = 0$

Dividing this by 4,

$$\alpha^4 + \alpha^2\omega^2\mu\epsilon - \frac{\omega^2\mu^2\sigma^2}{4} = 0$$

Adding and subtracting $\left(\dfrac{\omega^2 \mu \epsilon}{2}\right)^2$

$$\alpha^4 + \alpha^2 \omega^2 \mu \epsilon + \left(\frac{\omega^2 \mu \epsilon}{2}\right)^2 = \frac{\omega^2 \mu^2 \sigma^2}{4} + \left(\frac{\omega^2 \mu \epsilon}{2}\right)^2$$

or,
$$\left(\alpha^2 + \frac{\omega^2 \mu \epsilon}{2}\right)^2 = \frac{\omega^2 \mu^2 \sigma^2}{4} + \frac{\omega^4 \mu^2 \epsilon^2}{4}$$

or,
$$\left(\alpha^2 + \frac{\omega^2 \mu \epsilon}{2}\right)^2 = \frac{\omega^4 \mu^2 \epsilon^2}{4}\left(1 + \frac{\sigma^2}{\omega^2 \epsilon^2}\right)$$

Taking square root on either side, we get

$$\alpha^2 + \frac{\omega^2 \mu \epsilon}{2} = \sqrt{\frac{\omega^4 \mu^2 \epsilon^2}{4}\left(1 + \frac{\sigma^2}{\omega^2 \epsilon^2}\right)}$$

or,
$$\alpha^2 = \frac{\omega^2 \mu \epsilon}{2}\sqrt{1 + \frac{\sigma^2}{\omega^2 \epsilon^2}} - \frac{\omega^2 \mu \epsilon}{2} \tag{5.12}$$

$$\alpha = \pm\omega\sqrt{\frac{\mu \epsilon}{2}\left(\sqrt{1 + \frac{\sigma^2}{\omega^2 \epsilon^2}} - 1\right)}$$

But α cannot be (–)ve. Hence

$$\boxed{\alpha = \omega\sqrt{\frac{\mu \epsilon}{2}\left(\sqrt{1 + \frac{\sigma^2}{\omega^2 \epsilon^2}} - 1\right)}}$$

Hence proved.

Now substituting Equation (5.12) in Equation (5.10), we get

$$\frac{\omega^2 \mu \epsilon}{2}\sqrt{1 + \frac{\sigma^2}{\omega^2 \epsilon^2}} - \frac{\omega^2 \mu \epsilon}{2} - \beta^2 = \omega^2 \mu \epsilon$$

or, $$+\beta^2 = \frac{\omega^2 \mu\epsilon}{2}\sqrt{1+\frac{\sigma^2}{\omega^2\epsilon^2}} + \frac{\omega^2\mu\epsilon}{2}$$

$$\beta = \omega\sqrt{\frac{\mu\epsilon}{2}\left(\sqrt{1+\frac{\sigma^2}{\omega^2\epsilon^2}}+1\right)}$$ Hence proved.

5.13 SUMMARY OF PROPAGATION CHARACTERISTICS OF EM WAVES IN A CONDUCTING MEDIUM

1. Propagation constant, $\gamma = \sqrt{-\omega^2\mu\epsilon + j\omega\mu\sigma}, \left(\frac{1}{m}\right)$

2. Phase shift constant, $\beta = \omega\sqrt{\frac{\mu\epsilon}{2}\left(\sqrt{1+\frac{\sigma^2}{\omega^2\epsilon^2}}+1\right)}$, (rad/m)

3. Attenuation constant, $\alpha = \omega\sqrt{\frac{\mu\epsilon}{2}\left(\sqrt{1+\frac{\sigma^2}{\omega^2\epsilon^2}}-1\right)}$, (dB/m)

4. Velocity of propagation of EM wave,

$$\upsilon = f\lambda = \frac{\omega}{\beta} = \frac{1}{\sqrt{\frac{\mu\epsilon}{2}\left(\sqrt{1+\frac{\sigma^2}{\omega^2\epsilon^2}}+1\right)}}, \text{(m/s)}$$

5. $\lambda = \frac{2\pi}{\beta}$ (m) or $\lambda = \frac{1}{f\sqrt{\frac{\mu\epsilon}{2}\left(\sqrt{1+\frac{\sigma^2}{\omega^2\epsilon^2}}+1\right)}}$

6. Intrinsic impedance, η

$$\frac{E}{H} = \sqrt{\frac{j\omega\mu}{\sigma + j\omega\epsilon}}\ \Omega$$

Problem 5.7 Earth has a conductivity of $\sigma = 10^{-2}$ mho/m, $\epsilon_r = 10$, $\mu_r = 2$. What are the conducting characteristics of the earth at

(a) $f = 50\,\text{Hz}$ (b) $f = 1\,\text{kHz}$ (c) $f = 1\,\text{MHz}$

(d) $f = 100\,\text{MHz}$ (e) $f = 10\,\text{GHz}$

Solution The parameters of earth are $\sigma = 10^{-2}$ mho/m, $\epsilon_r = 10$, $\mu_r = 2$.

$$\frac{\sigma}{\omega\epsilon} = \frac{10^{-2}}{2\pi f \times 10 \times \frac{1}{36\pi} \times 10^{-9}}$$

$$= \frac{10^{-2} \times 18}{10^{-8} \times f} = \frac{18 \times 10^6}{f}$$

(a) At $f = 50$ Hz

$$\frac{\sigma}{\omega\epsilon} = \frac{18 \times 10^6}{50} = 0.36 \times 10^6 = 3.6 \times 10^5$$

So this is >> 1. Hence it behaves like a good conductor.

(b) At $f = 1$ kHz,

$$\frac{\sigma}{\omega\epsilon} = \frac{18 \times 10^6}{f} = \frac{18 \times 10^6}{10^3} = 18 \times 10^3$$

This is >> 1. Hence it behaves like a good conductor.

(c) At $f = 1$ MHz,

$$\frac{\sigma}{\omega\epsilon} = \frac{18 \times 10^6}{10^6} = 18$$

It behaves like a moderate conductor.

(d) At $f = 100$ MHz,

$$\frac{\sigma}{\omega\epsilon} = \frac{18 \times 10^6}{100 \times 10^6} = 0.18$$

Earth behaves like a quasi-dielectric.

(e) At $f = 10$ GHz,

$$\frac{\sigma}{\omega\epsilon} = \frac{18 \times 10^6}{10 \times 10^9} = 18 \times 10^{-4}$$

that is, $\frac{\sigma}{\omega \epsilon} << 1.$

Earth behaves like a good dielectric.

Problem 5.8 A medium like copper conductor which is characterised by the parameters $\sigma = 5.8 \times 10^7$ mho/m, $\epsilon_r = 1, \mu_r = 1$ supports a uniform plane wave of frequency 60 Hz. Find the attenuation constant, propagation constant, intrinsic impedance, wavelength and phase velocity of the wave.

Solution Let us obtain the ratio

$$\frac{\sigma}{\omega \epsilon} = \frac{5.8 \times 10^7}{2\pi \times 60 \times 8.854 \times 10^{-12}} = 173 \times 10^{14}$$

This is >> 1. Therefore, it is a very good conductor.

Attenuation constant

$$= \alpha = \left(\frac{\omega \mu \sigma}{2}\right)^{\frac{1}{2}} = 117.2\,\mathrm{m}^{-1}$$

Phase constant

$$= \beta = \left(\frac{\omega \mu \sigma}{2}\right)^{\frac{1}{2}} = 117.2\left(\frac{1}{m}\right)$$

Propagation constant

$$= \gamma = \alpha + j\beta = 117.2 + j117.2\left(\frac{1}{m}\right)$$

Wavelength

$$= \lambda = \frac{2\pi}{\beta} = \frac{2\pi}{117.2} = 0.0536\,\mathrm{m}$$

Intrinsic impedance, η

$$= \sqrt{\frac{j\omega\mu}{\sigma}} = (2.022 + j2.022) \times 10^{-6}\,\Omega$$

Phase velocity of wave,

$$\upsilon = \lambda f = 0.0536 \times 60 = 3.216\,\mathrm{m/s}$$

$\alpha = 117.2\,\mathrm{m}^{-1}$ $\quad\quad$ $\beta = 117.2\,\mathrm{m}^{-1}$

$$\gamma = 117.2 + j117.2\ \text{m}^{-1} \qquad \lambda = 5.36\ \text{cm}$$

$$\eta = (2.022 + j2.022)\ \mu\Omega \qquad \upsilon = 3.216\ \text{m/s}$$

5.14 CONDUCTORS AND DIELECTRICS

As some media behave like good conductors at one frequency range and like good dielectrics at some other frequency range, the conventional definitions of conductors and dielectrics are not satisfactory in communication through EM waves.

The displacement current density,

$$\mathbf{J}_d = \dot{\mathbf{D}} = \epsilon\,\dot{\mathbf{E}} = j\omega\epsilon\,\mathbf{E}$$

and the conduction current density,

$$\mathbf{J}_c = \sigma\mathbf{E}$$

$$\frac{\mathbf{J}_c}{\mathbf{J}_d} = \frac{\sigma}{\omega\epsilon}$$

Definition of a good conductor

If $\dfrac{\sigma}{\omega\epsilon} >> 1$, the medium is a good conductor.

Definition of a good dielectric

If $\dfrac{\sigma}{\omega\epsilon} << 1$, the medium is a good dielectric.

Dissipation factor of a dielectric material $(\mathbf{D}_f)$ is defined as

$$D_f \equiv \frac{\sigma}{\omega\epsilon}$$

When D_f is small for a specific type of dielectric material, the dissipation factor is practically the same as its power factor which is equal to sin ϕ. Here, $\phi = \tan^{-1} D_f$.

5.15 WAVE PROPAGATION CHARACTERISTICS IN GOOD DIELECTRICS

Attenuation constant, α is given by

$$\alpha = \omega\sqrt{\frac{\mu\epsilon}{2}\left(\sqrt{1+\frac{\sigma^2}{\omega^2\epsilon^2}} - 1\right)} \tag{5.13}$$

For dielectrics, $\frac{\sigma}{\omega \in} << 1$

Expanding $\left(1+\frac{\sigma^2}{\omega^2 \epsilon^2}\right)^{\frac{1}{2}}$ by Binomial series, higher order terms can be neglected.

$$\sqrt{1+\frac{\sigma^2}{\omega^2 \epsilon^2}} \approx \left(1+\frac{\sigma^2}{2\omega^2 \epsilon^2}\right)$$

Now Equation (5.13) becomes

$$\alpha \approx \omega \sqrt{\frac{\mu\epsilon}{2}\left[\left(1+\frac{\sigma^2}{2\omega^2 \epsilon^2}\right)-1\right]}$$

$$= \omega \sqrt{\frac{\mu\epsilon\sigma^2}{4\omega^2 \epsilon^2}}$$

$$= \omega \sqrt{\frac{\mu\sigma^2}{4\omega^2 \epsilon}}$$

$$\boxed{\alpha = \frac{\sigma}{2}\sqrt{\frac{\mu}{\epsilon}}}$$

The phase shift constant, β is given by

$$\beta = \omega \sqrt{\frac{\mu\epsilon}{2}\left(\sqrt{1+\frac{\sigma^2}{\omega^2 \epsilon^2}}+1\right)}$$

$$\approx \omega \sqrt{\frac{\mu\epsilon}{2}\left(1+\frac{\sigma^2}{2\omega^2 \epsilon^2}+1\right)}$$

that is,

$$\beta \approx \omega \sqrt{\frac{\mu\epsilon}{2}\left(2+\frac{\sigma^2}{2\omega^2 \epsilon^2}\right)}$$

$$= \omega \sqrt{\mu\epsilon\left(1+\frac{\sigma^2}{4\omega^2 \epsilon^2}\right)}$$

$$\beta \approx \omega\sqrt{\mu\epsilon}\left(1+\frac{\sigma^2}{8\omega^2\epsilon^2}\right)$$

The velocity of propagation is

$$\upsilon = \frac{\omega}{\beta} = \frac{\omega}{\omega\sqrt{\mu\epsilon}\left(1+\frac{\sigma^2}{8\omega^2\epsilon^2}\right)}$$

$$\approx \frac{1}{\sqrt{\mu\epsilon}\left(1+\frac{\sigma^2}{8\omega^2\epsilon^2}\right)}$$

$$\upsilon \approx \frac{1}{\sqrt{\mu\epsilon}}\left(1-\frac{\sigma^2}{8\omega^2\epsilon^2}\right)$$

Intrinsic or characteristic impedance of general medium, η

$$\eta = \sqrt{\frac{j\omega\mu}{\sigma + j\omega\epsilon}} = \sqrt{\frac{j\omega\mu}{j\omega\epsilon}\left(\frac{1}{1+\frac{\sigma}{j\omega\epsilon}}\right)} = \sqrt{\frac{\mu}{\epsilon}\left(\frac{1}{1+\frac{\sigma}{j\omega\epsilon}}\right)}$$

$$\eta \approx \sqrt{\frac{\mu}{\epsilon}}\left(1+j\frac{\sigma}{2\omega\epsilon}\right)$$

5.16 SUMMARY OF THE PROPAGATION CHARACTERISTICS OF EM WAVES IN GOOD DIELECTRICS

1. Attenuation constant, $\alpha \approx \frac{\sigma}{2}\sqrt{\frac{\mu}{\epsilon}}$, dB/m

2. Phase shift constant, $\beta \approx \omega\sqrt{\mu\epsilon}\left(1+\frac{\sigma^2}{8\omega^2\epsilon^2}\right)$, rad/m

3. Velocity of the wave, $\upsilon = \frac{\omega}{\beta} \approx \frac{1}{\sqrt{\mu\epsilon}}\left(1 - \frac{\sigma^2}{8\omega^2\epsilon^2}\right)$, m/s

4. Intrinsic impedance, $\eta \approx \sqrt{\frac{\mu}{\epsilon}}\left(1 + j\frac{\sigma}{2\omega\epsilon}\right)$, Ω

5.17 WAVE PROPAGATION CHARACTERISTICS IN GOOD CONDUCTORS

The propagation constant, γ is given by

$$\gamma = \sqrt{-\omega^2\mu\epsilon + j\omega\mu\sigma}$$

$$= \sqrt{j\omega\mu\sigma}\sqrt{\left(1 + j\frac{\omega\epsilon}{\sigma}\right)}$$

$$\approx \sqrt{j\omega\mu\sigma} \quad \left[\text{as} \frac{\omega\epsilon}{\sigma} << 1 \text{ for good conductors}\right]$$

$$= \sqrt{\omega\mu\sigma}\,\angle 45°$$

Thus, $\gamma = \alpha + j\beta = \sqrt{\omega\mu\sigma}\,\angle 45°$

or, $\alpha = \beta = \sqrt{\frac{\omega\mu\sigma}{2}}$

The velocity of the wave in a conductor is

$$\upsilon = \frac{\omega}{\beta} = \sqrt{\frac{2\omega}{\mu\sigma}}$$

and the intrinsic impedance of the conductor is

$$\eta = \sqrt{\frac{j\omega\mu}{\sigma + j\omega\epsilon}}$$

$$= \sqrt{\frac{\mu}{\epsilon}\left(\frac{1}{1 + \frac{\sigma}{j\omega\epsilon}}\right)}$$

$$= \sqrt{\frac{j\omega\mu\epsilon}{\sigma\epsilon}}$$

$$\eta = \sqrt{\frac{j\omega\mu}{\sigma}}$$

5.18 SUMMARY OF CHARACTERISTICS OF WAVE PROPAGATION IN GOOD CONDUCTORS

$$\alpha = \sqrt{\frac{\omega\mu\sigma}{2}}, \text{dB/m}$$

$$\beta = \sqrt{\frac{\omega\mu\sigma}{2}}, \text{rad/m}$$

$$\upsilon = \sqrt{\frac{2\omega}{\mu\sigma}}, \text{m/s}$$

$$\eta = \sqrt{\frac{j\omega\mu}{\sigma}}, \Omega$$

5.19 DEPTH OF PENETRATION, δ (m)

Definition The depth of penetration is defined as that depth at which the wave attenuates to $\frac{1}{e}$ or approximately 37 per cent of its original amplitude. Depth of penetration is also called **skin depth**. It is a measure of depth to which an EM wave can penetrate the medium.

The depth of penetration

$$\delta = \frac{1}{\alpha}, \alpha \text{ being attenuation constant}$$

The conductivity of the medium attenuates the wave during propagation. At radio frequencies, the rate of attenuation is more in good conductors.

Proof of $\delta = \frac{1}{\alpha}$

Let the wave attenuation be represented by

$$E = E_0 e^{-\alpha z}$$

z being the direction of propagation.

At $$z = \delta$$

$$E = E_0 e^{-\alpha\delta} \tag{5.14}$$

As per the definition of δ, we have

$$E = E_0 e^{-1} \quad \text{at } z = \delta \tag{5.15}$$

From Equations (5.14) and (5.15), we have

$$E_0 e^{-1} = E_0 e^{-\alpha\delta}$$

$$\alpha\delta = 1$$

or, $$\boxed{\delta = \frac{1}{\alpha}}$$ Hence proved.

Problem 5.9 Find the depth of penetration, δ of an EM wave in copper at f = 60 Hz and f = 100 MHz. For copper, $\sigma = 5.8 \times 10^7$ mho/m, $\mu_r = 1, \epsilon_r = 1$.

Solution For copper, at f = 60 Hz,

$$\frac{\sigma}{\omega\epsilon} = \frac{5.8 \times 10^7}{2\pi \times 60 \times 8.854 \times 10^{-12}}$$

$$= 174 \times 10^{14} >> 1.$$

Therefore, at f = 60 Hz, copper is a very good conductor.

The depth of penetration,

$$\delta = \frac{1}{\alpha} = \sqrt{\frac{2}{\omega\mu\sigma}} = \sqrt{\frac{2}{2\pi \times 60 \times 4\pi \times 10^{-7} \times 5.8 \times 10^7}}$$

or, $$\boxed{\delta = 8.53 \times 10^{-3} \text{ m}}$$

At f = 100 MHz,

$$\frac{\sigma}{\omega\epsilon} = \frac{5.8 \times 10^7}{2\pi \times 8.854 \times 10^{-4}} = 0.10425 \times 10^{11}$$

$$= 10.425 \times 10^9 >> 1$$

Copper is a very good conductor at f = 100 MHz.

The depth of penetration,

$$\delta = \frac{1}{\alpha} = \sqrt{\frac{2}{\omega\mu\sigma}}$$

$$= \sqrt{\frac{2}{2\pi \times 100 \times 10^{6} \times 4\pi \times 10^{-7} \times 5.8 \times 10^{7}}}$$

or,

$$\boxed{\delta = 6.608 \times 10^{-6} \text{ m}}$$

5.20 POLARISATION OF A WAVE

Definition Polarisation of a wave is defined as the direction of the electric field at a given point as a function of time.

The polarisation of a composite wave is the direction of the electric field.

Types of Polarisations

These are of three types, namely

(a) Linear polarisation

(b) Circular polarisation

(c) Elliptical polarisation

Linear Polarisation

A wave is said to be linearly polarised if the electric field remains along a straight line as a function of time at some point in the medium. Linear polarisation of a wave is again of three types, namely

(i) Horizontal polarisation

(ii) Vertical polarisation

(iii) Theta polarisation

When a wave travels in *z*-direction with $\tilde{\mathbf{E}}$ and $\tilde{\mathbf{H}}$ fields lying in *xy*-plane, if $\tilde{E}_y = 0$ and $\tilde{E}_x$ is present, it is said to be ***x*-polarised** or **horizontally polarised.**

If $\tilde{E}_y$ is only present and $\tilde{E}_x = 0$ the wave is said to be **vertically (*y*-polarised) polarised.** On the other hand, if $\tilde{E}_x$ and $\tilde{E}_y$ are present and are in phase then the wave is said to be θ-polarised. This is given by

$$\theta = \left(\tan^{-1}\frac{E_y}{E_x}\right)$$

Circular Polarisation

A wave is said to be circularly polarised when the electric field traces a circle. If $\tilde{E}_x$ and $\tilde{E}_y$ have equal magnitudes and a 90 degree phase difference, the locus of the resultant $\tilde{E}$ is a circle and the wave is **circularly polarised.**

Let **E** of a uniform plane wave travelling in the z-direction be represented by

$$\mathbf{E}_{(z)} = E_c\, e^{-j\beta z}$$

and in time varying form

$$\tilde{\mathbf{E}}_{(z,t)} = \text{Re}\{E_c e^{-j\beta z} e^{j\omega t}\}$$

As the wave moves in z-direction, $\tilde{\mathbf{E}}$ and $\tilde{\mathbf{H}}$ lie in x-y plane.

Here, $\mathbf{E}_c$ is a complex vector, that is, $\mathbf{E}_c$ can be written as

$$\mathbf{E}_c = \mathbf{E}_1 + j\,\mathbf{E}_2$$

where $\mathbf{E}_1$ and $\mathbf{E}_2$ are real vectors.

At some point in space (say $z = 0$) $\tilde{\mathbf{E}}_{(z,\,t)}$ becomes

$$\tilde{\mathbf{E}}_{(0,\,t)} = \text{Re}\,[(\mathbf{E}_1 + j\mathbf{E}_2)\,e^{j\omega t}]$$

$$= E_1 \cos\omega t - E_2 \sin\omega t$$

As per the definition of circular polarisation, the electric vector at $z = 0$ is expressed as

$$\mathbf{E}_c = E_k\,\mathbf{a}_x + jE_k\,\mathbf{a}_y$$

$$\tilde{E}\,(0,\,t) = E_k \cos\omega t\,\mathbf{a}_x - E_k \sin\omega t\,\mathbf{a}_y$$

that is,

$$\tilde{E}_x = E_k \cos\omega t$$

$$\tilde{E}_y = -E_k \sin\omega t$$

So,

$$\boxed{\tilde{E}_x^2 + \tilde{E}_y^2 = E_k^2}$$

This represents a circle.

Elliptical Polarisation

If $\tilde{E}_x$ and $\tilde{E}_y$ are not equal in magnitude and they differ by 90° phase, then the tip of the resultant electric vector traces an ellipse. The wave is said to be **elliptically polarised**.

Here $\mathbf{E}_c$ can be written as

$$\mathbf{E}_c = a\mathbf{a}_x + jb\mathbf{a}_y \qquad \text{where } a \text{ and } b \text{ are constants}$$

$$\tilde{\mathbf{E}}(0, t) = a\cos\omega t\mathbf{a}_x - b\sin\omega t\mathbf{a}_y$$

$$\tilde{E}_x = a\cos\omega t$$

$$\tilde{E}_y = -b\sin\omega t$$

or,

$$\frac{\tilde{E}_x^2}{a^2} + \frac{\tilde{E}_y^2}{b^2} = 1$$

This is the equation of an ellipse and hence the wave is said to be elliptically polarised.

5.21 SOURCES OF DIFFERENT POLARISED EM WAVES

Horizontal dipole produces horizontally polarised waves.

Vertical dipole produces vertically polarised waves.

Inclined dipole produces θ-polarised waves.

Circular slots produce circularly polarised waves.

Elliptical slots produce elliptically polarised waves.

5.22 DIRECTION COSINES OF A VECTOR FIELD

Definition The direction cosine of a vector field is defined as the cosine of the angle made by the vector with the required coordinate axis.

Consider a vector **E** which is arbitrarily oriented with respect to the Cartesian coordinate axes. Assume **E** makes angles θ_x, θ_y and θ_z with x, y and z-axes (Fig. 5.3). Then the component of a vector in a given direction is the projection of the vector, **E** on a line in that direction, that is,

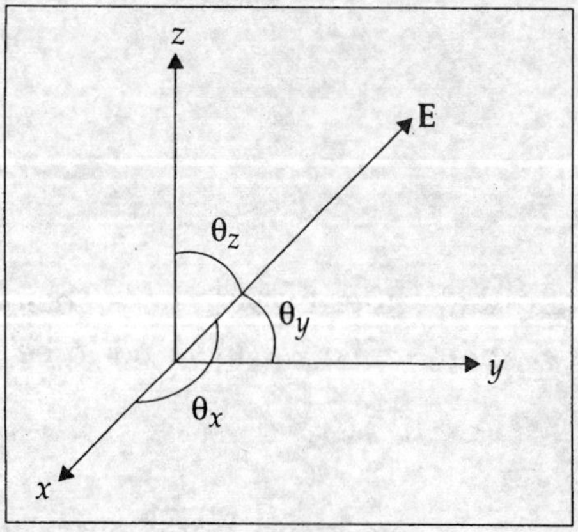

Fig. 5.3 Arbitrarily oriented vector, **E**

$$E_x = \mathbf{E}.\mathbf{a}_x = E\cos\theta_x$$

$$E_y = \mathbf{E}.\mathbf{a}_y = E\cos\theta_y$$

$$E_z = \mathbf{E}.\mathbf{a}_z = E\cos\theta_z$$

$\cos\theta_x$, $\cos\theta_y$ and $\cos\theta_z$ are known as direction cosines of the vector along the coordinate axes.

Problem 5.10 A vessel under sea water requires a minimum signal level of 20μ V/m. What is the depth in the sea that can be reached by a 4.0 MHz plane wave from an aeroplane? The wave has an electric field intensity of 100 V/m. The propagation is vertical into the sea. For sea water, $\sigma = 4$ mho/m, $\mu_r = 1$, $\epsilon_r = 81$.

Solution At $f = 4.0\,\text{MHz} = 4.0\times10^6$ Hz

For sea water

$$\sigma = 4.0\ \text{mho/m}$$

$$\mu_r = 1$$

$$\epsilon_r = 81$$

$$\frac{\sigma}{\omega\epsilon} = \frac{4}{2\pi\times4\times10^6\times8.854\times10^{-12}\times81}$$

$$= \frac{10^6}{2253.06} = 221.9 >> 1$$

Hence sea water is a good conductor.

Intrinsic impedance of sea water,

$$\eta_2 = \sqrt{\frac{\omega\mu}{\sigma}}\angle 45^\circ = \sqrt{\frac{2\pi\times10^6\times4\times4\pi\times10^{-7}}{4}}\angle 45^\circ$$

$$= 2.8\angle 45^\circ\Omega$$

η_1 (free space) $= 377\Omega$

α_1 (free space) $= 0$

$$\beta_1\ (\text{free space}) = \frac{\omega}{\upsilon} = \frac{2\pi\times4\times10^6}{3\times10^8} = 8.377\times10^{-2}\,(\text{m}^{-1})$$

Transmission coefficient

$$= \frac{2\eta_2}{\eta_1+\eta_2}$$

$$\eta_2 = 2.8\ \angle 45^\circ = 1.9798 + j1.9798$$

Transmission coefficient

$$= \frac{3.9596 + j3.9596}{377 + 1.9798 + j1.9798}$$

$$= \frac{5.6\angle 45^\circ}{378.9798 + j1.9798} = 14.77 \times 10^{-3} \angle 44.70$$

The transmitted electric field

$$= 14.77 \times 10^{-3} \times 100 = 1.477 \text{ V/m}$$

Propagation takes place in the form of $e^{-\alpha_2 x}$. Therefore, the distance at which the signal becomes 20μ V/m is found out from

$$1.477 e^{-\alpha_2 d} = 20 \times 10^{-6}$$

or,
$$e^{-\alpha_2 d} = \frac{20 \times 10^{-6}}{1.477} = 13.54 \times 10^{-6}$$

or,
$$-\alpha_2 d = \ln 13.54 \times 10^{-6}$$
$$= -6 \ln 13.54$$
$$= -15.6338$$
$$d = \frac{+15.6338}{\alpha_2}$$

But
$$\alpha_2 = \sqrt{\frac{\omega\mu\sigma}{2}}$$
$$= \sqrt{\frac{2\pi \times 4 \times 10^6 \times 4 \times 4\pi \times 10^{-7}}{2}}$$
$$= 8\pi \times \sqrt{10^{-1}}$$
$$= 25 \times 0.316$$
$$= 7.905\ (\text{m}^{-1})$$

$$d = \frac{15.6338}{7.905} = 1.977 \text{m}$$

$$\boxed{d = 1.977 \text{m}}$$

that is, the signal will reach the vessel when it is at a depth of 1.977 m.

Problem 5.11 The electric field of a plane wave propagating in a medium is given by $\mathbf{E} = 4.0e^{-\alpha x}\cos(10^9 \pi t) - \beta x)\,\mathbf{a}_z$ V/m. The medium is characterised by $\epsilon_r = 49$, $\mu_r = 4$ and $\sigma = 4$ mho/m. Find the magnetic field of the wave.

Solution Here $\omega = \pi \times 10^9$

The ratio $\quad \dfrac{\sigma}{\omega\epsilon} = \dfrac{36\times 4}{49} = 2.938$

This is neither << 1 nor >> 1

$$|\eta| = \left|\sqrt{\frac{j\omega\mu}{\sigma + j\omega\epsilon}}\right| = \left|\sqrt{\frac{j\pi\times 10^9 \times 4\times 4\pi\times 10^{-7}}{4 + j\pi\times 10^9 \times 49 \times \dfrac{1}{36\pi\times 10^9}}}\right|$$

$$= \left|\sqrt{\frac{j16\pi^2\times 10^2\times 36}{4\times 36 + j49}}\right| = \frac{240\pi\ \angle 45^\circ}{\sqrt{144 + j49}}$$

$$= \frac{240\pi\ \angle 45^\circ}{(20736 + 240)\ \angle 18^\circ} = \frac{240\pi\ \angle 45^\circ}{\sqrt{152.10}\ \angle 9.39^\circ}$$

$$= 61.13\ \angle 35.604^\circ$$

$$\boxed{\mathbf{H} = -0.065\cos(10^9\pi t - \beta x - 35.604^\circ)\,\mathbf{a}_y}$$

Problem 5.12 An elliptical polarised wave has an electric field of

$$\mathbf{E} = \sin(\omega t - \beta z)\,\mathbf{a}_x + 2\sin(\omega t - \beta z + 75^\circ)\,\mathbf{a}_y \text{ V/m.}$$

Find the power per unit area conveyed by the wave in free space.

Solution

$$E_x = \sin(\omega t - \beta z), \text{ V/m}$$

$$E_y = 2\sin(\omega t - \beta z + 75^\circ), \text{ V/m}$$

According to Poynting theorem, we have

$$\mathbf{P}_{\text{inst}} = \mathbf{E}\times\mathbf{H}$$

$$P_{av} = \frac{1}{2}\frac{E^2}{\eta_0} = \frac{1}{2}\frac{(E_x^2 + E_y^2)}{\eta_0}$$

$$= \frac{1}{2}\left(\frac{1+4}{377}\right)$$

$$\boxed{P_{av} = 0.00663 \text{ w/m}^2 = 6.63 \text{ mw/m}^2}$$

5.23 WAVES ON A PERFECT CONDUCTOR—NORMAL INCIDENCE

1. When a wave in air is incident on a perfect conductor normally, it is entirely reflected.
2. As neither **E** nor **H** can exist in a perfect conductor, none of the energy is transmitted through it.
3. As there are no losses within a perfect conductor, no energy is absorbed in it.
4. When an EM wave travelling in one medium is incident upon a second medium, it is partially reflected and partially transmitted.

Total fields of a wave at any point after reflection with normal incidence on a perfect conductor

Resultant electric field,

$$\tilde{E}_R(z, t) = 2E_i \sin\beta z \sin\omega t$$

E_i is the amplitude of the electric field of the incident wave, z = direction of propagation.

Resultant magnetic field,

$$\tilde{H}_R(z, t) = 2H_i \cos\beta z \cos\omega t$$

H_i is the amplitude of the magnetic field of the incident wave.

Let the electric field of the incident wave be

$$E_{\text{incident}} = E_i e^{-j\beta z}, \beta = \frac{2\pi}{\lambda}$$

Then the electric field of the reflected wave is

$$E_{\text{reflected}} = E_r e^{j\beta z}$$

The boundary condition is

$$E_{\tan 1} = E_{\tan 2} = 0 \text{ at } z = 0$$

This requires that

$$E_R = (E_i e^{-j\beta z} + E_r e^{j\beta z}) = 0$$

At $z = 0$

$$E_i + E_r = 0$$

$$E_i = -E_r$$

This means, the amplitudes of incident and reflected electric field strengths are equal but with a phase reversal on reflection.

Now $$\mathbf{E}_R(z) = E_i e^{-j\beta z} + E_r e^{j\beta z}$$
$$= E_i (e^{-j\beta z} - e^{j\beta z})$$
$$= 2jE_i \sin\beta z$$

In time varying form

$$\tilde{\mathbf{E}}_R(z, t) = \text{Re}\{-2jE_i \sin\beta z e^{j\omega t}\}$$

$$\tilde{\mathbf{E}}_R(x, t) = -2E_i \sin\beta z \sin\omega t$$

This obviously represents a standing wave. The variation of E_R is shown in Fig. 5.4.

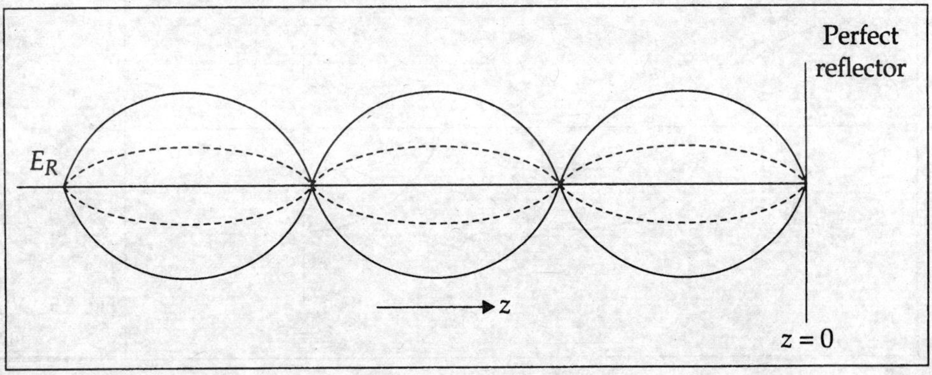

Fig. 5.4 Standing waves

Conclusions

1. The magnitude of electric field varies sinusoidally with distance from the reflecting plane.
2. $E_R = 0$ at the surface of the conductor ($z = 0$) and also at $z = \frac{n\lambda}{2}$, $n = 1, 2, 3, \ldots$
3. $(E_R)_{max} = 2E_i$.
4. E_R is maximum at $z = m\frac{\lambda}{4}$, $m = 1, 3, 5, \ldots$

Resultant magnetic field, $\mathbf{H}_R$

Let us write H_R as $H_R(z) = H_i e^{-j\beta z} + H_r e^{j\beta z}$

At the surface of a perfect conductor,

$$H_{\tan 1} - H_{\tan 2} = J_s$$

As J_s is not specified, we cannot use this boundary condition. Suppose $H_i = -H_r$. It leads to identical directions of incident and reflected powers. This cannot be true. Therefore, H_i and H_r should be the same at $z = 0$,

$$H_i = +H_r$$

$$H_R(x) = H_i\,(e^{-j\beta z} + e^{j\beta z})$$
$$= 2H_i \cos\beta z$$

In time varying form,

$$\tilde{H}_R(z, t) = \text{Re}\,\{2H_i \cos\beta z\, e^{j\omega t}\}$$

$$\tilde{H}_R(z, t) = 2H_i \cos\beta z \cos\omega t$$

This also represents a standing wave. The variation of the H_R is shown in Fig. 5.5.

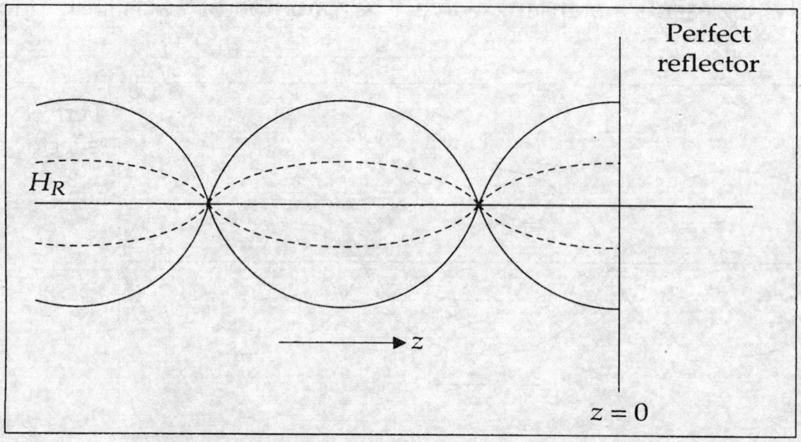

Fig. 5.5 Variation of resultant magnetic field

Conclusions

1. Magnitude of H_R varies cosinusoidally with distance.
2. $H_R = 0$ at $z = m\,\dfrac{\lambda}{4}$, $m = 1, 3, 5, \ldots$
3. $(H_R)_{\max} = 2H_i$
4. H_R is maximum at $z = n\dfrac{\lambda}{2}, n = 1, 2, 3, \ldots$ from the surface.

5.24 WAVES ON DIELECTRIC—NORMAL INCIDENCE

When an EM wave is incident normally on the surface of a dielectric, reflection and transmission take place.

For a perfect dielectric, $\sigma = 0$. Hence, there is no loss or no absorption of energy in it.

Reflection coefficient It is defined as the ratio of reflected wave and incident wave.

That is, the reflection coefficient

$$\equiv \frac{\text{reflected wave}}{\text{incident wave}}$$

Reflection coefficient for

$$E = \Gamma_E \equiv \frac{E_r}{E_i}$$

Reflection coefficient for

$$H = \Gamma_H \equiv \frac{H_r}{H_i}$$

where E_r = reflected electric field

E_i = incident electric field

H_r = reflected magnetic field

H_i = incident magnetic field

Transmission coefficient It is defined as the ratio of transmitted wave and incident wave.

Transmission coefficient

$$\equiv \frac{\text{transmitted wave}}{\text{incident wave}}$$

Transmission coefficient for E is

$$T_E \equiv \frac{E_t}{E_i}$$

and transmission coefficient for H is

$$T_H \equiv \frac{H_t}{H_i}$$

Expressions for reflection and transmission coefficients are:

$$\Gamma_E = \frac{E_r}{E_i} = \frac{\eta_2 - \eta_1}{\eta_2 + \eta_1}$$

$$\Gamma_H = \frac{H_r}{H_i} = \frac{\eta_1 - \eta_2}{\eta_1 + \eta_2}$$

$$T_E = \frac{E_t}{E_i} = \frac{2\eta_2}{\eta_1 + \eta_2}$$

$$T_H = \frac{H_t}{H_i} = \frac{2\eta_1}{\eta_1 + \eta_2}$$

where η_1 and η_2 are intrinsic impedances of medium 1 and medium 2 respectively.

Proof Let $\in_1, \mu_1, \eta_1$ be the permittivity, permeability and intrinsic impedance of medium 1. $\in_2, \mu_2, \eta_2$ are the values for medium 2. We know that

$$E_i = \eta_1 H_i$$

$$E_r = -\eta_1 H_r$$

$$E_t = \eta_2 H_t$$

At the boundary of a dielectric, the tangential components of E and H are continuous, that is,

$$E_i + E_r = E_t$$

$$H_i + H_r = H_t$$

From the above equations, we have

$$\eta_1 (H_i + H_r) = E_i - E_r = \eta_1 \; H_t$$

$$= \eta_1 H_t = (E_i - E_r)$$

$$= \frac{\eta_1}{\eta_2} \; E_t = \frac{\eta_1}{\eta_2} \; (E_i + E_r)$$

$$\eta_2 \; (E_i - E_r) = \eta_1 \; (E_i + E_r)$$

or, $$\eta_2 E_i - \eta_1 E_i = \eta_1 E_r + \eta_2 E_r$$

So, $$\frac{E_r}{E_i} = \Gamma_E = \frac{\eta_2 - \eta_1}{\eta_2 + \eta_1}$$ Hence proved.

Now consider, $$E_t = E_i + E_r$$

$$\frac{E_t}{E_i} = 1 + \frac{E_r}{E_i}$$

$$= 1 + \frac{\eta_2 - \eta_1}{\eta_2 + \eta_1}$$

$$= \frac{\eta_2 + \eta_1 + \eta_2 - \eta_1}{\eta_1 + \eta_2}$$

So, $$\frac{E_t}{E_i} = T_E = \frac{2\eta_2}{\eta_1 + \eta_2}$$ Hence proved.

From the above equations

$$\frac{H_r}{H_i} = -\frac{E_r}{E_i} = -\frac{\eta_2 - \eta_1}{\eta_2 + \eta_1}$$

$$\frac{H_r}{H_i} = \Gamma_H = \frac{\eta_1 - \eta_2}{\eta_1 + \eta_2}$$

Hence proved.

Similarly, consider

$$H_i + H_r = H_t$$

or, $$\frac{H_t}{H_i} = 1 + \frac{H_r}{H_i}$$

$$= 1 + \frac{\eta_1 - \eta_2}{\eta_1 + \eta_2}$$

$$\frac{H_t}{H_i} = T_H = \frac{2\eta_1}{\eta_1 + \eta_2}$$ Hence proved.

5.25 OBLIQUE INCIDENCE OF A PLANE WAVE ON A BOUNDARY PLANE

Reflection and transmission of a wave depend on

1. the type of polarisation of a wave
2. the medium of the boundary.

General polarisations, namely, parallel and perpendicular, are considered.

Parallel Polarisation

It is defined as the polarisation in which the electric field of the wave is parallel to the **plane of incidence**. Parallel polarisation is also called **vertical polarisation**.

Perpendicular Polarisation

It is defined as the polarisation in which the electric field of the wave is perpendicular to the **plane of incidence**. Perpendicular polarisation is also called **horizontal polarisation.**

Plane of Incidence

It is a plane which contains the incident, reflected and transmitted rays and is normal to the boundary.

It is described in Fig. 5.6 in which *x*-*y* is the plane of incidence.

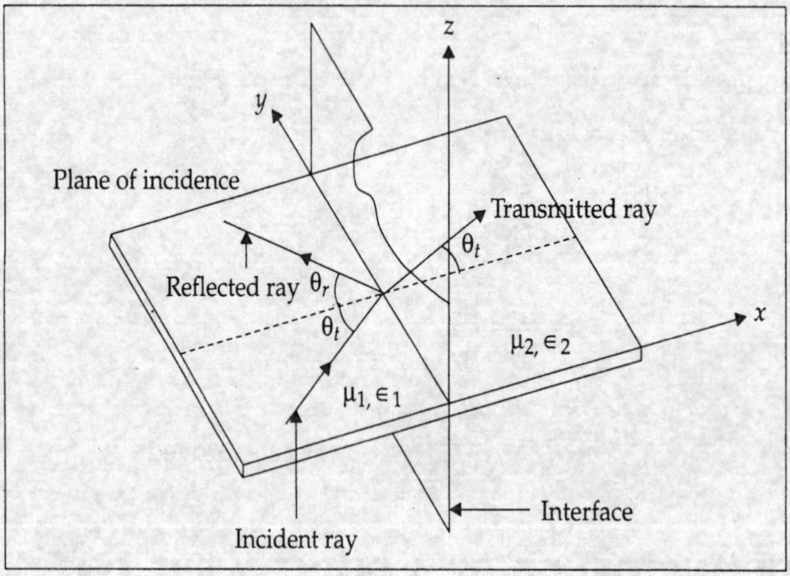

Fig. 5.6 Plane of incidence

5.26 OBLIQUE INCIDENCE OF WAVE ON PERFECT CONDUCTOR

When a wave is incident on a perfect conductor, it is reflected back into the same medium. The resultant fields depend on the type of polarisation.

Parallel Polarisation

The incident and reflected electric fields are shown in Fig. 5.7.

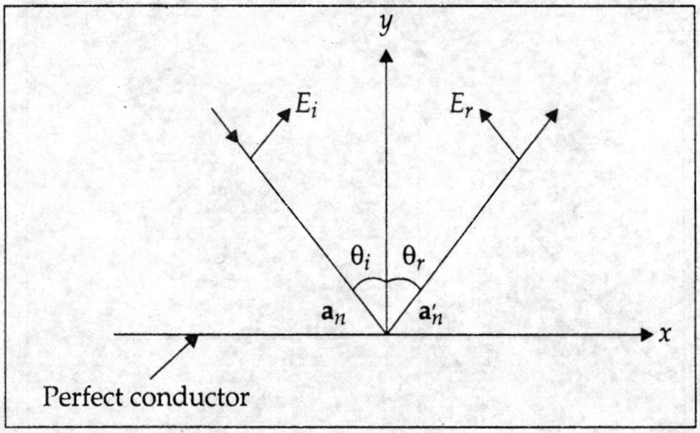

Fig. 5.7 Incident and reflected electric fields in parallel polarisation

The incident and reflected magnetic fields are shown in Fig. 5.8.

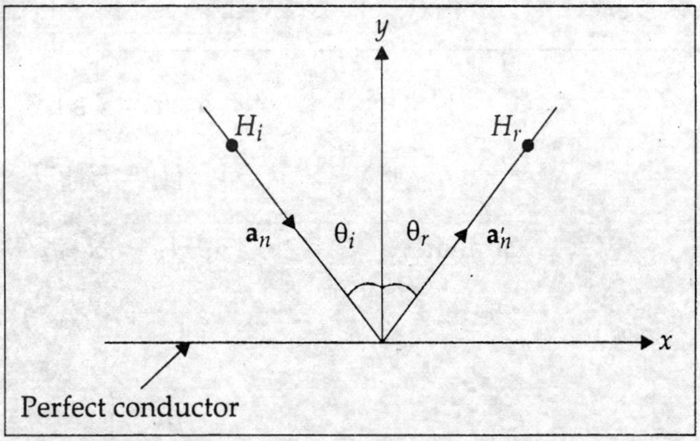

Fig. 5.8 Incident and reflected magnetic fields in parallel polarisation

The incident magnetic field is given by

$$H_I = H_i \, e^{-\gamma(\mathbf{a}_n \cdot \mathbf{r})}$$

where $\mathbf{a}_n$ = unit vector normal to the plane

$\mathbf{r} = (x, y, z)$ is a radius vector on the plane

$$\mathbf{a}_n \cdot \mathbf{r} = x \cos \theta_x + y \cos \theta_y + z \cos \theta_z$$

where

$$\theta_x = \left(\frac{\pi}{2} - \theta_i\right)$$

$$\theta_y = (\pi - \theta_i)$$

$$\theta_z = \frac{\pi}{2}$$

θ_x, θ_y and θ_z are the angles made by a unit vector normal to the plane with x, y and z-axes (Fig. 5.9).

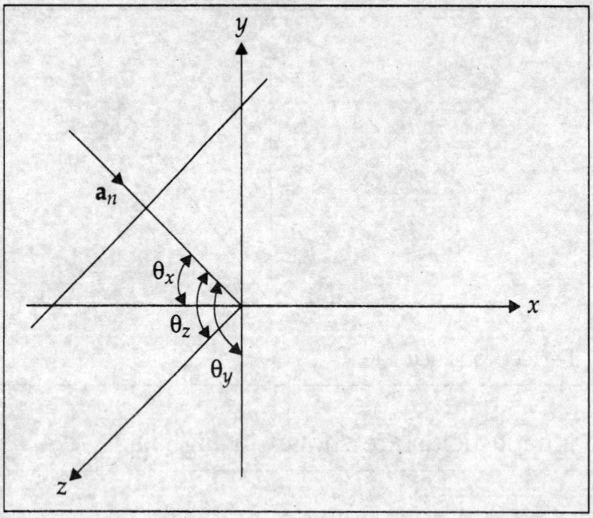

Fig. 5.9 Unit vector normal to the plane

$$\mathbf{a}_n.\mathbf{r} = x \sin\theta_i - y \cos\theta_i$$

$$H_I = H_i e^{-\gamma(x\sin\theta - y\cos\theta)}$$

Similarly,
$$H_R = H_r e^{-\gamma(\mathbf{a}'_n.\mathbf{r})}$$

Here,
$$\mathbf{a}'_n.\mathbf{r} = x \cos\theta'_x + y\cos\theta'_y + z\cos\theta'_z$$

$$\theta'_x = \left(\frac{\pi}{2} - \theta_r\right)$$

$$\theta'_y = \theta_r$$

$$\theta'_z = \frac{\pi}{2}$$

θ'_x, θ'_y and θ'_z are the angles made by the unit vector normal to the plane with x, y and z-axes (Fig. 5.10).

$$H_R = H_r e^{-\gamma(x\sin\theta + y\cos\theta)}$$

$$H_T = H_I + H_R$$

$$= H_i e^{-\gamma(x\sin\theta - y\cos\theta)} + H_r e^{-\gamma(x\sin\theta + y\cos\theta)}$$

At the surfaces of the conductor, $E_i = -E_r$. In order to satisfy the direction of power flow, H_i must be equal to H_r that is, $H_i = H_r$. Now

$$H_T = H_i[e^{-\gamma(x\sin\theta - y\cos\theta)} + e^{-\gamma(x\sin\theta + y\cos\theta)}]$$

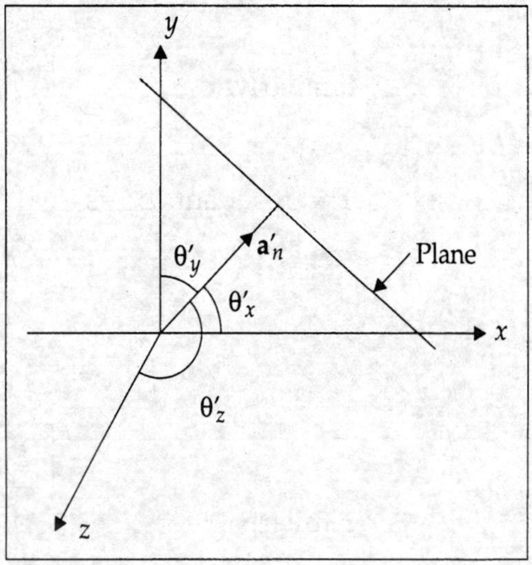

Fig. 5.10 Unit vector normal to the plane

$$= H_i e^{-\gamma x \sin\theta} (e^{\gamma y \cos\theta} + e^{-\gamma y \cos\theta})$$

Here, γ = propagation constant $= \alpha + j\beta$

But in free space, $\alpha = 0$.

Hence $\qquad \gamma = j\beta$

The expression for H_T is given by

$$H_T = H_i e^{-j\beta x \sin\theta} (e^{j\beta y \cos\theta} + e^{-j\beta y \cos\theta})$$

$$\boxed{H_T = 2H_i \cos(\beta y \cos\theta)\, e^{-j\beta x \sin\theta}}$$

or, $$\boxed{H_T = 2H_i \cos(\beta_y y) e^{-j\beta_x x}}$$

where, $$\beta_y = \beta \cos\theta,\ \beta_x = \beta \sin\theta$$

Conclusions

1. The maximum value of H_T is $2H_i$.
2. The maximum occurs at $y = 0$, at $y = \dfrac{\lambda_y}{2}$ and at its even multiples.
3. The minimum is zero and it occurs at $y = \dfrac{\lambda_y}{4}$ and at its odd multiples.

In order to find the resultant electric field, we can make use of the relations between the magnetic and electric fields.

$$E_I = \eta H_I$$

$$E_x = \eta \cos\theta\, H_I, \text{ for the incident wave}$$

$$= -\eta \cos\theta\, H_R, \text{ for the reflected wave}$$

Similarly, $$E_y = \eta \sin\theta\, H_I, \text{ for the incident wave}$$

$$= \eta \cos\theta\, H_R, \text{ for the reflected wave}$$

The resultant E_x is

$$E_x = \eta \cos\theta\, H_I - \eta \cos\theta\, H_R$$

$$= \eta \cos\theta\, [H_I - H_R]$$

$$= \eta \cos\theta\, [H_i\, e^{-j\beta\,(x\sin\theta - y\cos\theta)} - H_r\, e^{-j\beta\,(x\sin\theta + y\cos\theta)}]$$

$$\boxed{E_{Tx} = 2j\eta \cos\theta H_i\, \sin\beta_y\, y e^{-j\beta_x x}}$$

Similarly, $$E_y = \eta \sin\theta\, H_I, \text{ for the incident wave}$$

$$= \eta \sin\theta\, H_R, \text{ for the reflected wave}$$

$$\boxed{E_{Ty} = 2\eta \sin\theta\, H_i\, \cos\beta_y\, y e^{-j\beta_x x}}$$

Perpendicular Polarisation

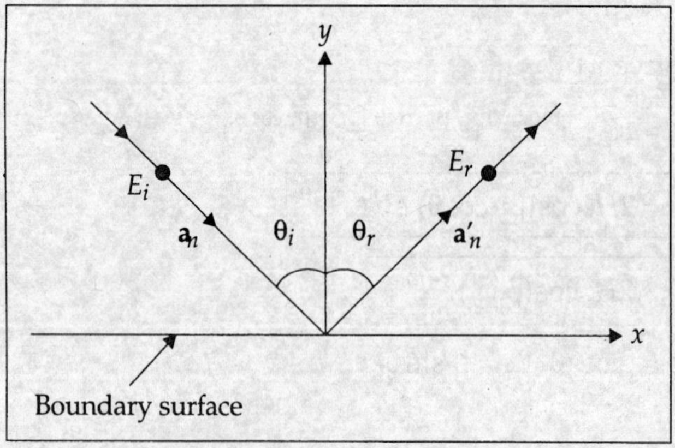

Fig. 5.11 Electric field in perpendicular polarisation

For the incident wave,

$$E_I = E_i\, e^{-j\,(\mathbf{a}_n \cdot \mathbf{r})}$$

$$\mathbf{a}_n \cdot \mathbf{r} = x\cos\left(\frac{\pi}{2} - \theta\right) + y\cos(\pi - \theta) + z\cos\frac{\pi}{2}$$

$$E_I = E_i\, e^{-j\beta\,(x\sin\theta - y\cos\theta)}$$

For the reflected wave,

$$E_R = E_r\, e^{-j\beta\,(\mathbf{a}'_n \cdot \mathbf{r})}$$

$$\mathbf{a}'_n.\mathbf{r} = x\cos\left(\frac{\pi}{2}-\theta\right)+y\cos\theta+z\cos\frac{\pi}{2}$$

$$= x\sin\theta + y\cos\theta$$

$$E_R = E_r e^{-j\beta(x\sin\theta+y\cos\theta)}$$

But $E_i = E_r$

$$E_T = E_i\,[e^{-j\beta(x\sin\theta-y\cos\theta)}+e^{-j\beta(x\sin\theta+y\cos\theta)}]$$

$$= 2jE_i\sin(\beta y\cos\theta)\,e^{-j\beta x\sin\theta}$$

or,

$$E_T = 2jE_i\sin\beta_y\,ye^{-j\beta_x x}$$

where

$$\beta_y = \beta\cos\theta,\ \beta_x = \beta\sin\theta$$

5.27 OBLIQUE INCIDENCE OF A PLANE WAVE ON DIELECTRIC

When a wave is incident on a dielectric, a part of it is reflected and a part of it is transmitted through the dielectric. If θ_i, θ_r and θ_t are the angles of the incident, reflected and transmitted rays, $\theta_i = \theta_r$. The angles θ_i and θ_t are related by **Snell's law**, that is,

$$\frac{\sin\theta_i}{\sin\theta_t} = \sqrt{\frac{\epsilon_{r2}}{\epsilon_{r1}}}$$

Parallel Polarisation

Consider Fig. 5.12.

The boundary condition on E is

$$E_{\tan 1} = E_{\tan 2}$$

$$(E_i - E_r)\cos\theta_i = E_t\cos\theta_t$$

Dividing both sides by E_i, we get

$$\left(1-\frac{E_r}{E_i}\right)\cos\theta_i = \frac{E_t}{E_i}\cos\theta_t$$

or,

$$\frac{E_t}{E_i} = \left(1-\frac{E_r}{E_i}\right)\frac{\cos\theta_i}{\cos\theta_t} \tag{5.16}$$

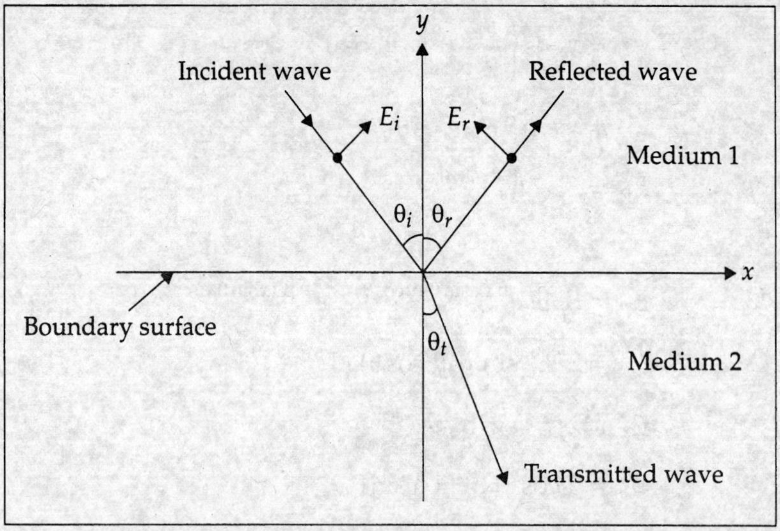

Fig. 5.12 Incident, reflected and transmitted rays

By the law of conservation of energy, incident energy is equal to the sum the of reflected and transmitted energies, that is,

$$\frac{1}{\eta_1} E_i^2 \cos\theta_i = \frac{1}{\eta_1} E_r^2 \cos\theta_i + \frac{1}{\eta_2} E_t^2 \cos\theta_t$$

$$1 = \frac{E_r^2}{E_i^2} + \frac{\eta_1}{\eta_2} \frac{E_t^2}{E_i^2} \frac{\cos\theta_t}{\cos\theta_i} \tag{5.17}$$

From Equations (5.16) and (5.17)

$$\frac{E_r^2}{E_i^2} = 1 - \frac{\eta_1}{\eta_2}\left(1 - \frac{E_r}{E_i}\right)^2 \left(\frac{\cos\theta_i}{\cos\theta_t}\right)^2 \frac{\cos\theta_t}{\cos\theta_i}$$

$$= 1 - \frac{\eta_1}{\eta_2}\left(1 - \frac{E_r}{E_i}\right)^2 \frac{\cos\theta_i}{\cos\theta_t}$$

$$\frac{E_r^2}{E_i^2} - 1 = -\frac{\eta_1}{\eta_2}\left(\frac{E_r}{E_i} - 1\right)^2 \frac{\cos\theta_i}{\cos\theta_t}$$

$$\left(\frac{E_r}{E_i} - 1\right)\left(\frac{E_r}{E_i} + 1\right) = -\frac{\eta_1}{\eta_2}\left(\frac{E_r}{E_i} - 1\right)^2 \frac{\cos\theta_i}{\cos\theta_t}$$

$$\left(\frac{E_r}{E_i} + 1\right) = \frac{\eta_1}{\eta_2}\left(1 - \frac{E_r}{E_i}\right) \frac{\cos\theta_i}{\cos\theta_t}$$

Simplifying this expression, we get

$$\frac{E_r}{E_i} = \frac{\eta_1 \cos\theta_i - \eta_2 \cos\theta_t}{\eta_1 \cos\theta_i + \eta_2 \cos\theta_t}$$

But
$$\eta_1 = \sqrt{\frac{\mu_1}{\epsilon_1}}, \ \eta_2 = \sqrt{\frac{\mu_2}{\epsilon_2}}$$

For most of the dielectrics,

$$\mu = \mu_0$$

$$\mu_1 = \mu_2 = \mu_0$$

$$\eta_1 = \sqrt{\frac{\mu_0}{\epsilon_1}}, \ \eta_2 = \sqrt{\frac{\mu_0}{\epsilon_2}}$$

Hence the reflection coefficient is

$$\frac{E_r}{E_i} = \frac{\sqrt{\epsilon_{r2}} \cos\theta_i - \sqrt{\epsilon_{r1}} \cos\theta_t}{\sqrt{\epsilon_{r2}} \cos\theta_i + \sqrt{\epsilon_{r1}} \cos\theta_t}$$

$$\frac{E_r}{E_i} = \frac{\sqrt{\epsilon_{r2}} \cos\theta_i - \sqrt{\epsilon_{r1} (1 - \sin^2\theta_t)}}{\sqrt{\epsilon_{r2}} \cos\theta_i + \sqrt{\epsilon_{r1} (1 - \sin^2\theta_t)}}$$

as
$$\frac{\sin^2\theta_i}{\sin^2\theta_t} = \frac{\epsilon_{r2}}{\epsilon_{r1}}$$

$$\epsilon_{r1} \sin^2\theta_i = \epsilon_{r2} \sin^2\theta_t$$

$$(1 - \sin^2\theta_t) = \left(1 - \frac{\epsilon_{r1}}{\epsilon_{r2}} \sin^2\theta_i\right)$$

$$\frac{E_r}{E_i} = \frac{\sqrt{\epsilon_{r2}} \cos\theta_i - \sqrt{\epsilon_{r1} \left(1 - \frac{\epsilon_{r1}}{\epsilon_{r2}} \sin^2\theta_i\right)}}{\sqrt{\epsilon_{r2}} \cos\theta_i + \sqrt{\epsilon_{r1} \left(1 - \frac{\epsilon_{r1}}{\epsilon_{r2}} \sin^2\theta_i\right)}}$$

or,
$$\frac{E_r}{E_i} = \frac{\frac{\epsilon_{r2}}{\epsilon_{r1}}\cos\theta_i - \sqrt{\left(\frac{\epsilon_{r2}}{\epsilon_{r1}} - \sin^2\theta_i\right)}}{\frac{\epsilon_{r2}}{\epsilon_{r1}}\cos\theta_i + \sqrt{\left(\frac{\epsilon_{r2}}{\epsilon_{r1}} - \sin^2\theta_i\right)}}$$

Perpendicular Polarisation

Consider Fig. 5.13 in which E is z-directed and x-y is the plane of incidence.

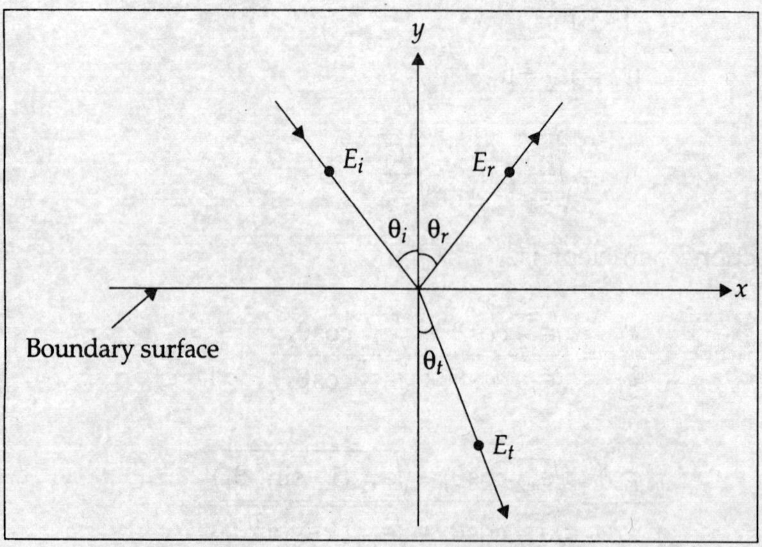

Fig. 5.13 Wave incidence with perpendicular polarisation

From the boundary condition, we have

$$E_i + E_r = E_t$$

$$\frac{E_r}{E_i} + 1 = \frac{E_t}{E_i}$$

From the law of conservation of energy, we have

$$\left(\frac{E_r}{E_i}\right)^2 = 1 - \frac{\eta_2}{\eta_1}\left(\frac{E_t}{E_i}\right)^2 \frac{\cos\theta_t}{\cos\theta_i}$$

$$= 1 - \frac{\eta_2}{\eta_1}\left(1 + \frac{E_r}{E_i}\right)^2 \frac{\cos\theta_t}{\cos\theta_i}$$

As $$\eta_1 = \sqrt{\frac{\mu_0}{\epsilon_1}}, \eta_2 = \sqrt{\frac{\mu_0}{\epsilon_2}}$$

On simplification of the above expression, we get

$$\frac{E_r}{E_i} = \frac{\cos\theta_i - \sqrt{\left(\frac{\epsilon_{r2}}{\epsilon_{r1}}\right) - \sin^2\theta_i}}{\cos\theta_i + \sqrt{\left(\frac{\epsilon_{r2}}{\epsilon_{r1}}\right) - \sin^2\theta_i}}$$

5.28 BREWSTER ANGLE

Definition Brewster angle is the angle of incidence at which there is no reflection.

Brewster angle for parallel polarisation

For parallel polarisation, we have

$$\frac{E_r}{E_i} = \frac{\frac{\epsilon_{r2}}{\epsilon_{r1}}\cos\theta_i - \sqrt{\left(\frac{\epsilon_{r2}}{\epsilon_{r1}}\right) - \sin^2\theta_i}}{\frac{\epsilon_{r2}}{\epsilon_{r1}}\cos\theta_i + \sqrt{\left(\frac{\epsilon_{r2}}{\epsilon_{r1}}\right) - \sin^2\theta_i}}$$

At Brewster angle, $\frac{E_r}{E_i} = 0$

$$\frac{\epsilon_{r2}}{\epsilon_{r1}}\cos\theta_i = \sqrt{\frac{\epsilon_{r2}}{\epsilon_{r1}} - \sin^2\theta_i}$$

$$\left(\frac{\epsilon_{r2}}{\epsilon_{r1}}\right)^2\cos^2\theta_i = \frac{\epsilon_{r2}}{\epsilon_{r1}} - \sin^2\theta_i$$

$$\left(\frac{\epsilon_{r2}}{\epsilon_{r1}}\right)^2 - \left(\frac{\epsilon_{r2}}{\epsilon_{r1}}\right)^2\sin^2\theta_i = \frac{\epsilon_{r2}}{\epsilon_{r1}} - \sin^2\theta_i$$

Simplifying this, we get

$$\sin^2\theta_i = \frac{\epsilon_{r2}}{\epsilon_{r1} + \epsilon_{r2}}$$

or,

$$\cos^2\theta_i = \frac{\epsilon_{r2}}{\epsilon_{r1} + \epsilon_{r2}} - 1 = \frac{\epsilon_{r1}}{\epsilon_{r1} + \epsilon_{r2}}$$

$$\tan\theta_i = \sqrt{\frac{\in_{r2}}{\in_{r1}}}$$

Brewster angle, $$\theta_i = \tan^{-1}\sqrt{\frac{\in_{r2}}{\in_{r1}}}$$

Brewster angle for perpendicular polarisation

We have $$\frac{E_r}{E_i} = \frac{\cos\theta_i - \sqrt{\left(\frac{\in_{r2}}{\in_{r1}}\right) - \sin^2\theta_i}}{\cos\theta_i + \sqrt{\left(\frac{\in_{r2}}{\in_{r1}}\right) - \sin^2\theta_i}}$$

At Brewster angle, $$\frac{E_r}{E_i} = 0$$

$$\cos^2\theta_i = \frac{\in_{r2}}{\in_{r1}} - \sin^2\theta_i$$

or, $$\cos^2\theta_i + \sin^2\theta_i = \frac{\in_{r2}}{\in_{r1}} = 1$$

$$\in_{r1} = \in_{r2}$$

Hence, the condition for no reflection in perpendicular polarisation is $\in_{r1} = \in_{r2}$.

5.29 TOTAL INTERNAL REFLECTION

Definition Total internal reflection is said to exist if

1. the angle of incidence is very high
2. medium 1 is denser than medium 2 and
3. $\sin\theta_i > \sqrt{\frac{\in_{r2}}{\in_{r1}}}$

For total internal reflection,

$$\frac{E_r}{E_i} = \frac{\cos\theta_i + j\sqrt{\sin^2\theta_i - \left(\frac{\in_2}{\in_1}\right)}}{\cos\theta_i - j\sqrt{\sin^2\theta_i - \left(\frac{\in_2}{\in_1}\right)}} \quad \text{for perpendicular polarisation}$$

$$\frac{E_r}{E_i} = \frac{\left(\frac{\epsilon_2}{\epsilon_1}\right)\cos\theta_i + j\sqrt{\sin^2\theta_i - \left(\frac{\epsilon_2}{\epsilon_1}\right)}}{\left(\frac{\epsilon_2}{\epsilon_1}\right)\cos\theta_i - j\sqrt{\sin^2\theta_i - \left(\frac{\epsilon_2}{\epsilon_1}\right)}} \quad \text{for parallel polarisation}$$

The concept of total internal reflection is often used in binocular optics. For these applications, glass prisms are used to shorten the instrument.

5.30 SURFACE IMPEDANCE

Definition It is defined as the ratio of the tangential electric field, E_t to the linear current density, J_s which flows due to the electric field, that is,

$$z_s \equiv \frac{E_t}{J_s}, \text{ohms}$$

For a flat thick conducting sheet, the current density (Fig. 5.14) is given by

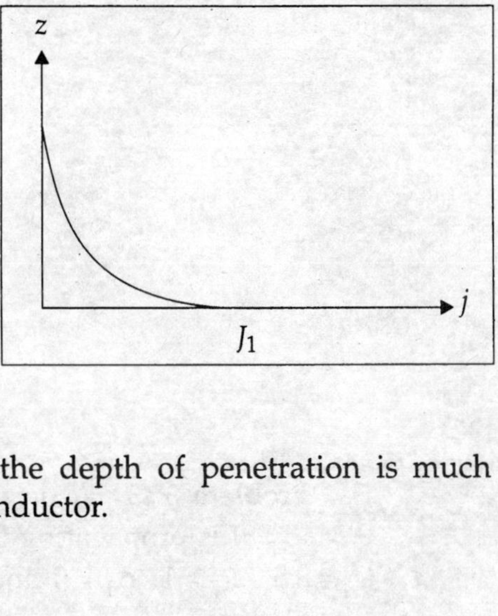

Fig. 5.14 Current density

$$J = J_1 e^{-\gamma z}$$

As the conductor considered is thick, the depth of penetration is much smaller compared to the thickness of the conductor.

The surface current density is

$$J_s = \int_0^\infty J\,dz$$

$$= J_1 \int_0^\infty e^{-\gamma z}\,dz$$

$$= J_1 \left[-\frac{e^{-\gamma z}}{\gamma}\right]_0^\infty$$

$$= \frac{J_1}{\gamma}$$

The current density at the surface, J_1 is given by

$$J_1 = \sigma E_t$$

From the above expressions, we have

$$z_s = \frac{E_t}{J_s} = \frac{J_1}{\sigma} \times \frac{\gamma}{J_1} = \frac{\gamma}{\sigma}$$

For a perfect conductor, σ is very high.

$$\gamma = \sqrt{-\omega^2 \mu \epsilon + j\omega\mu\sigma}$$

$$= \sqrt{j\omega\mu\,(\sigma + j\omega\epsilon)}$$

$$\approx \sqrt{j\omega\mu\sigma}$$

$$z_s = \frac{\sqrt{j\omega\mu\sigma}}{\sigma} = \sqrt{\frac{j\omega\mu}{\sigma}}$$

$$= \sqrt{\frac{\omega\mu}{\sigma}} \angle 45^\circ$$

$$z_s = R_s + jX_s$$

$$= \sqrt{\frac{\omega\mu}{2\sigma}} + j\sqrt{\frac{\omega\mu}{2\sigma}}$$

$$R_s = \sqrt{\frac{\omega\mu}{2\sigma}} \text{ and } X_s = \sqrt{\frac{\omega\mu}{2\sigma}}$$

Problem 5.13 A perpendicularly polarised wave is incident at an angle of $\theta_i = 15^\circ$. It is propagating from medium 1 to medium 2. Medium 1 is defined by $\epsilon_{r_1} = 8.5$, $\mu_{r_1} = 1$, $\sigma_1 = 0$ and medium 2 is free space. If $E_i = 1.0\,\text{mV/m}$, determine E_r, H_i, H_r.

Solution The intrinsic impedance of medium 1 is given by

$$\eta_1 = \sqrt{\frac{\mu_1}{\epsilon_1}} = \sqrt{\frac{\mu_0}{\epsilon_0}} \times \sqrt{\frac{\mu_{r_1}}{\epsilon_{r_1}}}$$

$$= 377 \times \sqrt{\frac{1}{8.5}}$$

$$\boxed{\eta_1 = 129\Omega}$$

As medium 2 is free space,

$$\eta_2 = 377\Omega$$

By Snell's law

$$\frac{\sin\theta_i}{\sin\theta_t} = \sqrt{\frac{\epsilon_2}{\epsilon_1}} = \sqrt{\frac{\epsilon_0}{\epsilon_0\, 8.5}}$$

$$\theta_t = 49°$$

The reflection coefficient for electric field is

$$\frac{E_r}{E_i} = \frac{\eta_2 \cos\theta_i - \eta_1 \cos\theta_i}{\eta_2 \cos\theta_i + \eta_1 \cos\theta_i} = 0.490$$

$$E_r = 0.490 \times 1.0 \times 10^{-3} = 0.490\,\text{mV/m}$$

$$\boxed{E_r = 0.490\,\text{mV/m}}$$

Now $$\frac{E_i}{H_i} = \eta_1$$

or, $$H_i = \frac{E_i}{\eta_1} = \frac{1.0 \times 10^{-3}}{129}$$

$$\boxed{H_i = 7.75\,\mu\text{A/m}}$$

and similarly, $$\frac{E_r}{H_r} = \eta_1$$

or, $$\boxed{H_r = 3.798\,\mu\text{A/m}}$$

5.31 POYNTING VECTOR AND FLOW OF POWER

When EM waves travel from one point to another, there will be energy flow across the surface involved.

Poynting Theorem It states that the cross product of **E** and **H** at any point is a measure of the rate of energy flow per unit area at that point, that is,

$$\mathbf{P} = \mathbf{E} \times \mathbf{H}\ \text{watts/m}^2$$

Poynting Vector, P is defined as

$$\mathbf{P} \equiv \mathbf{E} \times \mathbf{H}\ \text{watts/m}^2$$

If **E** and **H** are instantaneous, **P** is also instantaneous.

Proof First Maxwell's equation is

$$\nabla \times \mathbf{H} = \dot{\mathbf{D}} + \mathbf{J} = \epsilon \dot{\mathbf{E}} + \mathbf{J}$$

or,
$$\mathbf{J} = \nabla \times \mathbf{H} - \epsilon \dot{\mathbf{E}}$$

that is,
$$\mathbf{E}.\mathbf{J} = \mathbf{E}.\nabla \times \mathbf{H} - \epsilon \mathbf{E}.\dot{\mathbf{E}}$$

But
$$\nabla.(\mathbf{E} \times \mathbf{H}) = \mathbf{H}.\nabla \times \mathbf{E} - \mathbf{E}.\nabla \times \mathbf{H}$$

$$\mathbf{E}.\mathbf{J} = \mathbf{H}.\nabla \times \mathbf{E} - \nabla.\mathbf{E} \times \mathbf{H} - \epsilon \mathbf{E}.\dot{\mathbf{E}}$$

As
$$\nabla \times \mathbf{E} = -\mu \dot{\mathbf{H}}, \text{ this becomes}$$

$$\mathbf{E}.\mathbf{J} = -\mu \mathbf{H}.\dot{\mathbf{H}} - \epsilon \mathbf{E}.\dot{\mathbf{E}} - \nabla.\mathbf{E} \times \mathbf{H}$$

Here
$$\mathbf{E}.\dot{\mathbf{E}} = \frac{1}{2}\frac{\partial}{\partial t} E^2$$

$$\mathbf{H}.\dot{\mathbf{H}} = \frac{1}{2}\frac{\partial}{\partial t} H^2$$

$$\mathbf{E}.\mathbf{J} = \nabla.\mathbf{E} \times \mathbf{H} - \frac{\partial}{\partial t}\left(\frac{1}{2} \epsilon H^2 + \frac{1}{2} \epsilon E^2\right)$$

Taking volume integral,

$$\int_{\upsilon} \mathbf{E}.\mathbf{J}\, d\upsilon = -\int_{\upsilon} (\nabla.\mathbf{E} \times \mathbf{H})\, d\upsilon - \frac{\partial}{\partial t}\int\left(\frac{1}{2} \epsilon E^2 + \frac{1}{2}\mu H^2\right) d\upsilon$$

By divergence theorem,

$$\int_{\upsilon} \nabla.\mathbf{E} \times \mathbf{H}\, d\upsilon = \oint_{S} (\mathbf{E} \times \mathbf{H}).d\mathbf{S}$$

$$\int_{\upsilon} \mathbf{E}.\mathbf{J}\, d\upsilon = -\oint_{S} (\mathbf{E} \times \mathbf{H}).d\mathbf{S} - \frac{\partial}{\partial t}\int_{\upsilon}\left(\frac{1}{2} \epsilon E^2 + \frac{1}{2}\mu H^2\right) d\upsilon$$

The left hand side represents energy dissipated in the volume.

The second term on the right hand side represents the rate at which the stored energy in magnetic and electric fields is changing. (–)ve sign indicates decrease. Therefore, by the law of conservation of energy, the rate of energy dissipation in the volume is equal to the rate at which the stored energy in static electric and magnetic fields in the volume is decreasing plus the rate at which the energy is entering the volume from outside. Therefore,

$$-\oint_S (\mathbf{E}\times\mathbf{H}).d\mathbf{S}$$

represents inward power flow, or

$$\oint_S (\mathbf{E}\times\mathbf{H}).d\mathbf{S}$$

represents outward power flow.

$$\oint_S (\mathbf{E}\times\mathbf{H}).d\mathbf{S} \text{ is in watts}$$

$\mathbf{E}\times\mathbf{H}$ represents power flow per unit area, or, $\mathbf{E}\times\mathbf{H}$ is in watts/m^2.

$$\mathbf{P} = \mathbf{E}\times\mathbf{H} \qquad \text{Hence proved.}$$

Problem 5.14 The magnetic field, **H** of a plane wave has a magnitude of 5 mA/m in a medium defined by $\epsilon_r = 4$, $\mu_r = 1$. Determine (a) the average power flow (b) the maximum energy density in the plane wave.

Solution (a) We have

$$\frac{E}{H} = \sqrt{\frac{\mu}{\epsilon}} = \sqrt{\frac{4\pi\times10^{-7}}{4\times10^{-9}\times\frac{1}{36}}} = 188.4\Omega$$

$$E = 188.4H = 188.4\times5\times10^{-3} = 942.0\text{ mV/m}$$

$$P_{av} = \frac{E^2}{2\times188.5} = \frac{942.0\times10^{-3}\times942\times10^{-3}}{377} = 2353.75\mu\text{ w/m}^2$$

(b) The maximum energy density of the wave is

$$W_E = \frac{1}{2}\epsilon E^2\times2 = \epsilon E^2 = \epsilon_0\epsilon_r E^2$$

$$= 4\times8.854\times10^{-12}\times942\times942\times10^{-6}$$

$$= 314.269\times10^5\times10^{-18}$$

$$= 31.42\times10^6\times10^{-18}$$

$$\boxed{W_E = 31.42\text{ PJ/m}^3}$$

Problem 5.15 A plane wave travelling in a medium of $\epsilon_r = 1$, $\mu_r = 1$ has an electric field intensity of $100\times\sqrt{\pi}$ V/m. Determine the energy density in the magnetic field and also the total energy density.

Solution The electric energy density is given by

$$W_E = \frac{1}{2} \in E^2 = \frac{1}{2} \in_0 \in_r E^2$$

$$= \frac{1}{2} \times 8.854 \times 10^{-12} \times 1 \times 100^2 \times \pi$$

$$= 13.9 \times 10^{-8} = 139 \times 10^{-9}$$

$$W_E = 139 \text{ nJ/m}^3$$

As the electric energy density is equal to that of the magnetic field for a plane travelling wave,

$$\boxed{W_H = 139 \text{ nJ/m}^3}$$

So the total energy density,

$$\boxed{W_T = 278 \text{ nJ/m}^3}$$

Problem 5.16 The conductivity of sea water, $\sigma = 5$ mho/m, $\in_r = 80$. What is the distance, an EM wave can be transmitted at 25 kHz and 25 MHz when the range corresponds to 90% of attenuation?

Solution If the wave is moving in x-direction, we have

$$e^{-\alpha x} = 0.1$$

that is,

$$\ln (e^{-\alpha x}) = \ln (0.1)$$

$$-\alpha x = \ln (0.1)$$

$$-\alpha x = -2.30$$

or,

$$x = \frac{2.30}{\alpha}$$

$$\alpha = \omega \sqrt{\frac{\mu \in}{2} \left(\sqrt{1 + \frac{\sigma^2}{\omega^2 \in^2}} - 1 \right)}$$

at $f = 25$ kHz, $\alpha = 0.702$,

at $f = 25$ MHz,

$$\alpha = 21.96$$

at $f = 25$ kHz, $x = \dfrac{2.3}{0.702}$

Hence $\boxed{x = 3.27 \text{ m}}$

at $f = 25\,\text{MHz}$,

$$x = \frac{2.3}{21.96}$$

or $\boxed{x = 0.104 \text{ m}}$

Problem 5.17 A plane wave with a frequency of 2 MHz is incident upon a copper conductor normally. The wave has an electric field amplitude of $E = 2\text{ mV/m}$. Copper has $\mu_r = 1$, $\epsilon_r = 1$ and $\sigma = 5.8\times10^7$ mho/m. Find the average power density absorbed by copper.

Solution Copper is a good conductor

$$\eta = \sqrt{\frac{\mu\omega}{\sigma}} = \sqrt{\frac{4\pi\times10^{-7}\times2\pi\times2\times10^6}{58\times10^6}}$$

$$= \frac{4\pi}{\sqrt{5.8}}\times10^{-4} = \frac{4\pi}{2.40}\times10^{-4} = 5.235\times10^{-4}\,\Omega$$

$$P_{av} = \frac{1}{2}\frac{E^2}{|\eta|} = \frac{\frac{1}{2}\times4\times10^{-6}}{5.235\times10^{-4}} = 0.382\times10^{-2}\text{ w/m}^2$$

$$\boxed{P_{av} = 3.82\,\text{mw/m}^2}$$

5.32 COMPLEX POYNTING VECTOR

It is defined as

$$\mathbf{P}_c \equiv \frac{1}{2}\,\mathbf{E}\times\mathbf{H}^*$$

where $\mathbf{H}^*$ is the complex conjugate of $\mathbf{H}$.

$$P_{av} = \frac{1}{2}\,\text{Re}\,\{\mathbf{E}\times\mathbf{H}^*\}$$

$$P_{\text{react}} = \frac{1}{2}\,\text{Im}\,(\mathbf{E}\times\mathbf{H}^*)$$

Complex Poynting vector is useful to find average power flow. We have,

$$P_{av} = \frac{1}{2}\,\text{Re}\,(EH^*)$$

$$= \frac{1}{2} \text{Re} (\eta H H^*)$$

If $$H = H_m e^{j\omega t}$$

$$H^* = H_m e^{-j\omega t}$$

$$\boxed{P_{av} = \frac{1}{2} H_m^2 \text{ Re} (\eta)}$$

POINTS/FORMULAE TO REMEMBER

- For a uniform plane wave propagating in x-direction, $E_x = 0$ and $H_x = 0$.
- Intrinsic impedance of free space is $120\pi\,\Omega$.
- Intrinsic impedance of a medium is $\eta = \frac{E}{H}$.
- The wave equations in a conductive medium are:

$$\nabla^2 \mathbf{E} = \mu \in \ddot{\mathbf{E}} + \mu\sigma\dot{\mathbf{E}}$$

$$\nabla^2 \mathbf{H} = \mu \in \ddot{\mathbf{H}} + \mu\sigma\dot{\mathbf{H}}$$

- The solution of uniform plane wave propagating in x-direction is

$$\mathbf{E} = f\,(x - \upsilon_0 t)$$

- Propagation constant is $\gamma = \sqrt{-\omega^2 \mu \in + j\omega\mu\sigma}$
- Attenuation constant in free space is zero.
- Phase constant in free space is $\beta = \omega\sqrt{\mu_0 \epsilon_0}$
- Phase velocity in free space, $\upsilon_p = \frac{\omega}{\beta}$
- Dissipation factor, $D_f = \frac{\sigma}{\omega\epsilon}$
- Attenuation constant in good conductors is $\alpha = \sqrt{\frac{\omega\mu\sigma}{2}}$
- Phase constant in good conductors is $\beta = \sqrt{\frac{\omega\mu\sigma}{2}}$
- α in good dielectrics is $\alpha = \frac{\sigma}{2}\sqrt{\frac{\mu}{\in}}$
- Depth of penetration, $\delta = \frac{1}{\alpha}$
- Poynting vector, $\mathbf{P} = \mathbf{E} \times \mathbf{H}$

- Complex Poynting vector, $\mathbf{P}_c = \frac{1}{2}\,\mathbf{E} \times \mathbf{H}^*$
- Reflection coefficient $= \dfrac{\text{reflected wave}}{\text{incident wave}}$
- Transmission coefficient $= \dfrac{\text{transmitted wave}}{\text{incident wave}}$
- Brewster angle, $\theta_b = \tan^{-1}\sqrt{\dfrac{\epsilon_2}{\epsilon_1}}$
- Snell's law is given by $\dfrac{\sin\theta_i}{\sin\theta_t} = \sqrt{\dfrac{\epsilon_2}{\epsilon_1}}$

OBJECTIVE QUESTIONS

1. Characteristic impedance of a medium is $\sqrt{\frac{\epsilon}{\mu}}$. (Yes/No)
2. Velocity of propagation of uniform plane wave and its phase velocity are identical. (Yes/No)
3. Brewster angle is the angle of reflection. (Yes/No)
4. $\mathbf{P} \times \mathbf{H}$ gives average power. (Yes/No)
5. $\mathbf{E} = e^{j2x}\,\mathbf{a}_y$ means that $\mathbf{E} = \sin(\omega t + 2x)\,\mathbf{a}_y$. (Yes/No)
6. **E** and **H** in good conductors are in time phase. (Yes/No)
7. Power density is represented by Poynting vector. (Yes/No)
8. Complex Poynting vector is $\mathbf{E} \times \mathbf{H}^*$. (Yes/No)
9. The units of Poynting vector are watts. (Yes/No)
10. Depth of penetration is nothing but α. (Yes/No)
11. β has the unit of radian. (Yes/No)
12. Unit of α and β are the same. (Yes/No)
13. Unit of the propagation constant is m^{-1}. (Yes/No)
14. Brewster angle is the same as critical angle. (Yes/No)
15. In circular polarisation, E_x and E_y components have the same magnitude. (Yes/No)
16. In elliptical polarisation, E_x and E_y components have the same magnitude. (Yes/No)
17. Horizontal polarisation is said to be linear polarisation. (Yes/No)
18. When an EM wave is incident on a perfect conductor normally, standing waves are produced. (Yes/No)
19. According to Snell's law, the angle of incidence and the angle of reflection are the same. (Yes/No)
20. Polarisation and the direction of propagation of an EM wave are one and the same. (Yes/No)
21. In perpendicular polarisation with oblique incidence on a dielectric, there exists Brewster angle. (Yes/No)

22. If $\mathbf{E} = \cos(6\times10^7 t - \beta z)\,\mathbf{a}_x$, β is ________.

23. If the attenuation of a plane wave in a medium is $22.5\times10^3\ \text{m}^{-1}$, the depth of penetration is ________.

24. If the depth of penetration of a plane wave in a medium is 2 mm, the attenuation constant is ________.

25. Brewster angle is given by ________.

Answers

1. No	2. Yes	3. No	4. No	5. Yes
6. No	7. Yes	8. No	9. No	10. No
11. No	12. No	13. Yes	14. No	15. Yes
16. No	17. Yes	18. Yes	19. No	20. No
21. No	22. 0.2 rad/m		23. 4 cm	24. $0.5\times10^3\ \text{m}^{-1}$
25. $\tan^{-1}\sqrt{\dfrac{\epsilon_2}{\epsilon_1}}$				

MULTIPLE CHOICE QUESTIONS

1. Poynting vector is given by
   (a) $\mathbf{E} \times \mathbf{H}$ (b) $\mathbf{E.H}$ (c) $\mathbf{H} \times \mathbf{E}$ (d) $\mathbf{H.E}$

2. Poynting vector gives
   (a) rate of energy flow (b) direction of polarisation
   (c) electric field (d) magnetic field

3. **E.H** of a uniform plane wave is
   (a) EH (b) 0 (c) ηE^2 (d) ηH^2

4. Absolute permeability of free space is
   (a) $4\pi \times 10^{-7}$ A/m (b) $4\pi \times 10^{-7}$ H/m
   (c) $4\pi \times 10^{-7}$ F/m (d) $4\pi \times 10^{-7}$ H/m^2

5. For a uniform plane wave in the *x*-direction
   (a) $E_x = 0$ (b) $H_x = 0$
   (c) $E_x = 0$ and $H_x = 0$ (d) $E_z = 0$

6. Depth of penetration in free space is
   (a) infinity (b) $1/\alpha$ (c) 0 (d) small

7. Complex Poynting vector, **P** is
   (a) $P = E \times H^*$ (b) $\mathbf{P} = \mathbf{E} \times \mathbf{H}^*$
   (c) $P = \frac{1}{2} \mathbf{E} \times \mathbf{H}^*$ (d) $\mathbf{E} \times \mathbf{H}$

8. Uniform plane wave is
   (a) longitudinal in nature (b) transverse in nature
   (c) neither transverse nor longitudinal (d) *x*-directed

9. The direction of propagation of EM wave is obtained from
   (a) $\mathbf{E} \times \mathbf{H}$ (b) $\mathbf{E.H}$ (c) $\mathbf{E}$ (d) $\mathbf{H}$

10. The velocity of an EM wave is
    (a) inversely proportional to β (b) inversely proportional to α
    (c) directly proportional to β (d) directly proportional to α

11. Velocity of the wave in an ideal conductor is
(a) zero (b) very large
(c) moderate (d) small

12. If $E = 2$ V/m of a wave in free space, (H) is
(a) $\frac{1}{60\pi}$ A/m (b) 60π A/m (c) 120π A/m (d) 240π A/m

13. If wet soil has $\sigma = 10^{-2}$ mho/m, $\epsilon_r = 15$, $\mu_r = 1$, $f = 60$ Hz, it is a
(a) good conductor (b) good dielectric
(c) semi-conductor (d) magnetic material

14. If wet soil has $\sigma = 10^{-2}$ mho/m, $\epsilon_r = 15$, $\mu_r = 1$, at 10 GHz, it is a
(a) good conductor (b) good dielectric
(c) semi-conductor (d) semi-dielectric

15. The cosine of the angle between two vectors is
(a) sum of the products of the directions of the two vectors
(b) difference of the products of the directions of the two vectors
(c) product of the products of the directions of the two vectors
(d) zero

16. If **E** is a vector, then $\nabla . \nabla \times \mathbf{E}$ is
(a) 0 (b) 1
(c) does not exist (d) none of these

17. Velocity of an EM wave in free space is
(a) independent of f (b) increases with increase in f
(c) decreases with increase in f (d) zero

18. The direction of propagation of an EM wave is given by
(a) the direction of **E** (b) the direction of **H**
(c) the direction of $\mathbf{E} \times \mathbf{H}$ (d) the direction of $\mathbf{E} . \mathbf{H}$

19. For uniform plane wave propagating in z-direction
(a) $\mathbf{E}_x = 0$ (b) $\mathbf{H}_x = 0$
(c) $\mathbf{E}_y = 0, \mathbf{H}_y = 0$ (d) $\mathbf{E}_z = 0, \mathbf{H}_z = 0$

20. For free space,
(a) $\sigma = \infty$ (b) $\sigma = 0$
(c) $J \neq 0$ (d) $\mu_r = 0$

21. Velocity of propagation of an EM wave is

(a) $\sqrt{\dfrac{\epsilon_0}{\mu_0}}$ (b) $\dfrac{\mu_0}{\epsilon_0}$ (c) $\dfrac{1}{\sqrt{\mu_0 \epsilon_0}}$ (d) $\dfrac{\epsilon_0}{\mu_0}$

22. The intrinsic impedance of the medium whose $\sigma = 0,\ \epsilon_r = 9,\ \mu_r = 1$ is

(a) $40\pi\Omega$ (b) 9Ω (c) $120\pi\Omega$ (d) $60\pi\Omega$

23. The wavelength of a wave with a propagation constant $= 0.1\pi + j0.2\pi$ is

(a) 10 m (b) 20 m (c) 30 m (d) 25 m

24. Electric field just above a conductor is always

(a) normal to the surface (b) tangential to source

(c) zero (d) ∞

Answers

1. (a)	2. (a)	3. (b)	4. (b)	5. (c)
6. (a)	7. (c)	8. (b)	9. (a)	10. (a)
11. (a)	12. (a)	13. (a)	14. (b)	15. (a)
16. (a)	17. (a)	18. (c)	19. (d)	20. (b)
21. (c)	22. (a)	23. (a)	24. (a)	

EXERCISE PROBLEMS

1. Find the depth of penetration for copper at 20 MHz. For copper
$$\sigma = 5.8\times10^7 \text{ mho/m}, \mu_r = 1$$
2. What is the velocity of propagation of a uniform plane wave in a medium whose $\in_r = 10$, $\mu_r = 3$?
3. Conduction current in a copper wire is 4.0 amps. Find the displacement current in the wire at 1 MHz and 10 MHz. For copper $\in_r = 1$, $\mu_r = 1$, $\sigma = 5.8\times10^7$ mho/m.
4. An EM wave in free space is incident normally on a dielectric whose $\in_r = 5.0$. Find the reflection and transmission coefficients.
5. If an electric field in free space is $\mathbf{E} = 2.0\cos(\omega t - \beta z)\,\mathbf{a}_x$ V/m, find the average power flowing across a square whose side is 2 m. The square is in $z = a$ constant plane.
6. Derive the condition under which the electric field $\mathbf{E} = k\cos(3\times10^8 t - z)\,\mathbf{a}_y$ exists in a source free dielectric medium. Here k is a constant, β is a constant.
7. If the electric and magnetic fields in a medium ($\mu_r = 1$ and $\in_r = 4$) are given by $\mathbf{E} = 10\cos(10^8 t - \beta z)\,\mathbf{a}_x$ and $\mathbf{H} = \dfrac{1}{25}\cos(10^8 t - \beta z)\,\mathbf{a}_y$ find out η, f and β.

CHAPTER 6

GUIDED WAVES

The propagation characteristics of guided waves are different from wave propagation characteristics in free space.

The main objective of this chapter is to provide the conceptual behaviour of EM waves in guided structures. The topics include:

- fields and propagation characteristics between parallel plates
- fields and propagation characteristics in hollow rectangular and circular waveguides
- TE, TM and TEM waves
- cavity resonators
- solved problems, points/formulae to remember, objective and multiple choice questions and exercise problems.

Do you know?

The Transverse Electromagnetic (TEM) wave does not exist in a hollow waveguide.

6.1 INTRODUCTION

In the previous chapter, propagation characteristics of uniform plane waves in free space were considered. These characteristics are very important when waves propagate between transmitter and receiver in free space. However, there are practical situations where propagation is by means of guided waves, that is, the EM waves may be guided between conducting planes or along a pair of wires or coaxial transmission lines. They may also be guided in wave-guide structures. In these cases, the medium of propagation is not free space.

6.2 WAVES BETWEEN PARALLEL PLATES

In these structures, the waves are no more uniform plane waves. Hence their propagation characteristics are described in terms of Transverse Electric (TE) and Transverse Magnetic (TM) waves.

6.3 DERIVATION OF FIELD EQUATIONS BETWEEN PARALLEL PLATES AND PROPAGATION PARAMETERS

Consider Fig. 6.1.

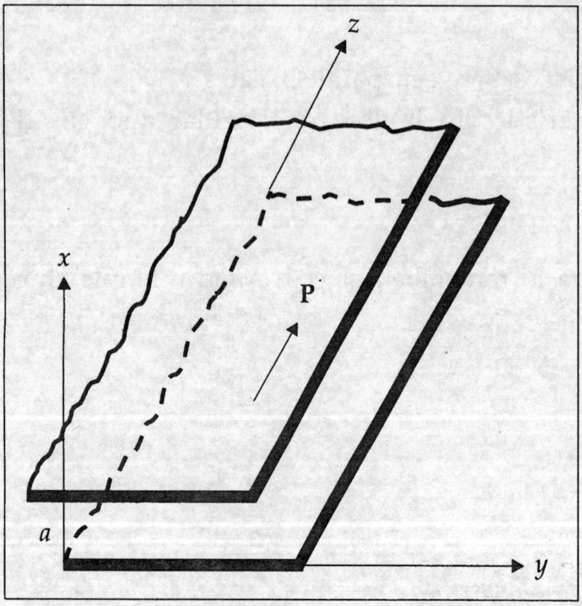

Fig. 6.1 Parallel conducting plates

Assume that:

(a) the plates are extended to infinity in y and z-directions

(b) the plates are perfectly conducting

(c) the plates are separated by a distance of 'a' metres in x-direction

(d) the space between the plates is air or free space

(e) the direction of power flow is z

(f) all field components in z-direction vary as $e^{-\gamma_p z}$ where $\gamma_p = \alpha_p + j\beta_p$. Here γ_p is the propagation constant, α_p is the attenuation constant and β_p is the phase constant. In time varying form the fields vary as

$$e^{-\gamma_p z} . e^{j\omega t} = e^{j(\omega t - \beta_p z)} \qquad [\text{as } \alpha_p = 0]$$

(g) the field is uniform in y-direction as the plates are extended to infinity in y-direction

As the medium between the plates is air, the first and second Maxwell's equations are given by

$$\nabla \times \mathbf{H} = j\omega\epsilon_0 \mathbf{E} \tag{6.1}$$

and

$$\nabla \times \mathbf{E} = -j\omega\mu_0 \mathbf{H} \tag{6.2}$$

The corresponding wave equations in air are:

$$\nabla^2 \mathbf{E} = -\omega^2 \mu_0 \epsilon_0 \mathbf{E} \tag{6.3}$$

and

$$\nabla^2 \mathbf{H} = -\omega^2 \mu_0 \epsilon_0 \mathbf{H} \tag{6.4}$$

Expanding Equation (6.1), we get

$$\nabla \times \mathbf{H} = \begin{vmatrix} \mathbf{a}_x & \mathbf{a}_y & \mathbf{a}_z \\ \dfrac{\partial}{\partial x} & \dfrac{\partial}{\partial y} & \dfrac{\partial}{\partial z} \\ H_x & H_y & H_z \end{vmatrix}$$

$$= \mathbf{a}_x \left[\frac{\partial H_z}{\partial y} - \frac{\partial H_y}{\partial z}\right] + \mathbf{a}_y \left[\frac{\partial H_x}{\partial z} - \frac{\partial H_z}{\partial x}\right] + \mathbf{a}_z \left[\frac{\partial H_y}{\partial x} - \frac{\partial H_x}{\partial y}\right]$$

$$= j\omega\epsilon_0 E_x \mathbf{a}_x + j\omega\epsilon_0 E_y \mathbf{a}_y + j\omega\epsilon_0 E_z \mathbf{a}_z$$

Equating the respective components on either side, we get

$$\left.\begin{aligned} \frac{\partial H_z}{\partial y} - \frac{\partial H_y}{\partial z} &= j\omega\epsilon_0 E_x \\ \frac{\partial H_x}{\partial z} - \frac{\partial H_z}{\partial x} &= j\omega\epsilon_0 E_y \\ \frac{\partial H_y}{\partial x} - \frac{\partial H_x}{\partial y} &= j\omega\epsilon_0 E_z \end{aligned}\right\} \tag{6.5}$$

Similarly, expanding Equation (6.2) and equating the respective components on both sides, we get

$$\left.\begin{aligned}\frac{\partial E_z}{\partial y}-\frac{\partial E_y}{\partial z}&=-j\omega\mu_0 H_x\\ \frac{\partial E_x}{\partial z}-\frac{\partial E_z}{\partial x}&=-j\omega\mu_0 H_y\\ \frac{\partial E_y}{\partial x}-\frac{\partial E_x}{\partial y}&=-j\omega\mu_0 H_z\end{aligned}\right\}\tag{6.6}$$

If $H=H^0$, at $z=0$, the field components can be written as

$$H_y = H_y^0 e^{-\gamma_p z}$$

and similarly

$$\left.\begin{aligned}\frac{\partial H_y}{\partial z} &= -\gamma_p H_y^0 e^{-\gamma_p z} = -\gamma_p H_y\\ \frac{\partial H_x}{\partial z} &= -\gamma_p H_x^0 e^{-\gamma_p z} = -\gamma_p H_x\\ \frac{\partial E_y}{\partial z} &= -\gamma_p E_y^0 e^{-\gamma_p z} = -\gamma_p E_y\\ \text{and}\quad \frac{\partial E_x}{\partial z} &= -\gamma_p E_x^0 e^{-\gamma_p z} = -\gamma_p E_x\end{aligned}\right\}\tag{6.7}$$

Substituting Equation (6.7) in Equations (6.5) and (6.6), we get

$$\left.\begin{aligned}\frac{\partial H_z}{\partial y}+\gamma_p H_y &= j\omega\epsilon_0 E_x\\ -\gamma_p H_x-\frac{\partial H_z}{\partial x} &= j\omega\epsilon_0 E_y\\ \frac{\partial H_y}{\partial x}-\frac{\partial H_x}{\partial y} &= j\omega\epsilon_0 E_z\\ \frac{\partial E_z}{\partial y}+\gamma_p E_y &= -j\omega\mu_0 H_x\\ -\gamma_p E_x-\frac{\partial E_z}{\partial x} &= -j\omega\mu_0 H_y\\ \frac{\partial E_y}{\partial x}-\frac{\partial E_x}{\partial y} &= -j\omega\mu_0 H_z\end{aligned}\right\}\tag{6.8}$$

But all terms containing derivatives with respect to y vanish as the fields do not vary with y. Hence Equation (6.8) becomes

$$\left.\begin{aligned} +\gamma_p H_y &= j\omega\epsilon_0 E_x \\ -\gamma_p H_x - \frac{\partial H_z}{\partial x} &= j\omega\epsilon_0 E_y \\ \frac{\partial H_y}{\partial x} &= j\omega\epsilon_0 E_z \\ \gamma_p E_y &= -j\omega\mu_0 H_x \\ -\gamma_p E_x - \frac{\partial E_z}{\partial x} &= -j\omega\mu_0 H_y \\ \frac{\partial E_y}{\partial x} &= -j\omega\mu_0 H_z \end{aligned}\right\} \tag{6.9}$$

As **E** can be written as

$$\mathbf{E} = E^0 e^{-\gamma_p z}$$

$$\frac{\partial \mathbf{E}}{\partial z} = -\gamma_p E^0 e^{-\gamma_p z} = -\gamma_p \mathbf{E}$$

and similarly

$$\left.\begin{aligned} \frac{\partial^2 \mathbf{E}}{\partial z^2} &= \gamma_p^2 \mathbf{E} \\ \frac{\partial \mathbf{H}}{\partial z} &= -\gamma_p \mathbf{H} \\ \frac{\partial^2 \mathbf{H}}{\partial z^2} &= +\gamma_p^2 \mathbf{H} \end{aligned}\right\} \tag{6.10}$$

The wave equations

$$\frac{\partial^2 \mathbf{E}}{\partial x^2} + \frac{\partial^2 \mathbf{E}}{\partial y^2} + \frac{\partial^2 \mathbf{E}}{\partial z^2} = -\omega^2 \epsilon_0 \mu_0 \mathbf{E}$$

and

$$\frac{\partial^2 \mathbf{H}}{\partial x^2} + \frac{\partial^2 \mathbf{H}}{\partial y^2} + \frac{\partial^2 \mathbf{H}}{\partial z^2} = -\omega^2 \epsilon_0 \mu_0 \mathbf{H}$$

now become

$$\frac{\partial^2 \mathbf{E}}{\partial x^2} + 0 + \gamma_p^2 \mathbf{E} = -\omega^2 \mu_0 \epsilon_0 \mathbf{H}$$

or,

$$\left.\begin{aligned}\frac{\partial^2 \mathbf{E}}{\partial x^2}+\gamma_p^2 \mathbf{E} &= -\omega^2 \mu_0 \epsilon_0 \mathbf{E}\\ \frac{\partial^2 \mathbf{H}}{\partial x^2}+\gamma_p^2 \mathbf{H} &= -\omega^2 \mu_0 \epsilon_0 \mathbf{H}\end{aligned}\right\} \tag{6.11}$$

Solving Equation (6.9) simultaneously, we get the field components as

$$\left.\begin{aligned}H_x &= \frac{-\gamma_p}{h_p^2}\frac{\partial H_z}{\partial x}\\ H_y &= \frac{-j\omega\epsilon_0}{h_p^2}\frac{\partial E_z}{\partial x}\\ E_x &= \frac{-\gamma_p}{h_p^2}\frac{\partial E_z}{\partial x}\\ E_y &= \frac{j\omega\mu_0}{h_p^2}\frac{\partial H_z}{\partial x}\end{aligned}\right\} \tag{6.12}$$

where

$$h_p^2 = \gamma_p^2 + \omega^2 \mu_0 \epsilon_0 \tag{6.13}$$

It is seen from Equation (6.12) that if $E_z = 0,\ H_z = 0$ as in the case of uniform plane wave, all the components will vanish. Therefore, there should be either **E** or **H** in the direction of propagation. In view of this, the solution is divided into two parts:

1. **Transverse Electric waves (TE waves)** In this $E_z = 0$ and $H_z \neq 0$.
2. **Transverse Magnetic waves (TM waves)** In this $H_z = 0$ and $E_z \neq 0$.

6.4 FIELD COMPONENTS FOR TE WAVES ($E_z = 0$)

TE wave means transverse electric wave for which there is no component of **E** in the direction of propagation, or, $E_z = 0$.

For this case, as $E_z = 0$, it is obvious from Equation (6.12) that $H_y = 0$ and $E_x = 0$. Substituting these values in Equation (6.11), we get the component of y as,

$$\frac{\partial^2 E_y}{\partial x^2}+\gamma_p^2 E_y = -\omega^2 \mu_0 \epsilon_0 E_y$$

that is, $$\frac{\partial^2 E_y}{\partial x^2} = -(\gamma_p^2 + \omega^2 \mu_0 \epsilon_0) E_y$$

or, $$\frac{\partial^2 E_y}{\partial x^2} = -h_p^2 E_y \tag{6.14}$$

or, $$\frac{\partial^2 E_y^0}{\partial x^2} = -h_p^2 E_y^0 \tag{6.15}$$

where $$h_p^2 = (\gamma_p^2 + \omega^2 \mu_0 \epsilon_0)$$

This equation has the solution in the form of

$$E_y^0 = A_1 \sin h_p x + A_2 \cos h_p x \tag{6.16}$$

where A_1 and A_2 are constants.

Using the boundary condition, $E_y = 0$ at $x = 0$, A_2 is zero.

So, $$E_y^0 = A_1 \sin h_p x e^{-\gamma_p z} \tag{6.17}$$

At $x = a$, $E_y = 0$ requires that

$$h_p = \frac{m\pi}{a}, \; m = 1, 2, 3, \ldots \tag{6.18}$$

Here if $m = 0$, all the field components vanish.

Thus, $$E_y^0 = A_1 \sin\left(\frac{m\pi}{a} x\right) e^{-\gamma_p z} \tag{6.19}$$

Substituting Equation (6.19) in Equation (6.18), the other components are obtained. The resultant expressions for the field components are:

$$E_y^0 = A_1 \sin\left(\frac{m\pi}{a} x\right)$$

$$H_z^0 = \frac{-m\pi}{j\omega\mu_0 a} A_1 \cos\left(\frac{m\pi}{a} x\right)$$

$$H_x^0 = \frac{-\gamma_p}{j\omega\mu_0} A_1 \sin\left(\frac{m\pi}{a} x\right)$$

These represent the fields for TE_m.

The field components in complete form are:

$$E_y = E_y^0 \, e^{-\gamma_p z}$$

$$H_z = H_z^0 \, e^{-\gamma_p z}$$

$$H_x = H_x^0 \, e^{-\gamma_p z}$$

The instantaneous electric and magnetic fields for TE_1 wave at some instant of time are shown in Fig. 6.2.

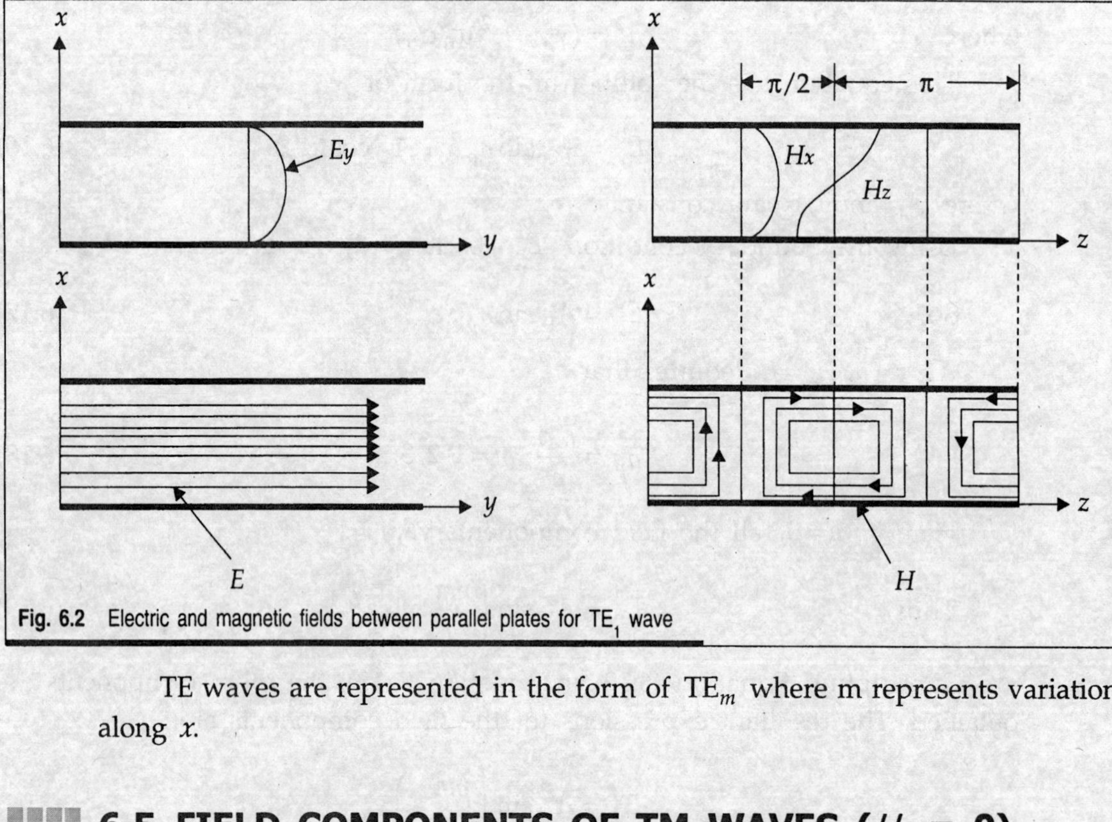

Fig. 6.2 Electric and magnetic fields between parallel plates for TE_1 wave

TE waves are represented in the form of TE_m where m represents variation along x.

6.5 FIELD COMPONENTS OF TM WAVES (H_z = 0)

TM wave means transverse magnetic wave for which there is no component of magnetic field in the direction of propagation, or, $H_z = 0$.

The wave equation for y-component of **H** from Equation (6.11) can be written as

$$\frac{\partial^2 H_y}{\partial x^2} + \gamma_p^2 H_y = -\omega^2 \mu_0 \in_0 H_y \tag{6.20}$$

or,
$$\frac{\partial^2 H_y}{\partial x^2} = -h_p^2 H_y \tag{6.21}$$

where
$$h_p^2 = \gamma_p^2 + \omega^2 \mu_0 \in_0$$

The solution of Equation (6.21) appears in the form of

$$H_y^0 = (A_3 \sin h_p x + A_4 \cos h_p x) \tag{6.22}$$

As the tangential component of **H** is not zero at the surface of a conductor, the boundary condition cannot be applied directly to H_y to obtain A_3 and A_4. From Equation (6.9) we have

$$\frac{\partial H_y^0}{\partial x} = j\omega \in_0 E_z^0 \tag{6.23}$$

Putting Equation (6.22) in Equation (6.23), and writing the equation for E_z, we get

$$E_z^0 = \frac{h_p}{j\omega \in_0} [A_3 \cos hx - A_4 \sin h_p x] \tag{6.24}$$

The boundary conditions are:

$$E_z^0 = 0 \text{ at } x = 0$$

$$E_z^0 = 0 \text{ at } x = a$$

Applying the first boundary condition, A_3 becomes zero. For the second boundary condition, $E_z = 0$ at $x = a$. E_z becomes zero if

$$h_p = \left(\frac{m\pi}{a}\right), m = 0, 1, 2, 3, \ldots$$

Here, $m = 0$ is also possible as there still exist some more components.

$$E_z^0 = \frac{-h_p}{j\omega \in_0} A_4 \sin\left(\frac{m\pi}{a} x\right) \tag{6.25}$$

From Equations (6.9) and (6.25) we get the remaining components H_y and E_x. As a whole, the expressions of field components for TM waves are:

$$E_z^0 = \frac{-h_p}{j\omega \in_0} A_4 \sin\left(\frac{m\pi}{a} x\right)$$

$$H_y^0 = A_4 \cos\left(\frac{m\pi}{a} x\right) \tag{6.26}$$

$$E_x^0 = \frac{\gamma_p A_4}{j\omega \in_0} \cos\left(\frac{m\pi}{a} x\right) \tag{6.27}$$

The field components in complete form for TM_m waves are:

$$E_z = E_z^0 e^{-\gamma_p z}$$

$$H_y = H_y^0 e^{-\gamma_p z}$$

$$E_x = E_x^0 e^{-\gamma_p z}$$

The field variations for TM_1 wave between parallel plates are shown in Fig. 6.3.

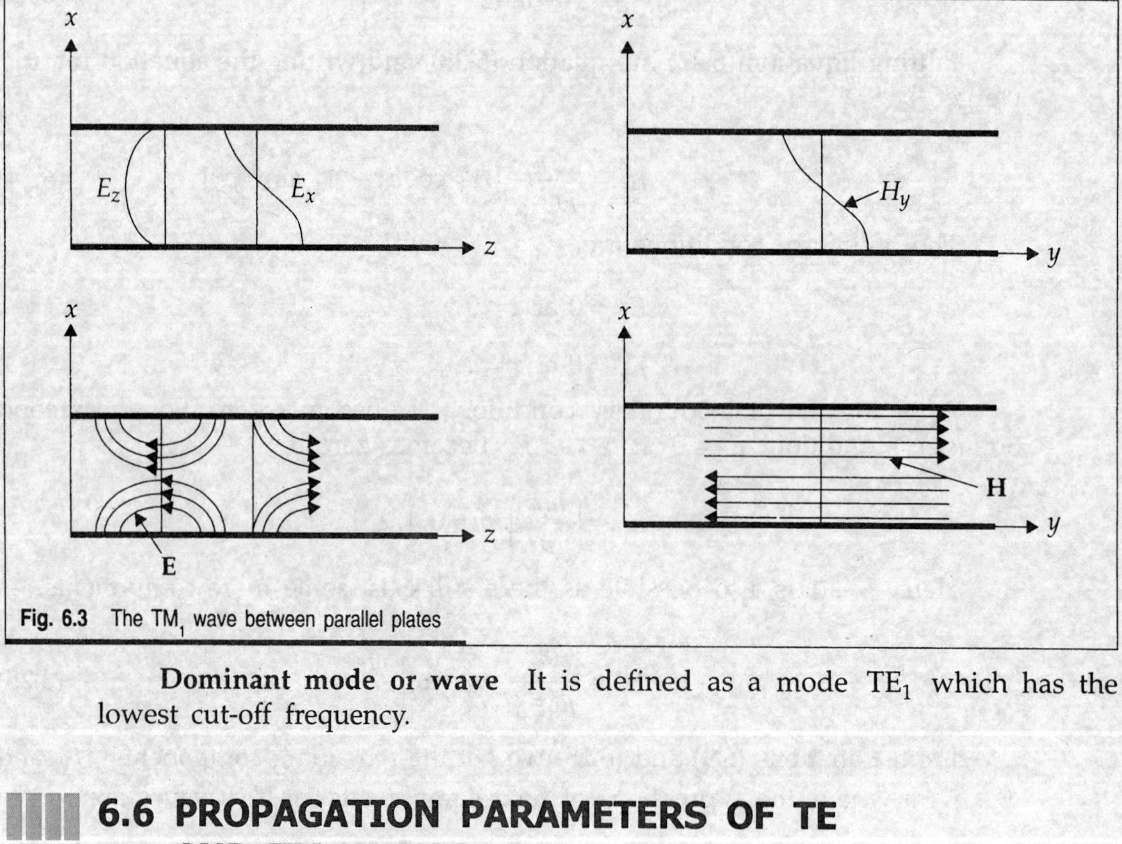

Fig. 6.3 The TM_1 wave between parallel plates

Dominant mode or wave It is defined as a mode TE_1 which has the lowest cut-off frequency.

6.6 PROPAGATION PARAMETERS OF TE AND TM WAVES

We have the expression for propagation of TE and TM waves as

$$\gamma_p = \sqrt{h_p^2 - \omega^2 \mu_0 \in_0}$$

where $$h_p = \left(\frac{m\pi}{a}\right)$$

If $\omega^2 \mu_0 \epsilon_0 > \left(\frac{m\pi}{a}\right)^2$, γ_p will be purely imaginary, or,

$$\gamma_p = j\beta_p$$

β_p is the phase constant

So, $$\beta_p = \sqrt{\omega^2 \mu_0 \epsilon_0 - \left(\frac{m\pi}{a}\right)^2}$$

If $\left(\frac{m\pi}{a}\right)^2 > \omega^2 \mu_0 \epsilon_0$, γ_p will be purely real, or,

$$\gamma_p = \alpha_p$$

α_p is the attenuation constant

So, $$\alpha_p = \sqrt{\left(\frac{m\pi}{a}\right)^2 - \omega^2 \mu_0 \epsilon_0}$$

For all frequencies less than ω_c,

where $$\omega_c = \frac{1}{\sqrt{\mu_0 \epsilon_0}}\left(\frac{m\pi}{a}\right)$$

γ_p is real and $\beta_p = 0$.. This means that fields will be attenuated exponentially in z-direction and there is no wave motion as $\beta_p = 0$.

At $f > f_c$, $\gamma_p = j\beta_p$, wave motion takes place and $\alpha_p = 0$. This f_c is known as cut-off frequency.

Cut-off frequency (f_c)

Definition Cut-off frequency, f_c is defined as a frequency below which there exists only attenuation and $\beta_p = 0$ and above which $\alpha_p = 0$ and β_p exists.

$$f_c = \frac{m}{2a\sqrt{\mu_0 \epsilon_0}}$$

6.7 GUIDE WAVELENGTH

Wavelength between the plates is $\lambda_p = \frac{2\pi}{\beta_p}$

$$= \frac{2\pi}{\sqrt{\omega^2 \mu_0 \in_0 - \left(\frac{m\pi}{a}\right)^2}}$$

$$= \sqrt{\frac{2\pi}{\omega^2 \mu_0 \in_0 - \omega_c^2 \mu_0 \in_0}}$$

$$= \frac{2\pi}{\omega\sqrt{\mu_0 \in_0}\left(1 - \frac{\omega_c^2}{\omega^2}\right)^{1/2}}$$

But $$\lambda = \frac{2\pi}{\beta} = \text{free space wavelength}$$

where $$\beta = \frac{\omega}{\upsilon_0}$$

$$\upsilon_0 = \frac{1}{\sqrt{\mu_0 \in_0}}$$

$$\frac{2\pi}{\omega\sqrt{\mu_0 \in_0}} = \lambda$$

Now $$\lambda_p = \frac{\lambda}{\left[1 - \left(\frac{f_c}{f}\right)^2\right]^{1/2}}$$

$$\lambda_p = \frac{\lambda}{\left[1 - \left(\frac{\lambda}{\lambda_c}\right)^2\right]^{1/2}} \tag{6.28}$$

where λ_c = cut-off wavelength

The propagation parameters are:

$$\gamma_p = \sqrt{\left(\frac{m\pi}{a}\right)^2 - \omega^2 \mu_0 \in_0}$$

$$\alpha_p = \sqrt{\left(\frac{m\pi}{a}\right)^2 - \omega^2\mu_0 \in_0}\,, \text{ if } \left(\frac{m\pi}{a}\right)^2 > \omega^2 \mu_0 \in_0$$

$$\beta_p = \sqrt{\omega^2\mu_0 \in_0 - \left(\frac{m\pi}{a}\right)^2}\,, \text{ if } \omega^2 \mu_0 \in_0 > \left(\frac{m\pi}{a}\right)^2$$

$$\lambda_p = \frac{2\pi}{\beta_p} = \frac{\lambda}{\left[1-\left(\frac{\lambda}{\lambda_C}\right)^2\right]^{1/2}}$$

$$\upsilon_p = \frac{\omega}{\beta_p}$$

$$f_c = \frac{m}{2a\sqrt{\mu_0 \in_0}}$$

For TE_1

$$f_c = \frac{1}{2a\sqrt{\mu_0 \in_0}}$$

$$\lambda_c = 2a$$

6.8 TRANSVERSE ELECTROMAGNETIC WAVE (TEM WAVE)

Definition 1 TEM wave is a wave for which there are no components of **E** and **H** in the direction of propagation.

Definition 2 TEM wave is a TM_m wave for $m=0$, or, $TEM = TEM_0$. TEM wave is called **principal wave**.

The field components of TEM wave are obtained from $m=0$. They are:

$$H_y = A_4$$

$$E_x = \frac{\gamma_p}{j\omega\in_0} A_4$$

The propagation parameters for TEM wave are

$$\gamma_p = j\omega \sqrt{\mu_0 \in_0}\left(\frac{1}{m}\right)$$

$$\alpha_p = 0$$
$$\beta_p = \omega\sqrt{\mu_0 \in_0} \text{ (rad/m)}$$
$$\upsilon_p = 1/\sqrt{\mu_0 \in_0} \text{ (m)}$$
$$\lambda_p = \lambda = \upsilon_0/f \text{ (m)}$$
$$f_c = 0$$
$$\lambda_c = \infty$$
$$\eta_p = \left|\frac{E_x}{H_y}\right| = \sqrt{\frac{\mu_0}{\in_0}} = 120\pi\Omega$$

TEM wave between parallel plates are given in Fig. 6.4.

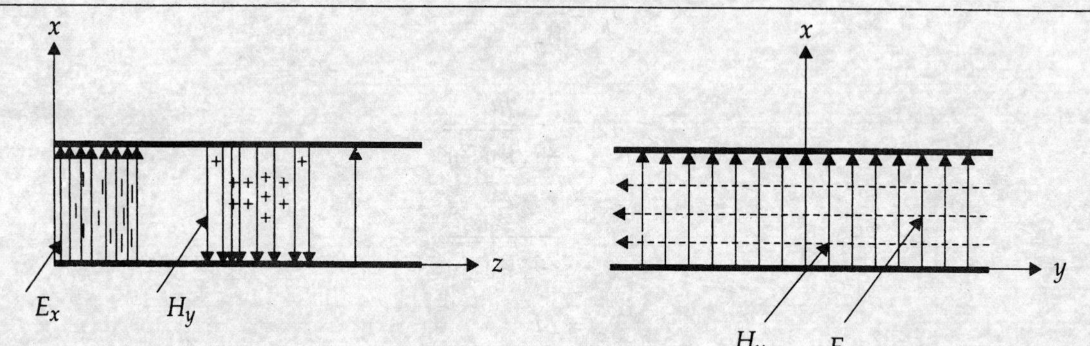

Fig. 6.4 TEM wave fields between parallel plates

6.9 VELOCITIES OF PROPAGATION

Free space velocity (υ_0) It is defined as the velocity of propagation of an EM wave in free space.

Phase velocity (υ_p) It is defined as the velocity of propagation of equiphase surfaces along the guide.

Group velocity (υ_g) It is defined as the velocity with which energy propagates in a waveguide.

The phase velocity in a waveguide is always greater than free space velocity.

In free space, velocity of propagation is equal to the phase velocity.

Group velocity in waveguide is less than free space velocity.

The velocities of propagation are related by

$$\upsilon_0^2 = \upsilon_g \upsilon_p$$

where $$\upsilon_p = \frac{\lambda_p}{\lambda}\upsilon_0$$

$$\upsilon_g = \frac{\lambda}{\lambda_p}\upsilon_0$$

6.10 ATTENUATION IN PARALLEL PLATE GUIDES

Attenuation factor in general

$$\alpha \equiv \frac{\text{power lost per unit length}}{2\times\text{power transmitted}}$$

Attenuation factor for TEM wave,

$$\alpha_p = \frac{1}{\eta_p a}\sqrt{\frac{\omega\mu_c}{2\sigma_c}}, \text{Nepers/m}$$

where a = plate separation, m

σ_c = conductor conductivity, mho/m

μ_c = conductor permeability, H/m

$\omega = 2\pi f$

η_p = intrinsic impedance, Ω

Attenuation factor is also given by

$$\alpha_p = \frac{R}{2z_0}$$

where R = resistance, Ω/m

z_0 = characteristic impedance, Ω

Attenuation of TE waves

$$\alpha_p = \frac{2m^2\pi^2\sqrt{\omega\mu_c/2\sigma_c}}{\omega\mu a^2\sqrt{\omega^2\mu\epsilon - \left(\frac{m\pi}{a}\right)^2}} \tag{6.29}$$

where $m = 1, 2, 3, \ldots$

$\omega = 2\pi f$

μ = permeability of medium between plates

ϵ = permittivity of medium between plates

6.11 WAVE IMPEDANCES

Wave impedances at any point between parallel plates are defined by the following relations:

$$z_{xy}^{+} \equiv \frac{E_x}{H_y}$$

$$z_{yx}^{+} \equiv -\frac{E_y}{H_x}$$

$$z_{yz}^{+} \equiv \frac{E_y}{H_z}$$

$$z_{zy}^{+} \equiv -\frac{E_z}{H_y}$$

$$z_{zx}^{+} \equiv \frac{E_z}{H_x}$$

$$z_{xz}^{+} \equiv -\frac{E_x}{H_z}$$

These wave impedances can be evaluated from field expressions. For example,

$$z_{yx}^{+} = \frac{j\omega\mu_0}{\gamma_p}$$

Problem 6.1 If a wave of 6 GHz is propagating between two parallel conducting plates separated by 30 mm, find the cut-off wavelength, guide wavelength for TE_1 mode.

Solution Frequency of the wave,

$$f = 6\,\text{GHz}$$

Plate separation,

$$a = 30\,\text{mm} = 0.03\,\text{m}$$

Cut-off wavelength for TE_1 is

$$\lambda_c = 2\times \text{(plane separation)}$$

$$= 2\times 0.03 = 0.06\ \text{m}$$

$$\boxed{\lambda_c = 6\ \text{cm}}$$

Free space wavelength,

$$\lambda = \frac{\upsilon_0}{f} = \frac{3\times 10^8}{6\times 10^9} = 0.05\ \text{m}$$

The guide wavelength

$$\lambda_p = \frac{\lambda}{\left[1-\left(\frac{\lambda}{\lambda_c}\right)^2\right]^{1/2}}$$

$$= \frac{0.05}{\left[1-\left(\frac{0.05}{0.06}\right)^2\right]^{1/2}}$$

$$= \frac{0.05}{(1-0.694)^{1/2}}$$

$$= \frac{0.05}{(0.3055)^{1/2}}$$

$$= \frac{0.05}{0.5527} = 0.0905$$

$$\boxed{\lambda = 9.05\ \text{cm}}$$

Problem 6.2 When a wave of 6 GHz propagates in parallel conducting plates separated by 3 cm, find the phase velocity, group velocity of the wave for the dominant wave.

Solution Frequency of the wave,

$$f = 6\ \text{GHz}$$

Plate separation, $a = 3\ \text{cm}$

Free space wavelength, $\lambda = \frac{\upsilon_0}{f} = \frac{3\times 10^8}{6\times 10^9}$

$$\lambda = 5\ \text{cm}$$

Cut-off wavelength, $\lambda_c = 2\times a = 2\times 3 = 6\ \text{cm}$

Phase velocity, $$\upsilon_p = \left(\frac{\lambda_p}{\lambda}\right)\upsilon_0$$

$$= \frac{\lambda}{\left[1-\left(\frac{\lambda}{\lambda_C}\right)^2\right]^{1/2}} \times \frac{1}{\lambda}\upsilon_0$$

$$= \frac{\upsilon_0}{\left[1-\left(\frac{\lambda}{\lambda_C}\right)^2\right]^{1/2}}$$

$$= \frac{3\times10^8}{\left[1-\left(\frac{5}{6}\right)^2\right]^{1/2}}$$

$$\boxed{\upsilon_p = 5.430\times10^8 \text{ m/s}}$$

Group velocity, $$\upsilon_g = \frac{\lambda}{\lambda_p}\upsilon_0$$

$$= \left[1-\left(\frac{\lambda}{\lambda_C}\right)^2\right]^{1/2}\upsilon_0$$

$$= \left[1-\left(\frac{5}{6}\right)^2\right]^{1/2} \times 3\times10^8$$

$$\boxed{\upsilon_g = 1.66\times10^8 \text{ m/s}}$$

Problem 6.3 When a wave of 6 GHz is to be propagated between two parallel conducting plates separated by 60 mm, find the modes that will propagate through the guide.

Solution The modes which have their cut-off frequencies less than the frequency of the wave will propagate.

Here, plate separation, $a = 60\text{ mm} = 6.0\text{ cm}$

$$f = 6\text{ GHz} = 6\times10^9\text{ Hz}$$

For TE_m $\quad f_c = \left(\frac{m}{2a}\right)\upsilon_0$

For TE_1 $\quad f_c = \frac{1}{2\times 6}\times 3\times 10^{10} = 2.5\,\text{GHz}$

As $f_c < f$, TE_1 propagates.

For TE_2 $\quad f_c = \frac{2}{2\times 6}\times 3\times 10^{10} = 5\,\text{GHz}$

It propagates.

For TE_3 $\quad f_c = \frac{3}{2\times 6}\times 3\times 10^{10} = 7.5\,\text{GHz}$

As $f_c > f$, it does not propagate.

For TE_4 $\quad f_c = \frac{4}{2\times 6}\times 3\times 10^{10} = 10\,\text{GHz}$

As $f_c > f$, this will not propagate.

6.12 WAVES IN RECTANGULAR WAVEGUIDES

A rectangular waveguide is a hollow metallic device with four sides closed and two sides open. It can be used as

1. a radiator
2. a high pass filter
3. a transmission line
4. a feed element to antennas

A hollow rectangular waveguide supports only **TE** and **TM** waves/modes and it does not support **TEM** mode.

6.13 DERIVATION OF FIELD EQUATIONS IN RECTANGULAR HOLLOW WAVEGUIDES

Field expressions can be obtained from the solutions of Maxwell's equations and wave equations.

Assumptions:

(a) Space inside the waveguide is free space or air.

(b) The walls of the waveguides are perfectly conducting.

(c) The direction of propagation of power is z.

(d) The dimension of the narrow wall is b metres.

(e) The dimension of the broad wall is a metres.

(f) The fields in z-direction vary as $e^{-\gamma_g z}$.

As the medium inside the waveguide is air, the first and second Maxwell's equations are given by

$$\nabla \times \mathbf{H} = j\omega\epsilon\, \mathbf{E}$$

$$\nabla \times \mathbf{E} = -j\omega\mu\, \mathbf{H}$$

Expanding these equations, we get

$$\nabla \times \mathbf{H} = \begin{vmatrix} \mathbf{a}_x & \mathbf{a}_y & \mathbf{a}_z \\ \frac{\partial}{\partial x} & \frac{\partial}{\partial y} & \frac{\partial}{\partial z} \\ H_x & H_y & H_z \end{vmatrix}$$

$$= \mathbf{a}_x \left[\frac{\partial H_z}{\partial y} - \frac{\partial H_y}{\partial z}\right] + \mathbf{a}_y \left[\frac{\partial H_x}{\partial z} - \frac{\partial H_z}{\partial x}\right] + \mathbf{a}_z \left[\frac{\partial H_y}{\partial x} - \frac{\partial H_x}{\partial y}\right]$$

$$= j\omega\epsilon\, E_x\, \mathbf{a}_x + j\omega\epsilon\, E_y\, \mathbf{a}_y + j\omega\epsilon\, E_z\, \mathbf{a}_z$$

Equating the respective components, we get

$$\left.\begin{aligned} \frac{\partial H_z}{\partial y} - \frac{\partial H_y}{\partial z} &= j\omega\epsilon\, E_x \\ \frac{\partial H_x}{\partial z} - \frac{\partial H_z}{\partial x} &= j\omega\epsilon\, E_y \\ \frac{\partial H_y}{\partial x} - \frac{\partial H_x}{\partial y} &= j\omega\epsilon\, E_z \end{aligned}\right\} \tag{6.30}$$

and

$$\left.\begin{aligned} \frac{\partial E_z}{\partial y} - \frac{\partial E_y}{\partial z} &= -j\omega\mu\, H_x \\ \frac{\partial E_x}{\partial z} - \frac{\partial E_z}{\partial x} &= -j\omega\mu\, H_y \\ \frac{\partial E_y}{\partial x} - \frac{\partial E_x}{\partial y} &= -j\omega\mu\, H_z \end{aligned}\right\} \tag{6.31}$$

As the fields are assumed to be varying in the form of $e^{-\gamma_g z}$, combining time variation, we get

$$H_y = H_y^0 e^{-\gamma_g z}$$

$$\frac{\partial H_y}{\partial z} = -\gamma_g H_y^0 e^{-\gamma_g z} = -\gamma_g H_y$$

Similarly,

$$\frac{\partial H_x}{\partial z} = -\gamma_g H_x$$

$$\frac{\partial H_y}{\partial z} = -\gamma_g H_y \tag{6.32}$$

and

$$\frac{\partial E_x}{\partial z} = -\gamma_g E_x$$

Substituting Equation (6.32) in Equations (6.30) and (6.31), we get

$$\frac{\partial H_z}{\partial y} + \gamma_g H_y = j\omega\epsilon E_x$$

$$-\gamma_g H_x - \frac{\partial H_z}{\partial x} = j\omega\epsilon E_y \tag{6.33}$$

$$\frac{\partial H_y}{\partial x} - \frac{\partial H_x}{\partial y} = j\omega\epsilon E_z$$

$$\frac{\partial E_z}{\partial y} + \gamma_g E_y = -j\omega\mu H_x$$

$$\frac{\partial E_z}{\partial x} + \gamma_g E_x = j\omega\mu H_y$$

$$\frac{\partial E_y}{\partial x} - \frac{\partial E_x}{\partial y} = -j\omega\mu H_z$$

The wave equations are:

$$\nabla^2 \mathbf{E} = -\omega^2 \mu\epsilon \mathbf{E}$$

$$\nabla^2 \mathbf{H} = -\omega^2 \mu\epsilon \mathbf{H}$$

These can be written as

$$\frac{\partial^2 \mathbf{E}}{\partial x^2}+\frac{\partial^2 \mathbf{E}}{\partial y^2}+\frac{\partial^2 \mathbf{E}}{\partial z^2}=-\omega^2\mu\in \mathbf{E}$$

$$\frac{\partial^2 \mathbf{H}}{\partial x^2}+\frac{\partial^2 \mathbf{H}}{\partial y^2}+\frac{\partial^2 \mathbf{H}}{\partial z^2}=-\omega^2\mu\in \mathbf{H}$$

The wave equations for E_z and H_z are given by

$$\frac{\partial^2 E_z}{\partial x^2}+\frac{\partial^2 E_z}{\partial y^2}+\gamma_g^2 E_z=-\omega^2\mu\in E_z$$

$$\frac{\partial^2 H_z}{\partial x^2}+\frac{\partial^2 H_z}{\partial y^2}+\gamma_g^2 H_z=-\omega^2\mu\in H_z \tag{6.34}$$

Equation (6.33) can be mathematically manipulated to get the following. Consider

$$\gamma_g H_x+\frac{\partial H_z}{\partial x}=-j\omega\in E_y \tag{6.35}$$

and

$$\frac{\partial E_z}{\partial y}+\gamma_g E_y=-j\omega\mu H_x \tag{6.36}$$

Equation (6.35) becomes

$$\gamma_g H_x=-j\omega\in E_y-\frac{\partial H_z}{\partial x} \tag{6.37}$$

But E_y from Equation (6.36) is

$$E_y=\frac{1}{\gamma_g}(-j\omega\mu H_x)-\frac{1}{\gamma_g}\frac{\partial E_z}{\partial y} \tag{6.38}$$

From Equations (6.37) and (6.38), we get

$$\gamma_g H_x=-j\omega\in\left(\frac{-j\omega\mu}{\gamma_g}H_x\right)+\frac{j\omega\in}{\gamma_g}\frac{\partial E_z}{\partial y}-\frac{\partial H_z}{\partial x}$$

$$=\frac{-\omega^2\mu\in}{\gamma_g}H_x+\frac{j\omega\in}{\gamma_g}\frac{\partial E_z}{\partial y}-\frac{\partial H_z}{\partial x}$$

or, $$H_x = \frac{-\omega^2\mu\epsilon}{\gamma_g^2} H_x + \frac{j\omega\epsilon}{\gamma_g^2}\frac{\partial E_z}{\partial y} - \frac{1}{\gamma_g}\frac{\partial H_z}{\partial x}$$

or, $$H_x \frac{(\gamma_g^2 + \omega^2\mu\epsilon)}{\gamma_g^2} = \frac{j\omega\epsilon}{\gamma_g^2}\frac{\partial E_z}{\partial y} - \frac{1}{\gamma_g}\frac{\partial H_z}{\partial x}$$

$$H_x = \frac{j\omega\epsilon}{h_g^2}\frac{\partial E_z}{\partial y} - \frac{\gamma_g}{h_g^2}\frac{\partial H_z}{\partial x}$$

where $$h_g^2 = \gamma_g^2 + \omega^2\mu\epsilon$$

Similarly, $$H_y = -\frac{\gamma_g}{h_g^2}\frac{\partial H_z}{\partial y} - \frac{j\omega\epsilon}{h_g^2}\frac{\partial E_z}{\partial x}$$

$$E_x = -\frac{\gamma_g}{h_g^2}\frac{\partial E_z}{\partial x} - \frac{j\omega\mu}{h_g^2}\frac{\partial H_z}{\partial y} \tag{6.39}$$

$$E_y = -\frac{\gamma_g}{h_g^2}\frac{\partial E_z}{\partial y} + j\omega\mu\frac{\partial H_z}{\partial x}$$

In the above equations, if $E_z = 0$ and $H_z = 0$, all the field components vanish. Hence, the wave cannot satisfy TEM wave characteristics. They are transverse magnetic (TM) and transverse electric (TE) waves. A typical rectangular waveguide is shown in Fig. 6.5.

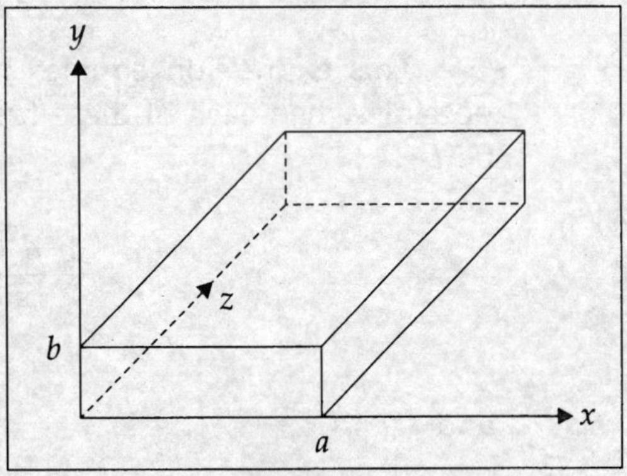

Fig. 6.5 A rectangular waveguide

Transverse Magnetic (TM) Waves in Rectangular Waveguide

TM waves are EM waves for which there is no component of **H** in the direction of propagation, that is, $H_z = 0$.

The wave equations given by Equation (6.34) can be easily solved using the method of product solution. In this method, two ordinary differential equations with known solutions are obtained. We know that,

$$E_z(x, y, z) = E_z^0(x, y)\, e^{-\gamma_g z}$$

If
$$E_z^0 = XY \tag{6.40}$$

where X is only a function of x and Y is only a function of y.

From Equations (6.34) and (6.35), we can write

$$Y\frac{d^2X}{dx^2} + X\frac{d^2Y}{dy^2} + \gamma_g^2 XY = -\omega^2\mu\epsilon XY$$

or,
$$Y\frac{d^2X}{dx^2} + X\frac{d^2Y}{dy^2} + (\gamma_g^2 + \omega^2\mu\epsilon)XY = 0$$

$$Y\frac{\partial^2 Y}{\partial x^2} + X\frac{\partial^2 Y}{\partial y^2} + h_g^2 XY = 0$$

$$\frac{1}{X}\frac{d^2X}{dx^2} + \frac{1}{Y}\frac{d^2Y}{dy^2} + h_g^2 = 0$$

$$\frac{1}{X}\frac{d^2X}{dx^2} + h_g^2 = -\frac{1}{Y}\frac{d^2Y}{dy^2} \tag{6.41}$$

This expression equates a function of x to another function of y. This is possible when each of these functions is equal to some constant. Let the constant be B

$$\frac{1}{X}\frac{d^2X}{dx^2} + h_g^2 = B^2 \tag{6.42}$$

and
$$\frac{1}{Y}\frac{d^2Y}{dy^2} = -B^2 \tag{6.43}$$

The general solution of these equations are

$$X = A_1 \cos Cx + A_2 \sin Cx$$

where
$$C^2 = h_g^2 - B^2$$

and
$$Y = A_3 \cos By + A_4 \sin By$$

$$E_z^0 = XY$$
$$= A_1 A_3 \cos Cx \cos By + A_1 A_4 \cos Cx \sin By$$
$$+ A_2 A_3 \sin Cx \cos By + A_2 A_4 \sin Cx \sin By$$

The constants A_1, A_2, A_3, A_4 are evaluated using the boundary conditions. Using the boundary condition

$$E_z^0 = 0 \text{ at } x = 0$$
$$E_z^0 = A_1 A_3 \cos By + A_1 A_4 \sin By$$

This is zero if $A_1 = 0$

$$E_z^0 = A_2 A_3 \sin Cx \cos By + A_2 A_4 \sin Cx \sin By \tag{6.44}$$

At $y = 0$, Equation (6.44) becomes

$$E_z^0 = A_2 A_3 \sin Cx$$

For this to vanish, A_2 or A_3 can be zero, while assuming $C \neq 0$. Keeping $A_2 = 0$ in Equation (6.44), E_z^0 becomes zero. Hence instead of $A_2 = 0$ assume $A_3 = 0$. Then Equation (6.44) becomes

$$E_z^0 = A_2 A_4 \sin Cx \sin By$$
$$= K \sin Cx \sin By \qquad [K = A_2 A_4]$$

At $x = a$,

$$E_z^0 = K \sin Ca \sin By$$

For this to vanish for all values of y (assuming $B \neq 0$) the constant C must be

$$C = \frac{m\pi}{a}, \; m = 1, 2, 3, \ldots$$

At $y = b$,

$$E_z^0 = K \sin \frac{m\pi}{a} x \sin Bb$$

For this to vanish for all values of x, B must be

$$B = \frac{n\pi}{b}, \; n = 1, 2, 3, \ldots$$

Hence, the final expression for E_z^0 is

$$E_z^0 = K \sin \frac{m\pi}{a} x \sin \frac{n\pi}{b} y \tag{6.45}$$

$$= K \sin Cx \sin By$$

From Equations (6.44) and (6.45), we get

$$\left.\begin{aligned} E_x^0 &= -\frac{\gamma_g K}{h_g^2} B \cos Cx \sin By \\ E_y^0 &= -\frac{\gamma_g K}{h_g^2} B \sin Cx \cos By \end{aligned}\right\} \tag{6.46}$$

$$H_x^0 = \frac{j\omega\in K}{h_g^2} B \sin Cx \cos By$$

$$H_y^0 = \frac{-j\omega\in K}{h_g^2} C \cos Cx \sin By$$

where $C = \frac{m\pi}{a},\ B = \frac{n\pi}{b}$

Therefore, the field components for a TM wave are:

TM_{mn} wavefield components

$$E_z = (K \sin Cx \sin By)\, e^{(-\gamma_g z + j\omega t)}$$

$$E_x = \left(-\frac{\gamma_g K}{h_g^2} C \cos Cx \sin By\right) e^{(-\gamma_g z + j\omega t)}$$

$$E_y = \left(-\frac{\gamma_g K}{h_g^2} B \sin Cx \cos By\right) e^{(-\gamma_g z + j\omega t)}$$

$$H_x = \left(\frac{j\omega\in K}{h_g^2} B \sin Cx \cos By\right) e^{(-\gamma_g z + j\omega t)}$$

$$H_y = \left(\frac{-j\omega\in K}{h_g^2} C \cos Cx \sin By\right) e^{(-\gamma_g z + j\omega t)}$$

$$C = \frac{m\pi}{a} \text{ and } B = \frac{n\pi}{b}$$

Field variations of TM_{11} mode/wave are shown in Fig. 6.6.

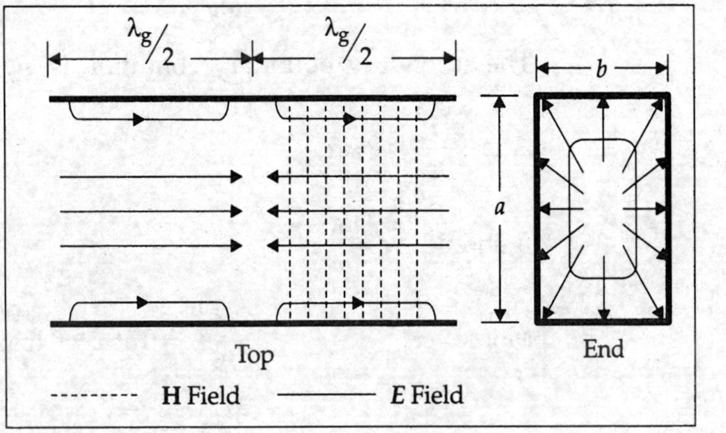

Fig. 6.6 Field variations of TM_{11} wave in hollow rectangular waveguide

Transverse Electric Waves

TE waves are EM waves for which there is no component of **E** in the direction of propagation, that is, $E_z = 0$.

The expressions for TE waves are derived in the same manner as in the case of TM waves. From Equation (6.39), we have

$$E_x = -\frac{\gamma_g}{h_g^2}\frac{\partial E_z}{\partial x} - \frac{j\omega\mu}{h_g^2}\frac{\partial H_z}{\partial y}$$

But $E_z = 0$

Hence, $$E_x = \frac{-j\omega\mu}{h_g^2}\frac{\partial H_z}{\partial y}$$

The first boundary condition is

$$E_x = 0 \text{ at } y = 0$$

that is, $$\frac{\partial H_z^0}{\partial y} = 0 \text{ at } y = 0$$

But $$H_z^0 = (A_1 \cos Cx + A_2 \sin Cx).(A_3 \cos By + A_4 \sin By)$$

$$\frac{\partial H_z^0}{\partial y} = (A_1 \cos Cx + A_2 \sin Cx).(-A_3 B \sin By + A_4 B \cos By)$$

As $$\frac{\partial H_z^0}{\partial y} = 0 \text{ at } y = 0,\ A_4 = 0$$

The secondary boundary condition is $E_x = 0$ at $y = b$

A is $\frac{\partial H_z^0}{\partial y} = 0$ at $y = b$

B should be $\frac{n\pi}{b}$

Moreover, $$\frac{\partial H_z^0}{\partial x} = 0 \text{ at } x = 0 \text{ leads to } A_2 = 0$$

and $$\frac{\partial H_z^0}{\partial x} = 0 \text{ at } x = a, \text{ leads to } c = \frac{m\pi}{a}$$

$$H_z^0 = A_1 A_3 \cos\left(\frac{m\pi}{a}x\right)\cos\left(\frac{n\pi}{b}y\right) \tag{6.47}$$

Substituting Equation (6.47) in Equation (6.39), we get

$$E_x^0 = \frac{j\omega\mu}{h_g^2} KB \cos Cx \sin By$$

$$E_y^0 = -\frac{j\omega\mu}{h_g^2} KC \sin Cx \cos By$$

$$H_x^0 = \frac{\gamma_g}{h_g^2} KC \sin Cx \cos By$$

$$H_y^0 = \frac{\gamma_g}{h_g^2} KB \cos Cx \sin By$$

The field expressions for TE waves in complete form are

$$H_z = (K \cos Cx \cos By)\, e^{(j\omega t - \gamma_g z)}$$

$$E_x = \left(\frac{j\omega\mu}{h_g^2} KB \cos Cx \sin By\right) e^{(j\omega t - \gamma_g z)}$$

$$E_y = \left(\frac{-j\omega\mu}{h_g^2} KC \sin Cx \cos By\right) e^{(j\omega t - \gamma_g z)}$$

$$H_x = \left(\frac{\gamma_g}{h_g^2} KC \sin Cx \cos By \right) e^{(j\omega t - \gamma_g z)}$$

$$H_y = \left(\frac{\gamma_g}{h_g^2} KB \sin Cx \cos By \right) e^{(j\omega t - \gamma_g z)}$$

where $$h_g^2 = \gamma_g^2 + \omega^2 \mu \in$$

$$C = \frac{m\pi}{a},\ B = \frac{n\pi}{b}$$

$$K = A_1 A_3$$

As TE_{10} is popular, its field equations are

$$H_z = K \cos \frac{\pi x}{a} e^{(j\omega t - \gamma_g z)}$$

$$H_x = \frac{\gamma_g a K}{\pi} \sin \frac{\pi x}{a} e^{(j\omega t - \gamma_g z)}$$

$$E_y = \frac{-j\omega\mu a K}{\pi} \sin \frac{\pi x}{a} e^{(j\omega t - \gamma_g z)}$$

$$E_x = 0,$$

$$H_y = 0$$

The field patterns are shown in Fig. 6.7.

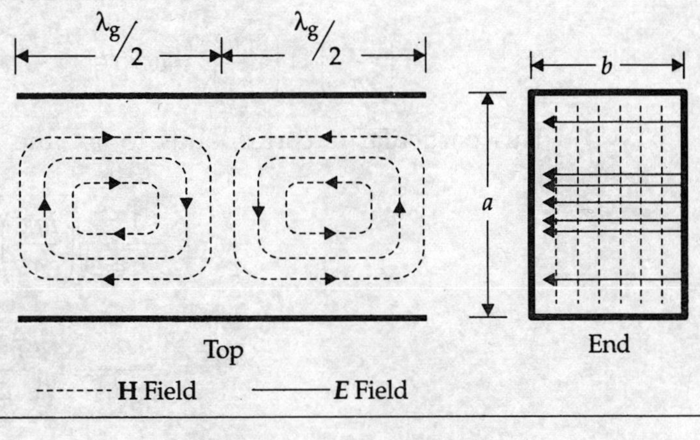

Fig. 6.7 Field patterns of TE_{10} mode

6.14 PROPAGATION PARAMETERS OF TE AND TM WAVES IN RECTANGULAR WAVEGUIDES

We have $$h_g^2 = B^2 + C^2,\ B = \frac{n\pi}{b}, C = \frac{m\pi}{a}$$

and $$h_g^2 = \gamma_g^2 + \omega^2 \mu \in$$

or, $$\gamma_g = \sqrt{h_g^2 - \omega^2 \mu \in}$$

$$\gamma_g = \sqrt{\left(\frac{m\pi}{a}\right)^2 + \left(\frac{n\pi}{b}\right)^2 - \omega^2 \mu \in}$$

If $$\left[\left(\frac{m\pi}{a}\right)^2 + \left(\frac{n\pi}{b}\right)^2\right] > \omega^2 \mu \in$$

$$\gamma_g = \alpha_g \text{ (purely real)}$$

$$\alpha_g = \sqrt{\left(\frac{m\pi}{a}\right)^2 + \left(\frac{n\pi}{b}\right)^2 - \omega^2 \mu \in}$$

If $$\left(\frac{m\pi}{a}\right)^2 + \left(\frac{n\pi}{b}\right)^2 < \omega^2 \mu \in,$$

$$\alpha_g = j\beta_g \text{ (purely imaginary)}$$

$$\beta_g = \sqrt{\omega^2 \mu \in - \left(\frac{m\pi}{a}\right)^2 - \left(\frac{n\pi}{b}\right)^2}$$

If $$\left(\frac{m\pi}{a}\right)^2 + \left(\frac{n\pi}{b}\right)^2 = \omega^2 \mu \in$$

This particular ω corresponds to ω_c, the angular cut-off frequency.

$$\omega_c = \frac{1}{\sqrt{\mu \in}} \sqrt{\left(\frac{m\pi}{a}\right)^2 + \left(\frac{n\pi}{b}\right)^2}$$

or, $$f_c = \frac{1}{2\pi\sqrt{\mu \in}} \sqrt{\left(\frac{m\pi}{a}\right)^2 + \left(\frac{n\pi}{b}\right)^2}$$

And cut-off wavelength is

$$\lambda_c = \frac{\upsilon}{f_c}, \upsilon = \frac{1}{\sqrt{\mu \in}}$$

$$\lambda_c = \frac{2}{\sqrt{\left(\frac{m}{a}\right)^2 + \left(\frac{n}{b}\right)^2}}$$

The velocity of wave propagation is

$$\upsilon_p = \frac{\omega}{\beta_g} = \frac{\omega}{\sqrt{\omega^2 \mu \in - \left(\frac{m\pi}{a}\right)^2 - \left(\frac{n\pi}{b}\right)^2}}$$

$$\lambda_g = \frac{2\pi}{\beta_g} = \frac{2\pi}{\sqrt{\omega^2 \mu \in - \left(\frac{m\pi}{a}\right)^2 - \left(\frac{n\pi}{b}\right)^2}}$$

Derivation of guide wavelength in terms of free space and cut-off wavelengths

The expression for guide wavelength is

$$\lambda_g = \frac{2\pi}{\beta_g} = \frac{2\pi}{\sqrt{\omega^2 \mu \in - \left(\frac{m\pi}{a}\right)^2 - \left(\frac{n\pi}{b}\right)^2}}$$

As $\left(\frac{m\pi}{a}\right)^2 + \left(\frac{n\pi}{b}\right)^2 = \omega_c^2 \mu \in$

$$\lambda_g = \frac{2\pi}{\sqrt{\omega^2 \mu \in - \omega_c^2 \mu \in}}$$

$$= \frac{2\pi}{\omega \sqrt{\mu \in} \left[1 - \left(\frac{\omega_c}{\omega}\right)^2\right]^{1/2}}$$

But $$\upsilon = \frac{1}{\sqrt{\mu \in}}, \lambda = \frac{\upsilon}{f}$$

$$\lambda_g = \frac{\upsilon}{f\sqrt{1-\left(\frac{\omega_c}{\omega}\right)^2}}$$

or, $$\lambda_g = \frac{\lambda}{\left[1-\left(\frac{\lambda}{\lambda_c}\right)^2\right]^{1/2}}$$

The summary of propagation parameters of TE and TM waves are:

1. Propagation constant,

$$\gamma_g = \sqrt{\left(\frac{m\pi}{a}\right)^2 + \left(\frac{n\pi}{b}\right)^2 - \omega^2\mu\in}, \left(\frac{1}{m}\right)$$

$$= \alpha_g \text{ if } \left(\frac{m\pi}{a}\right)^2 + \left(\frac{n\pi}{b}\right)^2 > \omega^2\mu\in$$

2. Phase constant,

$$\beta_g = \sqrt{\omega^2\mu\in - \left(\frac{m\pi}{a}\right)^2 - \left(\frac{n\pi}{b}\right)^2}, (\text{rad/m})$$

3. Cut-off frequency,

$$f_c = \frac{1}{2\pi\sqrt{\mu\in}}\sqrt{\left(\frac{m\pi}{a}\right)^2 + \left(\frac{n\pi}{b}\right)^2}, \text{ (Hz)}$$

4. Cut-off wavelength,

$$\lambda_c = \frac{2}{\sqrt{\left(\frac{m}{a}\right)^2 + \left(\frac{n}{b}\right)^2}}, (\text{m})$$

5. Phase velocity,

$$\upsilon_p = \frac{\omega}{\sqrt{\omega^2\mu\in - \left(\frac{m\pi}{a}\right)^2 - \left(\frac{n\pi}{b}\right)^2}}, \text{ (m/s)}$$

6. Guide wavelength,

$$\lambda_g = \frac{2\pi}{\sqrt{\omega^2\mu\epsilon - \left(\frac{m\pi}{a}\right)^2 - \left(\frac{n\pi}{b}\right)^2}}, \text{ (m)}$$

or,

$$\lambda_g = \frac{\lambda}{\left[1-\left(\frac{\lambda}{\lambda_c}\right)^2\right]^{1/2}}, \text{ (m)}$$

Dominant wave/mode It is defined as a wave which has the lowest cut-off frequency. This is represented by TE_{10}.

In TE_{mn} or TM_{mn} waves, m represents the number of half-period variations of the field along x-axis and n represents the number of half-period variations of the field along y-axis. Here, the broad wall is along the x-axis and the narrow wall is along the y-axis.

Propagation and field equation of dominant mode, TE_{10}

1. Cut-off frequency for dominant mode, TE_{10}

$$f_c \text{ of } TE_{10} = \frac{1}{2a\sqrt{\mu\epsilon}} = \frac{\upsilon}{2a}$$

2. Cut-off wavelength for dominant mode, TE_{10}

$$\lambda_c \text{ of } TE_{10} = 2a$$

Here
λ_c = cut-off wavelength (m)
a = broad wall dimension (m)
m, n = integers = 1, 2, 3, ...

$$H_z = K\cos\frac{\pi x}{a}\, e^{(j\omega t - \gamma_g z)}$$

$$H_x = \frac{\gamma_g aK}{\pi}\sin\frac{\pi x}{a}\, e^{(j\omega t - \gamma_g z)}$$

$$E_y = \frac{-j\omega\mu aK}{\pi}\sin\frac{\pi x}{a}$$

$$E_x = 0,\ H_y = 0$$

$$\beta_g = \sqrt{\omega^2 \mu \in - \left(\frac{\pi}{a}\right)^2}$$

$$\upsilon_g = \frac{\omega}{\sqrt{\omega^2 \mu \in - \left(\frac{\pi}{a}\right)^2}}$$

$$\lambda_g = \frac{2\pi}{\sqrt{\omega^2 \mu \in - \left(\frac{\pi}{a}\right)^2}}$$

Transverse Electromagnetic Waves

In **TEM** wave, both **E** and **H** are entirely transverse to the direction of propagation, that is, if the direction of propagation is along z, $E_z = 0$, and $H_z = 0$.

TEM wave is called principal wave. Its cut-off frequency is zero and it exists in **two conductor** transmission lines or in free space.

Characteristics of TEM waves

1. TEM $\equiv \text{TM}_{00}T$
2. For TEM, $E_z = 0, H_z = 0$
3. Its cut-off frequency, $f_c = 0$
4. It exists only in **two conductor** transmission lines or in free space.
5. It does not exist in hollow waveguides.
6. $\lambda_g = \lambda$
7. $\beta_g = \beta$
8. $\alpha = 0$
9. $\eta = \eta_0$
10. $\lambda_c = \infty$

6.15 TEM WAVE DOES NOT EXIST IN HOLLOW WAVEGUIDES

Proof Method 1 Consider TM wavefield equations given by

$$E_z = (K \sin Cx \sin By) e^{(j\omega t - \gamma_g z)}$$

$$E_x = \left(-\frac{\gamma_g}{h_g^2} KC \sin Cx \cos By \right) e^{(j\omega t - \gamma_g z)}$$

$$E_y = \left(-\frac{\gamma_g}{h_g^2} KB \sin Cx \cos By \right) e^{(j\omega t - \gamma_g z)}$$

$$H_x = \left(\frac{j\omega \in K}{h_g^2} B \sin Cx \cos By \right) e^{(j\omega t - \gamma_g z)}$$

$$H_y = \left(\frac{-j\omega \in}{h_g^2} KC \cos Cx \sin By \right) e^{(j\omega t - \gamma_g z)}$$

As TEM = TM_{00}, that is, when $m=0$, $n=0$, all the above field components vanish. This itself indicates that there exists no TEM waves in hollow waveguides.

Method 2 Assume that TEM wave exists within a hollow waveguide. Then, the magnetic field lines must be in the transverse plane. Also we know that

$$\nabla.\mathbf{B} = \nabla.\mu\mathbf{H} = 0$$

or,

$$\nabla.\mathbf{H} = 0$$

This requires that the lines of **H** be closed loops. Hence, if a TEM wave exists (by hypothesis) inside the waveguide, the lines of **H** are closed loops in a plane perpendicular to the axis of the guide. It may be noted that the direction of propagation is along the axis.

By Maxwell's first equation, we have

$$\oint \mathbf{H}.d\mathbf{L} = \int_s (\dot{\mathbf{D}} + \mathbf{J}).d\mathbf{S}$$

that is, the magnetomotive force around the closed loop of **H** lines is equal to the sum of axial displacement and conduction currents. As the space inside the guide is air or free space,

$$\mathbf{J} = \sigma\mathbf{E} = 0 \qquad [\text{as } \sigma = 0]$$

that is, conduction current is zero. Hence the axial current must be a displacement current. If there exists displacement current in the axial direction which is the direction of propagation of EM energy, there should be a component of **E** $\left(\text{as } \frac{\partial \mathbf{D}}{\partial t} = \in \frac{\partial \mathbf{E}}{\partial t} = \text{displacement current density}\right)$ in the axial direction. The presence of **E** along the axial or direction of propagation indicates the absence of

TEM wave. Therefore, we conclude that there exists no TEM in hollow waveguides of any shape.

6.16 EXCITATION METHODS FOR DIFFERENT TE AND TM WAVES/MODES

The excitation of TE_{10} by a probe is shown in Fig. 6.8.

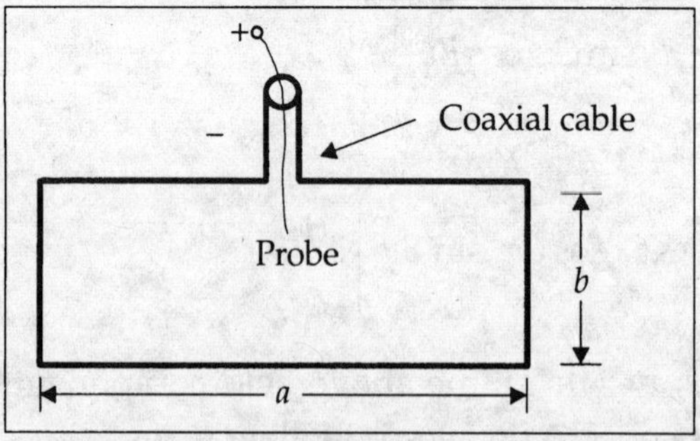

Fig. 6.8 Excitation of TE_{10} by a probe

The excitation of TE_{10} by a loop is shown in Fig. 6.9.

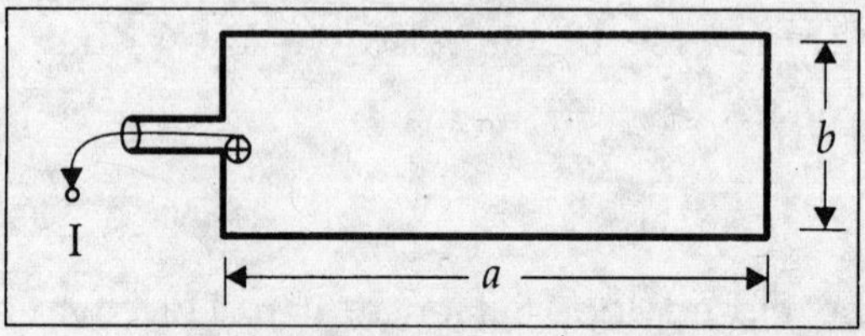

Fig. 6.9 Excitation of TE_{10} by a loop

The excitation method of TM_{11} by a probe and a loop are shown in Figs. 6.10 and 6.11.

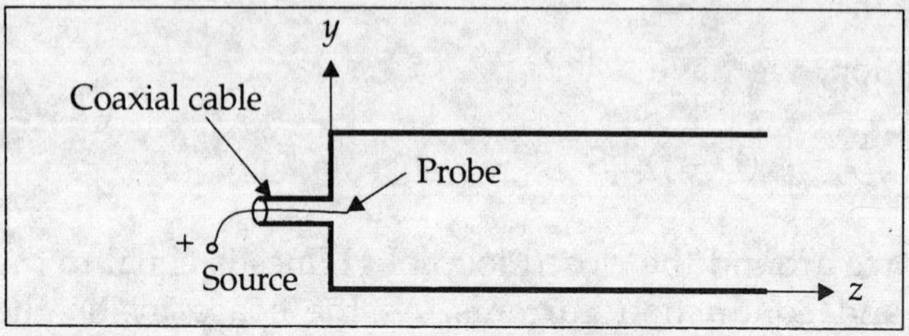

Fig. 6.10 Excitation method of TM_{11} by a probe

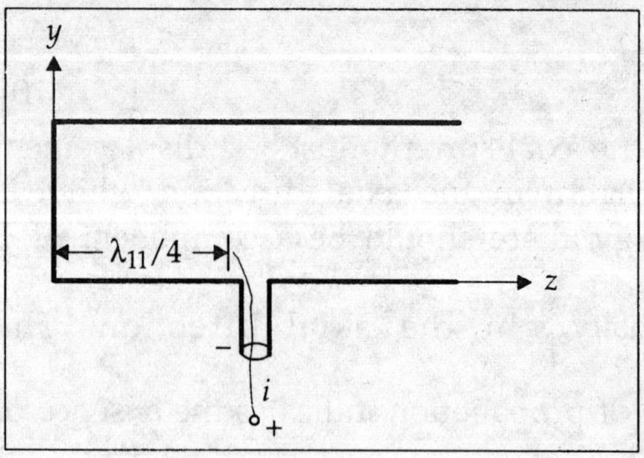

Fig. 6.11 Excitation method of TM_{11} by a loop

6.17 EVANESCENT WAVE OR MODE

This is defined as a wave TE_{mn} or TM_{mn} in which the operating frequency is less than the cut-off frequency and wave propagation does not take place. For evanescent wave, the TM_{mn} wave impedance is purely capacitive and this causes only reactive power or energy storage.

6.18 WAVE IMPEDANCE IN WAVEGUIDE

For TE_{mn} wave, wave impedance is defined as

$$\eta_{mn}^{TE} = \frac{E_x}{H_y} = -\frac{E_y}{H_x} = \frac{\omega\mu}{\beta_g}$$

or,

$$\eta_{mn}^{TE} = \frac{\eta}{\sqrt{1-\left(\frac{f_c}{f}\right)^2}}$$

where η = intrinsic impedance of the unbounded medium.

The wave impedance is purely resistive and average power flow occurs in the waveguide when $f > f_c$.

For non-propagating TM_{mn} wave, the wave impedance

$$\eta_{mn}^{TM} \equiv \frac{E_x}{H_y} = \frac{-j\alpha}{\omega\epsilon}$$

when $f < f_c$ for a particular TM_{mn} mode.

The variation of magnitude of wave impedances of TE_{mn} and TM_{mn} for $f < f_c$ is shown in Fig. 6.12.

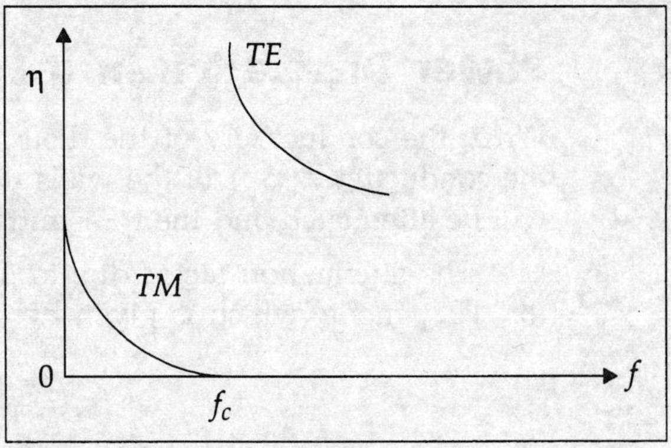

Fig. 6.12 Magnitude of TE_{mn} and TM_{mn} wave impedances ($f < f_c$)

The wave impedance for the propagating modes is found to be

$$\eta_{mn}^{TM} = \frac{\gamma_g}{j\omega\epsilon}$$

or,

$$\eta_{mn}^{TM} = \eta\sqrt{1-\left(\frac{f_c}{f}\right)^2}$$

From the above equations, we get

$$\sqrt{\eta_{mn}^{TE}\,\eta_{mn}^{TM}} = \eta = \sqrt{\frac{\mu}{\epsilon}}$$

6.19 POWER TRANSMITTED IN A LOSSLESS WAVEGUIDE

The average power transmitted in z-direction is found by integration of the z component of the complex Poynting vector over a transverse cross-section of the waveguide, that is,

$$P_{av} = \frac{1}{2}\,\text{Re}\iint_{\text{cross-section}} \mathbf{E}\times\mathbf{H}^{*}.\mathbf{a}_z\,dS$$

where $\mathbf{H}^{*}$ = complex conjugate of $\mathbf{H}$

Power transmission takes place through TE_{10} wave. Using the corresponding component values of **E** and **H***, P_{av} is given by

$$P_{av} = \frac{ab\,|\mathbf{E}|^2}{4\eta_{10}^{TE}} \text{ watts}$$

In an ideal guide, P_{av} is independent of z-direction.

Power Dissipation in a Lossy Waveguide

When the conductivity of the dielectric (σ_d) in the waveguide is non-zero and the conductivity (σ_c) of the walls is not infinite, wave in the propagating mode will be attenuated and the transmitted power will decrease exponentially with z.

The attenuation factor due to dielectric loss is the indication of power loss for TE_{10} mode, which is given by

$$\alpha_d = \frac{\omega\mu_d\,\sigma_d}{2\beta_g} = \frac{1}{2}\,\eta_{10}^{TE}\,\sigma_d,\ \text{Neper/m}$$

The attenuation factor due to wall loss indicates power loss and is given by

$$\alpha_w = \frac{P_{\text{loss}}(z)}{2P_{av}(z)} = \frac{P_{\text{loss}}}{2p_{10}}$$

where P_{loss} = power flow into the first 1 m of the inner surface of the wall

$$P_{10} = P_{av}(z)\,e^{2\alpha_w z}$$

and simplified expression for α_w of TE_{10} mode is given by

$$\alpha_w = \frac{R_{sc}}{\eta}\left(\sqrt{\frac{f}{f_c}}\right) \times \frac{a + 2b\left(\frac{f_c}{f}\right)^2}{ab\sqrt{1-\left(\frac{f_c}{f}\right)^2}} \text{ Np/m}$$

where R_{sc} = surface resistance at cut-off frequency of TE_{10}, Ω

$$\eta = \sqrt{\frac{\mu}{\in}}\ \Omega$$

The total attenuation factor is

$$\alpha_t = \alpha_d + \alpha_w, \text{ np/m}$$

Problem 6.4 Find the cut-off frequencies for TE_{12} mode in a hollow rectangular waveguide whose dimensions are:

(a) $a = 2.286\,\text{cm}, b = 1.016\,\text{cm}$

(b) $a = 1.016\,\text{cm}, b = 2.286\,\text{cm}$

(c) $a = 1\,\text{cm}, b = 1\,\text{cm}$

(d) $a = 10\,\text{cm}, b = 10\,\text{cm}$

Solution The cut-off frequency for TE_{12} mode is given by

$$f_c = \frac{\upsilon_0}{2}\sqrt{\left(\frac{m}{a}\right)^2 + \left(\frac{n}{b}\right)^2}$$

(a) $a = 2.286\,\text{cm}, b = 1.016\,\text{cm}$

$$f_c = \frac{\upsilon_0}{2}\sqrt{\left(\frac{1}{a}\right)^2 + \left(\frac{2}{b}\right)^2}$$

$$= \frac{3\times10^{10}}{2}\sqrt{\left(\frac{1}{2.286}\right)^2+\left(\frac{2}{1.016}\right)^2}$$

$$= 30.215\,\text{GHz}$$

(b) $a = 1.016\,\text{cm},\ b = 2.286\,\text{cm}$

$$f_c = \frac{3\times10^{10}}{2}\sqrt{\left(\frac{1}{1.016}\right)^2+\left(\frac{2}{2.286}\right)^2}$$

$$= 19.75\,\text{GHz}$$

(c) $a = 1\,\text{cm},\ b = 1\,\text{cm}$

$$f_c = \frac{3\times10^{10}}{2}\sqrt{1^2+2^2}$$

$$= 33.54\,\text{GHz}$$

(d) $a = 10$ cm, $b = 10$ cm

$$f_c = \frac{3\times10^{10}}{2}\sqrt{\left(\frac{1}{10}\right)^2+\left(\frac{2}{10}\right)^2}$$

$$= 3.354\ \text{GHz}$$

Problem 6.5 A rectangular waveguide with dimensions 3 × 2 cm operates at 10 GHz. Find f_c, λ_c, λ, λ_g, β_g, υ_p of TE_{10} mode.

Solution $a = 3\,\text{cm},\ b = 2\,\text{cm},\ f = 10\,\text{GHz}$

For TE_{10} mode,

$$f_c = \frac{\upsilon_0}{2a} = \frac{3\times10^{10}}{2\times3} = 0.5\times10^{10}$$

$$\boxed{f_c = 5\,\text{GHz}}$$

$$\boxed{\lambda_c = 2a = 2\times3 = 6\,\text{cm}}$$

$$\boxed{\lambda = \frac{3\times10^{10}}{10\times10^9} = 3\,\text{cm}}$$

$$\lambda_g = \frac{\lambda}{\left[1-\left(\frac{\lambda}{\lambda_c}\right)^2\right]^{1/2}}$$

$$= \frac{3}{\left[1 - \left(\frac{3}{6}\right)^2\right]^{1/2}}$$

$$= \frac{3}{(1 - 0.5^2)^{1/2}}$$

$$= \frac{3}{(0.75)^{1/2}} = 3.464$$

$$\boxed{\lambda_g = 3.464 \text{ cm}}$$

$$\boxed{\beta_g = \frac{2\pi}{\lambda_g} = 1.8138 \text{ rad/cm}}$$

$$\upsilon_p = \frac{\omega}{\beta_g} = 34.64 \times 10^9$$

$$\boxed{\upsilon_p = 3.4 \times 10^8 \text{ m/s}}$$

Problem 6.6 Find the broad wall dimension of a rectangular waveguide when the cut-off frequency for TE_{10} mode is (a) 3 GHz, (b) 30 GHz.

Solution (a) $f_c = 3$ GHz for TE_{10} mode

$$f_c = \frac{\upsilon_0}{2a}$$

a = broad wall dimension

or,

$$a = \frac{\upsilon_0}{2f_c} = \frac{3 \times 10^{10}}{2 \times 3 \times 10^9}$$

$$\boxed{a = 0.5 \times 10 = 5 \text{ cm}}$$

(b) $f_c = 30$ GHz for TE_{10} mode

$$a = \frac{\upsilon_0}{2f_c} = \frac{3 \times 10^{10}}{2 \times 30 \times 10^9} = 0.5 \text{ cm}$$

or,

$$\boxed{a = 0.5 \text{ cm}}$$

Problem 6.7 A hollow rectangular waveguide operates at $f = 1$ GHz and it has the dimensions of 5 × 2 cm. Check whether TE_{21} mode propagates or not.

Solution The propagation constant is given by

$$\gamma_g = \sqrt{\left(\frac{m\pi}{a}\right)^2 + \left(\frac{n\pi}{b}\right)^2 - \omega^2 \mu_0 \epsilon_0}$$

Here $a = 5\,\text{cm} = 0.05\,\text{m}$

$b = 2\,\text{cm} = 0.02\,\text{m}$

$f = 1\,\text{GHz} = 10^9\,\text{Hz}$

For TE_{21} $m = 2, n = 1$

$\mu_0 = 4\pi \times 10^{-7}\,\text{H/m}$

$\epsilon_0 = 8.854 \times 10^{-12}\,\text{F/m}$

γ_g for TE_{21} mode is

$$\gamma_g = \sqrt{\left(\frac{2\pi}{0.05}\right)^2 + \left(\frac{\pi}{0.02}\right)^2 - 4\pi^2 \times 10^{18} \times 4\pi \times 10^{-7} \times 8.854 \times 10^{-12}}$$

$$= 200.1 + j0,\ \text{NP/m}$$

$$= \alpha_g \text{ and } \beta_g = 0$$

$$\gamma_g = \alpha_g = 200.1\,\text{Np/m}$$

As γ_g is purely real, there is no propagation of TE_{21} mode.

6.20 WAVEGUIDE RESONATORS

A waveguide resonator is a resonator at high frequencies. It is made up of a rectangular waveguide with its open ends closed by shorts (Fig. 6.13).

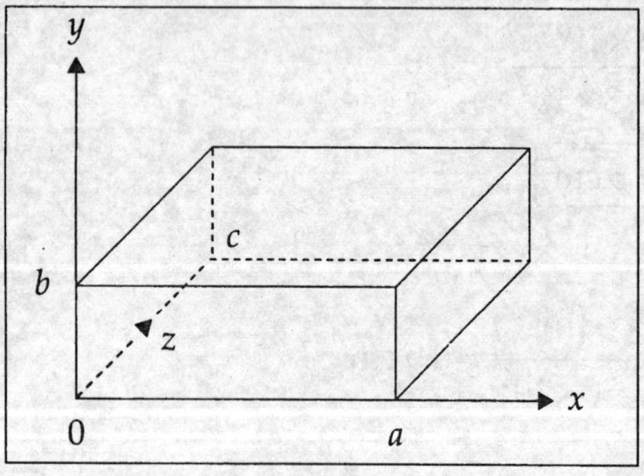

Fig. 6.13 Rectangular cavity

It is used for energy storage. As there is no propagation through the shorted ports, standing waves exist inside the cavity. These resonators are used for various applications, particularly in klystrons and wave metres.

Features of Resonators

1. A rectangular cavity (Fig. 6.13) is a rectangular waveguide whose open ends are shorted. In this type of structure, standing waves, TE and TM waves exist.
2. Resonators are mainly used for energy storage. At high frequencies RLC circuit elements are inefficient when used as resonators. This is because the dimensions of the circuits are of the order of operating wavelength. Because of this, radiation takes place which is undesirable.
3. The EM resonator cavities find extensive applications in klystron tubes, band pass filters, wave metres and microwave ovens.

TM Mode ($H_z = 0$)

If $$E_z = (x, y, z) = X(x)\, Y(y)\, Z(z),$$

We can write,
$$X(x) = C_1 \cos Ax + C_2 \sin Ax$$
$$Y(y) = C_3 \cos By + C_4 \sin By$$
$$Z(z) = C_5 \cos Cz + C_6 \sin Cz$$

where $$K_r^2 = A^2 + B^2 + C^2 = \omega^2 \mu \in = \beta_r^2$$

and $$A = \left(\frac{m\pi}{a}\right), B = \left(\frac{n\pi}{b}\right), C = \left(\frac{\ell\pi}{c}\right)$$

$$m = 1, 2, 3, \ldots, n = 1, 2, 3, \ldots, \ell = 0, 1, 2, 3, \ldots$$

Here, we have three boundary conditions to solve the constants, $C_1, C_2, \ldots$ etc.

$$E_z = 0 \text{ at } x = 0 \text{ and at } x = a$$
$$E_z = 0 \text{ at } y = 0 \text{ and at } y = b$$
$$E_x = 0,\ E_y = 0 \text{ at } z = 0 \text{ and at } z = c$$

After simplification, E_{zr} becomes

$$E_z = E_m \sin\left(\frac{m\pi x}{a}\right) \sin\left(\frac{n\pi y}{b}\right) \cos\left(\frac{\ell\pi z}{c}\right)$$

where $$E_m = C_2\, C_4\, C_5$$

The resonant frequency is given by

$$f_r = \frac{\beta_r}{2\pi\sqrt{\mu \in}} = \frac{\beta_r\, \upsilon}{2\pi}$$

or,
$$f_r = \frac{\upsilon}{2}\sqrt{\left(\frac{m}{a}\right)^2 + \left(\frac{n}{b}\right)^2 + \left(\frac{\ell}{c}\right)^2}$$

and
$$\lambda_r = \frac{\upsilon}{f_r} = \frac{2}{\sqrt{\left(\frac{m}{a}\right)^2 + \left(\frac{n}{b}\right)^2 + \left(\frac{\ell}{c}\right)^2}}$$

TE Mode ($E_z = 0$)

As in the case of TM mode, H_z is expressed

$$H_z\ (x, y, z) = X(x)\ Y(y)\ Z(z)$$

where
$$X(x) = p_1 \cos Bx + p_2 \sin Bx$$
$$Y(y) = p_3 \cos Ay + p_4 \sin Ay$$
$$Z(z) = p_5 \cos Cz + p_6 \sin Cz$$

The set of boundary conditions are

$$H_z = 0 \text{ at } z = 0 \text{ and at } z = c$$
$$\frac{\partial H_z}{\partial x} = 0 \text{ at } x = 0 \text{ and at } x = a$$
$$\frac{\partial H_z}{\partial y} = 0 \text{ at } y = 0 \text{ and at } y = b$$

Using the boundary conditions, and simplifying, we get

$$H_z = H_m \cos\left(\frac{m\pi x}{a}\right)\cos\left(\frac{n\pi y}{b}\right)\sin\left(\frac{\ell\pi z}{c}\right)$$

where
$$m = 0, 1, 2, 3, \ldots$$
$$n = 0, 1, 2, 3, \ldots$$
$$l = 1, 2, 3, \ldots$$

Dominant Mode

Dominant mode is defined as the mode which has the lowest resonant frequency for a given cavity size (a, b, c).

The waves are represented by TE_{mnl}, TM_{mnl}.

Degenerate Mode

Modes having the same resonant frequency are called degenerate modes. Ideally, the walls of the resonant cavity have infinite conductivity. But practically, cavity walls have finite conductivity. As a result, some stored energy is lost.

Quality Factor, *Q*

Quality factor is defined as

$$Q \equiv 2\pi \frac{\text{average stored energy}}{\text{loss of energy in a cycle}}$$

$$Q = 2\pi \frac{W_{av}}{W_L} = \omega \frac{W_{av}}{W_L}$$

where $\omega = 2\pi f$

W_L = average power loss in a cycle

W_{av} = average stored energy

Quality factor, for dominant mode, TE_{101} is

$$Q = \frac{(a^2 + c^2)\, abc}{\delta\, [2b\,(a^3 + c^3) + ac\,(a^2 + c^2)]}$$

where δ = depth of penetration in cavity walls

$$\delta = \frac{1}{\sqrt{\pi f_{101} \mu_0 \sigma_c}} \text{ (m)}$$

6.21 SALIENT FEATURES OF CAVITY RESONATORS

1. A completely closed metallic structure forms a cavity and it is called cavity resonator.
2. It stores energy.
3. TE and TM modes exist in the cavity.
4. In TE mode, $E_z = 0$ (z is propagation direction) and E_x, E_y, H_x, H_y and H_z are present.
5. In TM mode, $H_z = 0$ and H_x, H_y, E_x, E_y and E_z are present.
6. In cavities, the electric and magnetic fields do not propagate along z-axis but they oscillate with time at a specified location.
7. The lowest order of TM_{mnl} mode is TM_{110}.
8. The resonant frequency of the lowest order TM mode is

$$f_{110} = \frac{\upsilon}{2}\sqrt{\frac{1}{a^2} + \frac{1}{b^2}}$$

9. The lowest order for TE_{mnl} is TE_{101}.
10. The resonant frequency of the lowest order TE mode is

$$f_{101} = \frac{\upsilon}{2}\sqrt{\frac{1}{a^2}+\frac{1}{c^2}}$$

Problem 6.8 A copper rectangular cavity resonator is structured by $3 \times 1 \times 4$ cm. Find its resonant frequency for TM_{110} mode.

Solution The dimensions of resonator are:

$$a = 3\,\text{cm} = 0.03\,\text{m}$$
$$b = 1\,\text{cm} = 0.01\,\text{m}$$
$$c = 4\,\text{cm} = 0.04\,\text{m}$$

For TM_{110}, the resonant frequency is

$$f_r = \frac{\upsilon}{2}\sqrt{\frac{1}{a^2}+\frac{1}{b^2}+\frac{1}{c^2}}$$

$$f_r = \frac{3\times10^8}{2}\sqrt{\frac{1}{(0.03)^2}+\frac{1}{(0.01)^2}+\frac{1}{(0.04)^2}}$$

$$f_r = \frac{3\times10^8}{2}\sqrt{\frac{10{,}000}{9}+10{,}000+\frac{10{,}000}{16}}$$

$$f_r = \frac{3\times10^8}{2}\times100\sqrt{\frac{169}{144}}$$

$$f_r = \frac{3\times10^{10}}{2}\times\frac{13}{12}$$

$$f_r = 1.625\times10^{10}\ \text{Hz}$$

$$\boxed{f_r = 16.25\,\text{GHz}}$$

Problem 6.9 A copper walled rectangular cavity resonator is structured by $3 \times 1 \times 4$ cm and operates at the dominant modes of TE and TM. Find its resonant frequency and quality factor. The conductivity of copper is 5.8×10^7 mho/m. There is air inside the cavity.

Solution For TM mode, the dominant mode is TM_{100}. Its resonant frequency is

$$f_r = \frac{\upsilon}{2}\sqrt{\frac{1}{a^2}+\frac{1}{b^2}}$$

Here $$\upsilon = \frac{1}{\sqrt{\mu_0 \in_0}} = 3\times 10^8 \text{ m/s}$$

$$a = 3\,\text{cm} = 0.03\,\text{m}$$

$$b = 1\,\text{cm} = 0.01\,\text{m}$$

$$f_r = \frac{3\times 10^8}{2}\sqrt{\left(\frac{1}{0.03}\right)^2 + \left(\frac{1}{0.01}\right)^2}$$

$$= \frac{3\times 10^8}{2}\sqrt{1111.1 + 10{,}000}$$

$$= 1.5\times 10^8 \times 105.4$$

$$\boxed{f_r = 15.8\,\text{GHz}}$$

For TE mode, the resonant frequency of the dominant, TE_{101} is

$$f_r = \frac{\upsilon}{2}\sqrt{\left(\frac{1}{a}\right)^2 + \left(\frac{1}{l}\right)^2}$$

$$= 1.5\times 10^8 \sqrt{1111.1 + \left(\frac{1}{0.04}\right)^2}$$

$$= 1.5\times 10^8 \sqrt{1111.1 + 625.0}$$

$$\boxed{f_r = 6.249\,\text{GHz}}$$

The quality factor, Q for TE_{101}

$$= \frac{(a^2 + c^2)\, abc}{\delta\,[2b\,(a^3 + c^3) + ac\,(a^2 + c^2)]}$$

where $$\delta = \frac{1}{\sqrt{\pi f_{101}\, \mu_0\, \sigma_c}}$$

$$= \frac{1}{\sqrt{\pi \times 6.249\times 10^9 \times 4\pi\times 10^{-7} \times 5.8\times 10^7}}$$

$$= 8.3598\times 10^{-7} \text{ m}$$

$$Q = \frac{(3^2+4^2)\times 3\times 1\times 4\times 10^{-2}}{8.3598\times 10^{-7}[2\times 1\,(3^3+4^3)+3\times 4\,(3^2+4^2)]}$$

$$= \frac{25\times 12\times 10^{-2}}{8.3598\times 10^{-7}[2\times 91]+12\times 25}$$

$$= \frac{300\times 10^{-2}}{8.3598\times 10^{-7}[182+300]}$$

$$= \frac{300\times 10^{-2}}{482}\times\frac{1}{8.3598}\times 10^{7}$$

$$\boxed{Q = 7{,}445}$$

Problem 6.10 A copper walled resonant cavity is dielectric ($\in_r = 4$) filled and its dimensions are 5 × 4 × 10 cm. Determine the resonant frequency of TE_{101} and its quality factor.

Solution The conductivity of copper,

$$\sigma_c = 5.8\times 10^{7} \text{ mho/m}$$

$$a = 5\,\text{cm}$$

$$b = 4\,\text{cm}$$

$$c = 10\,\text{cm}$$

The resonant frequency of TE_{101} is

$$f_r = \frac{\upsilon}{2}\sqrt{\left(\frac{1}{a}\right)^2+\left(\frac{1}{c}\right)^2}$$

$$= \frac{\upsilon_0}{2\sqrt{\in_r}}\sqrt{\left(\frac{1}{a}\right)^2+\left(\frac{1}{c}\right)^2}$$

$$= 0.75\times 10^{8}\sqrt{\left(\frac{1}{a}\right)^2+\left(\frac{1}{c}\right)^2}$$

$$= 0.75\times 10^{8}\sqrt{\left(\frac{1}{0.05}\right)^2+\left(\frac{1}{0.1}\right)^2}$$

$$= 0.75 \times 10^8 \sqrt{400 + 100}$$

$$\boxed{f_r = 1.677\,\text{GHz}}$$

The quality factor, Q for this mode is

$$Q = \frac{(a^2 + c^2)\,abc}{\delta\,[2b\,(a^3 + c^3) + ac\,(a^2 + c^2)]}$$

$$\delta = \frac{1}{\sqrt{\pi f_{101}\,\mu_0\,\sigma_c}}$$

$$= \frac{1}{\sqrt{\pi \times 1.677 \times 10^9 \times 4\pi \times 10^{-7} \times 5.8 \times 10^7}}$$

$$= 0.032275 \times 10^{-4}$$

$$= 32.275 \times 10^{-7}\ \text{(m)}$$

and

$$\frac{(a^2 + c^2)\,abc}{2b\,(a^3 + c^3) + ac\,(a^2 + c^2)}$$

$$= \frac{(25 + 100)\,5 \times 4 \times 10}{2 \times 4\,(125 + 1000) + 5 \times 10\,(25 + 100)}$$

$$= \frac{125 \times 200}{9{,}000 + 6{,}250} = \frac{25{,}000}{15{,}250}$$

$$= 1.6393\,\text{cm}$$

$$= 1.6393 \times 10^{-2}\ \text{m}$$

$$Q = \frac{1.6393 \times 10^{-2}}{32.275 \times 10^{-7}}$$

$$= 0.05079 \times 10^5$$

$$\boxed{Q = 5{,}079}$$

6.22 CIRCULAR WAVEGUIDES

The waveguides of circular cross-section (Fig. 6.14) are used to transmit EM waves from one point to another. Unlike rectangular waveguides, the circular waveguides do not have unique orientation as it is perfectly symmetrical around the axis.

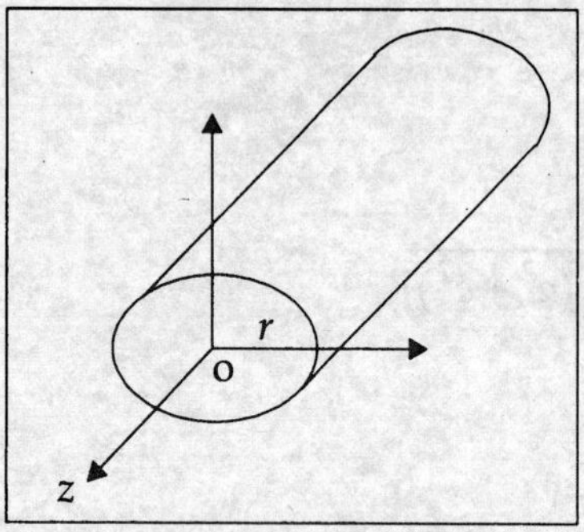

Fig. 6.14 Circular waveguide

6.23 SALIENT FEATURES OF CIRCULAR WAVEGUIDES

1. It is easy to manufacture.
2. They are used in rotational coupling.
3. Rotation of polarisation exists and this can be overcome by rotating modes symmetrically.
4. TM_{01} mode is preferred to TE_{01} as it requires a smaller diameter for the same cut-off wavelength.
5. TE_{01} does not have practical application.
6. For $f > 10$ GHz, TE_{01} has the lowest attenuation per unit length of the waveguide.
7. The main disadvantage is that its cross-section is larger than that of a rectangular waveguide for carrying the same signal.
8. The space occupied by circular waveguides is more than that of a rectangular waveguide.
9. The determination of fields here consists of differential equations of certain type. Their solutions involve Bessel functions.
10. Here also TE and TM modes exist.

    For TM wave, the solution of axial component

    $$E_{z,nm} = J_n(K_c r)(A \cos n\phi + B \sin \phi)$$

    and for TE wave, it is

    $$H_{z,nm} = J_n(K_c r)(A' \cos n\phi + B' \sin n\phi)$$

    where $J_n(K_c r)$ = Bessel function of the first kind

$$r = \text{the radius of the guide}$$

$$K_c = \text{the cut-off wave number}$$

$$A, B, A', B' = \text{constants}$$

The solutions for the Bessel function are obtained for certain values of K_c where these values of K_c are known as eigen values. If K_c is to produce solution of the Bessel function, $(K_c r)$ must be the roots of the Bessel function. Then

$$J_n(K_c r) = 0$$

The propagation parameters for nm^{th} mode TM waves are:

Phase constant,

$$\beta_{nm} = (K^2 - K_{c,nm}^2)^{1/2}$$

where $$K_{c,nm} = \frac{p_{nm}}{r}$$

and $$\beta_{nm} = \left[K^2 - \left(\frac{p_{nm}}{r}\right)^2\right]^{1/2}$$

where p_{nm} = the roots of the Bessel function

$$K = \text{free space wave number} = \frac{2\pi}{\lambda}$$

The cut-off wavelength for TM wave,

$$\lambda_c = \frac{2\pi}{K_{c,nm}} = \frac{2\pi r}{p_{nm}}$$

The roots of the Bessel function for TM mode are shown in Table 6.1.

Table 6.1 Roots of Bessel Function (TM Mode)

Order n	*First order* p_{n1}	*Second order* p_{n2}	*Third order* p_{n3}
0	2.405	5.520	8.654
1	3.832	7.016	10.174
2	5.135	8.417	11.620

The roots of the Bessel function for TE mode are shown in Table 6.2.

Table 6.2 Roots of Bessel Function (TE Mode)

Order n	*First order* p'_{n1}	*Second order* p'_{n2}	*Third order* p'_{n3}
0	3.832	7.016	10.174
1	1.841	5.331	8.536
2	3.054	6.706	9.970

For circular waveguides, TE_{11} is the dominant mode. The propagation parameters for TE_{nm} mode are:

$$\beta_{nm} = [K^2 - K'^2_{c,nm}]^{1/2}$$

where

$$K'^2_{c,nm} = \frac{p'_{nm}}{r}$$

$$\beta_{nm} = \left[K^2 - \left(\frac{p'_{nm}}{r}\right)^2\right]$$

$$\lambda_{c,nm} = \frac{2\pi r}{p'_{nm}}$$

Guide wavelength,

$$\lambda_g = \frac{\lambda}{\left[1 - \left(\frac{\lambda}{\lambda_{c,nm}}\right)^2\right]^{1/2}}$$

Problem 6.11 If the radius of a circular waveguide $r = 1.27$ cm, $f = 10$ GHz, find the cut-off wavelength for the dominant mode and phase constant. Assume that the waveguide is air-filled. Take $p'_{11} = 1.841$.

Solution For the dominant mode,

$$m = 1,\ n = 1$$

For TE_{11}, we have

$$\lambda_{c,11} = \frac{2\pi r}{\rho'_{11}},\ p'_{11} = 1.841$$

$$\boxed{\lambda_{c,11} = \frac{2\pi \times 1.27}{1.841} = 4.32\,\text{cm}}$$

The phase constant,

$$\beta_{11} = \left[K^2 - \frac{(p'_{11})^2}{r^2}\right]^{1/2}$$

Here $$K = \frac{2\pi}{\lambda} = \frac{2\pi f}{\upsilon_0} = \frac{\omega}{\upsilon_0}$$

$$\beta_{11} = \frac{2\pi \times 10^{10}}{3\times 10^8} = 209$$

$$\beta_{11} = \left[209^2 - \left(\frac{1.841}{0.0127}\right)^2\right]^{1/2}$$

$$\boxed{\beta_{11} = 150\,\text{rad/m}}$$

Problem 6.12 Determine the size of the circular waveguide required to propagate TE_{11} mode if $\lambda_c = 8\,\text{cm}\,(\rho'_{11} = 1.841)$.

Solution We have $$\lambda_{c,11} = \frac{2\pi r}{p'_{11}}$$

or, $$r = \frac{\lambda_{c,11} \times p'_{11}}{2\pi}$$

$$= \frac{8 \times 1.841}{2\pi} = 2.350\,\text{cm}$$

Radius of the guide $\boxed{= 2.35\,\text{cm}}$

POINTS/FORMULAE TO REMEMBER

- EM wave propagates in a waveguide with multiple reflections.
- TE and TM waves exist in a waveguide.
- TEM wave does not exist in a hollow waveguide.
- TEM wave exists between parallel plates.
- The dominant mode has the lowest cut-off frequency.
- The propagation constant between parallel plates is $\gamma_p = \sqrt{\left(\frac{m\pi}{a}\right)^2 - \omega^2 \mu_0 \epsilon_0}$.
- The cut-off frequency is a frequency below which the wave is attenuated completely.
- Between the parallel plates, $f_c = \frac{m}{2a\sqrt{\mu_0 \epsilon_0}}$.
- $\lambda_p = \frac{\lambda}{\sqrt{\left[1 - \left(\frac{\lambda}{\lambda_0}\right)^2\right]}}$
- λ_c for $TE_1 = 2a$
- Group velocity, phase velocity and free space velocity are related by $\upsilon_0^2 = \upsilon_p \upsilon_g$.
- Attenuation factor between parallel plates is $\alpha_p = \frac{1}{\eta a}\sqrt{\frac{\omega\mu_c}{2\sigma_c}}$ Nepers/m.
- Attenuation of TE waves between parallel plates is $\alpha_p = \frac{2m^2\pi^2\sqrt{\omega\mu_c/2\sigma_c}}{\omega\mu a^2\sqrt{\omega^2\mu\epsilon - \left(\frac{m\pi}{a}\right)^2}}$.
- A rectangular waveguide is used as a radiator, a high pass filter, a transmission line and feed element to an antenna.
- $\lambda_g = \frac{\lambda}{\sqrt{1 - \left(\frac{\lambda}{\lambda_c}\right)^2}}$

- TEM $= TM_{00}$
- For TEM, $E_z = 0, H_z = 0$
- $\lambda_g = \lambda, \beta_g = \beta, \eta = \eta_0, \alpha = 0$ for TEM
- Typically, wave impedance for TE_{mn} wave $\eta_{mn}^{TE} = \dfrac{E_x}{H_y}$.
- Circular waveguides can be used to produce circular polarisation.
- TEM wave has zero cut-off frequency.
- The field components of TE wave between parallel plates are E_y, H_x, H_z.
- The field components of TM wave between parallel plates are E_x, E_z, H_y.
- TEM wave between parallel plates has only E_x and H_y components.
- The field components of TE in a hollow rectangular waveguide are E_x, E_y, H_x, H_y and H_z.
- The field components of TM in a hollow rectangular waveguide are E_x, E_y, E_z, H_x and H_y.
- The resonant frequency TE_{101} in a cavity resonator is $f_{101} = \dfrac{\upsilon}{2}\sqrt{\dfrac{1}{a^2} + \dfrac{1}{c^2}}$.
- The resonant frequency of TM_{110} in a cavity resonator is $f_{110} = \dfrac{\upsilon}{2}\sqrt{\dfrac{1}{a^2} + \dfrac{1}{b^2}}$.

OBJECTIVE QUESTIONS

1. TEM mode exists in parallel plate structure. (Yes/No)
2. TEM mode exists in a hollow rectangular waveguide. (Yes/No)
3. The quality factor is nothing but figure of merit of a cavity resonator. (Yes/No)
4. If quality factor is high, the power loss in the walls of a waveguide is less. (Yes/No)
5. The phase constant, β_{mn} of TE wave is the same as that of TM wave in a rectangular waveguide. (Yes/No)
6. The wavelength, λ_{mn} of TM wave is the same as that of TE wave in a rectangular waveguide. (Yes/No)
7. Phase velocity, υ_p in TE wave is the same as that of TM wave in a rectangular waveguide. (Yes/No)
8. Group velocity, υ_g in TE wave is the same as that of TM wave in a rectangular waveguide. (Yes/No)
9. A rectangular waveguide can be excited for TE_{10} mode by a loop. (Yes/No)
10. The presence of modes depends on the shape of the waveguide. (Yes/No)
11. The presence of modes depends on the dimensions of the waveguide. (Yes/No)
12. The presence of modes depends on the medium inside the waveguide. (Yes/No)
13. The presence of modes depends on the operating frequency. (Yes/No)
14. A rectangular waveguide can be used as an antenna. (Yes/No)
15. A rectangular waveguide can be used as a transmission line. (Yes/No)
16. A calibrated cavity resonator is nothing but a frequency meter at microwave frequency. (Yes/No)
17. In cavity resonator, standing waves are absent. (Yes/No)
18. A cavity resonator at microwave frequency is similar to a low frequency resonant circuit. (Yes/No)
19. Each mode of TE wave in a rectangular waveguide has a different cut-off frequency. (Yes/No)

20. In a hollow rectangular waveguide, TE mode has all the components of **E** and **H**. (Yes/No)

21. In a hollow rectangular waveguide, TM mode has all the components of **E** and **H**. (Yes/No)

22. The phase velocity in a rectangular waveguide is greater than that in free space. (Yes/No)

23. The group velocity for each TE_{mn} mode in rectangular waveguide is different. (Yes/No)

24. The phase velocity for each TE_{mn} mode is the same as in a rectangular waveguide. (Yes/No)

25. Guide wavelength varies with each TE and TM mode. (Yes/No)

26. Guide wavelength depends on waveguide dimensions. (Yes/No)

27. Evanescent mode is a mode which does not propagate. (Yes/No)

28. For evanescent mode, the propagation constant is equal to attenuation constant. (Yes/No)

29. Wave impedance in a waveguide is smaller than that of free space. (Yes/No)

30. If the impedance is resistive in a waveguide, the average power flow exists. (Yes/No)

31. $\eta_{mn}^{TE}\,\eta_{mn}^{TM} = \eta^2$ (Yes/No)

32. The cut-off frequencies of TE_{10} and TE_{01} are the same. (Yes/No)

33. If the operating frequency is 10 GHz and cut-off frequency for a TE_{10} in a rectangular waveguide is 12 GHz, wave propagation takes place. (Yes/No)

34. In cavity resonators, there exists no reflected waves. (Yes/No)

35. In rectangular waveguides, if the propagation constant is purely real, propagation takes place. (Yes/No)

36. Phase velocity is nothing but the velocity of propagating energy in a waveguide. (Yes/No)

37. Phase velocity is the speed of the constant phase points on the travelling wave. (Yes/No)

38. Propagation constant is the same for TE and TM modes in a waveguide. (Yes/No)

39. If the conductivity of the wall of a waveguide is high, the skin depth is small. (Yes/No)

40. The resonant frequency depends on the length of the cavity resonator. (Yes/No)

41. TEM wave means ________.

42. TE wave means ________.

43. TM wave means ________.

44. In TE mode, the electric field in the direction of propagation, z is equal to ________.

45. When an EM wave is propagating in z-direction, the magnetic field in z-direction for TM wave is ________.

46. The cut-off frequency of TEM mode is ________.

47. The cut-off wavelength of TEM mode is ________.

48. Cavity resonator is obtained from a rectangular waveguide by ________.

49. The cut-off frequency for the dominant mode in a rectangular waveguide is ________.

50. The quality factor of a resonating system is ________.

51. The propagation constant, γ_{mn} for TE_{mn} at operating frequency less than cut-off frequency is ________.

52. At an operating frequency greater than the cut-off frequency of TE_{mn} wave, the propagation constant is ________.

53. The rectangular waveguide can be excited for TE_{10} mode by ________.

54. The power flow in z-direction in a waveguide requires the presence of the components of ________.

55. The number of modes in rectangular waveguides are ________.

56. The modes in a waveguide are ________.

57. A waveguide is usually made of ________.

58. The field components in a waveguide are obtained using ________.

59. In TEM mode in a hollow rectangular waveguide, ________ components of **E** and **H** are present.

60. The phase velocity in a rectangular waveguide is ________ when $f = f_c$.

61. The relation between phase, group velocities and free space velocity is ________.

62. If the wave impedance is capacitive in a waveguide, the average power flow is ________.

63. For the best cavity resonator, the quality factor is ________.

Answers

1. Yes	2. No	3. Yes	4. Yes	5. Yes
6. Yes	7. Yes	8. Yes	9. Yes	10. Yes
11. Yes	12. Yes	13. Yes	14. Yes	15. Yes
16. Yes	17. No	18. Yes	19. Yes	20. No
21. No	22. Yes	23. Yes	24. No	25. Yes
26. Yes	27. Yes	28. Yes	29. Yes	30. Yes
31. Yes	32. No	33. No	34. No	35. No
36. No	37. Yes	38. Yes	39. Yes	40. Yes

41. Transverse electromagnetic wave 42. Transverse electric wave
43. Transverse magnetic wave 44. Zero 45. Zero 46. Zero
47. Infinity 48. Closing both ends of the waveguide with conducting plates
49. Lowest

50. $Q = \omega_r \dfrac{W_{av}}{p_d}$, ω_r is angular resonant frequency, W_{av} is the time-averaged stored energy (*J*) and P_d is the power dissipation

51. α_{mn} 52. Imaginary 53. A probe 54. E_x and H_y 55. Infinity
56. Nothing but discrete characteristic patterns
57. Copper or brass 58. Maxwell's equations 59. No
60. Infinity 61. $\upsilon_p \, \upsilon_g = \upsilon_0^2$ 62. Zero 63. Infinity

MULTIPLE CHOICE QUESTIONS

1. When a wave is propagating in *x*-direction, TE wave has E_x equal to

   (a) zero (b) E_y (c) E_z (d) $E_y + E_z$

2. For a wave propagating in *z*-direction in a hollow rectangular waveguide, TEM wave has

   (a) $E_z = 0$ (b) $H_z = 0$
   (c) all components of *E* and *H* are zero (d) $E_x = 0, H_z = 0$

3. The cut-off frequency of TEM wave is

   (a) zero (b) f (c) infinity (d) that of TE_{10}

4. If $a = 2\,\text{cm}, b = 1\,\text{cm}$ for a waveguide, the cut-off frequency for TE_{10} mode is

   (a) 0.75 GHz (b) 7.5 GHz (c) 1.5 GHz (d) 1.5 MHz

5. If the cut-off frequency for the dominant mode in a rectangular waveguide is 4 GHz, the dimension of the broad wall along *x*-axis is

   (a) 4 cm (b) 3.75 cm (c) 5 cm (d) 37.5 cm

6. If $f = 10$ GHz, the broad dimension of a rectangular waveguide, $a = 2$ cm, the cut-off frequency for TE_{10} is

   (a) 0.75 GHz (b) 7.5 GHz (c) 10 GHz (d) 5 GHz

7. At $f = 10$ GHz, $a = 2$ cm, $b = 1$ cm, the cut-off frequency for TE_{01} is

   (a) 10 GHz (b) 1.5 GHz (c) 15 GHz (d) 7.5 GHz

8. If in a rectangular waveguide the attenuation constant is 2 dB/m and the phase constant is 2 rad/m, the propagation constant is

   (a) $2 + j2\,\text{dB/m}$ (b) $2 + j2\,\text{np/m}$
   (c) 2 dB/m (d) 2 np/m

9. If $\beta = 300\,\text{rad/m}$ for TE_{21} mode in a rectangular waveguide, the guide wavelength for TE_{21} is

   (a) 0.209 cm (b) 0.209 m (c) 2.09 m (d) 20.9 cm

10. The cut-off frequency for TE_{10} is 5 GHz at an operating frequency of 6 GHz. The phase constant for TE_{10} is

(a) 69.5 rad/m (b) 6.95 rad/m
(c) 0.833 rad/m (d) 1.2 rad/m

11. If the resonant frequency is 6 GHz, the time averaged stored energy is 10 Joules and the power dissipation is 20 watts in a resonator, the quality factor is
(a) 188.49 (b) 188.49×10^9
(c) 200 (d) 188

12. The ideal value of quality factor is
(a) ∞ (b) 0 (c) 100 (d) very small

13. If $\eta_{mn}^{TM} = 100$, $\eta_{mn}^{TE} = 150$, the intrinsic impedance is
(a) 122.47Ω (b) 1.5Ω (c) 0.333Ω (d) 12.247Ω

14. If the phase constant of TM_{21} is 320 rad/m, $\alpha = 0$ at an operating frequency of 10 GHz, the characteristic impedance of TM_{21} is
(a) 32Ω (b) 539Ω (c) 300Ω (d) 53.9Ω

15. The cut-off wavelength for the dominant mode in a rectangular waveguide having dimensions 4×3 cm is
(a) 12 cm (b) 8 cm (c) 1.33 cm (d) 0.75 cm

16. The cut-off wavelength for TE_{32} in a waveguide is 10 GHz. The guide wavelength for TE_{32} mode is
(a) 3 cm (b) 3.2 cm (c) 6 cm (d) 3 m

17. Skin depth in the copper wall of a waveguide operating at $f = 6.25$ GHz is
(a) 8.39 mm (b) 83.9 μm (c) 83.9 mm (d) 839 mm

18. The mode of propagation in an air-filled waveguide of 2×1 cm dimensions operating at 10 GHz is
(a) TE_{01} (b) TE_{21} (c) TE_{10} (d) TE_{11}

19. If the propagation constant in a waveguide is $105 + j10$, np/m the attenuation constant is
(a) 105 np/m (b) 105 dB/m (c) 10.5 np/m (d) 115 np/m

20. If $\gamma_{mn} = \alpha_{mn}$, in a waveguide
(a) propagation takes place (b) no propagation takes place
(c) energy is increased (d) energy is decreased

Answers

1. (a)	2. (c)	3. (a)	4. (b)	5. (b)
6. (b)	7. (c)	8. (a)	9. (b)	10. (a)
11. (b)	12. (a)	13. (a)	14. (b)	15. (b)
16. (a)	17. (b)	18. (c)	19. (a)	20. (b)

EXERCISE PROBLEMS

1. A rectangular waveguide operates at 1 GHz and has dimensions of 5 × 2 cm. Find the distance from the source end at which the electric field of TE_{21} wave becomes 0.5% of its starting amplitude at $z = 0$. The amplitude of the electric field in z-direction at $z = 0$ is 10 kV/m.

2. A rectangular waveguide with dimensions of 2.5 × 1.0 cm operates below 15.0 GHz. What are the TE modes that can be propagated if the waveguide is filled with material whose relative permittivity, $\epsilon_r = 4$ and $\mu_r = 1$, $\sigma = 0$? Find their cut-off frequencies and cut-off wavelengths.

3. A hollow rectangular waveguide operates at 10 GHz. If the cut-off frequency of TM_{21} mode is 6 GHz, find its phase velocity, wavelength, phase constant.

4. A hollow rectangular waveguide has the dimensions of $a = 2.286$ cm and $b = 1.016$ cm. Find the cut-off frequencies of TM_{11} and TM_{20}.

5. Find the cut-off wavelength of TE_{10} and TE_{20} mode in the hollow rectangular waveguide of dimensions $a = 3\,\text{cm}, b = 1.5\,\text{cm}$.

6. In a hollow square waveguide, the cut-off frequency for TE_{10} mode is (a) 10 GHz (b) 20 GHz (c) 30 GHz. Find the dimensions of the waveguide.

7. Find the radius of the circular waveguide required to propagate TE_{11} mode if $\lambda_c = 10\,\text{cm}$. ($\rho'_{11} = 1.84$).

8. Find the cut-off frequency of TE_{21} mode in a circular waveguide of radius 4 cm. (Take $\rho'_{21} = 3.054$).

CHAPTER

7

TRANSMISSION LINES

Transmission lines are nothing but guided conducting structures which are used in power distribution at low frequencies, in communications and computer networks at higher frequencies.

The main aim of this chapter is to provide the overall concepts of Transmission Line theory. They include:

- transmission lines and applications
- equivalent circuits
- primary and secondary constants
- lossless and distortionless lines
- loading of lines and RF lines
- reflection coefficient and VSWR
- stubs and smith charts
- solved problems, points/formulae to remember, objective and multiple choice questions and exercise problems.

Do you know?

Human beings and animals consist of several ideal transmission lines which are noiseless, attenuationless and distortionless. These transmission lines are powered by the heart and controlled by the brain through parallel processing.

7.1 TRANSMISSION LINES

A transmission line is a means of transfer of information from one point to another. Usually it consists of two conductors. It is used to connect a source to a load. The source may be a transmitter and the load may be a receiver.

7.2 TYPES OF TRANSMISSION LINES

The various types of transmission lines are

(a) Two-wire parallel lines
(b) Coaxial lines
(c) Twisted wires
(d) Parallel plates or planar lines
(e) Wire above conducting line
(f) Microstrip lines
(g) Optical fibres

Typical configurations of the above are shown in Fig. 7.1.

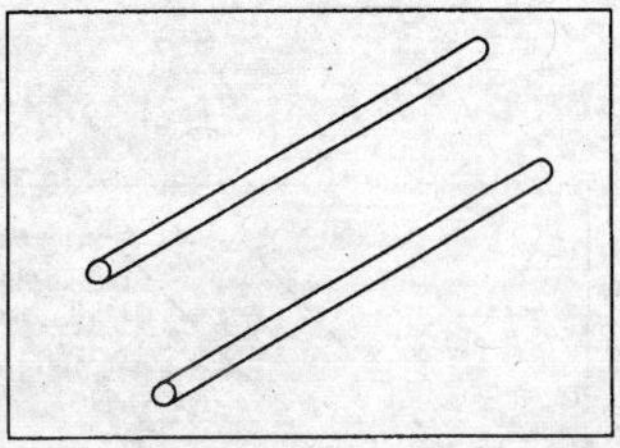

(*a*) Two-wire parallel line

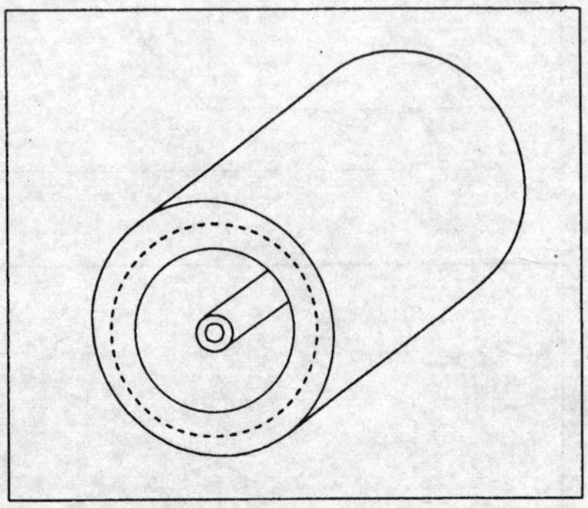

(*b*) Coaxial line

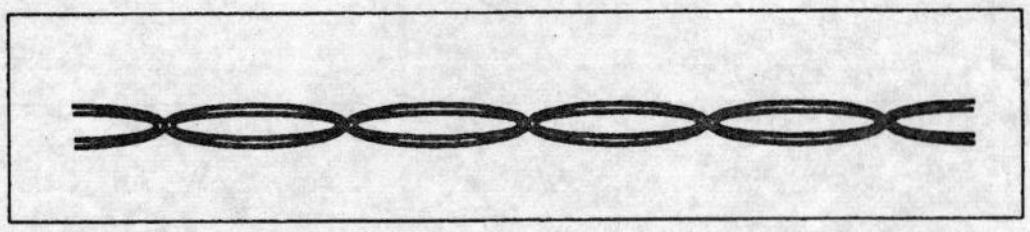

(*c*) Twisted pair of lines

Fig. 7.1 Types of transmission lines

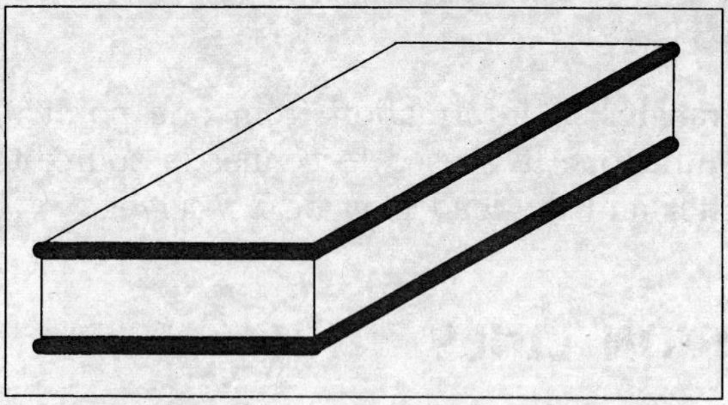

(*d*) Planar line

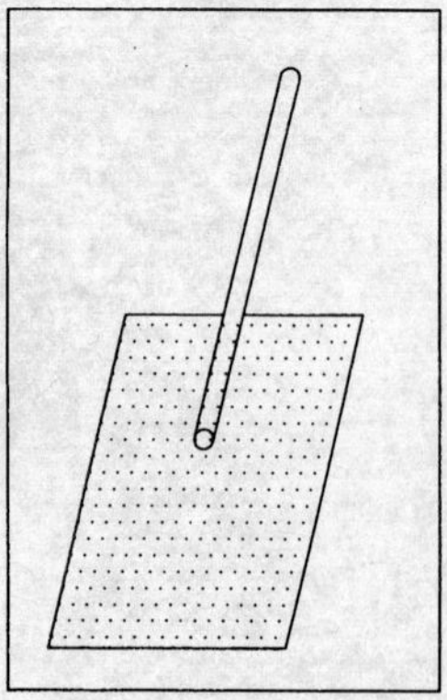

(*e*) Wire above conducting plane

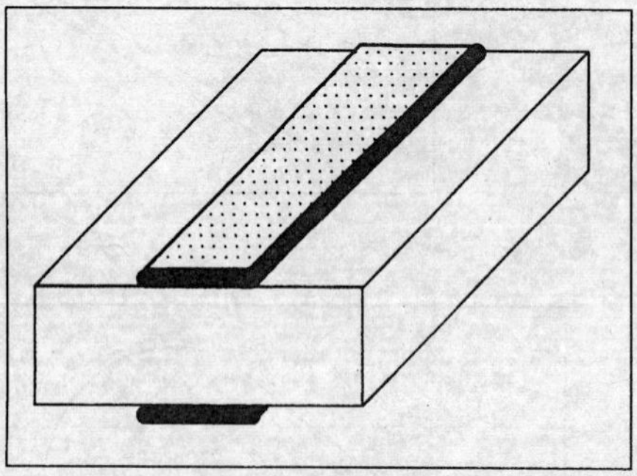

(*f*) Microstrip line

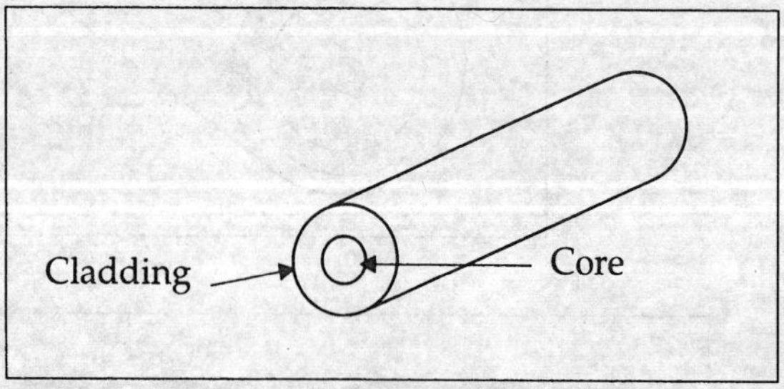

(*g*) A typical step-index optical fibre

Fig. 7.1 Types of transmission lines

7.3 APPLICATIONS OF TRANSMISSION LINES

Applications of transmission lines are:

1. They are used to transfer energy from one circuit to another.
2. They can be used as circuit elements like inductors, capacitors and so on.
3. They can be used as impedance matching devices.
4. They can be used as stubs.
5. They can be used as measuring devices.
6. Coaxial cables are frequently used in laboratories and to connect televisions to TV antennas.
7. Microstrips are used in integrated circuits in which metallic strips connecting electronic elements are deposited on dielectric substrates.
8. Twisted pairs and coaxial cables are used in computer networks such as Ethernet and Internet.
9. Pair of parallel lines are used in telephony and power transmission.
10. Planar lines are used to connect transmitters and antennas.
11. Optical fibres are used to transmit information over long and short distances with negligible attenuation.

7.4 EQUIVALENT CIRCUIT OF A PAIR OF TRANSMISSION LINES

The equivalent circuit of a transmission line is a distributed network. This consists of cascaded sections and each section consists of a series Resistance R, series Induction L, shunt Capacitance C, and shunt conductance G. One section of the equivalent circuit is shown in Fig. 7.2. Here R is expressed in ohm/unit length, L in Henry/unit length, C in Farad/unit length and G in Mho per unit length.

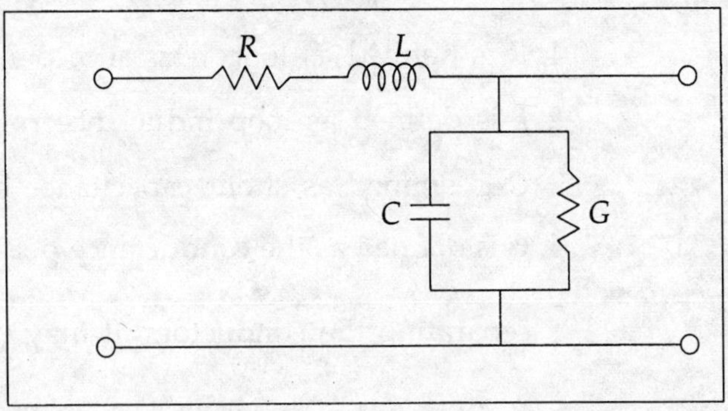

Fig. 7.2 Equivalent circuit of a two-conductor transmission line

Electric and Magnetic Fields in Parallel Plate and Coaxial Lines

In parallel plate transmission, if z is the direction of propagation, the electric and magnetic field distributions are shown in Fig. 7.3.

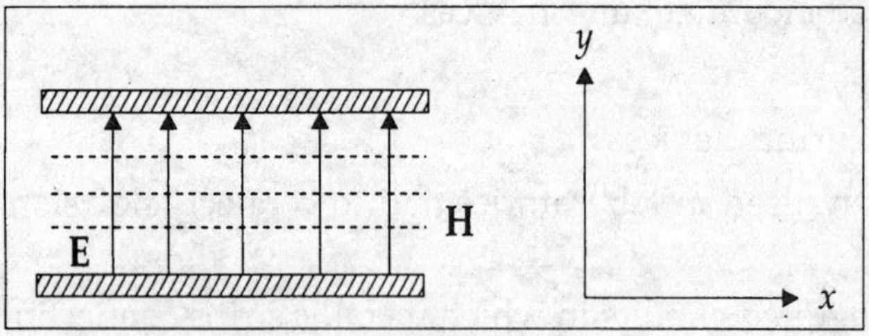

Fig. 7.3 **E** and **H** in a parallel plate transmission line

The **E** and **H** fields in a coaxial line are shown in Fig. 7.4.

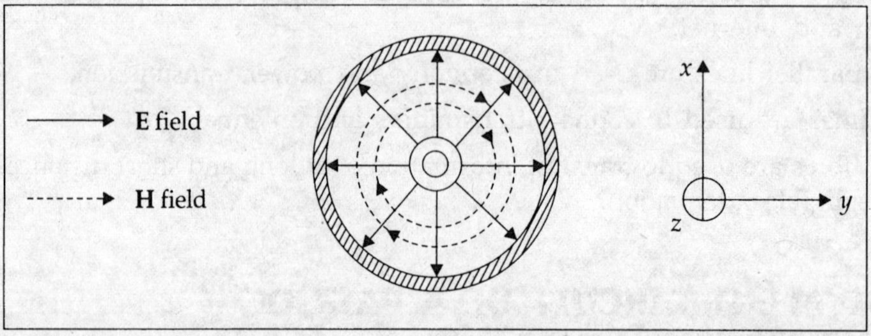

Fig. 7.4 **E** and **H** fields in a coaxial line

7.5 PRIMARY CONSTANTS

The R (Ω /Km), L (H/Km), C (F/Km) and G (mho/Km) are known as primary constants.

Salient aspects of primary constants:

1. R is defined as loop resistance per unit length of line.
2. L is defined as loop inductance per unit line length.
3. C is defined as shunt capacitance between two wires per unit length.
4. G is defined as the conductance per unit length due to the dielectric medium separating the conductors. It may be noted that $\left(G \neq \frac{1}{R}\right)$.
5. R, L, C, G are distributed along the length of the line.

6. For each line, the conductors are characterised by σ_c and $\mu_c = \mu_0, \epsilon_c = \epsilon_0$, and the dielectric medium, which is basically homogeneous, separating the conductor is characterised by $\sigma_d, \mu_d, \epsilon_d$.
7. R, L, C and G depend on the geometry of transmission line, characteristics of the dielectric material and in some cases on the frequency.

   The relations are presented for quick reference.

For parallel wires

$$R_{dc} = \frac{2}{\sigma_c \pi a^2}, \Omega/\text{m} \ (f < 10\,\text{kHz})$$

$$R_{ac} = \frac{1}{\pi \sigma_c \delta a}, \Omega/\text{m} \ (f > 10\,\text{kHz})$$

$$L = \left[\mu_r + 9.21 \log_{10} \frac{d}{a}\right] \times 10^{-7}\ \text{H/m}$$

$$C = \frac{\pi \epsilon_d}{\ln\left(\frac{d}{a}\right)}\ \text{F/m}$$

$$G = \frac{C}{\epsilon_d} \sigma_d\ \text{mho/m}$$

where σ_c = conductivity of conductors

a = radius of wire

$\delta = \dfrac{2}{\sqrt{\pi f \mu_c \sigma_c}}$ = skin depth

d = spacing between wires

ϵ_d = dielectric constant of the dielectric material

μ_r = relative permeability of the conductor material
= 1 for non-magnetic material

Internal inductance (L_i) is due to internal flux linkages in the conductors. It is

$$L_i = \frac{R_{ac}}{2\pi f}\ (\text{H/m}) \text{ for } f > 10\text{kHz}$$

$$= \frac{\mu_0}{4\pi} \text{ (H/m) for } f < 10 \text{ kHz}$$

External inductance (L_e) is due to flux linkages with the flux external to the wire.

$$L_e = \frac{\mu_d}{\pi} \times \frac{1}{ln\left(\frac{d}{a}\right)} \text{ H/m} \qquad (\text{for } d >> a)$$

For coaxial cable

$$R_{dc} = \frac{1}{\sigma_c \pi} \times \left[\frac{1}{a^2} + \frac{1}{t\,(b+t)}\right]$$

$$R_{ac} = \frac{1}{2\pi\sigma_c \delta}\left[\frac{1}{a} + \frac{1}{b}\right] \text{ for } t >> \delta$$

$$L_i = \frac{R_{ac}}{2\pi f} \text{ (H/m) for } f > 10 \text{ kHz}$$

$$= \frac{\mu_0}{4\pi} \text{ (H/m) for } f < 10 \text{ kHz}$$

$$L = L_i + L_e$$

$$C = \pi \epsilon_d \left[\frac{2}{\ln\left(\frac{b}{a}\right)}\right], \text{ (F/m)}$$

$$G = \frac{C}{\epsilon_d} \sigma_d \text{ (mho/m)}$$

where a = inner radius
b = outer radius
t = outer thickness

For parallel plates

$$R_{dc} = \frac{2}{\sigma_c wt}$$

$$R_{ac} = \frac{2}{\pi\sigma_c \delta w} \quad \text{for } t >> \delta$$

$$L_i = \frac{R_{ac}}{2\pi f}, \text{ (H/m) for } f > 10\,\text{kHz}$$

$$= \frac{\mu_0}{4\pi}, \text{ H/m for } f < 10\,\text{kHz}$$

$$L_e = \frac{\mu_d d}{w} \text{ (H/m)}$$

$$C = \frac{\pi^2 \epsilon_d d}{w} \text{ (F/m)}$$

$$G = \frac{C}{\epsilon_d} \sigma_d \text{ (mho/m)}$$

where w = width

d = separation

7.6 TRANSMISSION LINE EQUATIONS

Consider Fig. 7.5 in which a line of length l, voltage and current at source end, V_s and I_s are shown. The voltage, V_L and current I_L at the load end are also shown.

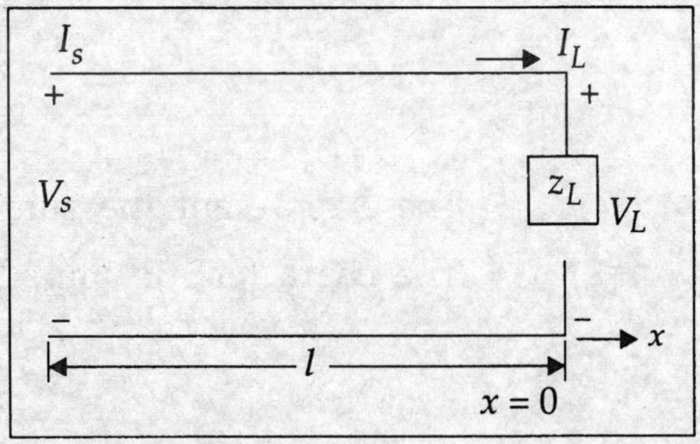

Fig. 7.5 A transmission line with load, Z_L

Let V and I be the voltage and current on the line at any arbitrary location. Assume V_f and I_f are for the forward wave and V_r and I_r are for the reflected wave. Then, we can write

$$\frac{dV}{dx} = -(R + j\omega L)\,I$$

$$\frac{dI}{dx} = -(G + j\omega C)\,V$$

Differentiating and combining, we get

$$\frac{d^2 V}{dx^2} = \gamma^2 V \tag{7.1}$$

$$\frac{d^2 I}{dx^2} = \gamma^2 I \tag{7.2}$$

where $\gamma^2 = (R + j\omega L)\,(G + j\omega C)$

or, $\gamma = \sqrt{(R + j\omega L)\,(G + j\omega C)}$

= propagation constant

And the series impedance

$$\boxed{Z = (R + j\omega L)}$$

The shunt admittance

$$\boxed{Y = (G + j\omega C)}$$

$$\boxed{\gamma \equiv \sqrt{ZY}}$$

The solutions of Equations (7.1) and (7.2) are either in exponential form or in hyperbolic function form. In the first form,

$$V = V_f\, e^{-\gamma x} + V_r\, e^{\gamma x}$$

$$I = I_f\, e^{-\gamma x} + I_r\, e^{\gamma x}$$

These represent the sum of the forward and reflected waves.

The characteristic impedance,

$$z_0 = \frac{V_f}{I_f} \text{ or } -\frac{V_r}{I_r}$$

z_0 is related to R, L, C and G and is given by

$$\boxed{z_0 \equiv \sqrt{\frac{Z}{Y}} = \sqrt{\frac{R + j\omega L}{G + j\omega C}}}$$

If the line is terminated in a load impedance of z_L, then

$$z_L = \frac{V}{I} = \frac{(V_f + V_r)}{(I_f + I_r)}$$

$$= \frac{V_f + V_r}{\frac{V_f}{z_0} - \frac{V_r}{z_0}}$$

$$= z_0 \frac{(V_f + V_r)}{(V_f - V_r)}$$

or,
$$z_L V_f - z_L V_r = z_0 V_f + z_0 V_r$$

Dividing both sides by V_f, we get

$$z_L - z_L \frac{V_r}{V_f} = z_0 + z_0 \frac{V_r}{V_f}$$

or,
$$\frac{V_r}{V_f} [z_L + z_0] = z_L - z_0$$

or,
$$\frac{V_r}{V_f} = \frac{z_L - z_0}{z_L + z_0}$$

The reflection coefficient,

$$\rho \equiv \frac{V_r}{V_f} = \frac{z_L - z_0}{z_L + z_0}$$

7.7 INPUT IMPEDANCE OF A TRANSMISSION LINE

Input impedance z_i is defined as the ratio of voltage and current at the sending end.

In hyperbolic function form, the solutions of Equations (7.1) and (7.2) are given by

$$V = A_1 \cosh \gamma x + B_1 \sinh \gamma x$$

$$I = A_2 \cosh \gamma x + B_2 \sinh \gamma x$$

The constants, A_1, A_2, B_1 and B_2 are found using the boundary conditions, that is,

$$V = V_L,\ I = I_L \text{ at } x = 0$$

$$V = V_s,\ I = I_s \text{ at } x = l$$

Then
$$A_1 = V_L,\ B_1 = -\sqrt{\frac{R+j\omega L}{G+j\omega C}}\, I_L$$

$$A_2 = I_L,\ B_2 = -\sqrt{\frac{G+j\omega C}{R+j\omega L}}\, V_L$$

$$V_s = V_L \cosh \gamma x - z_0 I_L \sinh \gamma x$$

$$I_s = I_L \cosh \gamma x - \frac{V_L}{z_0} \sinh \gamma x$$

z_L is chosen at $x = 0$ and if $l = -x$, we have

$$V_s = V_L \cosh \gamma l + z_0 I_L \sinh \gamma l \tag{7.3}$$

$$I_s = I_L \cosh \gamma l + \frac{V_L}{z_0} \sinh \gamma l \tag{7.4}$$

Here, l is measured from the load end. From Equations (7.3) and (7.4), we have

$$z_i = \frac{V_s}{I_s} = \frac{V_L \cosh \gamma l + z_0 I_L \sinh \gamma l}{I_L \cosh \gamma l + \left(\frac{V_L}{z_0}\right) \sinh \gamma l}$$

If the line is short circuited at the receiving end, $z_L = 0$ and $V_L = 0$. Then the input impedance is

$$(z_i)_{sc} = z_0 \tanh \gamma l$$

Now if $z_L = \infty$, $I_L = 0$, then the input impedance is

$$(z_i)_{oc} = z_o \coth \gamma l$$

or,
$$(z_i)_{sc}\,(z_i)_{oc} = z_0^2$$

7.8 SECONDARY CONSTANTS

The secondary constants are

1. Propagation constant, γ and
2. Characteristic impedance, z_0

Propagation constant

Definition 1 $$\gamma \equiv \sqrt{\text{series impedance} \times \text{shunt admittance}}$$

$$\equiv \sqrt{ZY}, \left(\frac{1}{m}\right)$$

Definition 2 $$\gamma \equiv \log_e \left(\frac{I_s}{I_L}\right) = \log_e \left(\frac{V_s}{V_L}\right), \text{nepers}$$

Definition 3 $$\gamma \equiv 20 \log_{10} \left(\frac{I_s}{I_L}\right) = 20 \log_{10} \left(\frac{V_s}{V_L}\right), \text{dB}$$

Definition 4 $\gamma \equiv \alpha + j\beta$

where α = attenuation constant, dB/m

β = phase constant, rad/m

It may be noted that

$$1 \text{ neper} = 8.686 \text{ dB}$$
$$1 \text{ rad} = 57.3°$$

Consider

$$\gamma = \alpha + j\beta = \sqrt{(R + j\omega L)(G + j\omega C)}$$

$$\gamma^2 = \alpha^2 + 2j\alpha\beta - \beta^2$$

$$= RG - \omega^2 LC + j\omega RC + j\omega LG$$

$$\alpha^2 - \beta^2 = RG - \omega^2 LC \tag{7.5}$$

$$|\gamma| = \sqrt{\alpha^2 + \beta^2}$$

$$\alpha^2 + \beta^2 = \sqrt{(R^2 + \omega^2 L^2)(G^2 + \omega^2 C^2)} \tag{7.6}$$

From Equations (7.5) and (7.6), we have

$$2\alpha^2 = (RG - \omega^2 LC) + \sqrt{(R^2 + \omega^2 L^2)(G^2 + \omega^2 C^2)}$$

or,
$$\alpha = \sqrt{\frac{1}{2}[(RG-\omega^2 LC)+\sqrt{(R^2+\omega^2 L^2)(G^2+\omega^2 C^2)}]} \quad (7.7)$$

From Equations (7.6) and (7.7), we get

$$2\beta^2 = (RG - \omega^2 LC) - \sqrt{(R^2 + \omega^2 L^2)(G^2 + \omega^2 C^2)}$$

or,
$$\beta = \sqrt{\frac{1}{2}\left[(\omega^2 LC - RG) + \sqrt{(R^2 + \omega^2 L^2)(G^2 + \omega^2 C^2)}\right]}$$

Characteristic impedance, z_0

Definition 1 The characteristic impedance, z_0 of a line is defined as the ratio of the forward voltage wave to the forward current wave at any point on the line,

that is,
$$z_0 \equiv \frac{V_f}{I_f}, \Omega$$

Definition 2 z_0 is also defined as the ratio of the square root of series impedance to the square root of shunt admittance, or,

$$z_0 \equiv \sqrt{\frac{Z}{Y}} \equiv \sqrt{\frac{R+j\omega L}{G+j\omega C}}$$

Definition 3 z_0 is defined as the minus of the ratio of the reflected voltage wave to the reflected current wave at any point on the line, or,

$$z_0 \equiv -\frac{V_r}{I_r}$$

7.9 LOSSLESS TRANSMISSION LINES

A transmission line is said to be lossless if the conductors of the line are perfect, or, $\sigma_c = \infty$ and the dielectric medium between the lines is lossless, or, $\sigma_d = 0$. Also, a line is said to be lossless, if

$$R = 0 = G$$

For lossless line,

$$\gamma = \alpha + j\beta = j\beta, \alpha = 0$$

$$= \sqrt{(R + j\omega L)(G + \omega C)}$$

As $R = 0, G = 0$

$$\gamma = j\omega\sqrt{LC} = j\beta \text{ and } \beta = \omega\sqrt{LC}$$

or, $\beta = \omega\sqrt{LC}$, $\alpha = 0$ for a lossless line

$$z_0 = \sqrt{\frac{R + j\omega L}{G + j\omega C}}$$

As $R = 0 = G$

$$z_0 = \sqrt{\frac{L}{C}}$$

The velocity of propagation in lossless line

$$\upsilon_p = \frac{\omega}{\beta} = \frac{1}{\sqrt{LC}}$$

7.10 DISTORTIONLESS LINE

A transmission line is said to be distortionless when the attenuation constant, α is frequency-independent and the phase constant, β is linearly dependent on the frequency or when

$$\frac{R}{L} = \frac{G}{C}$$

Consider $\gamma = \sqrt{(R + j\omega L)(G + j\omega C)}$

$$= \sqrt{RG\left(1 + \frac{j\omega L}{R}\right)\left(1 + \frac{j\omega C}{G}\right)}$$

If $\dfrac{R}{L} = \dfrac{G}{C}$

$$= \sqrt{RG}\left(1 + \frac{j\omega C}{G}\right)$$

$$\alpha = \sqrt{RG}, \beta = \omega\sqrt{LC}$$

Consider

$$z_0 = \sqrt{\frac{R + j\omega L}{G + j\omega C}}$$

$$= \sqrt{\frac{R\,(1 + j\omega L/R)}{G\,(1 + j\omega C/G)}}$$

$$= \sqrt{\frac{R}{G}} = \sqrt{\frac{L}{C}} = R_0 + jX_0, \text{ that is, } X_0 = 0$$

$$z_0 = R_0 = \sqrt{\frac{R}{G}} = \sqrt{\frac{L}{C}} \text{ for a distortionless line}$$

The velocity of propagation for distortionless line is

$$\upsilon_p = \frac{\omega}{\beta} = \frac{\omega}{\omega\sqrt{LC}} = \frac{1}{\sqrt{LC}}$$

The overall transmission line characteristics are shown in Table 7.1.

Table 7.1 Propagation Parameters for Different Types of Lines

Parameter	*General transmission line*	*Lossless line*	*Distorstionless line*
γ	$\sqrt{(R + j\omega L)(G + j\omega C)}$	$j\omega\sqrt{LC}$	$\sqrt{RG} + j\omega\sqrt{LC}$
z_0	$\sqrt{\frac{R + j\omega L}{G + j\omega C}}$	$\sqrt{\frac{L}{C}}$	$\sqrt{\frac{L}{C}}$ or $\sqrt{\frac{R}{G}}$
α	$\left[\frac{1}{2}(RG - \omega^2 LC) + \sqrt{(R^2 + \omega^2 L^2)}\sqrt{(G^2 + \omega^2 C^2)}\right]^{1/2}$	0	$\sqrt{\frac{R}{G}}$

Parameter	*General transmission line*	*Lossless line*	*Distorstionless line*
β	$\left[\frac{1}{2}(\omega^2 LC - RG) + \frac{1}{2}\{(R^2+\omega^2 L^2)^{1/2} (G^2+\omega^2 C^2)^{1/2}\}\right]^{1/2}$	$\omega\sqrt{LC}$	$\omega\sqrt{LC}$
υ_p	$\frac{\omega}{\beta}$	$\frac{1}{\sqrt{LC}}$	$\frac{1}{\sqrt{LC}}$

7.11 PHASE AND GROUP VELOCITIES

Energy is propagated along a transmission line in the form of Transverse Electromagnetic wave (TEM wave). The phase velocity, υ_p for TEM wave is

$$\upsilon_p = \frac{1}{\sqrt{\mu\epsilon}}$$

For a transmission line, $\mu_r = 1$, but ϵ_r may be different. Then

$$\upsilon_p = \frac{\upsilon_0}{\sqrt{\epsilon_r}}$$

$$\upsilon_0 = \text{free space velocity}$$

For a lossless and distortionless transmission line,

$$\upsilon_p = \frac{1}{\sqrt{LC}}$$

Individual waves propagate with the same **phase velocity** if β is proportional to ω. If β is not proportional to ω and if the wave components travel with different velocities, the envelope of the wave travels with a velocity, known as **group velocity**, υ_g, that is,

$$\upsilon_g \equiv \frac{\partial\omega}{\partial\beta}$$

This is also the velocity at which energy is propagated along the line.

7.12 LOADING OF LINES

Introduction of inductance in series with the line is called loading and such lines are called **loaded lines.**

Effect of loading

This is shown in Fig. 7.6.

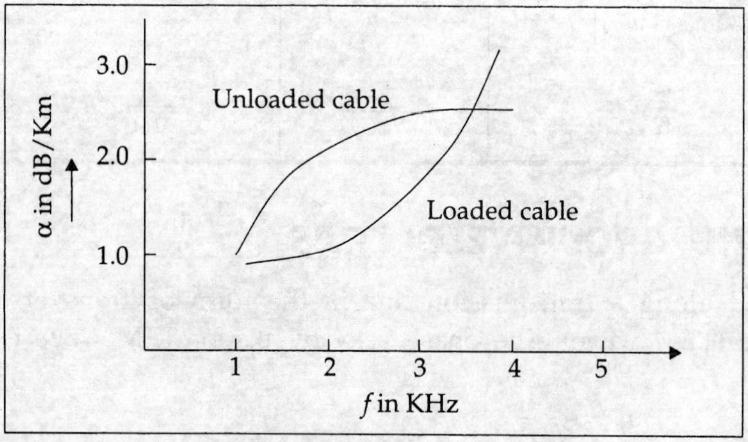

Fig. 7.6 Effect of loading on the cable

Types of loading

1. **Continuous loading** Here loading is done by winding a type of iron around the conductor. This increases inductance but it is expensive.
2. **Patch loading** This type of loading uses sections of continuously loaded cable separated by sections of unloaded cable. Hence cost is reduced.
3. **Lumped loading** Here, loading is introduced at uniform intervals. It may be noted that hysteresis and eddy current losses are introduced by loading and hence design should be optimal.

7.13 INPUT IMPEDANCE OF LOSSLESS TRANSMISSION LINE

For lossless line,

$$\gamma = j\beta \qquad [\text{as } \alpha = 0]$$

and as $\tanh j\beta l = j\tan\beta l$ and $z_0 = R_0$, z_i becomes

$$z_i = R_0 \left[\frac{z_L + jR_0 \tan\beta l}{R_0 + jz_L \tan\beta l}\right] \qquad \text{(for a lossless line)}$$

For shorted line, $z_L = 0$

$$Z_{sc} = z_0 \tanh \gamma l$$

For open circuited line, $z_L = \infty$

$$Z_{oc} = z_i = z_0 \coth \gamma l$$

and

$$(z_i)_{sc}\,(z_i)_{oc} = z_0^2$$

For matched line, $z_L = z_0$

$$z_i = z_0$$

Input impedances of a transmission line for different cases are given in Table 7.2.

Table 7.2 Input Impedance for Different Loads

Type of line	*Input impedance,* z_i
Lossy line	$z_0 \left[\dfrac{z_L + z_0 \tanh \gamma l}{z_0 + z_L \tanh \gamma l}\right]$
Lossless line	$z_0 \left[\dfrac{z_L + jz_0 \tan \beta l}{z_0 + jz_L \tan \beta l}\right]$
Lossy line with shorted load $(z_L = 0)$	$z_0 \tanh \beta l$
Lossy line with open circuited load $(z_L = \infty)$	$z_0 \coth \beta l$
Lossy line with matched load $(z_L = z_0)$	z_0
Lossless line with shorted load	$j\, z_0 \tan \beta l$
Lossless line with open circuited load	$-\, j\, z_0 \cot \beta l$
Lossless line with matched load	z_0

The variation of input impedance of lossless line when shorted and open circuited are shown in Fig. 7.7.

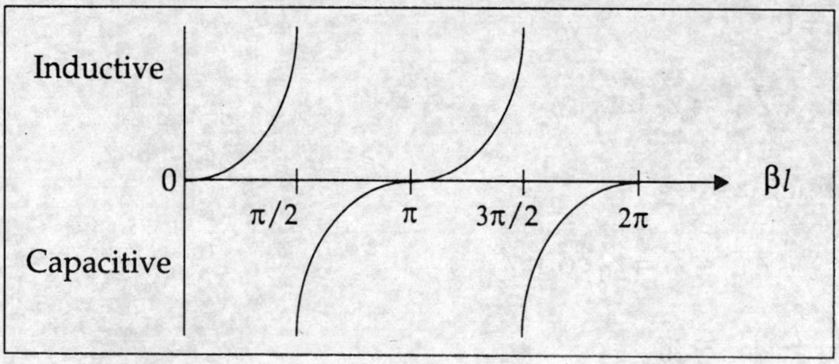

(a) Shorted load

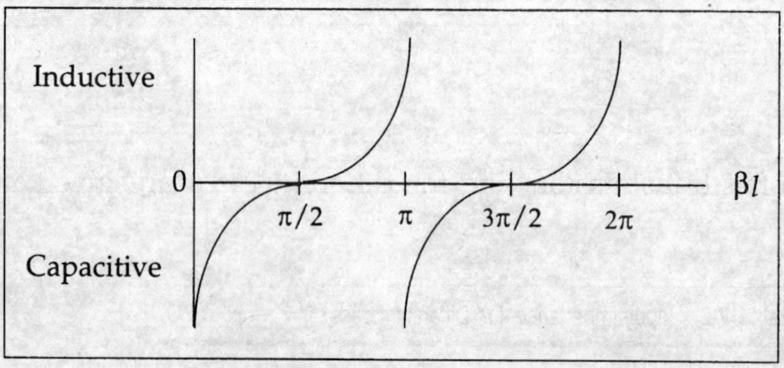

(b) Open circuited load

Fig. 7.7 Input impedance variation

7.14 RF LINES

At radio frequencies,

$$\omega L >> R$$

$$\omega C >> G$$

Then,

$$Z = R + j\omega L \approx j\omega L$$

$$Y = G + j\omega C \approx j\omega C$$

$$z_0 = \sqrt{\frac{R + j\omega L}{G + j\omega C}} \approx \sqrt{\frac{L}{C}}$$

and

$$\gamma = \sqrt{(R + j\omega L)\,(G + j\omega C)}$$

$$\gamma \approx j\omega\,\sqrt{LC}$$

As $\gamma = \alpha + j\beta$, $\alpha \approx 0$, $\beta \approx \omega\sqrt{LC}$, it is not enough if α is small compared to β. Hence let us consider,

$$\gamma = \sqrt{(R + j\omega L)\,(G + j\omega C)}$$

$$\gamma = j\omega\sqrt{LC}\sqrt{\left(1 + \frac{R}{j\omega L}\right)\left(1 + \frac{G}{j\omega C}\right)}$$

$$\approx j\omega\sqrt{LC}\left(1+\frac{R}{2j\omega L}\right)\left(1+\frac{G}{2j\omega C}\right)$$

$$\approx \frac{R}{2\sqrt{L/C}}+\frac{G}{2}\sqrt{L/C}+j\omega\sqrt{LC}$$

$$\alpha \approx \frac{1}{2}\left(\frac{R}{z_0}+Gz_0\right)$$

$$\beta = \omega\sqrt{LC}$$

For *RF* lines, the input impedance is

$$z_i = z_0\left[\frac{z_L + jz_0 \tan\beta l}{z_0 + jz_L \tan\beta l}\right]$$

7.15 RELATION BETWEEN REFLECTION COEFFICIENT, LOAD AND CHARACTERISTIC IMPEDANCES

$$\rho = \frac{z_L - z_0}{z_L + z_0}$$

where ρ = reflection coefficient
z_L = load impedance
z_0 = characteristic impedance

Proof For lossless *RF* lines, we have

$$\gamma = j\beta$$

$$V_{\text{incident}} = V_i\, e^{j\beta l}$$

Similarly,

$$V_{\text{reflected}} = V_r\, e^{-j\beta l}$$

At the load, $(l=0)$

$$V_L = V_i + V_r$$

$$z_0 = \frac{V_{\text{incident}}}{I_{\text{incident}}} = \frac{-V_{\text{reflected}}}{I_{\text{reflected}}}$$

I at any point from load is

$$I = I_{\text{incident}} + I_{\text{reflected}}$$

Load current, I_L

$$I_L = I_{\text{incident}} - I_{\text{reflected}}$$

$$= \frac{V_{\text{incident}}}{z_0} - \frac{V_{\text{reflected}}}{z_0}$$

$$I_L = \frac{V_i - V_r}{z_0}$$

But by definition, the reflection coefficient, ρ is

$$\rho = \frac{V_r}{V_i}$$

and

$$z_L = \frac{V_L}{I_L} = \frac{(V_i + V_r)}{(V_i - V_r)} z_0$$

$$V_i z_L - z_L V_r = z_0 V_i + z_0 V_r$$

Dividing both sides by V_i, we get

$$z_L - z_L \frac{V_r}{V_i} = z_0 + z_0 \frac{V_r}{V_i}$$

$$\rho = \frac{V_r}{V_i} = \frac{z_L - z_0}{z_L + z_0}$$

Hence proved.

If

$$z_L = z_0, \rho = 0$$

$$z_L = 0, \rho = -1$$

$$z_L = \infty, \rho = 1$$

$$z_L = \text{purely reactive}, |\rho| = 1$$

7.16 RELATION BETWEEN REFLECTION COEFFICIENT AND VOLTAGE STANDING WAVE RATIO (VSWR)

VSWR is defined as

$$\text{VSWR} \equiv \frac{V_{\text{max}}}{V_{\text{min}}}$$

$$= \frac{1+|\rho|}{1-|\rho|}$$

Proof We can write

$$V_{max} = |V_i| + |V_r|$$

$$= |V_i|[1 + |\rho|]$$

Similarly,

$$V_{min} = |V_i| - |V_r| = |V_i|(1 - |\rho|)$$

$$\text{VSWR} = S = \frac{V_i\,(1+|\rho|)}{V_i\,(1-|\rho|)}$$

$$\text{VSWR} = \frac{1+|\rho|}{1-|\rho|}$$ Hence proved.

S ranges between 1 and ∞ or $1 < S < \infty$

We can also write $|\rho|$ as

$$|\rho| = \frac{S-1}{S+1}$$

When the line is terminated by purely resistive load,

$$S = \frac{z_0}{R_L} \text{ if } z_0 > R_L$$

$$S = \frac{R_L}{z_0} \text{ if } R_L > z_0$$

7.17 LINES OF DIFFERENT LENGTH—$\frac{\lambda}{8}, \frac{\lambda}{4}, \frac{\lambda}{2}$ LINES

By selecting a terminated line of suitable length, it is possible to produce the equivalent of a pure resistance, inductance and capacitance or any desired combination thereof.

Equivalent circuits for shorted and open lines are shown in Table 7.3.

Table 7.3 Equivalent Circuits for Shorted and Open Lines

Length of line	*Equivalent circuit element*	
	Shorted line	*Open line*
$l < \frac{\lambda}{4}$	Inductor	Capacitor
$l = \frac{\lambda}{4}$	Tank circuit	Series-resonant circuit
$\frac{\lambda}{4} < l < \frac{\lambda}{2}$	Capacitor	Inductor
$l = \frac{\lambda}{2}$	Series-resonant circuit	Tank circuit

In Table 7.4, lines of different length with short and open ends along with their equivalent circuits are shown.

Table 7.4 Equivalent Circuit and Impedance

Transmission line	*Equivalent circuit*	*Input impedance*
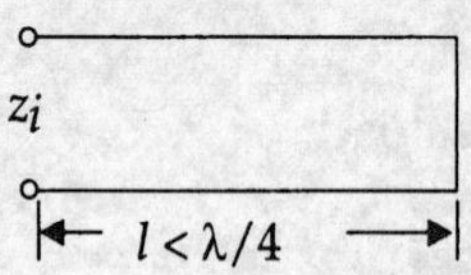		$z_i = +jz_0 \tan \beta l$
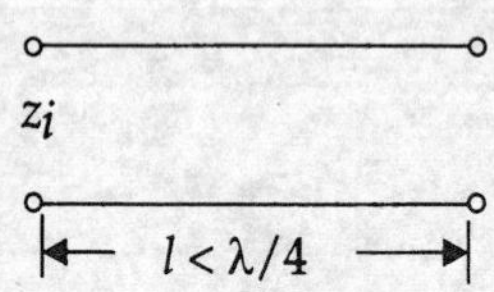	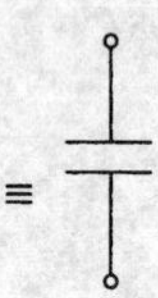	$z_i = -jz_0 \cot \beta l$
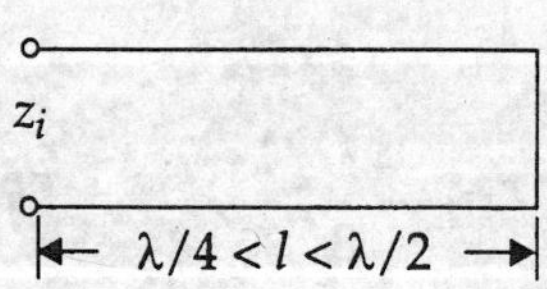	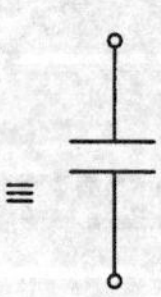	$z_i = +jz_0 \tan \beta l$
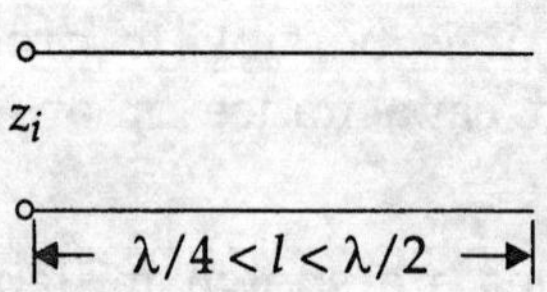	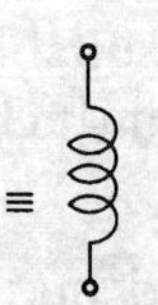	$z_i = -jz_0 \cot \beta l$

Transmission line	*Equivalent circuit*	*Input impedance*
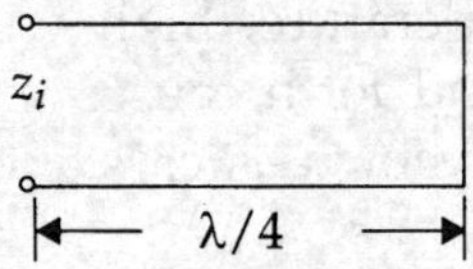	≡ 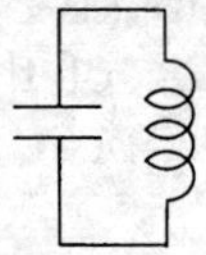	$z_i \approx \frac{2z_0^2}{Rl} = \frac{z_0^2}{\tanh \alpha l}$
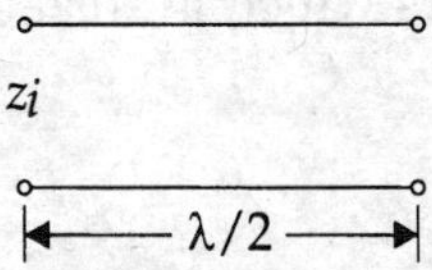	≡ 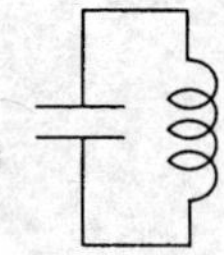	$z_i \approx \frac{2z_0^2}{Rl} = \frac{z_0^2}{\tanh \alpha l}$

7.18 LOSSES IN TRANSMISSION LINES

Losses in transmission lines are of three types:

Losses

Copper losses | Dielectric losses | Radiation losses

Copper Loss

These losses occur because of the following reasons:

1. **I^2R power loss** This is due to dissipation as a result of heating in pure resistance. Its features are:

(a) Copper loss is small if there are no current loops, that is, the line should be properly terminated without producing standing waves.

(b) Copper loss is more if z_0 is small. This is because, if z_0 is low, current is high. If the current is high, copper loss is more.

2. **Skin effect** Its features are:

(a) When an AC signal at high frequency is applied to the transmission lines, the current is confined to the surface (skin) of the conductors. This is known as skin effect. This effect reduces the cross-sectional area of the conductor with increase in frequency.

(b) Decrease in cross-sectional area increases the resistance.

(c) Increased resistance increases power losses.

3. **Crystallisation** Its features are:

(a) Copper losses increase due to ageing of the transmission line. Losses are more when the line is subjected to high temperature, high winds and moisture. Moreover, bending of the line back and forth causes the line to become brittle and cracks appear. This effect is known as crystallisation of the conductors.

(b) Crystallisation increases resistance in the conductors which in turn increases copper losses.

Dielectric Losses

These losses exist due to improper characteristics of dielectric.

Salient features:

(a) These are due to $I^2 R$ power dissipation because of the heating of the solid dielectric material between conductors in transmission lines. These losses are proportional to the voltage across the dielectric.

(b) With increased frequencies, solid dielectric properties worsen and hence transmission lines with solid dielectrics have limited applications.

(c) Lines with air dielectric are used at high frequencies, as air dielectric loss is very small.

Radiation Losses

Salient features:

(a) These losses are high when the spacing between the lines is high as the transmission line acts as an antenna. Therefore, radiation losses are more in parallel-wire lines than in coaxial lines.

(b) At high frequency, λ will be small and hence the transmission lines are not useful at high frequencies.

Problem 7.1 A transmission line with air as dielectric has $z_0 = 50\,\Omega$ and a phase constant of 3.0 rad/m at 10 MHz. Find the inductance and capacitance of the line.

Solution A line with air dielectric is lossless as $\sigma = 0$.

$$R = 0 = G \text{ and } \alpha = 0$$

$$z_0 = R_0 = 50\,\Omega = \sqrt{\frac{L}{C}}$$

and

$$\beta = \omega\sqrt{LC} = 2\pi f\sqrt{LC}$$

$$\frac{R_0}{\beta} = \frac{\sqrt{L/C}}{2\pi f\sqrt{LC}} = \frac{1}{\omega C}$$

or,
$$C = \frac{\beta}{\omega R_0} = \frac{3}{2\pi\times 10^7\times 50} = 0.9549\times 10^{-9}$$

$$\boxed{C = 954.9 \text{ pF/m}}$$

As
$$\sqrt{\frac{L}{C}} = R_0 = 50\Omega$$

$$L = 50^2 C = 2500\times 954.9\times 10^{-12}$$

$$\boxed{L = 2387.25 \text{ nH/m}}$$

Problem 7.2 A lossy cable which has $R = 2.25\,\Omega/\text{m}$, $L = 1.0\,\mu\text{H/m}$, $C = 1\,\text{pF/m}$, and $G = 0$ operates at $f = 0.5$ GHz. Find the attenuation constant of the line.

Solution The propagation constant is given by

$$\gamma = \alpha + j\beta = \sqrt{ZY}$$

$$Z = R + j\omega L$$

$$Y = G + j\omega C = j\omega C \quad [\text{As } G = 0]$$

$$\gamma = \sqrt{(R + j\omega L)(j\omega C)}$$

$$= \sqrt{(2.25 + j2\pi\times 500\times 10^6\times 1.0\times 10^{-6})\times j2\pi\times 500\times 10^6\times 1\times 10^{-12}}$$

$$= 1.125\times 10^{-3} + j3.142\,(\text{m}^{-1})$$

$$\boxed{\alpha = 1.125\times 10^{-3}\,(\text{m}^{-1})}$$

Problem 7.3 A transmission line in which no distortion is present has the following parameters: $z_0 = 50\,\Omega$, $\alpha = 0.020\,\text{m}^{-1}$, $\upsilon = 0.6\upsilon_0$. Determine *R*, *L*, *G*, *C* and wavelength at 0.1 GHz.

Solution For a distortionless line the condition is

$$RC = GL \text{ or } G = \frac{RC}{L}$$

and
$$z_0 = \sqrt{\frac{L}{C}}$$

$$\alpha = \sqrt{RG} = R\sqrt{\frac{C}{L}}$$

$$= \frac{R}{z_0}$$

and hence

$$R = \alpha z_0$$

$$= 20 \times 10^{-3} \times 50 = 1.0\,\Omega/\text{m}$$

$$L = \frac{z_0}{\upsilon} = \frac{50}{0.6 \times 3 \times 10^8} = 277\,\text{nH/m}$$

$$G = \frac{\alpha^2}{R} = \frac{400 \times 10^{-6}}{1.0}$$

$$G = 400\mu\ \text{mho/m}$$

$$C = \frac{1}{\upsilon z_0} = \frac{1}{0.6 \times 3 \times 10^8 \times 50} = 111.1\,\text{pF/m}$$

$$\lambda = \frac{\upsilon}{f} = \frac{0.6 \times 3 \times 10^8}{0.1 \times 10^9} = 1.8\,\text{m}$$

Problem 7.4 For a transmission line which is terminated in a normalised impedance z_n, VSWR = 2. Find the normalised impedance magnitude.

Solution Normalised impedance, z_n

$$z_n = \frac{z_L}{z_0}$$

$$S = \text{VSWR} = 2 = \frac{1 + |\rho|}{1 - |\rho|}$$

or,
$$|\rho| = \frac{S-1}{S+1} = \frac{2-1}{2+1} = \frac{1}{3}$$

We have

$$|\rho| = \left|\frac{z_L - z_0}{z_L + z_0}\right|$$

$$= \left|\frac{\frac{z_L}{z_0} - 1}{\frac{z_L}{z_0} + 1}\right| = \frac{1}{3}$$

$$\boxed{|z_n| = 2}$$

Problem 7.5 A lossless transmission line used in a TV receiver has a capacitance of 50 pF/m and an inductance of 200 nH/m. Find the characteristic impedance for sections of a line 10 metre long and 500 metre long.

Solution The characteristic impedance of a lossless transmission line is

$$z_0 = \sqrt{\frac{L}{C}}$$

The inductance, L of the line

$$= 200\,\text{nH/m}$$

For 10 m line,

$$L = 200 \times 10^{-9} \times 10 = 2000 \times 10^{-9}$$

$$= 2 \times 10^{-6}\ \text{H}$$

$$C = 50\,\text{pF/m}$$

For 10 m line,

$$C = 50 \times 10^{-12} \times 10 = 5 \times 10^{-10}\ \text{F}$$

The characteristic impedance, z_0

$$z_0 = \sqrt{\frac{L}{C}} = \sqrt{\frac{2 \times 10^{-6}}{5 \times 10^{-10}}}$$

$$= \sqrt{0.4 \times 10^4}$$

$$\boxed{z_0 = 63.245\,\Omega}$$

The inductance of 500 m line

$$L = 200 \times 10^{-9} \times 500$$

$$= 10{,}0000 \times 10^{-9}$$

$$L = 10^{-4}\ \text{H}$$

The capacitance of 500 m line

$$C = 50 \times 10^{-12} \times 500$$

$$= 25000 \times 10^{-12} = 25 \times 10^{-9}\ \text{F}$$

The characteristic impedance, z_0 of 500 m line

$$z_0 = \sqrt{\frac{L}{C}} = \sqrt{\frac{10^{-4}}{25 \times 10^{-9}}}$$

$$= \sqrt{0.04 \times 10^{-4} \times 10^{9}}$$

$$= \sqrt{0.4 \times 10^{-5} \times 10^{9}}$$

$$= \sqrt{0.4 \times 10^{4}}$$

$$= 0.632 \times 10^{2}$$

$$\boxed{z_0 = 63.2\ \Omega}$$

Problem 7.6 A two-wire open air line, whose diameter is 2.588 mm, is used in several applications. The wires are spaced at 290 mm between the centres. Find out the characteristic impedance of the line.

Solution Radius of the wire

$$r = \frac{d}{2} = \frac{2.588}{2} = 1.294\ \text{mm}$$

Spacing between the wires is

$$s = 290\ \text{mm}$$

The characteristic impedance of the two-wire open air line is

$$z_0 = 276 \log_{10} \frac{s}{r}$$

$$= 276 \log_{10} \frac{290}{1.294}$$

$$= 276 \log_{10} 224.11$$

$$= 276 \times 2.350$$

$$\boxed{z_0 = 648.7\ \Omega}$$

Problem 7.7 A copper coaxial line has an outside tubing of thickness 1.8 mm and its outside diameter is 30 mm. The thickness of the inner tubing is 1.0 mm and its outside diameter is 8 mm. Find the characteristic impedance of the line.

Solution Diameter of the outside conductor is

$$d_1 = 30 - 2 \times 1.8 = 26.4 \text{ mm}$$

Diameter of the inner conductor is

$$d_2 = 8 - 2 \times 1.0 = 6.0 \text{ mm}$$

For a coaxial cable, z_0 is

$$z_0 = 138 \log \left(\frac{d_1}{d_2} \right)$$

$$= 138 \log \left(\frac{26.4}{6} \right)$$

$$= 138 \log_{10} 4.40$$

$$\boxed{z_0 = 88.79 \Omega}$$

Problem 7.8 If a signal of 30 MHz is transmitted through a coaxial cable which has a capacitance of 30 pF/m and an inductance of 500 nH/m. (a) Find the time delay for a cable 1 m long, (b) propagation velocity, and (c) propagation delay over a cable length of 10 m.

Solution

(a) The time delay for 1 m long cable is

$$t_d = \sqrt{LC}$$

$$= \sqrt{(500 \times 10^{-9})\,(30 \times 10^{-12})}$$

$$= \sqrt{15 \times 10^{-18}}$$

$$\boxed{t_d = 3.87 \text{ n sec}}$$

(b) Velocity of propagation,

$$\upsilon_p = \frac{1\text{ metre}}{3.87\times10^{-9}}$$

$$\boxed{\upsilon_p = 2.5839\times10^{8} \text{ m/s}}$$

(c) The inductance of 10 metre cable is

$$L = 500\times10^{-9}\times10 = 5\times10^{-6} \text{ H}$$

The capacitance of 10 metre cable is

$$C = 30\times10^{-12}\times10 = 3\times10^{-10} \text{ F}$$

The time delay,

$$t_d = \sqrt{LC} = \sqrt{5\times10^{-6}\times3\times10^{-10}}$$

$$= 3.87\times10^{-8}$$

or

$$\boxed{t_d = 38.7 \text{ ns}}$$

Problem 7.9 The attenuation coefficient of a transmission line is 0.2 mNp/m. Find the attenuation coefficient in (a) dB/m (b) dB/mile.

Solution

(a) Attenuation coefficient in Np/m is

$$\alpha = 0.2\times10^{-3} \text{ Np/m}$$

Attenuation coefficient in dB/m is

$$\alpha = 8.686\times0.2\times10^{-3}$$

$$\boxed{\alpha = 1.7372\times10^{-3} \text{ dB/m}}$$

(b) 1 mile = 1609 m

The attenuation coefficient, α in dB/mile is

$$\alpha = 1.7372\times10^{-3}\times1609$$

$$\boxed{\alpha = 2.795 \text{ dB/mile}}$$

Problem 7.10 A lossless transmission line is terminated in a load impedance of $30 - j\,23\,\Omega$. Find the phase constant and the reflection coefficient of a line of length 50 m. Characteristic impedance, $z_0 = 50\Omega$. Wavelength on the line = 0.45 m.

Solution

$$z_L = (30 - j23)\ \Omega$$

$$z_0 = 50\Omega$$

$$\lambda = 0.45\text{m}$$

$$l = 50\text{m}$$

Phase constant,

$$\beta = \frac{2\pi}{\lambda} = 13.9626 \text{ rad/m}$$

Reflection coefficient,

$$\rho = \frac{z_L - z_0}{z_L + z_0}$$

$$= \frac{30 - j23 - 50}{30 - j23 + 50}$$

$$= \frac{-(20 + j23)}{(80 - j23)}$$

$$\boxed{\rho = -(0.1431 + j0.3071)}$$

Problem 7.11 A coaxial cable has z_0 of 75Ω and a capacitance of 70 pF/m. Find its inductance per metre. If the radius of the inner conductor is 0.292 mm and the relative permittivity of the dielectric is 2.3, determine the radius of the outer conductor.

Solution Radius of the inner conductor,

$$r_i = 0.292 \text{ mm}$$

We have

$$z_0 = \sqrt{\frac{L}{C}}$$

or,

$$\frac{L}{C} = z_0^2$$

$$L = z_0^2 C = 75^2 \times 70 \times 10^{-12}$$

$$\boxed{L = 0.3937 \mu\text{H/m}}$$

For a coaxial cable, z_0 is also given by

$$z_0 = \frac{138}{\sqrt{\epsilon_r}} \log\left(\frac{r_0}{r_i}\right)$$

or,
$$\log\left(\frac{r_0}{r_i}\right) = \frac{z_0\sqrt{\epsilon_r}}{138}$$

$$= \frac{75 \times \sqrt{2.3}}{138} = 0.8242$$

$$\frac{r_0}{r_i} = \text{Anti log } 0.8242 = 6.671$$

$$r_0 = 0.292 \times 6.671$$

$$\boxed{r_0 = 1.9479 \, \text{mm}}$$

where r_0 = radius of the outer conductor.

Problem 7.12 A lossless transmission line of length 100 m has an inductance of 28μH and a capacitance of 20 nF. Find (a) propagation velocity (b) phase constant at an operating frequency of 100 kHz (c) characteristic impedance of the line.

Solution Length of transmission line,

$$l = 100 \, \text{m}$$

Inductance of the line

$$= 28 \, \mu\text{H}$$

Inductance/metre
$$= \frac{28 \times 10^{-6}}{100} = 0.28 \, \mu\text{H/m}$$

Capacitance of the line $= 20$ nF

Capacitance per metre

$$= \frac{20 \times 10^{-9}}{100} = 0.20 \, \text{nF/m}$$

(a) Propagation velocity,

$$\upsilon_p = \frac{1}{\sqrt{LC}}$$

$$= \frac{1}{\sqrt{0.28\times10^{-6}\times0.2\times10^{-9}}}$$

$$\boxed{\upsilon_p = 1.336\times10^{8}\ \text{m/s}}$$

(b) Phase constant,

$$\beta = \frac{\omega}{\upsilon_p}$$

$$= \frac{2\pi\times100\times10^{3}}{1.336\times10^{8}}$$

$$= \frac{2\pi}{1.336}\times\frac{10^{5}}{10^{8}}$$

$$\boxed{\beta = 4.702\times10^{-3}\ \text{rad/m}}$$

(c) Characteristic impedance of the transmission line,

$$z_0 = \sqrt{\frac{L}{C}} = \sqrt{\frac{0.28\times10^{-6}}{0.2\times10^{-9}}} = 37.42\Omega$$

$$\boxed{z_0 = 37.42\Omega}$$

Problem 7.13 The dielectric material between two conductors of a lossless coaxial cable has $\epsilon_r = 4$ and $\mu_r = 1$. Diameter of the inner conductor is 2 mm. Characteristic impedance of the 10 m long cable is 50Ω. Determine the diameter of the outer conductor of the coaxial cable.

Solution The expression for inductance and capacitance of a coaxial cable is

$$L = \frac{\mu}{2\pi}\,\ell n\,\frac{d_0}{d_i}$$

and $$C = \frac{2\pi\epsilon}{\ell n\left(\dfrac{d_0}{d_i}\right)}$$

where μ = permeability = $\mu_0\,\mu_r$

ϵ = permittivity = $\epsilon_0\epsilon_r$

d_0 = diameter of outer conductor

d_i = diameter of inner conductor

The expression for the characteristic impedance, z_0 is

$$z_0 = \sqrt{\frac{L}{C}}$$

$$= \sqrt{\frac{\frac{\mu}{2\pi}\ell n\left(\frac{d_0}{d_i}\right)}{\frac{2\pi\epsilon}{\ell n\frac{d_0}{d_i}}}}$$

$$= \frac{1}{2\pi}\sqrt{\frac{\mu}{\epsilon}}\ln\frac{d_0}{d_i}$$

$$\ell n\frac{d_0}{d_i} = \frac{z_0 \times 2\pi}{\sqrt{\frac{\mu}{\epsilon}}}$$

Here $$\frac{\mu}{\epsilon} = \frac{\mu_r\mu_0}{\epsilon_r\epsilon_0} = \frac{4\pi\times10^{-7}}{4\times8.854\times10^{-12}}$$

$$\sqrt{\frac{\mu}{\epsilon}} = 1.8836\times10^2$$

$$\ell n\frac{d_0}{d_i} = \frac{50\times2\pi}{1.8836\times10^2} = 1.6678$$

$$\frac{d_0}{d_i} = e^{1.6678} = 5.300$$

$$d_0 = D_i \times 5.3 = 2\times5.3$$

$$\boxed{d_0 = 10.6 \text{ mm}}$$

Problem 7.14 A transmission line is lossless and is 25 m long. It is terminated in a load of $z_L = 40 + j30\,\Omega$ at a frequency of 10 MHz. The inductance and capacitance of the line are $L = 300$ nH/m, $C = 40$ pF/m. Find the input impedance at the source and at the mid-point of the line.

Solution The length of transmission line = 25 m.

Load impedance,

$$z_L = 40 + j30$$

Inductance,

$$L = 300 \text{ nH/m}$$

Capacitance,

$$C = 40 \text{ pF/m}$$

Characteristic impedance,

$$z_0 = \sqrt{\frac{L}{C}} = \sqrt{\frac{300\times10^{-9}}{40\times10^{-12}}} = 75\Omega$$

Phase constant, β

$$\beta = \omega\sqrt{LC}$$

$$= 2\pi f\sqrt{LC}$$

$$= 2\pi\times10\times10^6\sqrt{300\times10^{-9}\times40\times10^{-12}}$$

$$= 2\pi\times10^7\times10^{-11}\times\sqrt{300\times400}$$

$$= 2\pi\times346.41\times10^{-4}$$

$$\boxed{\beta = 0.2176 \text{ rad/m}}$$

Input impedance at the source end is

$$z_i = z_0\left[\frac{z_L + jz_0\tan\beta l}{z_0 + jz_L\tan\beta l}\right]$$

$$= 75\left[\frac{40 + j30 + j75\tan(0.2176\times25)}{75 + j(40 + j30)\tan(0.2176\times25)}\right]$$

$$\boxed{z_i = 37.116 - j23.165\Omega}$$

Similarly, input impedance at 12.5 m from source end is

$$z_i = 75\left[\frac{40 + j30 + j75\tan(0.2176\times12.5)}{75 + j(40 + j30)\tan(0.2176\times12.5)}\right]$$

$$\boxed{z_i = 33.154 + j3.198\Omega}$$

7.19 SMITH CHART AND APPLICATIONS

Smith chart is a polar plot of the reflection coefficient in terms of normalised impedance, $r + jx$. In other words, it is a graphical plot of normalised resistance and reactance in the reflection coefficient plane.

It was designed by Phillip H. Smith in 1939.

Construction of Smith Chart

It is constructed within a circle of unit radius ($|\rho| \le 1$) as in Fig. 7.8.

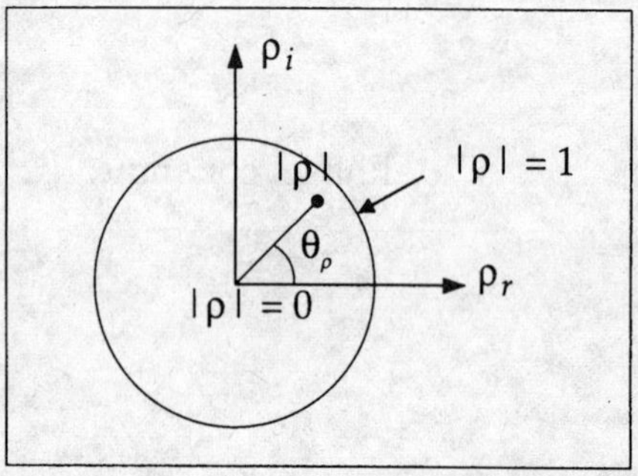

Fig. 7.8 Construction of Smith chart on unit circle

Smith chart provides the relation between reflection coefficient, ρ load, z_L and characteristic impedance, z_0. Here, the impedances are always normalised with respect to characteristic impedance,

that is, $$\rho = \frac{z_L - z_0}{z_L + z_0}$$

or, $$\rho = |\rho| \angle \theta_\rho = \rho_r + j\rho_i$$

Smith charts are constructed in terms of normalised impedances (z_L / z_0) to avoid construction of one chart for each z_0. Normalised load impedance is

$$z_n = \frac{z_L}{z_0} = r + jx$$

$$\rho = \rho_r + j\rho_i = \frac{z_n - 1}{z_n + 1}$$

or, $$z_n = \frac{(1+\rho_r) + j\rho_i}{(1-\rho_r) - j\rho_i}$$

$$r = \frac{1 - \rho_r^2 - \rho_i^2}{(1-\rho_r)^2 + \rho_i^2}$$

$$x = \frac{2\rho_i}{(1-\rho_r)^2+\rho_i^2}$$

or, $$\left[\rho_i - \frac{r}{r+1}\right]^2 + \rho_i^2 = \left(\frac{1}{1+r}\right)^2 \quad (7.8)$$

and $$(\rho_i - 1)^2 + \left(\rho_i - \frac{1}{x}\right)^2 = \left(\frac{1}{x}\right)^2 \quad (7.9)$$

The Equations (7.8) and (7.9) are the equations of circles like

$$(x-k)^2 + (y-h)^2 = a^2$$

Equation (7.8) is known as r-circle and Equation (7.9) is known as x-circle.

The resistance circle has centre at $(\rho_r, \rho_i) = \left(\frac{r}{1+r}, 0\right)$ and radius of $\left(\frac{1}{1+r}\right)$ as in Fig. 7.9.

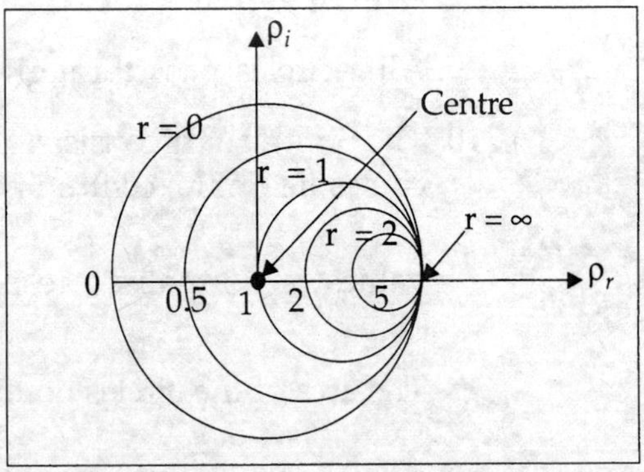

Fig. 7.9 *r*-circles

The reactance circle has centre at

$$(\rho_r, \rho_i) = \left(1, \frac{1}{x}\right)$$

and radius $= \left(\frac{1}{x}\right)$ as shown in Fig. 7.10.

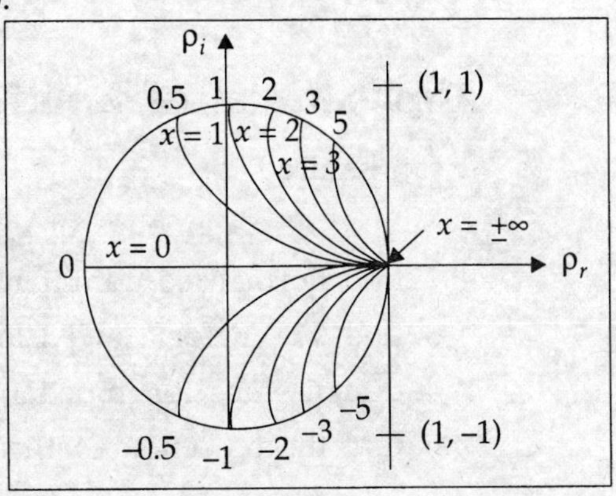

Fig. 7.10 The reactance circles

Salient features of r and x-circles

1. The centres of all r-circles lie on ρ_r-axis where ρ_r is the real part of the reflection coefficient. The centre straight line is called ρ_r axis.
2. The $r = 0$ circle having unity radius and centre at the origin is the largest.
3. The r-circles become progressively smaller as r increases from 0 to ∞ ending at the point $(\rho_r = 1, \rho_i = 1)$ for open circuit.
4. All circles pass through the $(\rho_r = 1, \rho_i = 0)$ point.
5. The centre of all x-circles lie on the $\rho_r = 1$ line.
6. The x-circles for $x > 0$ (inductive reactance) lie above ρ_r-axis and those for $x < 0$ (capacitive reactance) lie below the ρ_r-axis.
7. The $x = 0$ circle becomes ρ_r-axis.
8. x-circle becomes progressively smaller as $|x|$ increases from 0 to ∞ ending at the point $(\rho_r = 1, \rho_i = 0)$ for open circuit.
9. All x-circles pass through the $(\rho_r = 1, \rho_i = 0)$ point.
10. The Smith chart consists of two sets of circles or arcs of circles. The complete circles lie on the centre line on the chart. These circles correspond to various values of normalised resistance $\left(r = \dfrac{R}{z_0}\right)$ along the line.
11. The arcs of the circles on either side of the straight line correspond to various values of normalised line reactance, $\left(jx = \dfrac{jX}{z_0}\right)$.
12. The circles are orthogonal to each other.
13. The perimeter of the outer rim of the chart is of $\dfrac{\lambda}{2}$ length.

Applications of Smith Chart

It can be used to:

1. find the parameters of mismatched transmission lines
2. find normalised admittance from normalised impedance or vice-versa
3. find VSWR for a given load impedance
4. design stubs for impedance matchings
5. find the reflection coefficient

6. locate a voltage maximum on the line
7. find the input impedance of a transmission line.

7.20 STUBS

A stub is a piece of transmission line. It can be short circuited at the far end or open circuited. It has a pure reactance or susceptance. It is used to cancel out reactance or susceptance of a transmission line. In other words, it is used for impedance matching.

In general, shorted stubs are more frequently used since open ended stubs tend to radiate. The design parameters of stubs are (1) stub length and (2) stub distance from the load. The matching of transmission lines is done by the design of a single stub or a double stub.

Design of Single Stub Matching

The design consists of the following steps:

1. Given z_L and z_0 of transmission line, normalise z_L, or, find

$$z_n = z_L / z_0 = r + jx$$

2. Mark the point on the Smith chart where r-circle and x-arc intersect.
3. Draw a circle with the radius equal to the distance from the centre of the chart to the point.
4. Draw a straight line from this point through the centre of the chart to the other half of the chart, that is, travel around it through a distance of $\frac{\lambda}{4}$ (or, straight through) to find the load admittance. As the stub is placed in parallel with the main line, it is easy to deal with admittances to make stub calculations.
5. Mark the point where the straight line intersects the circle drawn. This point gives the normalised admittance, $(g + jb)$.
6. Note the point where the circle cuts the centre straight line of the chart to the right side of the centre. Read the value of that point. It gives VSWR straight away.
7. Note the point nearest to the load at which the normalised admittance is $1 \pm jb$. This is the point where $\pm jb$ intersects the $r = 1$ circle. Moreover, this is the point at which a stub, designed to tune out $\pm jb$ component, will be placed.

8. Note the distance travelled round the circumference of the chart. This is the distance of the stub from the load.
9. Move clockwise round the perimeter of the chart and find the point at which the susceptance tunes out $\pm jb$ susceptance of the line. For example, if the line admittance is $1 + jb$, the required susceptance is $-jb$.
10. Note the distance in wavelengths from ∞, $j\infty$ of the chart to the new point (Example: susceptance is $-b$). This gives the length of the stub required.

A typical stub is shown in Fig. 7.11.

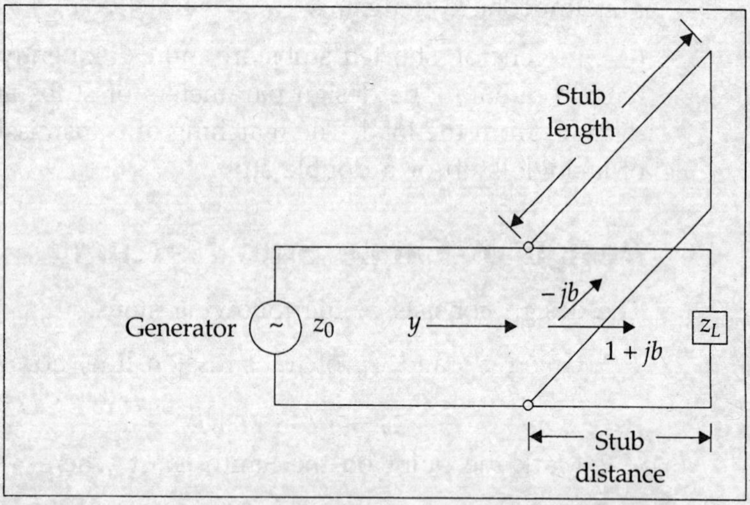

Fig. 7.11 Shorted stub connected to a transmission line

Problem 7.15 Find the input impedance of 75Ω lossless transmission line of length 0.1λ when the load is a short.

Solution Load impedance,

$$z_L = 0,\ \Omega\ \text{(short)}$$

Characteristic impedance,

$$z_0 = 75\Omega$$

Length of the line $= 0.1\lambda$

The normalised load impedance is

$$z_n = \frac{z_L}{z_0} = \frac{0}{75} = 0$$

Steps Involved

1. Start from the point P_{sc} at the left of the rim of the chart. P_{sc} is intersection of $r = 0$, $x = 0$.

2. Move clockwise from P_{sc} through the perimeter of the chart by 0.1λ towards generator. Mark point P_s (Refer Fig. 7.12).

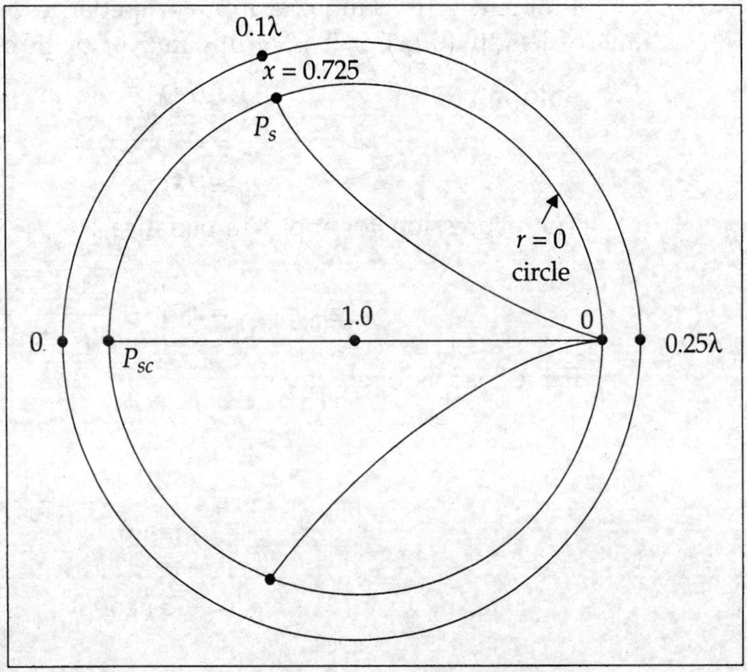

Fig. 7.12 Lossless transmission line where load is short

3. At P_s, $r = 0$ and $x = 0.725$, that is, z_{in} = normalised input impedance $= 0 + j0.725$.
4. The input impedance,

$$z_i = z_0 \; z_{in}$$

$$= 75\,(j0.725)$$

$$\boxed{z_i = j54.375, \Omega}$$

Analytical method The expression for z_i (for lossless line) is

$$= z_0 \left[\frac{z_L + jz_0 \tan \beta\ell}{z_0 + jz_L \tan \beta\ell}\right]$$

If $\qquad z_L = 0$, short circuit

$$z_i = jz_0 \tan\beta\ell$$

$$= j75 \tan\left(\frac{2\pi}{\lambda}\right)(0.1\lambda)$$

$$= j75 \tan\,(0.2\pi)$$

$$= j75 \tan 36°$$

$$\boxed{z_i = 54.375\Omega}$$

Problem 7.16 Find the input impedance of a 75Ω lossless transmission line of length (0.1λ) if it is terminated in open circuit.

Solution

$$l = 0.1\lambda$$

$$z_L = \infty \text{ (open)}$$

$$z_0 = 75\,\Omega$$

The expression for input impedance is

$$z_{i(\text{lossless})} = z_0 \left[\frac{z_L + jz_0 \tan \beta\ell}{z_0 + jz_L \tan \beta\ell}\right]$$

If the load is open circuit,

$$z_L = \infty$$

$$z_i = \frac{z_0}{j \tan \beta\ell} = -z_0\, j \cot \beta\ell$$

$$= -75j \cot \beta\ell = -75j \cot\left(\frac{2\pi}{\lambda}\right)(0.1\lambda)$$

$$= -j75 \cot(0.2\pi) = -j75 \times 1.3764$$

$$\boxed{z_i = -j103.23\Omega}$$

Smith chart method

1. Start from the point P_{oc} at the right of the rim of the chart.

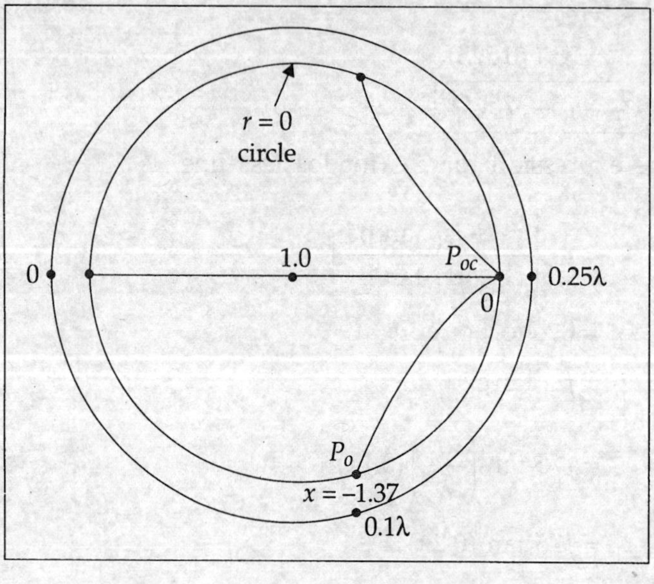

Fig. 7.13 Lossless transmission line terminated in open circuit

2. Move clockwise from P_{oc} through the perimeter of the chart by 0.1λ towards generator. Mark point P_o (Refer Fig. 7.13).
3. At P_o $r=0$, $x=-1.37$, that is, the normalised input impedance.

$$z_{\text{in}} = 0 - j1.37$$

4. Input impedance,

$$z_i = z_0 \ z_{in} = 75\,(-j1.37)$$

$$\boxed{z_i = -j103.23,\ \Omega}$$

Problem 7.17 A transmission line of length 0.40λ has a characteristic impedance of 100Ω and is terminated in a load impedance of $200 + j180\Omega$. Find the

(a) voltage reflection coefficient

(b) voltage standing wave ratio

(c) input impedance of the line

Solution The data is

$$z_L = 200 + j180\ \Omega$$

$$z_0 = 100\ \Omega$$

$$l = 0.4\lambda$$

$$z_n = \frac{z_L}{z_0} = \frac{200 + j180}{100} = 2 + j1.8$$

(a) (i) Refer Fig. 7.14. Locate the point $2.0 + j1.8$ on the Smith chart. It is represented by A.

(ii) Draw a circle of radius equal to OA. This OA is $|\rho|$ and it is equal to 0.591.

(iii) Draw a straight line OA and extend it to B. Read 0.207 in wavelength towards the generator.

(iv) The phase angle of the reflection coefficient is given by

$$(0.250 - 0.207) \times 4\pi = 0.043 \times 4\pi = 31°$$

Hence the reflection coefficient is

$$\boxed{\rho = 0.591\ \angle 31°}$$

(b) The reflection coefficient $|\rho| = 0.591$ circle meets the positive real axis op_{oc} at $r = 4$, that is,

$$\boxed{\text{VSWR} = S = 4}$$

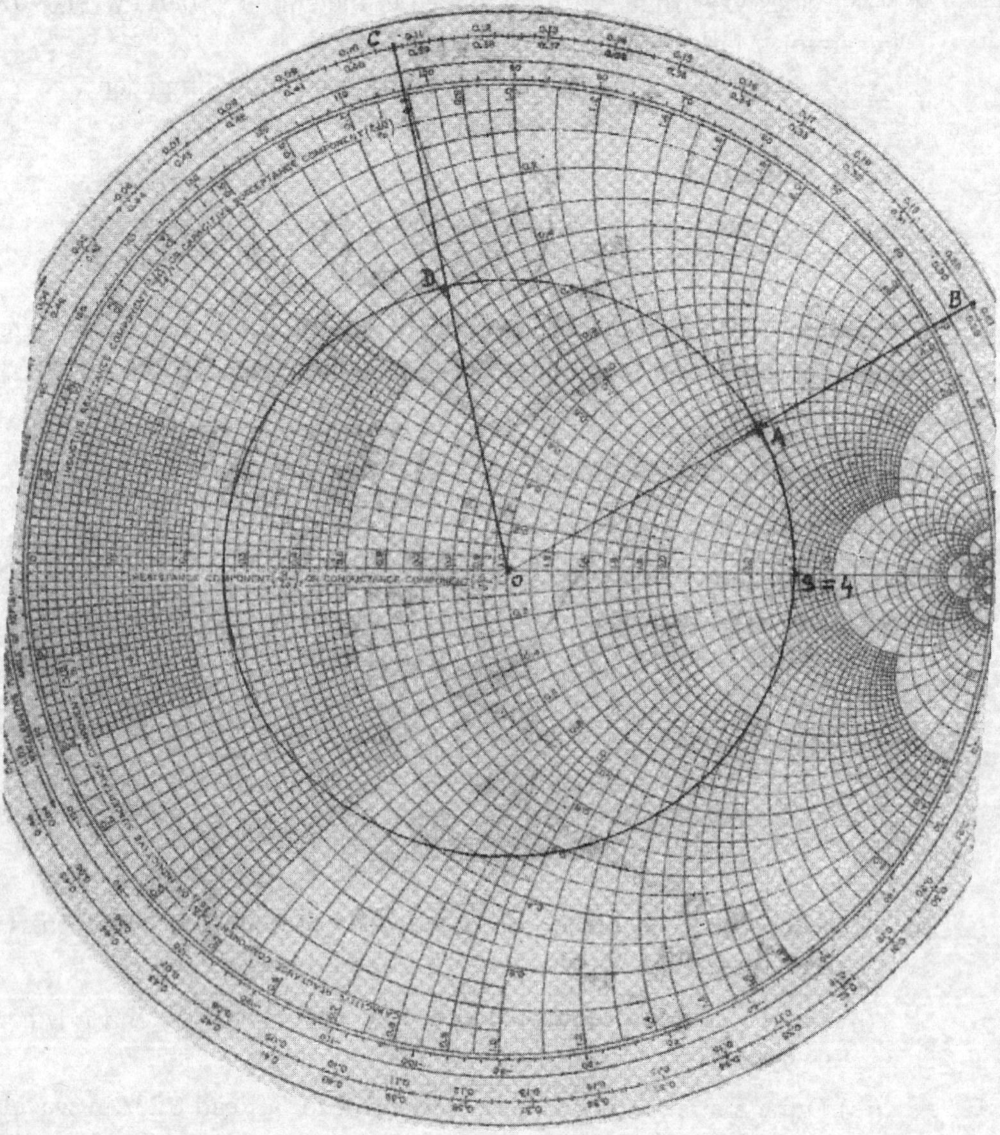

Fig. 7.14

(c) To determine z_i,

(i) Move B at 0.207 by a total of 0.40 wavelengths towards generator. Movement is made from 0.40 to 0.50 to 0.107. This point is represented by C.

(ii) Draw a line joining the centre and point C. This intersects circle at D.

(iii) Read the values of r and x at this point, that is, $r = 0.4$ and $x = 0.72$.

Normalised input impedance

$$= 0.4 + j0.72$$

Input impedance $= z_0\ (0.4 + j0.72) = 100\ (0.4 + j0.72)$

$$\boxed{z_i = 40 + j72,\ \Omega}$$

Problem 7.18 Design a stub to match a transmission line which is connected to a load impedance of $z_L = (450 - j600)\ \Omega$. The characteristic impedance of the line is $300\ \Omega$. The operating frequency is 20 MHz.

Solution $f = 20\ \text{MHz}$

$$\lambda = \frac{\upsilon}{f} = \frac{3\times 10^8}{20\times 10^6} = 15\ \text{m}$$

$$z_L = (450 - j600)\ \Omega$$

$$z_0 = 300\Omega$$

1. The normalised load impedance is

$$z_n = \frac{z_L}{z_0} = \frac{450 - j600}{300} = 1.5 - j2.0$$

2. Identify the point of intersection in Fig. 7.15, $r = 1.5$ and $x = -2.0$.
3. Draw a circle with a radius of OA. It cuts the centre line at 4.8. Therefore, VSWR = 4.8.
4. Draw a line OA and extend it to B. This point, B represents normalised admittance, y_n, that is, $y_n = 0.22 + j0.35$.
5. The drawn circle cuts $r = 1$ circle at C. This point corresponds to $1 + j1.7$.
6. The distance of D to E on the rim of the chart is the stub distance from the load.

   The stub distance $= (0.181 - 0.053)\ \lambda = 0.128\lambda$

   $= 0.128 \times 15 = 1.92\ \text{m}$

7. As the load has a susceptance of $+j1.7$, the stub is required to provide a susceptance of $-j1.7$. Therefore, mark a point by moving clockwise on the lower half of the chart. It is marked by F. Its distance from the short circuit admittance point is given by

   $$0.3342 - 0.25 = 0.0842\lambda$$

   Stub length $= 0.0842\lambda$

   $= 0.0842 \times 15 = 1.263\ \text{m}$

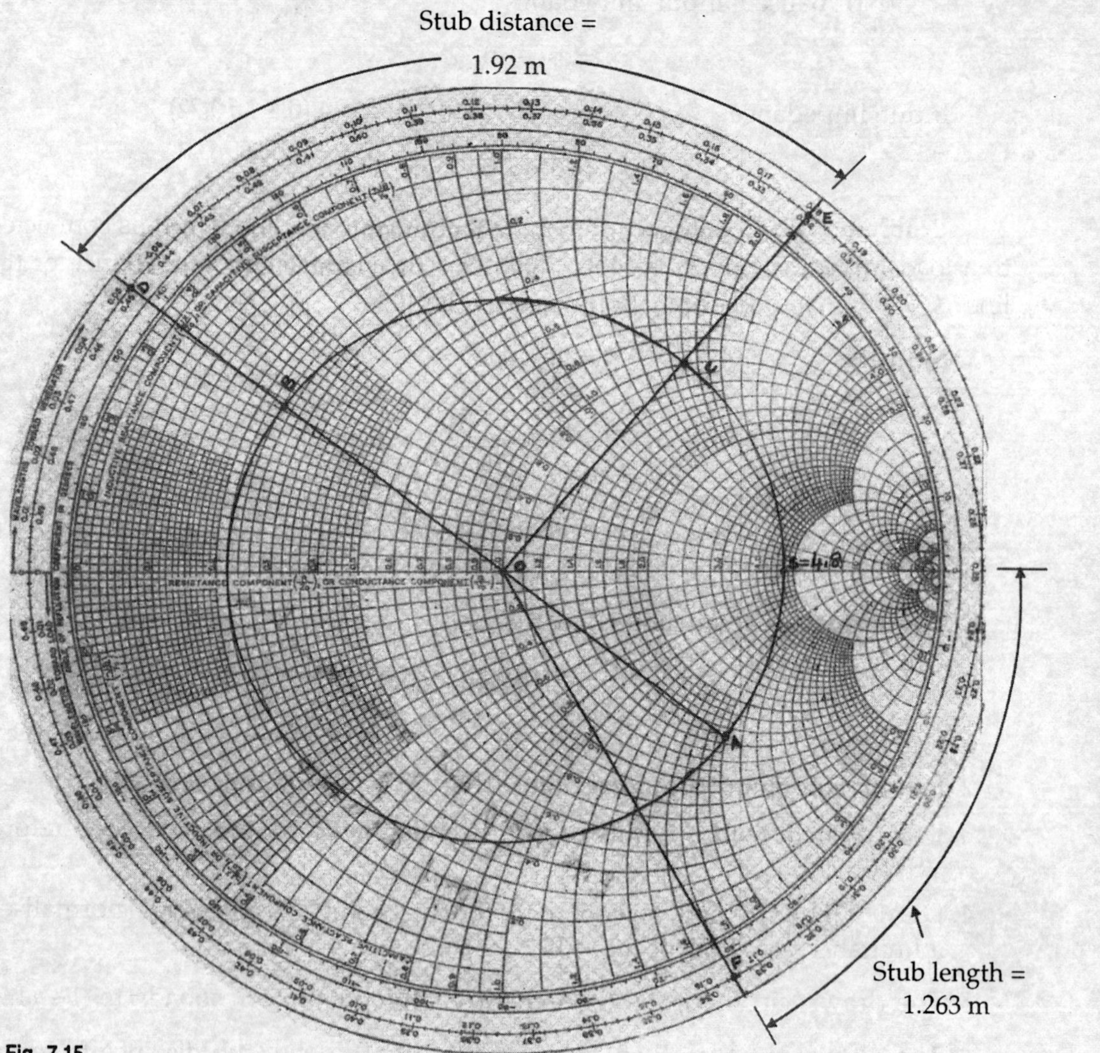

Fig. 7.15

The designed stub parameters are:

Stub length	= 1.263 m
Stub distance	= 1.92 m

7.21 DOUBLE STUBS

For the design of any device, it is convenient to have more parameters in designer's control for more freedom. For this purpose, to match the load with the transmission line, a second stub of adjustable position is included. A typical double stub is shown in Fig. 7.16.

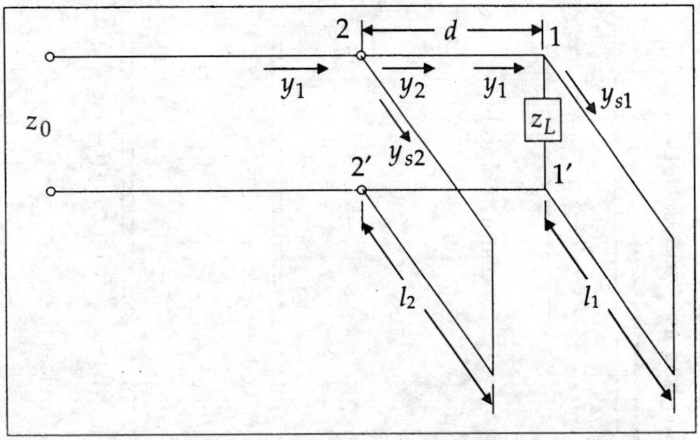

Fig. 7.16 Double stub transmission line

It is also possible to use a triple stub tuner for more design convenience.

In single stub matching the stub is placed on the line at a specified point. Its location varies with z_L and frequency. This creates some difficulties as the specified point may occur at an undesirable location. In such cases, double stubs are used. Here the distance between them is fixed such as $\frac{\lambda}{16}, \frac{\lambda}{8}, \frac{3\lambda}{16}, \frac{3\lambda}{8}$ and so on and the lengths of the two stubs are adjusted to match the load.

Design Methodology

Step 1 Fix the distance between the two stubs and keep stub 1 at the location of the load.

Step 2 Draw the circle corresponding to the normalised conductance, $g = 1$.

Step 3 Obtain the normalised distance of $\frac{d}{\lambda}$.

Step 4 Rotate the circle in anticlockwise direction by $\frac{d}{\lambda}$ wavelengths towards the load and draw. The point which represents y_1 is located here.

Step 5 Locate $y_L = g_L + jb_L$.

Step 6 Draw $g = g_L$ circle. This intersects the rotated $g = 1$ circle at one or two points where $y_L = g_L + jb_1$.

Step 7 Locate the corresponding y_2 points on the $g = 1$ circle. $y_2 = 1 + jb_2$.

Step 8 Find the stub length l_1 between the points representing y_l and y_L.

Step 9 Find the stub length l_2 from the angle between the point representing $-jb_2$ and p_{sc}.

The stub distances from the load need not be found as d is fixed.

In Table 7.5 given on the next page, propagation characteristics of EM waves in free space, in waveguides and in transmission lines are compared.

Table 7.5 Comparison between the Propagation Characteristics of EM Waves in Free Space, Waveguides and Transmission Lines

S. no.	*Parameter*	*Free space*	*Waveguides*	*Transmission lines*
1.	Characteristic impedance	$Z_0 = \frac{\lvert\mathbf{E}\rvert}{\lvert\mathbf{H}\rvert} = \sqrt{\frac{\mu_0}{\epsilon_0}}$	$Z_0 = \frac{V}{I} = \frac{\pi b\eta}{2a\sqrt{1-(f_c^2/f^2)}}$	$Z_0 = \frac{V_f}{I_f} = \sqrt{Z/Y} = \sqrt{\frac{L}{C}}$ for distortionless line
2.	Propagation constant	$\gamma = j\omega\sqrt{\mu\epsilon}$	$\gamma_g = \sqrt{\left(\frac{m\pi}{a}\right)^2 + \left(\frac{n\pi}{b}\right)^2 - \omega^2\mu\epsilon}$	$\gamma_l = \sqrt{ZY}$ $= j\omega\sqrt{LC}$ for lossless line
3.	Attenuation constant	$\alpha = 0$	$\alpha_g = \sqrt{\left(\frac{m\pi}{a}\right)^2 + \left(\frac{n\pi}{b}\right)^2 - \omega^2\mu\epsilon}$	$\alpha_l = \sqrt{\frac{1}{2}(RG-\omega^2LC) + \frac{1}{2}\sqrt{(R^2+\omega^2L^2)(G^2+\omega^2C^2)}}$ $\alpha_l = 0$ for lossless line
4.	Phase constant	$\beta = \omega\sqrt{\mu\epsilon}$	$\beta_g = \sqrt{\omega^2\mu\epsilon - \left(\frac{m\pi}{a}\right)^2 - \left(\frac{n\pi}{b}\right)^2}$	$\beta_l = \sqrt{\frac{1}{2}(\omega^2LC-RG) + \frac{1}{2}\sqrt{(R^2+\omega^2L^2)(G^2+\omega^2C^2)}}$ $= \omega\sqrt{LC}$ for lossless line
5.	Velocity of propagation	$\upsilon_0 = \frac{1}{\sqrt{\mu_0\epsilon_0}}$	$\upsilon_g = \frac{\upsilon_0^2}{\upsilon_p}$	$\upsilon_l = \frac{\upsilon_0^2}{\upsilon_p}$
6.	Phase velocity	$\upsilon_p = \upsilon_0$	$\upsilon_g = \frac{\omega}{\beta_g}$	$\upsilon_p = \sqrt{LC}$ $= \frac{\omega}{\beta_l}$ for lossless line
7.	Wavelength	$\lambda = \frac{2\pi}{\omega\sqrt{\mu_0\epsilon_0}}$	$\lambda_g = \frac{\lambda}{\sqrt{1-\left(\frac{\lambda}{\lambda_c}\right)^2}}$	$\lambda_l = \frac{2\pi}{\beta_l}$

POINTS/FORMULAE TO REMEMBER

- Transmission lines transfer information from one point to another.
- They are characterised by the presence of distributed constants R, L, G and C.
- Transmission lines are used as impedance matching devices, stubs.
- Coaxial cables are used to connect TVs to antennas.
- Microstrips are used in integrated circuits.
- Twisted pairs and coaxial cables are used in computer networks such as Ethernet.
- Optical fibres are characterised by negligible attenuation and infinite bandwidth.
- Equivalent circuit of a uniform transmission line is shown in Fig. 7.17.

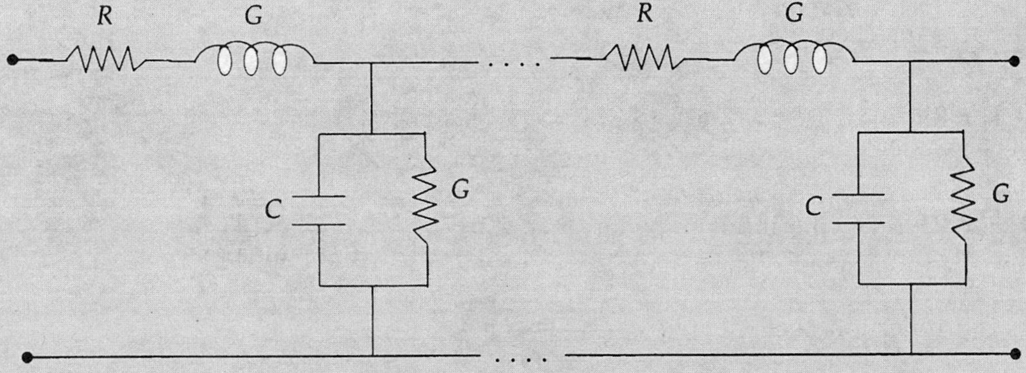

Fig. 7.17

- The characteristic impedance of a pair of transmission lines is $Z_0 = \sqrt{Z/Y}$.
- The propagation constant of a transmission line is $\gamma_l = \sqrt{ZY} = \alpha_l + j\beta_l$.
- For a lossless transmission line $R = 0$, $G = 0$, $Z_0 = \sqrt{\dfrac{L}{C}}$.
- For a distortionless transmission line,

$$\frac{R}{L} = \frac{G}{C}$$

$$\alpha_l = \sqrt{RG},\ \beta_l = \omega\sqrt{LC}$$

$$Z_0 = \sqrt{\frac{R}{G}} = \sqrt{\frac{L}{C}} = R_0$$

- For a lossless and distortionless transmission line, $\upsilon_p = \dfrac{1}{\sqrt{LC}}$.
- Loading of transmission lines is the introduction of inductance in series with the line.
- Input impedance of transmission line is $Z_i = Z_0\left[\dfrac{Z_L + Z_0\, j\tanh\gamma l}{Z_0 + Z_L\, j\tanh\gamma l}\right]$.
- Input impedance of a lossless transmission line is $Z_i = R_0\left[\dfrac{Z_L + jR_0\tan\beta l}{R_0 + jZ_L\tan\beta l}\right]$.
- For a matched line $Z_i = Z_0$.
- For RF lines, $\omega L >> R$, $\omega C >> G$, $\alpha = \dfrac{1}{2}\left(\dfrac{R}{Z_0} + GZ_0\right)$.
- For RF lines, the input impedance is $Z_i = Z_0\left[\dfrac{Z_L + jZ_0\tan\beta l}{Z_0 + jZ_L\tan\beta l}\right]$.
- The reflection coefficient, $\rho = \dfrac{Z_L - Z_0}{Z_L + Z_0}$
- VSWR $= \dfrac{1+|\rho|}{1-|\rho|}$
- VSWR varies between 1 and ∞.
- For resistance terminated load,

$$S = \frac{Z_0}{R_L} \text{ if } Z_0 > R_L$$

$$S = \frac{R_L}{Z_0} \text{ if } R_L > Z_0$$

- The losses in transmission lines are due to copper, dielectric and radiation.
- Smith chart is useful to measure transmission line parameters.

- The design parameters of a single stub are its length and its distance from the load end.
- Velocity of propagation in lossless line is $\upsilon_p = \frac{1}{\sqrt{LC}}$.
- VSWR is defined as $\frac{V_{\max}}{V_{\min}}$.
- Characteristic impedance is defined as $\frac{V_f}{I_f}$ or $-\frac{V_r}{I_r}$.
- Reflection coefficient is defined as $\frac{V_r}{V_f}$.

OBJECTIVE QUESTIONS

1. Guided waves require conductors for their existence. (Yes/No)
2. Microstrip can be used as a transmission line. (Yes/No)
3. A transmission line can be used as a resonant circuit. (Yes/No)
4. A transmission line can be used as a filter. (Yes/No)
5. A transmission line can be used as a waveshaping network. (Yes/No)
6. The normalised input impedance of a transmission line, $z_i = \frac{1}{z_L} \cdot z_L$ is normalised load impedance. (Yes/No)
7. The input impedance of a transmission line is a function of its length. (Yes/No)
8. VSWR $= \frac{1+|\rho|}{1-|\rho|}$ (Yes/No)
9. VSWR $= \frac{1+\rho}{1-\rho}$ (Yes/No)
10. VSWR $= \frac{1-|\rho|}{1+|\rho|}$ (Yes/No)
11. A stub can be shorted at one end. (Yes/No)
12. A stub can be open ended at one end. (Yes/No)
13. The magnitude of voltage along the perfectly matched transmission line is

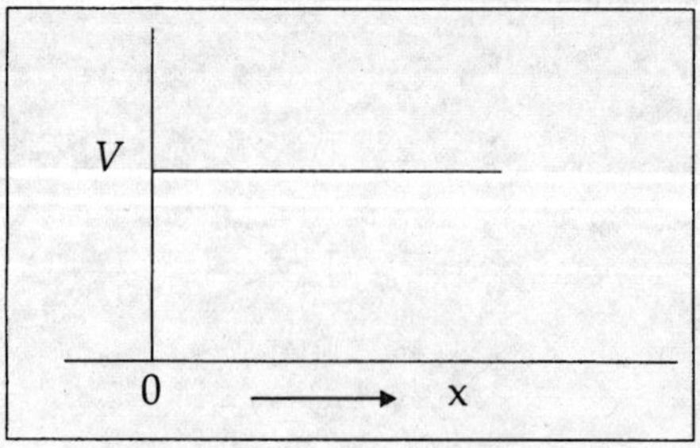

(Yes/No)

Fig. 7.18

14. The inductance of a transmission line depends on the permeability of the medium between the conductors. (Yes/No)

15. Capacitance of a transmission line depends on the dielectric constant of the medium between the conductors. (Yes/No)

16. The resistance in a transmission line depends on $\in_r$ of the medium between the conductors. (Yes/No)

17. The line resistance is expressed in units of ohm/m. (Yes/No)

18. Step function is a long pulse. (Yes/No)

19. Impulse function is a very short pulse. (Yes/No)

20. Principal wave is ________.

21. To exist, TEM wave requires ________.

22. A line is said to be a uniform transmission line ________.

23. The length of a microstrip line is usually ________.

24. The analysis of a transmission line can be made by ________.

25. A transmission line is said to be lossless ________.

26. The input impedance of a transmission line is ________ if there is a short circuit at the receiving end of the line.

27. For quarter-wavelength line, the input impedance at the sending end is ________.

28. The input impedance of one-half wavelength long line is ________.

29. A transmission line is said to have a discontinuity at a point if ________.

30. The reflection coefficient of a transmission line for a load impedance of z_L is ________.

31. The transmission coefficient of a transmission line for a load impedance of z_L is ________.

32. The transmission coefficient of a transmission line is ________.

33. VSWR = ________

34. A stub is a ________.

35. VSWR of a perfectly matched transmission line is ________.

36. The input impedance of a transmission line is ________.

37. The characteristic impedance, z_0 of a transmission line is ________ of forward voltage to current.

38. The transit time in a transmission line is ________.

39. The transit time is given by ________.

40. If VSWR $=2$, the magnitude of reflection coefficient is ________.

41. The characteristic impedance of a lossless transmission line is ________.

42. The propagation constant in a transmission line is ________.

43. The propagation constant in a lossless transmission line is ________.

44. Tracking of forward and backward waves along a transmission line is done by ________.

45. Lattice diagram is a ________.

46. Microstrip transmission lines are used in ________.

47. The phase constant of a transmission line is ________.

48. The velocity of propagation in a transmission line is ________.

49. If z_L is pure resistance and is greater then z_0, VSWR is ________.

50. If z_L is purely resistive and is less than z_0, VSWR is ________.

51. A practical transmission line has propagation constant equal to ________.

Answers

1. Yes	2. Yes	3. Yes	4. Yes	5. Yes
6. Yes	7. Yes	8. Yes	9. No	10. No
11. Yes	12. Yes	13. Yes	14. Yes	15. Yes
16. No	17. Yes	18. Yes	19. Yes	20. TEM wave

21. Two or more conductors

22. If the conductors are separated by the same dielectric, if they have same cross-sectional area along the length of the line and if they have same dimensions

23. A few mm 24. Smith chart

25. If the conductors have $\sigma = \infty$ and if the dielectric between the conductors has $\sigma = 0$

26. $z_o \tanh rl$ 27. $z_i = z_0^2 / z_L$ 28. $z_i = z_L$

29. z_0 changes at the point 30. $\rho = \dfrac{(z_L - z_0)}{(z_L + z_0)}$ 31. $T = \dfrac{2z_L}{z_L + z_o}$

32. $1 + \rho$ 33. $\dfrac{V_{\max}}{V_{\min}}$ 34. A piece of transmission line 35. One

36. $z_i = \dfrac{V_s}{I_s}$ 37. The ratio

38. The time taken for the wave to travel from one end to the other

39. $t_t = \dfrac{l}{\upsilon}$, l = length of line, υ is the velocity of propagation of the wave

40. 1/3 41. $z_0 = \sqrt{\dfrac{L}{C}}$ 42. $\sqrt{ZY}$ 43. $j\omega\sqrt{LC}$

44. Lattice diagram 45. A time-distance diagram 46. PCBs

47. $\beta = \omega\sqrt{LC}$ 48. $\upsilon_l = \dfrac{1}{\sqrt{LC}}$ 49. $\dfrac{z_L}{z_0}$ 50. $\dfrac{z_0}{z_L}$ 51. $\alpha + j\beta$

MULTIPLE CHOICE QUESTIONS

1. If the load impedance in a transmission line is $100 + j200\Omega$ and characteristic impedance is 100Ω, the normalised load impedance is
   (a) $1 + j2\Omega$ (b) $10{,}000 + j20{,}000\Omega$
   (c) $1 + j200\Omega$ (d) $100 + j2\Omega$

2. If the normalised load impedance of a transmission line is $3 + j4\Omega$, the normalised admittance is
   (a) $0.6 - j0.8$ mho (b) $1 - j1.0$ mho
   (c) $1 + j1.0$ mho (d) $0.6 - 0.8$ mho

3. If $z_0 = 50\Omega$, $z_L = 50 + 100j$ for a quarter wave transmission line, its input impedance is
   (a) $\dfrac{2{,}500}{50 + j100}$ (b) $\dfrac{50}{50 + j100}$ (c) $\dfrac{50 + j100}{2{,}500}$ (d) $\dfrac{50 + j100}{50}$

4. If the load impedance of one half-wavelength is $50 + j150\Omega$, its input impedance is
   (a) $50 - j150\Omega$ (b) $50 + j150\Omega$
   (c) $50 + j100\Omega$ (d) $1 + j1.5\Omega$

5. If the load impedance in a transmission line is z_L and z_0 is the characteristic impedance, reflection coefficient is
   (a) $\dfrac{(z_L - z_0)}{(z_L + z_0)}$ (b) $\dfrac{(z_L + z_0)}{(z_L - z_0)}$ (c) $\dfrac{z_L}{z_0}$ (d) $\dfrac{z_0}{z_L}$

6. If reflection coefficient in a transmission line for a given load is $0.5 + j0.5$, VSWR is
   (a) 1 (b) ∞ (c) 2 (d) $-\infty$

7. If maximum and minimum voltages on a transmission line are 4 V and 2 V respectively, VSWR is
   (a) 0.5 (b) 2 (c) 1 (d) 8

8. If the sending voltage and currents on a transmission line are 200 V and 2 amp for a given load the input impedance is
   (a) 100Ω (b) ∞ (c) 0.01Ω (d) 200Ω

9. If the voltage and current at the receiving end of a transmission line for a given load are 3.0 V and 200 mA respectively, the load impedance is given by

(a) 600Ω (b) 150Ω (c) 1500Ω (d) 15Ω

10. A lossless transmission line characterised by a distributed inductance of 300 nH/m and capacitance of 50 PF/m operates at no load. Its characteristic impedance is

(a) 60Ω (b) 600Ω (c) 77.45Ω (d) 774.5Ω

11. The velocity of propagation of a wave along a transmission line of length 100 m is 2.8×10^8 m/s. The delay on the transmission line is

(a) 3.57 ns (b) 357 ns (c) 0.357 ns (d) 2.8 μs

12. A 100 km long transmission line has an inductance of 27 mH. Its distributed inductance per metre is

(a) 27 mH (b) 2.7μH (c) 0.27μH (d) 27μH

13. A 100 km long transmission line has a capacitance of 20 nF. Its distributed capacitance per metre is

(a) 0.20 PF (b) 20 PF (c) 20 nH (d) 0.20 nF

14. A 100 m long lossless transmission line is operating at 100 kHz. If the velocity propagation of the wave on the line is 1.4×10^8 m/s, the propagation constant is

(a) 4.487 rad/m (b) 4.487 m rad/m

(c) 44 m rad/m (d) 0.4487 m rad/m

15. A voltage of 2 cos 10^5 t volts is applied to a parallel plate transmission line. If the plate separation is 2 mm, the electric field between the plates at the source end is

(a) 1,000 cos $10^5\,t$ V/m (b) cos $10^5\,t$ V/m

(c) 10 cos $10^5\,t$ V/m (d) cos $10^5\,t$ mV/m

16. A sinusoidal voltage with a wavelength of 100 cm is applied to a transmission line along which the velocity of propagation of the wave is 2.9×10^8 m/s. The frequency of the source is

(a) 2.9 MHz (b) 2.9 GHz

(c) 0.29 GHz (d) 0.29 MHz

17. If the magnitude of the reflection coefficient on a transmission line for a given load is 1/3, VSWR is

(a) 3 (b) 2 (c) 1.3 (d) 0.5

18. A 50Ω transmission line is connected to a load impedance yielding a VSWR of unity. The load impedance is

(a) 50Ω (b) 100Ω (c) 1Ω (d) 0Ω

19. A load impedance of 100Ω is connected to a 50Ω line. VSWR of unity is obtained by connecting

(a) another 50Ω in series with z_L (b) another 50Ω in parallel to z_L

(c) another 100Ω in parallel to z_z (d) another 100Ω in series with z_L

20. If the reflection coefficient at a point on a transmission line is – 0.5, the transmission coefficient is

(a) 0.5 (b) – 0.5 (c) 1.0 (d) 0

Answers

1. (a)	2. (a)	3. (a)	4. (b)	5. (a)
6. (b)	7. (b)	8. (a)	9. (b)	10. (c)
11. (b)	12. (c)	13. (a)	14. (b)	15. (a)
16. (c)	17. (b)	18. (a)	19. (c)	20. (a)

EXERCISE PROBLEMS

1. A parallel wire transmission line is made of copper. The separation between the wires is 1.0 m in air. The conductor radius is 2.0 mm. Find *L*, *C* and *G*.

2. A copper parallel wire transmission line operates at 1 MHz. For copper $\mu_c = \mu_0$, $\epsilon_c = \epsilon_0$, $\sigma_c = 5.8 \times 10^7$ mho/m. The radius of the wire, $a = 2.0$ mm. Find R_{dc} and R_{ac}.

3. A transmission line is characterised by
$$R = 10^{-2}\ \Omega/\text{m},\ \ G = 1.0\,\mu\ \text{mho/m and } C = 1.0\ \text{nF/m}.$$
Determine the characteristic impedance if the operating frequency is 1.59 kHz.

4. A transmission line has
$$R = 0.01\,\Omega/\text{m},\ \ G = 1.0\,\mu\ \text{mho/m},\ \ L = 10\,\mu\text{H/m},\ \ C = 1.0\,\mu\ \text{F/m}.$$
Find the attenuation constant, phase constant and phase velocity.

5. A telephone line of 100 km has $R = 4\Omega/\text{km}$, $L = 3\,\text{mH/km}$, $G = 1.0\mu$ mho/km, $C = 0.015\mu$ F/km. It operates at 796 Hz and is connected to a load impedance of 200Ω. Find the series impedance, shunt admittance, characteristic impedance and reflection coefficient.

6. A 100 km telephone line has $R = 4\Omega/\text{km}$, $L = 3\,\text{mH/km}$, $G = 1.0\mu$ mho/m and $C = 15\,\text{nF/m}$. It operates at $f = 796\,\text{Hz}$. Find the attenuation and phase constants.

7. A coaxial cable has air dielectric. Its outer radius is 2.0 mm and inner radius is 1.0 mm. Find its characteristic impedance.

8. A pair of transmission lines are separated by 1.0 m and each wire has a radius of 1 cm. Find z_0.

9. A uniform transmission line operating at 10 kHz has z_0 of 50Ω and propagation constant, $\gamma = (0.1 + j0.1)\ \text{m}^{-1}$. Find *R* and *L*.

10. A uniform transmission line operating at 1 kHz has $z_0 = 75\Omega$ and a propagation constant of $(0.1 + j0.2)\ \text{m}^{-1}$. Find *G* and *C*.

11. For a uniform transmission line, the open and short circuit impedances are given by $z_{oc} = 50 + j25\Omega$, $z_{sc} = 60 - j20\Omega$. Find z_0 of the line.

CHAPTER

8

RADIATION AND ANTENNAS

Antenna is a radiator and sensor of EM waves. It is also a transducer and impedance matching device. It can be designed to direct EM energy in desired directions and suppress it in unwanted directions.

The main aim of this chapter is to provide the fundamentals of antennas. They include:

- antenna definition, functions, properties and parameters
- basic antenna elements
- radiation mechanism, radiated power and radiation resistance of current elements, dipoles and monopoles
- directional characteristics of basic elements
- solved problems, points/formulae to remember, objective and multiple choice questions and exercise problems.

Do you know?

The Mobile telephone, TV, radio, satellite, or any type of modern communication is not possible without one type of antenna or the other and no radar works without an antenna.

EMF theory is essential for solving all types of antenna and free space problems.

Broadcast antennas transmit at various radio frequencies, depending on the channel, from about 550 kHz for AM radio upto 800 MHz for some UHF television stations. Frequencies for FM radio and VHF television lie in between these two.

8.1 GENERAL SOLUTION OF MAXWELL'S EQUATIONS

The general Maxwell's equations and retarded potentials are repeated here for the sake of continuity and clarity.

$$\nabla \times \mathbf{H} = \dot{\mathbf{D}} + \mathbf{J}$$
$$\nabla \times \mathbf{E} = -\dot{\mathbf{B}}$$
$$\nabla . \mathbf{D} = \rho_v$$
$$\nabla . \mathbf{B} = 0$$

where $\mathbf{H}$ = magnetic field, (A/m)

$\mathbf{D}$ = displacement electric flux density (c/m^2)

$$\dot{\mathbf{D}} = \frac{\partial \mathbf{D}}{\partial t} = \text{displacement electric current density (A/m}^2)$$

$\mathbf{J}$ = conduction current density (A/m^2)

$\mathbf{E}$ = electric field (V/m)

$\mathbf{B}$ = magnetic flux density (wb/m^2)

$$\dot{\mathbf{B}} = \frac{\partial \mathbf{B}}{\partial t} = \text{magnetic current density (V/m}^2)$$

From these equations, it is possible to obtain general solutions for **E** and **H** fields in terms of potentials and also expressions relating potentials and their sources.

8.2 EXPRESSIONS FOR E AND H IN TERMS OF POTENTIALS

By definition, vector magnetic potential, **A** is given by

$$\nabla \times \mathbf{A} = \mathbf{B} = \mu \mathbf{H}$$

or,

$$\nabla \times \dot{\mathbf{A}} = \mu \dot{\mathbf{H}}$$

But

$$\mu \dot{\mathbf{H}} = -\nabla \times \mathbf{E}$$

$$\nabla \times \dot{\mathbf{A}} = -\nabla \times \mathbf{E}$$

or,

$$\nabla \times (\mathbf{E} + \dot{\mathbf{A}}) = 0$$

This is true only if $(\mathbf{E} + \dot{\mathbf{A}})$ represents the gradient of a scalar. Hence,

$$\mathbf{E} + \dot{\mathbf{A}} = -\nabla V$$

$$\mathbf{E} = -\nabla V - \dot{\mathbf{A}}$$

And we know

$$\mathbf{B} = \nabla \times \mathbf{A}$$

or,
$$\mathbf{H} = \frac{1}{\mu} \nabla \times \mathbf{A}$$

Therefore, the general expressions for **E** and **H** are given by

$$\mathbf{E} = -\nabla V - \dot{\mathbf{A}}$$
$$\mathbf{H} = \frac{1}{\mu} \nabla \times \mathbf{A}$$

Expressions Relating Potentials and Their Sources

Consider
$$\nabla \times \mathbf{A} = \mathbf{B}$$
$$\nabla \times \nabla \times \mathbf{A} = \mu \nabla \times \mathbf{H}$$

or,
$$\nabla \times \nabla \times \mathbf{A} = \mu \in \dot{\mathbf{E}} + \mu \mathbf{J}$$

or,
$$\nabla \nabla . \mathbf{A} - \nabla^2 \mathbf{A} = \mu \in [-\nabla \dot{V} - \dot{\mathbf{A}}] + \mu \mathbf{J}$$
$$= -\mu \in \nabla \dot{V} - \mu \in \dot{\mathbf{A}} + \mu \mathbf{J}$$

Curl of **A** is known and its divergence is not specified.

Helmholtz Theorem

This states that any vector like **A** has unique meaning only if its curl and divergence are specified.

Hence, the divergence of **A** is

$$\nabla . \mathbf{A} = -\mu \in \dot{V}$$

This is known as Lorentz gauge condition.

Using Lorentz gauge condition, we get

$$\nabla(-\mu \in \dot{V}) - \nabla^2 \mathbf{A} = -\mu \in \nabla \dot{V} - \mu \in \dot{\mathbf{A}} + \mu \mathbf{J}$$

or,
$$\nabla^2 \mathbf{A} - \mu \in \dot{\mathbf{A}} = -\mu \mathbf{J}$$

Now consider
$$\nabla . \mathbf{D} = \rho_\upsilon$$

or,
$$\nabla . \mathbf{E} = \rho_\upsilon / \in$$

$$\nabla.(-\nabla V - \dot{\mathbf{A}}) = \rho_\upsilon / \in$$

or,
$$\nabla^2 V + \nabla.\dot{\mathbf{A}} = -\rho_\upsilon / \in$$

$$\nabla^2 V - \mu \in \ddot{V} = -\rho_\upsilon / \in$$

The final expressions are:

$$\nabla^2 \mathbf{A} - \mu \in \ddot{\mathbf{A}} = -\mu \mathbf{J}$$

$$\nabla^2 V - \mu \in \ddot{V} = -\frac{\rho_\upsilon}{\in}$$

For sinusoidal fields, these equations become

$$\nabla^2 \mathbf{A} + \omega^2 \mu \in \mathbf{A} = -\mu \mathbf{J}$$

$$\nabla^2 V + \omega^2 \mu \in V = -\frac{\rho_\upsilon}{\in}$$

8.3 RETARDED POTENTIALS

The potentials for static fields are

$$V(r) = \frac{1}{4\pi\in} \int_\upsilon \frac{\rho_\upsilon(r)}{r} d\upsilon$$

$$\mathbf{A}(r) = \frac{\mu}{4\pi} \int_\upsilon \frac{\mathbf{J}(r)}{r} d\upsilon$$

But for time varying fields, they can be written as,

$$V(r, t) = \frac{1}{4\pi\in} \int_\upsilon \frac{\rho_\upsilon(r, t)}{r} d\upsilon$$

$$\mathbf{A}(r, t) = \frac{\mu}{4\pi} \int_\upsilon \frac{\mathbf{J}(r, t)}{r} d\upsilon$$

The potentials are usually established due to time varying field only after some amount of propagation time. This propagation time depends on the distance between the point of the potentials from their sources and velocity of propagation of EM fields. As a result, the potentials are retarded by a time, r/υ_0. These potentials are known as **retarded potentials**, and they are expressed as

$$V(r, t) = \frac{1}{4\pi\epsilon} \int_{\upsilon} \frac{\rho_\upsilon(r, t - r/\upsilon_0)}{r} \, d\upsilon$$

$$\mathbf{A}(r, t) = \frac{\mu}{4\pi} \int_{\upsilon} \frac{\mathbf{J}(r, t - r/\upsilon_0)}{r} \, d\upsilon$$

It is well known that a uniform plane wave propagating in r direction has a phase variation represented by $e^{-j\beta r}$. Including this phase factor, the above potentials are given by

$$V(r, t) = \frac{1}{4\pi\epsilon} \int_{\upsilon} \frac{\rho_\upsilon(r, t - r/\upsilon_0)}{r} e^{-j\beta r} \, d\upsilon$$

$$\mathbf{A}(r, t) = \frac{\mu}{4\pi} \int_{\upsilon} \frac{\mathbf{J}(r, t - r/\upsilon_0)}{r} e^{-j\beta r} \, d\upsilon$$

8.4 ANTENNA DEFINITION

Antenna or Aerial means the same.

An antenna is defined in the following ways:

1. An antenna is a piece of conducting wire or rod with excitation.
2. An antenna is a source or radiator of Electromagnetic waves.
3. An antenna is a sensor of Electromagnetic waves.
4. An antenna is a transducer.
5. An antenna is an impedance matching device.

8.5 FUNCTIONS OF AN ANTENNA

1. It is used as a transducer, that is, it converts electrical energy into EM energy on the transmitting side and it converts EM energy into electrical energy on the receiving side.
2. It is used as an impedance matching device, that is, it matches the transmitter and free space on the transmitting side and it matches free space and the receiver on the receiving side.
3. It radiates in the desired directions and suppresses in the unwanted directions.

4. It is used as a radiator of EM waves.
5. It is used as a sensor of EM waves.

8.6 PROPERTIES OF AN ANTENNA

1. It has identical impedance when used for transmitting and receiving purposes.
2. It has identical directional characteristics when it is used for transmitting and receiving purposes.
3. It has the same effective length when it is used for transmitting and receiving purposes.

These properties can be proved using reciprocity theorem.

8.7 ANTENNA PARAMETERS

1. **Antenna Impedance, Z_a** It is defined as the ratio of input voltage to input current, that is,

$$Z_a \equiv \frac{V_i}{I_i}\ \Omega$$

Z_a is a complex quantity and it is written as

$$Z_a = R_a + jX_a$$

Here, the reactive part X_a results from fields surrounding the antenna. The resistive part, R_a is given by

$$R_a = R_l + R_r$$

Here R_l represents losses in the antenna. R_r is called radiation resistance.

2. **Radiation Resistance, R_r** R_r is defined as the fictitious or hypothetical resistance that would dissipate an amount of power equal to the radiated power.

or,
$$R_r \equiv \frac{\text{power radiated}}{I_{\text{RMS}}^2}$$

3. **Directional Characteristics** These are also called radiation characteristics or radiation pattern. These are of two types:

   (a) **Field strength pattern** It is the variation of the absolute value of field strength as a function of θ.

   That is, $\boxed{E\ vs\ \theta}$ is called field strength pattern.

(*b*) **Power pattern** It is the variation of radiated power with θ.

That is, $\boxed{P \text{ } vs \text{ } \theta}$ is called Power pattern.

More generally an antenna radiation pattern is a three dimensional variation of the radiation field.

4. **Effective Length of Antenna, (L_{eff})** It is used to indicate the effectiveness of the antenna as a radiator or receiver of EM energy.

L_{eff} of transmitting antenna It is that length of an equivalent linear antenna that has a constant current along its length and which radiates the same field strength as the actual antenna.

Refer Fig. 8.1.

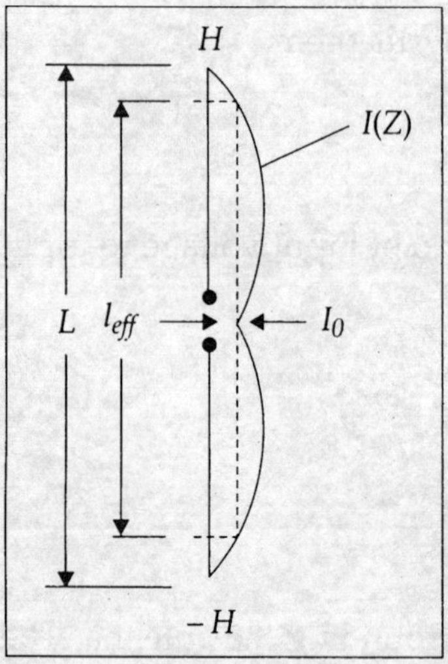

Fig. 8.1 Definition of effective length of transmitting antenna

L_{eff} of transmitting antenna is defined mathematically as

$$l_{eff}\,(Tx) \equiv \frac{1}{I_0} \int_{-H}^{H} I\,(z)\,dz$$

L_{eff} of receiving antenna It is defined as the ratio of the open circuit voltage developed at the terminals of the antenna under the received field strength, **E**, that is,

$$l_{eff}\,(\text{rec}) \equiv \frac{V_{0C}}{E}$$

Effective length of an antenna is always less than the actual length.

5. **Radiation Intensity, (*RI*)** It is defined as the power radiated in a given direction per unit solid angle, that is,

$$RI = r^2 P = \frac{r^2 E^2}{\eta_0} \text{ watts/unit solid angle}$$

Here η_0 = intrinsic impedance of the medium, (Ω)

r = radius of the sphere, (m)

P = power radiated-instantaneous

E = electric field strength, (V/m)

$RI = RI\,(\theta, \phi)$ is a function of θ and ϕ

6. **Directive Gain, (g_d)** It is defined as the ratio of radiation intensity in that direction to the average radiation intensity, that is,

$$g_d \equiv \frac{RI}{(RI)_{av}} = \frac{RI}{w_r/4\pi}$$

or,

$$g_d = \frac{4\pi\,(RI)}{w_r}$$

w_r = radiated power

7. **Directivity, D** It is defined as the ratio of the maximum radiation intensity to the average radiation intensity, that is,

$$D \equiv (g_d)_{\max}$$

$$D \text{ in dB} = 10 \log (g_d)_{\max}$$

8. **Power Gain, (g_p)** It is defined as the ratio of radiated power to the total input power, that is,

$$g_p \equiv \frac{4\pi\,(RI)}{w_t}$$

where $w_t = w_r + w_l$,

w_l = ohmic losses in the antenna

9. **Antenna Efficiency, (η)** It is defined as the ratio of radiated power to the input power, that is,

$$\eta \equiv \frac{w_r}{w_t} = \frac{w_r}{w_r + w_l} = \frac{g_p}{g_d}$$

10. **Effective Area** It is defined as

$$A_e \equiv \frac{\lambda^2}{4\pi}\,g_d$$

or,

$$A_e \equiv \frac{w_R}{P}$$

where w_R = received power (watt)

p = power flow per square metre (watts/m^2) for the incident wave

11. **Antenna Equivalent Circuit** It is a series R_a, L_a and C_a circuit (Fig. 8.2).

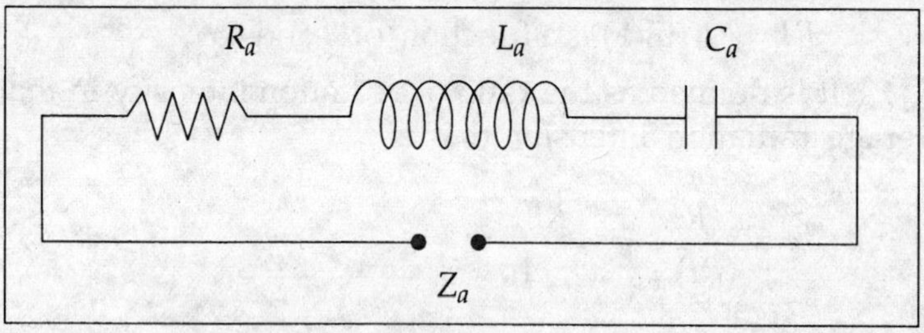

Fig. 8.2 Antenna equivalent circuit

The main difference between the antenna equivalent circuit and an *RLC* circuit is that R_a, L_a and C_a vary with frequency. As a result, the antenna conductance peak appears not at resonant frequency but at a frequency slightly away from f_r (Fig. 8.3).

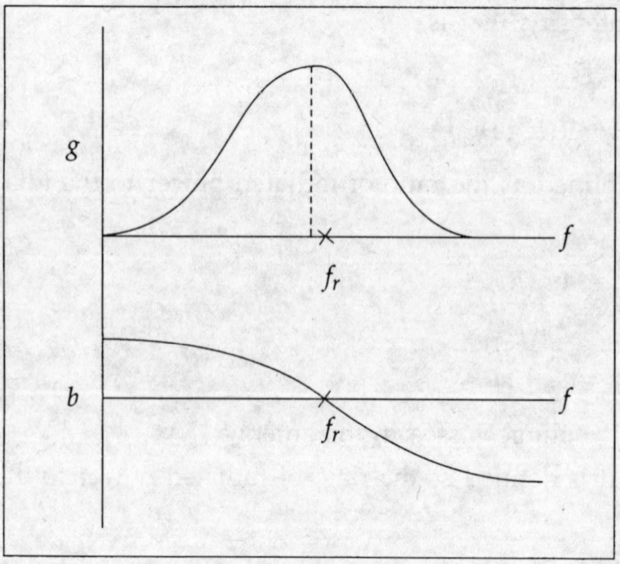

Fig. 8.3 Antenna conductance and susceptance variation

8.8 BASIC ANTENNA ELEMENTS

The basic antenna elements are:

1. Alternating current element or Hertzian dipole
2. Short dipole
3. Short monopole

4. Half wave dipole
5. Quarter wave monopole

1. **Alternating Current Element or Hertzian Dipole** It is a very short linear antenna in which the current along its length is approximately constant.

2. **Short Dipole** It is a linear antenna whose length is less than $\frac{\lambda}{4}$ and the approximate current distribution is triangular.

3. **Short Monopole** It is a linear antenna whose length is less than $\frac{\lambda}{8}$ and the approximate current distribution is triangular.

4. **Half Wave Dipole** It is a linear antenna whose length is $\frac{\lambda}{2}$ and the current distribution is sinusoidal.

5. **Quarter Wave Monopole** It is a linear antenna whose length is $\frac{\lambda}{4}$ and the current distribution is sinusoidal.

8.9 RADIATION MECHANISM

When a transmitting antenna is excited with an alternating voltage, the initial motion of a wave which is propagated through space is started by the balanced motion of charges in the antenna. The transmitting antenna has characteristics similar to those of a resonant circuit.

When energy is supplied to it, resonant oscillations occur in the antenna and violent variations in charge form an electric vector. The same violent motions of charges create a magnetic field about the antenna in the same manner as a magnetic field expands and collapses about a resonant circuit tank coil.

If energy is continuously applied to the antenna, energy moves away from the antenna into space in the form of EM waves. It may be noted that it is not the original antenna charges themselves that move through space but rather the motion they create. The charges around the antenna are set in motion first and they, in turn, set other charges further separated from the antenna into motion. This disturbance fans out from the antenna into space.

Wave motion of charges forms an electric field and a magnetic field which result from the motion of charges. These electric and magnetic fields are perpendicular to each other. The motion of these fields in the form EM waves has no boundaries and expand spherically. The EM energy density decreases with distance as they propagate.

8.10 RADIATION FIELDS OF AN ALTERNATING CURRENT ELEMENT (OR OSCILLATING ELECTRIC DIPOLE)

The concept of an alternating current element, $I\,dl \cos \omega t$ is of theoretical interest. But the theory developed for this can be extended to practical antennas. To derive radiation fields of antenna elements including current element, the concept of retarded vector magnetic potential is very useful.

Derivation of radiation fields consists of the following steps:

1. Write expression for retarded vector magnetic potential.
2. Write expressions for the components of **A** in Cartesian coordinates.
3. Express **A** in the components of the spherical coordinate system.
4. Obtain the components of **H** from $\mu\mathbf{H} = \nabla \times \mathbf{A}$.
5. Obtain the component of **E** from $\dot{\mathbf{E}} = \dfrac{\partial \mathbf{E}}{dt} = \dfrac{1}{\epsilon_0} \nabla \times \mathbf{H}$ (as $\mathbf{J} = 0$ for space).

Consider an alternating current element at the origin of a spherical coordinate system (Fig. 8.4).

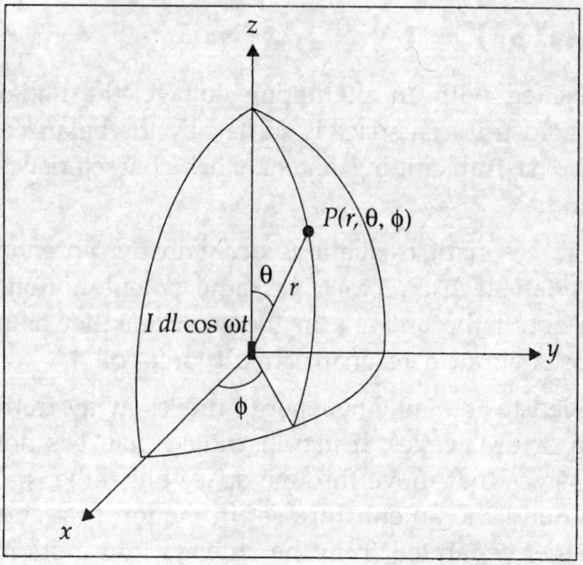

Fig. 8.4 Alternating current element at the origin

The vector magnetic potential, $\mathbf{A}(r, t)$ is given by

$$\mathbf{A}(r, t) = \frac{\mu}{4\pi} \int_{\upsilon} \frac{\mathbf{J}(r, t - r/\upsilon_0)}{r}\, d\upsilon$$

As the element is z-directed, **A** is also z-directed.

The volume integral in the above equation can be simplified by taking integration over the cross-sectional area of the element and an integration along its length. We know,

$$\int_s \mathbf{J}.\mathbf{ds} = I$$

and

$$\int_0^{dl} I\,dL = I\,dl$$

$$\mathbf{A} = A_z\,\mathbf{a}_z = \frac{\mu}{4\pi}\frac{I\,dl\cos\omega\,(t - r/\upsilon_0)}{r}\,\mathbf{a}_z$$

This means **A** has only z-component and $A_x = 0,\ A_y = 0$.

Changing Cartesian components into spherical coordinate components, we have

$$\left.\begin{aligned} A_r &= A_z\cos\theta \\ A_\theta &= -A_z\sin\theta \\ A_\phi &= 0 \end{aligned}\right\}$$

But we know

$$\mathbf{H} = \frac{1}{\mu}\,\nabla\times\mathbf{A}$$

$$H_r = \frac{1}{\mu}\,(\nabla\times\mathbf{A})_r$$

$$H_\theta = \frac{1}{\mu}\,(\nabla\times\mathbf{A})_\theta$$

and

$$H_\phi = \frac{1}{\mu}\,(\nabla\times\mathbf{A})_\phi$$

$$H_r = \frac{1}{\mu r\sin\theta}\left[\frac{\partial}{\partial\theta}(\sin\theta\,A_\phi) - \frac{\partial A_\theta}{\partial\phi}\right]$$

As $A_\phi = 0$ and $A_\theta \neq f(\phi)$, $H_r = 0$.

Similarly,
$$H_\theta = \frac{1}{\mu}\left[\frac{1}{r\sin\theta}\frac{\partial A_r}{\partial\theta} - \frac{1}{r}\frac{\partial(rA_\phi)}{\partial r}\right] = 0 \quad [\text{as } A_r \neq f(\phi)]$$

$$H_\phi = \frac{1}{\mu r}\left[\frac{\partial}{\partial r}(rA_\theta) - \frac{\partial A_r}{\partial\theta}\right]$$

Here
$$A_\theta = -A_z\sin\theta$$
$$= -\frac{\mu}{4\pi}\frac{I\,dl\cos\omega(t - r/\upsilon_0)}{r}\sin\theta$$
$$A_r = A_z\cos\theta$$
$$= +\frac{\mu}{4\pi}\frac{I\,dl\cos\omega(t - r/\upsilon_0)}{r}\cos\theta$$

Substituting the expressions of A_θ, A_r and simplifying, we get

$$H_\phi = \frac{I\,dl\sin\theta}{4\pi}\left[\frac{-\omega\sin\omega\left(t - \frac{r}{\upsilon_0}\right)}{r\upsilon_0} + \frac{\cos\omega\left(t - \frac{r}{\upsilon_0}\right)}{r^2}\right]$$

From Maxwell's first equation, we have

$$\nabla\times\mathbf{H} = \dot{\mathbf{D}} = \epsilon\,\dot{\mathbf{E}}$$
$$\mathbf{E} = \frac{1}{\epsilon_0}\int(\nabla\times\mathbf{H})\,dt$$

From this, we get

$$E_\phi = 0 \qquad [\text{as } \mathbf{H} = H_\phi\,\mathbf{a}_\phi]$$

$$E_\theta = \frac{I\,dl\sin\theta}{4\pi\epsilon_0}\left[-\frac{\omega\sin\omega t_d}{r\upsilon_0^2} + \frac{\cos\omega t_d}{r^2\upsilon_0} + \frac{\sin\omega t_d}{\omega r^3}\right]$$

and
$$E_r = \frac{2I\,dl\cos\theta}{4\pi\epsilon_0}\left[\frac{\cos\omega t_d}{r^2\upsilon_0} + \frac{\sin\omega t_d}{\omega r^3}\right]$$

where
$$t_d = \left(t - \frac{r}{\upsilon_0}\right)$$

The total field components of an alternating current element are

$$H_\phi = \frac{I\,dl \sin\theta}{4\pi}\left[-\frac{\omega \sin \omega t_d}{r\upsilon_0} + \frac{\cos \omega t_d}{r^2}\right]$$

$$E_\theta = \frac{I\,dl \sin\theta}{4\pi\epsilon_0}\left[-\frac{\omega \sin \omega t_d}{r\upsilon_0^2} + \frac{\cos \omega t_d}{r^2\upsilon_0} + \frac{\sin \omega t_d}{\omega r^3}\right]$$

$$E_r = \frac{2I\,dl \cos\theta}{4\pi\epsilon_0}\left[\frac{\cos \omega t_d}{r^2\upsilon_0} + \frac{\sin \omega t_d}{\omega r^3}\right]$$

$$H_r = 0,\ H_\theta = 0 \text{ and } E_\phi = 0$$

8.11 RADIATED POWER AND RADIATION RESISTANCE OF A CURRENT ELEMENT

The derivation of the expression for radiated power consists of the following steps:

1. Obtain field components as in Section 8.10.
2. Obtain expression for radiated power using Poynting vector.
3. Obtain average radiated power.
4. Obtain total power radiated from $P_T = \oint P_{av}\, ds$.
5. Identify expression for radiation resistance.

Poynting vector is

$$\mathbf{P} = \mathbf{E}\times\mathbf{H} \text{ watts/m}^2$$

$$P_\theta = -E_r H_\phi$$

$$P_\theta = -\left[\frac{2I\,dl\cos\theta}{4\pi\epsilon_0}\left(\frac{\cos\omega t_d}{r^2\upsilon_0} + \frac{\sin\omega t_d}{\omega r^3}\right)\right] \times \left[\frac{I\,dl\sin\theta}{4\pi}\left(-\frac{\omega\sin\omega t_d}{r\upsilon_0} + \frac{\cos\omega t_d}{r^2}\right)\right]$$

$$= \frac{2I^2\,dl^2\cos\theta\sin\theta}{16\pi^2\epsilon_0}\left[\frac{\sin^2\omega t_d}{r^4\upsilon_0} - \frac{\cos^2\omega t_d}{r^4\upsilon_0} - \frac{\sin\omega t_d\cos\omega t_d}{\omega r^5} + \frac{\omega\sin\omega t_d\cos\omega t_d}{r^3\upsilon_0}\right]$$

But $$2 \sin\theta \cos\theta = \sin 2\theta$$

$$\sin \omega t_d \cos \omega t_d = \frac{\sin 2\omega t_d}{2}$$

$$\sin^2 \omega t_d - \cos^2 \omega t_d = \cos 2\omega t_d$$

Using these identities, P_θ becomes

$$P_\theta = \frac{I^2\, dl^2 \sin 2\theta}{16\pi^2 \epsilon_0}\left[-\frac{\cos 2\omega t_d}{r^4\, \upsilon_0} - \frac{\sin 2\omega t_d}{2\omega r^5} + \frac{\omega \sin 2\omega t_d}{2r^3\, \upsilon_0^2}\right]$$

P_θ represents the instantaneous power flow in θ-direction. But the average value of $\cos 2\omega t_d$ or $\sin 2\omega t_d$ over a cycle is zero. Hence $(P_\theta)_{av} = 0$ at any value of r. This means power in θ-direction surges back and forth.

Similarly, $$P_r = E_\theta H_\phi$$

Using the expressions of E_θ and H_ϕ, we get

$$P_r = \frac{I^2\, dl^2 \sin^2\theta}{16\pi^2 \epsilon_0}\left[\frac{\sin \omega t_d \cos \omega t_d}{\omega r^5} + \frac{\cos^2 \omega t_d}{r^4\, \upsilon_0} - \frac{\omega \sin \omega t_d \cos \omega t_d}{r^3\, \upsilon_0^2} - \frac{\sin^2 \omega t_d}{r^4\, \upsilon_0} - \frac{\omega \sin \omega t_d \cos \omega t_d}{r^3\, \upsilon_0^2} + \frac{\omega^2 \sin^2 \omega t_d}{r^2\, \upsilon^3}\right]$$

$$P_r = \frac{I^2\, dl^2 \sin^2\theta}{16\pi^2 \epsilon_0}\left[\frac{\sin 2\omega t_d}{2\omega r^5} + \frac{\cos 2\omega t_d}{r^4\, \upsilon_0} - \frac{\omega \sin 2\omega t_d}{r^3\, \upsilon_0^2} + \frac{\omega^2 (1 - \cos 2\omega t_d)}{2r^2\, \upsilon_0^3}\right]$$

It is obvious that the average value of P_r is

$$P_{r(av)} = \frac{\omega^2 I^2\, dl^2 \sin^2\theta}{32\pi^2 r^2 \epsilon_0\, \upsilon_0^3}$$

or, $$P_{r(av)} = \frac{\eta_0}{2}\left(\frac{\omega I\, dl \sin\theta}{4\pi r\, \upsilon_0}\right)^2 \text{ watt/m}^2$$

where $$\eta_0 = \sqrt{\frac{\mu_0}{\epsilon_0}} = 120\pi\ \Omega$$

The total power radiated

$$P_T = \oint_{\text{surface}} P_{r(av)}\, ds$$

$P_{r(av)}$ is independent of ϕ and hence the elemental area ds on the spherical shell is

$$ds = 2\pi r^2 \sin\theta \, d\theta$$

Now P_T becomes

$$P_T = \int_0^{\pi} \frac{\eta_0}{2}\left(\frac{\omega I \, dl \sin\theta}{4\pi r \upsilon_0}\right) 2\pi r^2 \sin\theta \, d\theta = \frac{\eta_0 \, \omega^2 I^2 dl^2}{16\pi \upsilon_0^2} \int_0^{\pi} \sin^3\theta \, d\theta$$

But $$\int_0^{\pi} \sin^3\theta \, d\theta = \left(-\cos\theta + \frac{1}{3}\cos^3\theta\right)_0^{\pi} = \frac{4}{3}$$

$$P_T = \frac{\eta_0 \, \omega^2 I^2 dl^2}{12\pi \upsilon_0^2}$$

Here I is the peak value of current.

As $$I = \sqrt{2} I_{eff}$$

$$I^2 = 2 I_{eff}^2$$

Equation for P_T becomes

$$P_T = \frac{\eta_0 \, \omega^2 \, dl^2 \, I_{eff}^2}{6\pi \upsilon_0^2}$$

or, $$P_T = 80\pi^2 \left(\frac{dl}{\lambda}\right)^2 I_{eff}^2 \text{ watts}.$$

This is in the form of $P = I^2 R$. Hence the coefficient of I_{eff}^2 has the dimensions of resistance and it is called **Radiation Resistance**.

Radiation Resistance of a Hertzian dipole

$$R_r = 80\pi^2 \left(\frac{dl}{\lambda}\right)^2 \Omega$$

8.12 RADIATION, INDUCTION AND ELECTROSTATIC FIELDS

The field components of current elements are:

$$H_\phi = \frac{I \, dl \sin\theta}{4\pi}\left[-\frac{\omega \sin \omega t_d}{r \upsilon_0} + \frac{\cos \omega t_d}{r^2}\right]$$

$$E_\theta = \frac{I\,dl \sin\theta}{4\pi\epsilon_0}\left[-\frac{\omega \sin \omega t_d}{r\upsilon_0^2} + \frac{\cos \omega t_d}{r^2 \upsilon_0} + \frac{\sin \omega t_d}{\omega r^3}\right]$$

$$E_r = \frac{2I\,dl \cos\theta}{4\pi\epsilon_0}\left[\frac{\cos \omega t_d}{r^2 \upsilon_0} + \frac{\sin \omega t_d}{\omega r^3}\right]$$

H_ϕ field consists of terms $\frac{1}{r}$ and $\frac{1}{r^2}$. $\frac{1}{r^2}$ term dominates over $\frac{1}{r}$ term at points close to the current element. When r is small, $\frac{1}{r^2}$ term is called **Induction Field**.

On the other hand, $\frac{1}{r}$ term dominates over $\frac{1}{r^2}$ term when r is large. This $\frac{1}{r}$ term is called **Radiation Field** or **distant field** or **far-field**.

The expression for E_θ consists of three terms, $\frac{1}{r}, \frac{1}{r^2}$, and $\frac{1}{r^3}$ and the expression for E_r consists of $\frac{1}{r^2}$ and $\frac{1}{r^3}$ terms. The $\frac{1}{r^3}$ term is called **Electrostatic Field**.

In short,

> $\frac{1}{r}$ term in **E** and **H** fields is called **Radiation Field**
>
> $\frac{1}{r^2}$ term is called **Induction Field**
>
> $\frac{1}{r^3}$ term is called **Electrostatic Field**

If the induction and radiation fields have equal amplitudes, then from the expression of H_ϕ, we have

$$\frac{I\,dl\,\omega \sin\theta}{4\pi r \upsilon_0} = \frac{I\,dl \sin\theta}{4\pi r^2}$$

or,

$$\frac{\omega}{r\upsilon_0} = \frac{1}{r^2} \text{ or } r = \frac{\upsilon_0}{\omega} = \frac{\lambda}{2\pi} \approx \frac{\lambda}{6}$$

> At a distance of $r = \frac{\lambda}{2\pi}$ induction and radiation fields have equal amplitudes.

8.13 HERTZIAN DIPOLE

Hertzian dipole is defined as an infinitesimal current element $I\,dl$ which does not exist in real life.

or

Hertzian dipole is a short linear antenna which, when radiating, is assumed to carry **constant current** along its length.

As Hertzian dipole and alternating current elements are virtually the same, the radiated power and radiation resistance are given by

$$P_T = 80\pi^2 \left(\frac{dl}{\lambda}\right)^2 I_{\text{eff}}^2 \text{ watts}$$

$$R_r = 80\pi^2 \left(\frac{dl}{\lambda}\right)^2 \Omega$$

8.14 DIFFERENT CURRENT DISTRIBUTIONS IN LINEAR ANTENNAS

The possible current distributions are:

1. Constant current along its length—valid in Hertzian dipole.
2. Triangular current distribution (Fig. 8.5).

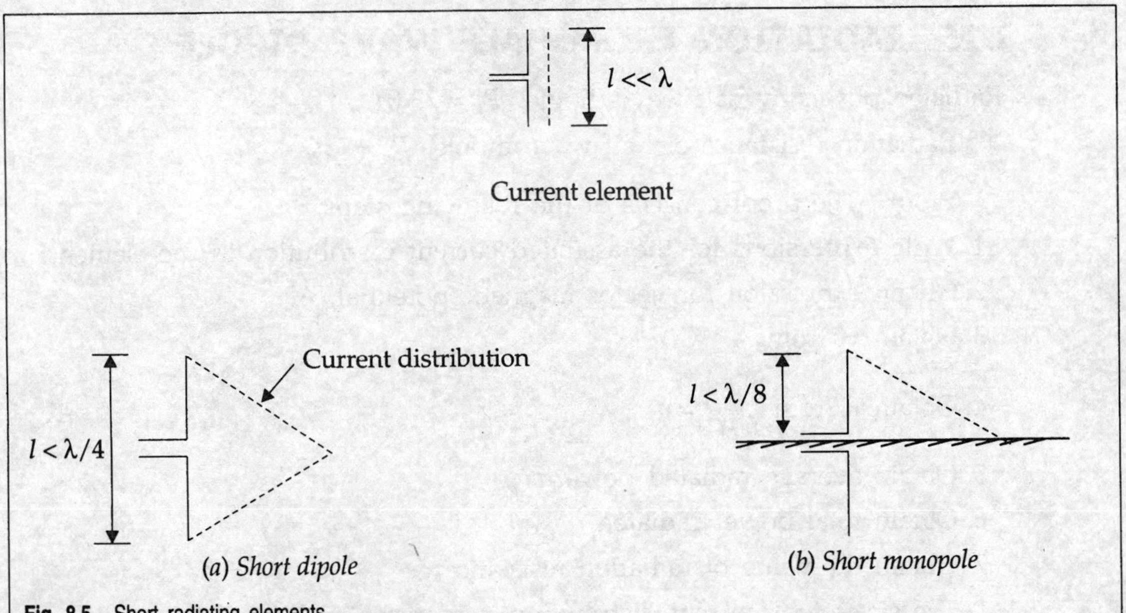

(a) Short dipole (b) Short monopole

Fig. 8.5 Short radiating elements

For triangular current distributions,

$$R_r\left(\text{Short dipole, } l < \frac{\lambda}{4}\right) = 20\pi^2\left(\frac{l}{\lambda}\right)^2\ \Omega$$

$$R_r\left(\text{Short monopole, } l < \frac{\lambda}{8}\right) = 10\pi^2\left(\frac{l}{\lambda}\right)^2\ \Omega$$

3. Sinusoidal current distribution (Fig. 8.6).

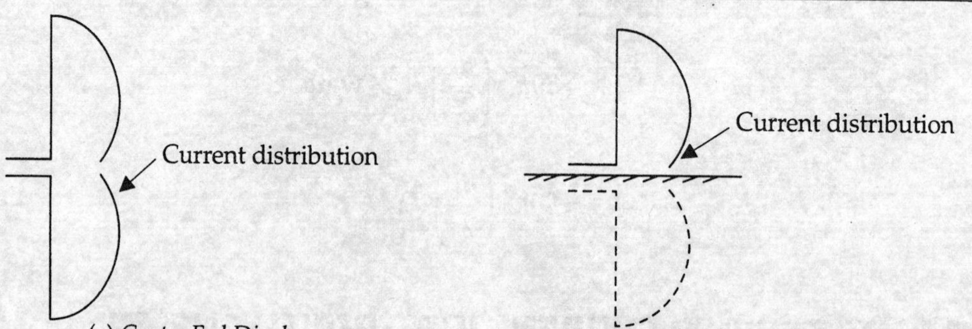

(*a*) *Centre Fed Dipole*

(*b*) Monopole

Fig. 8.6 Sinusoidal current distribution in dipole and monopole

4. Exact current distribution—This can be determined using the method of moment technique. The method is briefly presented in Chapter 9.

8.15 RADIATION FROM HALF WAVE DIPOLE

Radiated power by half wave dipole, $P_T = 73.0 I_{\text{eff}}^2$

Radiation resistance of half wave dipole, $R_r = 73\Omega$

Proof The proof consists of the following steps:

1. Write expressions for the assumed current distribution in the element.
2. Obtain expression for vector magnetic potential, **A**.
3. Obtain **H** from **A**.
4. Obtain **E** from $\left(\frac{E}{H}\right) = \eta_0$
5. Obtain average radiated power, P_{av}
6. Obtain total power radiated.
7. Obtain the value of radiation resistance.

The sinusoidal current distribution is represented by Fig. 8.7.

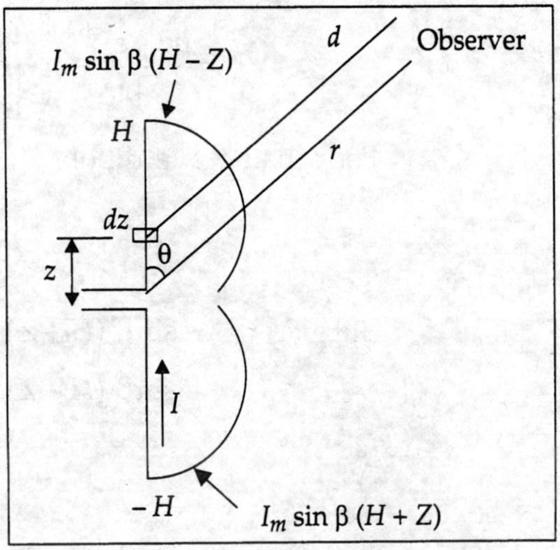

Fig. 8.7 Sinusoidal current distribution

$$I = I_m \sin\beta\,(H - Z) \text{ for } z > 0$$
$$= I_m \sin\beta\,(H + Z) \text{ for } z < 0$$

Here I_m = current maximum

The vector potential at a point, P due to the current element $I\,dz$ is given by

$$d\mathbf{A} = dA_z\mathbf{a}_z = \frac{\mu I e^{-j\beta d}\,dz}{4\pi d}\mathbf{a}_z$$

Here d is the distance from the current element to the point P. Let r be the distance of P from the origin. The total vector potential at P due to all current elements is given by

$$A_z = \frac{\mu}{4\pi}\int_{-H}^{H}\frac{Ie^{-j\beta d}}{d}\,dz$$

$$= \frac{\mu}{4\pi}\int_{-H}^{0}\frac{I_m \sin\beta(H+Z)}{d}e^{-j\beta d}\,dz + \frac{\mu}{4\pi}\int_{0}^{H}\frac{I_m \sin\beta(H-Z)}{d}e^{-j\beta d}\,dz$$

It is of interest here to consider radiation fields. d in the denominator can be approximated to r. But in the numerator, d is in the phase term and it is given by

$$d = r - z\cos\theta$$

Now A_z becomes

$$A_z = \frac{\mu}{4\pi}\int_{-H}^{0}\frac{I_m \sin\beta(H+Z)}{r}e^{-j\beta(r - z\cos\theta)}\,dz + \frac{\mu}{4\pi}\int_{0}^{H}\frac{I_m \sin\beta(H-Z)}{r}e^{-j\beta(r - z\cos\theta)}\,dz$$

$$= \frac{\mu I_m e^{-j\beta r}}{4\pi r}\left[\int_{-H}^{0} \sin\beta(H+Z)\, e^{j\beta z\cos\theta}\, dz + \int_{0}^{H} \sin\beta(H-Z)\, e^{j\beta z\cos\theta}\, dz\right]$$

For a half wave dipole,

$$H = \frac{\lambda}{4}$$

But

$$\sin\beta\,(H+Z) = \sin\beta\,\cos\beta z + \cos\beta H\,\sin\beta z$$

$$\sin\beta\,(H-Z) = \sin\beta H\,\cos\beta H - \cos\beta H\,\sin\beta z$$

As

$$\beta = \frac{2\pi}{\lambda},\ \sin\beta H = \sin\frac{2\pi}{\lambda}\times\frac{\lambda}{4} = 1,\ \cos\beta H = \cos\frac{\pi}{2} = 0$$

$$\sin\beta\,(H+Z) = \sin\beta\,(H-Z) = \cos\beta z$$

Using these values, A_z becomes

$$A_z = \frac{\mu I_m e^{-j\beta r}}{4\pi r}\left[\int_{-H}^{0} \cos\beta z\, e^{+j\beta z\cos\theta} + \int_{0}^{H} \cos\beta z\, e^{+j\beta z\cos\theta}\, dz\right]$$

But

$$\int_{-H}^{0} \cos\beta z e^{+j\beta z\cos\theta}\, dz = \int_{0}^{H} \cos\beta z\, e^{-j\beta z\cos\theta}\, dz$$

$$A_z = \frac{I_m\mu}{4\pi r}\, e^{-j\beta r}\left[\int_{0}^{\lambda/4} \cos\beta z\,(e^{j\beta z\cos\theta} + e^{-j\beta z\cos\theta})\, dz\right]$$

$$= \frac{I_m\mu}{4\pi r}\, e^{-j\beta r}\left[\int_{0}^{\lambda/4} \cos\{\beta z\,(e^{j\beta z\cos\theta} + e^{-j\beta z\cos\theta})\} + \cos\{\beta z\,(1-\cos\theta)\}\right] dz$$

$$= \frac{I_m\mu}{4\pi r}\, e^{-j\beta r}\left[\frac{\sin\{\beta z\,(1+\cos\theta)\}}{\beta\,(1+\cos\theta)} + \frac{\sin\{\beta z\,(1-\cos\theta)\}}{\beta\,(1-\cos\theta)}\right]_0^{\lambda/4}$$

$$= \frac{\mu I_m}{4\pi\beta r}\, e^{-j\beta r}\left[\frac{(1-\cos\theta)\cos\left(\frac{\pi}{2}\cos\theta\right) + (1+\cos\theta)\cos\left(\frac{\pi}{2}\cos\theta\right)}{\sin^2\theta}\right]$$

$$A_z = \frac{\mu I_m}{2\pi\beta r} e^{-j\beta r} \left[\frac{\cos\frac{\pi}{2}\cos\theta}{\sin^2\theta} \right]$$

But we have

$$\mu H_\phi = \frac{1}{r}\left[\frac{\partial}{\partial r}(rA_\theta) - \frac{\partial}{\partial\theta}A_r\right]$$

$$= \frac{1}{r}\left[\frac{\partial}{\partial r} r\,(-A_z \sin\theta) - \frac{\partial}{\partial\theta}(A_z\cos\theta)\right]$$

$$\mu H_\phi = -\sin\theta\,\frac{\partial A_z}{\partial r}$$

Hence

$$\mu H_\phi = -\frac{\partial}{\partial r}\left(\frac{\mu I_m e^{-j\beta r}}{2\pi\beta r}\,\frac{\cos\left(\frac{\pi}{2}\cos\theta\right)}{\sin^2\theta}\right)\sin\theta$$

$$H_\phi = \frac{jI_m e^{-j\beta r}}{2\pi r}\,\frac{\cos\left(\frac{\pi}{2}\cos\theta\right)}{\sin\theta}$$

We also know that

$$E_\theta = \eta_0 H_\phi,\ \ \eta_0 = 120\pi\Omega$$

$$E_\theta = \frac{j120\pi I_m e^{-j\beta r}}{2\pi r}\left[\frac{\cos\left(\frac{\pi}{2}\cos\theta\right)}{\sin\theta}\right]$$

$$= \frac{j60 I_m e^{-j\beta r}}{r}\left[\frac{\cos\left(\frac{\pi}{2}\cos\theta\right)}{\sin\theta}\right]$$

The magnitude of **E** for the radiation field is

$$E_\theta = \frac{60 I_m}{r}\left[\frac{\cos\left(\frac{\pi}{2}\cos\theta\right)}{\sin\theta}\right]\text{V/m}$$

E_θ and H_ϕ are in time phase. Hence the maximum value of Poynting vector is

$$P_m = (E_\theta)_{\max}\,(H_\phi)_{\max}$$

$$= \frac{60I_m}{r}\left[\frac{\cos\left(\frac{\pi}{2}\cos\theta\right)}{\sin\theta}\right] \times \frac{I_m}{2\pi r}\left(\frac{\cos\left(\frac{\pi}{2}\cos\theta\right)}{\sin\theta}\right)$$

$$= \frac{30I_m^2}{\pi r^2}\left[\frac{\cos^2\left(\frac{\pi}{2}\cos\theta\right)}{\sin^2\theta}\right]$$

The average value of Poynting vector is one half of the peak value.

$$P_{av} = \frac{15I_m^2}{\pi r^2}\left[\frac{\cos^2\left(\frac{\pi}{2}\cos\theta\right)}{\sin^2\theta}\right]$$

or,

$$P_{av} = \frac{\eta_0 I_m^2}{8\pi^2 r^2}\left[\frac{\cos^2\left(\frac{\pi}{2}\cos\theta\right)}{\sin^2\theta}\right]$$

Therefore total power radiated through a spherical surface half wave dipole is

$$P_T = \oint P_{av}\,ds = \frac{\eta_0 I_m^2}{8\pi^2 r^2}\int_0^\pi \frac{\cos^2\left(\frac{\pi}{2}\cos\theta\right)}{\sin^2\theta}\,2\pi r^2 \sin\theta\,d\theta$$

$$= \frac{\eta_0 I_m^2}{4\pi}\int_0^\pi \frac{\cos^2\left(\frac{\pi}{2}\cos\theta\right)}{\sin\theta}\,d\theta$$

The numerical evaluation of the integral $\int_0^\pi \frac{\cos^2\left(\frac{\pi}{2}\cos\theta\right)}{\sin\theta}\,d\theta$ by Simpson's or the Trapezoidal rule gives a value of 1.218.

$$P_T = \frac{\eta_0 I_m^2}{4\pi}$$

$$= \frac{120\pi I_m^2}{4\pi} \times 1.218 = 36.54 I_m^2$$

As $I_m = \sqrt{2} I_{\text{eff}}$; P_T becomes

$$P_T = 36.54 \times 2 \times I_{\text{eff}}^2$$

$$\boxed{P_T = 73.08\,\Omega\, I_{\text{eff}}^2, \text{ watts}}$$

The coefficient of I_{eff}^2 is nothing but radiation resistance. That is,

$$\boxed{R_r = 73.08\,\Omega}$$

8.16 RADIATION FROM QUARTER WAVE MONOPOLE

Radiated power of quarter wave monopole, $P_T = 36.5 I_{\text{eff}}^2$, watts.

Radiation resistance, $R_r = 36.5\Omega$

Proof Consider Fig. 8.8 in which a monopole with current distribution is shown.

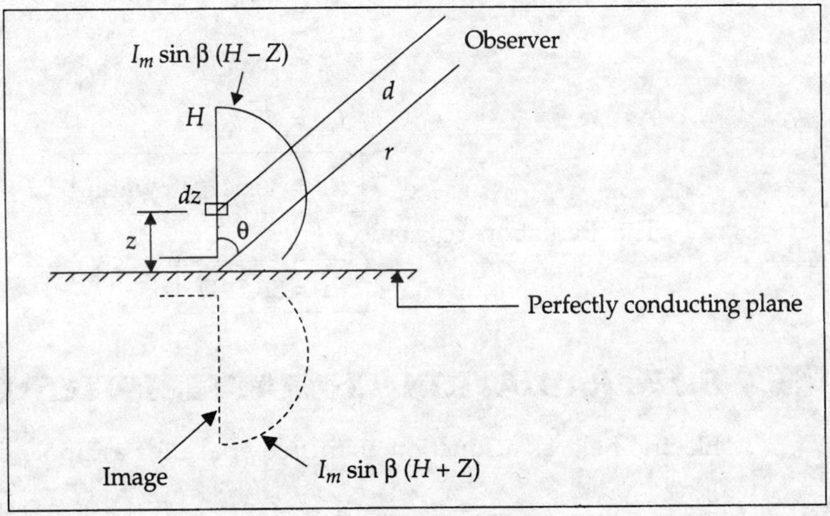

Fig. 8.8 Monopole

Obtain P_{av} exactly as described in half wave dipole, that is,

$$P_{av} = \frac{\eta_0 I_m^2}{8\pi^2 r^2} \left[\frac{\cos^2\left[\frac{\pi}{2}\cos\theta\right]}{\sin^2\theta} \right]$$

As the monopole is fed by a perfectly conducting plane at one end, it radiates only through a hemispherical surface. Therefore, the total radiated power is

$$P_T = \oint P_{av}\, ds$$

$$= \frac{\eta_0 I_m^2}{8\pi^2 r^2} \int_0^{\pi/2} \frac{\cos^2\left(\frac{\pi}{2}\cos\theta\right)}{\sin^2\theta} 2\pi r^2 \sin\theta\, d\theta$$

$$= \frac{\eta_0 I_m^2}{4\pi} \int_0^{\pi/2} \frac{\cos^2\left(\frac{\pi}{2}\cos\theta\right)}{\sin^2\theta} d\theta$$

Numerical evaluation of the integral $\int_0^{\pi/2} \frac{\cos^2\left(\frac{\pi}{2}\cos\theta\right)}{\sin\theta} d\theta$ by Simpson's or the Trapezoidal rule gives a value of 0.609.

$$P_T = \frac{\eta_0 I_m^2}{4\pi} \times 0.609$$

$$= 18.27 I_m^2$$

As $I_m = \sqrt{2}\, I_{eff}$

$$\boxed{P_T = 36.54 I_{eff}^2, \text{ watts}}$$

The Radiation resistance,

$$\boxed{R_r = 36.54\Omega}$$

8.17 RADIATION CHARACTERISTICS OF DIPOLES

Electric field as a function θ in free space for a dipole of length $2H$ is given by

$$E_\theta = \frac{j60 I_m e^{-j\beta r}}{r}\left[\frac{\cos(\beta H \cos\theta) - \cos\beta H}{\sin\theta}\right]$$

The amplitude of E_θ is

$$|E_\theta| = \frac{60 I_m}{r}\left[\frac{\cos(\beta H \cos\theta) - \cos\beta H}{\sin\theta}\right]$$

The normalised (E_θ) is

$$|E_\theta|_n = \frac{\dfrac{60I_m}{r}\left[\dfrac{\cos(\beta H\cos\theta)-\cos\beta H}{\sin\theta}\right]}{\dfrac{60I_m}{r}}$$

$$|E_\theta|_n = \left[\frac{\cos(\beta H\cos\theta)-\cos\beta H}{\sin\theta}\right]$$

The radiation patterns are the variation of $|E_\theta|_n$ with θ. These patterns for different lengths of dipole are shown in Fig. 8.9.

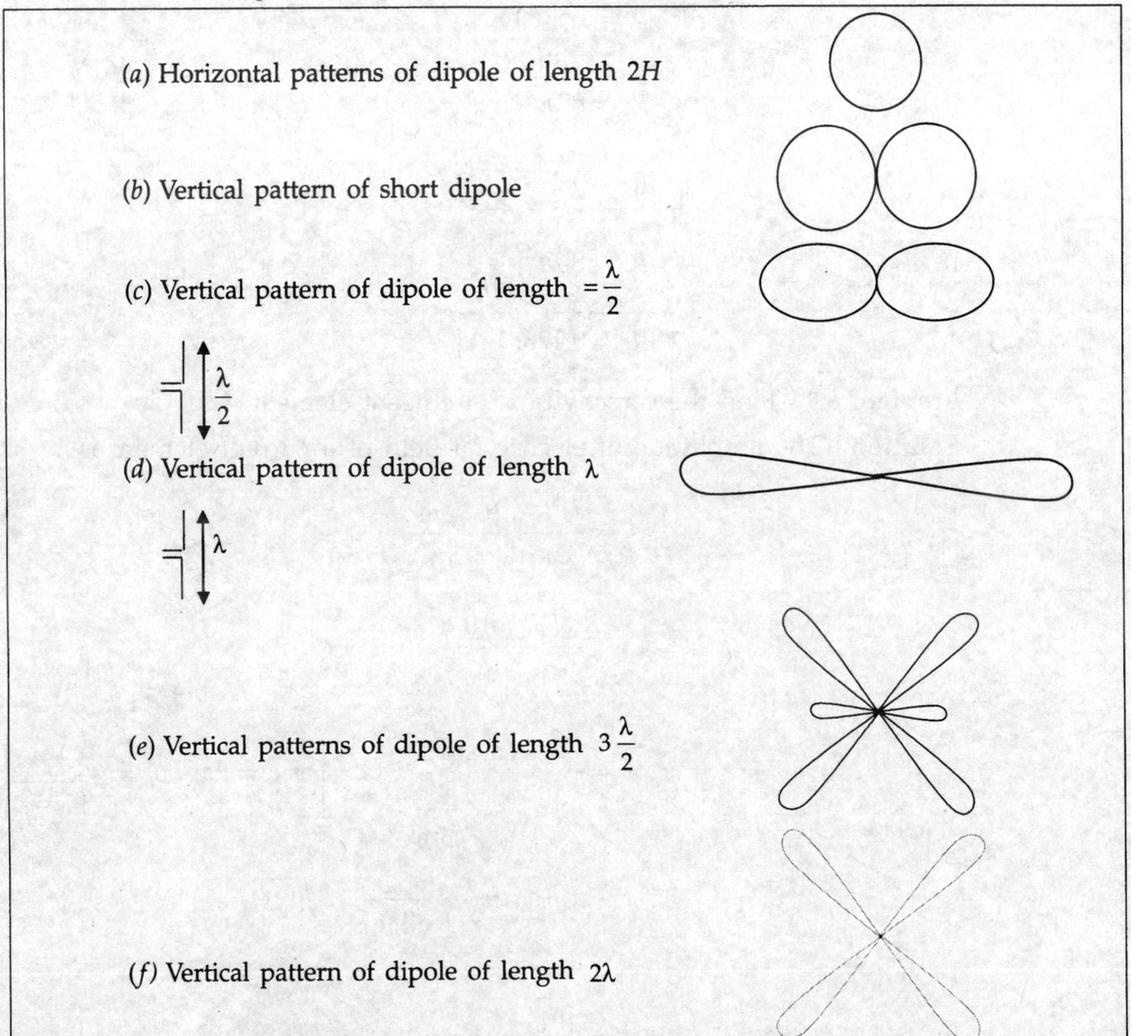

Fig. 8.9 Radiation patterns of dipoles

Problem 8.1 Find the radiation resistance of a Hertzian dipole of length $\frac{\lambda}{40}, \frac{\lambda}{60}, \frac{\lambda}{80}$.

Solution The radiation resistance of a Hertzian dipole of length *dl* is

$$R_r = 80\pi^2 \left(\frac{dl}{\lambda}\right)^2 \Omega$$

If $dl = \frac{\lambda}{40}$

$$R_r = 80\pi^2 \left(\frac{\lambda}{40} \times \frac{1}{\lambda}\right)^2$$

$$\boxed{R_r = 0.493\Omega}$$

If $dl = \frac{\lambda}{60}$

$$R_r = 80\pi^2 \frac{1}{60^2}$$

or,

$$\boxed{R_r = 0.219\Omega}$$

If $dl = \frac{\lambda}{80}$

$$R_r = 80\pi^2 \frac{1}{80^2}$$

or,

$$\boxed{R_r = 0.123\Omega}$$

Problem 8.2 Find the directivity of a current element, *I dl*.

Solution The amplitude of electric far-field of a current element is

$$E = \frac{I\,dl \sin\theta}{4\pi\epsilon_0} \times \frac{\omega}{r\upsilon_0^2}$$

$$= \frac{\omega I\,dl \sin\theta}{4\pi\epsilon_0 r\upsilon_0^2}$$

$$= \frac{2\pi f\ I\,dl \sin\theta}{4\pi\epsilon_0 r \dfrac{1}{\mu_0\epsilon_0}} \qquad \left[\text{as } \omega = 2\pi f,\ \upsilon_0 = \frac{1}{\sqrt{\mu_0\epsilon_0}}\right]$$

$$= \frac{I\,dl \sin\theta}{2r} \times f\sqrt{\mu_0\epsilon_0} \times \frac{\sqrt{\mu_0\epsilon_0}}{\epsilon_0}$$

But

$$\eta_0 = \sqrt{\frac{\mu_0}{\epsilon_0}}, \quad \lambda = \frac{\upsilon_0}{f}$$

or,
$$\frac{1}{\lambda}=\frac{f}{\upsilon_0}=f\sqrt{\mu_0\epsilon_0}$$

$$E=\frac{60\pi Idl\sin\theta}{\lambda r}$$

Maximum radiation occurs at $\theta=\pi/2$

$$E_{\max}=\frac{60\pi Idl}{\lambda r}$$

The radiated power of the current element is

$$w_r=80\pi^2\left(\frac{dl}{\lambda}\right)^2 I^2 \text{ watts}$$

If w_r is assumed to be 1 watt, then

$$I=\frac{\lambda}{\sqrt{80}\,\pi dl}\text{ ampere}$$

$$E_{\max}=\frac{60}{r\sqrt{80}}\text{ V/m}$$

The maximum radiation intensity is given by

$$RI=\frac{r^2E_{\max}^2}{\eta_0}$$

$$=\frac{r^2}{120\pi}\frac{60^2}{r^2\times 80}$$

$$RI=\frac{3}{8\pi}$$

The maximum directive gain, g_d (max)

$$g_d(\max)=\frac{4\pi\,(RI)}{w_r}$$

$$=4\pi\times\frac{3}{8\pi} \qquad \text{(as } w_r=1\text{ watt)}$$

$$=\frac{3}{2}=1.5$$

The directivity of current element

$$D = g_d\,(\max) = 1.5$$

or

$$D_{(\text{dB})} = 10 \log_{10} 1.5 = 1.76 \text{ dB}$$

Problem 8.3 Find the directivity of a half wave dipole.

Solution For a half wave dipole,

$$E_{\max} = \frac{60I}{r}$$

But $w_r = 73I^2 = \text{watts}$

For $w_r = 1\,\text{w},$

$$I = \frac{1}{\sqrt{73}}$$

$$E_{\max} = \frac{60}{r} \times \frac{1}{\sqrt{73}}$$

$$g_d\,(\max) = \frac{4\pi\,(RI)}{w_r}$$

$$= 4\pi\,(RI)$$

$$= 4\pi \times \frac{r^2 E^2}{\eta_0} \qquad \left[\text{as } RI = r^2 \frac{E^2}{\eta_0}\right]$$

$$= \frac{4\pi \times r^2}{\eta_0} \frac{60^2}{r^2} \frac{1}{73}$$

$$= \frac{4\pi \times 60 \times 60}{120\pi} \frac{1}{73}$$

$$= \frac{120}{73} = 1.644$$

$$g_d\,(\max) = D = 1.644$$

Problem 8.4 An antenna whose radiation resistance is 300Ω operates at a frequency of 1 GHz and with a current of 3 amperes. Find the radiated power.

Solution Radiated power,

$$w_r = I^2 R_r$$

$$= 3^2 \times 300$$

$$= 9 \times 300$$

$$\boxed{w_r = 2{,}700 \text{ watts}}$$

Problem 8.5 What is the effective area of a half wave dipole operating at 500 MHz?

Solution The effective area of an antenna is

$$A_e = \frac{\lambda^2}{4\pi} g_d$$

As $$f = 500 \text{ MHz}$$

$$\lambda = \frac{3 \times 10^8}{500 \times 10^6}$$

$$= \frac{3}{5} = 0.6 \text{ m}$$

Directivity of half wave dipole is

$$(g_d)_{\max} = D = 1.644$$

$$A_e = \frac{0.6^2}{4\pi} \times 1.644$$

$$\boxed{A_e = 0.047 \text{ m}^2}$$

Problem 8.6 Find the effective area of a Hertzian dipole operating at 100 MHz.

Solution As $f = 100$ MHz

$$\lambda = \frac{3 \times 10^8}{10^8} = 3 \text{ m}$$

Directivity of the Hertzian dipole,

$$D = 1.5$$

$$A_e = \text{effective area}$$

$$= \frac{\lambda^2 D}{4\pi} = \frac{3^2 \times 1.5}{4\pi} = 1.07 \text{ m}^2$$

$$\boxed{A_e = 1.07 \text{ m}^2}$$

Problem 8.7 An EM wave of 1 GHz is radiated by an antenna to cover a distance of 100 km. Determine the time taken by the wave to travel the above distance.

Solution The time taken by the EM wave is

$$t = \frac{d}{\upsilon_0}$$

where $d = 100\,\text{km}$

υ_0 = velocity of propagation

$$t = \frac{100 \times 1{,}000}{3 \times 10^8}$$

$$= \frac{10^5}{3 \times 10^8}$$

$$= 0.333 \times 10^{-3}$$

$$\boxed{t = 333\mu\ \text{sec}}$$

Problem 8.8 The directivity of an antenna is 30 and it operates at a frequency of 100 MHz. Find its maximum effective aperture.

Solution $D = 30$

$f = 100\,\text{MHz}$

$$\lambda = \frac{\upsilon_0}{f} = \frac{3 \times 10^8}{100 \times 10^6} = 3\,\text{m}$$

Maximum effective aperture

$$= \frac{\lambda^2}{4\pi} D$$

$$= \frac{3^2 \times 30}{4\pi} = \frac{270}{4\pi}$$

$$\boxed{A_{em} = 21.48\,\text{m}^2}$$

POINTS/FORMULAE TO REMEMBER

- Radiation intensity, $RI = \frac{r^2 E^2}{\eta_0}$ watts/unit solid angle.
- Directive gain, $g_d = \frac{4\pi \times (RI)}{w_r}$.
- Directivity, $D = (g_d)_{max}$.
- Power gain, $g_p = \frac{4\pi \times (RI)}{w_t}$.
- Antenna efficiency, $\eta = \frac{g_p}{g_d}$.
- Effective area, $A_e = \frac{\lambda^2}{4\pi} g_d$ or $A_e = \frac{\text{received power}}{\text{power flow of incident waves}}$.
- Far-field is represented by $\frac{1}{r}$ field term.
- Induction field is represented by $\frac{1}{r^2}$ field term.
- Radiation resistance of a Hertzian dipole is $80\pi^2 \left(\frac{dl}{\lambda}\right)^2 \Omega$.
- Electrostatic field is represented by $\frac{1}{r^3}$ term.
- The far-field and induction field have equal magnitudes at $r = \frac{\lambda}{2\pi}$.
- Radiation resistance of half wave dipole is 73Ω.
- Radiation resistance of quarter wave monopole is 36.5Ω.
- Horizontal pattern of vertical dipole is a circle.
- Radiated power flow of a vertical dipole is in the radial direction.

OBJECTIVE QUESTIONS

1. An antenna is a transducer. (Yes/No)
2. An antenna is a sensor of EM waves. (Yes/No)
3. An antenna acts as an impedance matching device. (Yes/No)
4. Effective length of a wire antenna is always greater than the actual length. (Yes/No)
5. Directive gain = Power gain for an antenna. (Yes/No)
6. The radiation fields are nothing but far-fields. (Yes/No)
7. The radiation pattern of vertical and horizontal dipoles are identical. (Yes/No)
8. The patterns of half wave dipole and quarter wave monopole are identical. (Yes/No)
9. The radiated fields of z-directed half wave dipole consists of $E_\theta, E_r, H_r, H_\theta,$ terms. (Yes/No)
10. The radiated fields of z-directed dipole consists of only E_θ, E_r and H_ϕ. (Yes/No)
11. Effective area of antenna is a function frequency. (Yes/No)
12. Electric and magnetic dipoles have the same physical structure. (Yes/No)
13. Magnetic dipole is a small current loop of wire. (Yes/No)
14. Input and radiation resistances are the same. (Yes/No)
15. The radiation resistance of a current element depends on frequency. (Yes/No)
16. The differential current element is nothing but Hertzian dipole. (Yes/No)
17. Delayed and retarded potentials mean the same. (Yes/No)
18. $\mathbf{J}dv = Id\mathbf{L}$ (Yes/No)
19. Radiated power of a current element depends on frequency. (Yes/No)
20. The units of scalar and vector magnetic potentials are the same. (Yes/No)
21. Electrostatic field contributes to the radiated power. (Yes/No)

22. Induction field does not contribute to radiation power. (Yes/No)

23. Electrostatic field does not contribute to radiation power. (Yes/No)

24. Array increases the directivity. (Yes/No)

25. Beam width is decreased by array. (Yes/No)

26. If the number of elements are increased, the beam width is reduced. (Yes/No)

27. Dipole is an omnidirectional antenna. (Yes/No)

28. Isotropic and omnidirectional antennas mean the same. (Yes/No)

29. Dipole and monopole mean the same except in length. (Yes/No)

30. Power gain and directive gain are the same. (Yes/No)

31. Radiation resistance of half-wave dipole is more than that of quarter monopole. (Yes/No)

32. Power gain and efficiency of antennas are the same. (Yes/No)

33. Effective area of receiving antenna depends on frequency. (Yes/No)

34. The effective area of an antenna is independent of the length of the antenna. (Yes/No)

35. The units of radiation intensity are __________.

36. Directivity is __________.

37. Efficiency of an antenna is __________.

38. Efficiency of an antenna in terms of directive and power gains is __________.

39. Effective area is __________.

40. The far-field is indicated by the presence of __________.

41. The induction field is indicated by the presence of __________.

42. The electrostatic field is indicated by the presence of __________.

43. The radiation resistance of an isolated half wave dipole is __________.

44. The radiation resistance of a quarter wave monopole is __________.

45. The current distribution in a half wave dipole is __________.

46. The current distribution in alternating current element is __________.

47. The current distribution in short dipoles is _________.

48. The directivity of current element is _________.

49. The directivity of half wave dipole is _________.

50. If a current element is x-directed, vector magnetic potential is _________.

51. Radiation resistance of a short monopole is _________.

52. Radiation resistance of a short dipole is _________.

53. At LF and VLF, polarisation often used is _________.

54. dB_i means _________.

55. dB_m means power gain in dB _________.

56. Antenna used in mobile communications is _________.

57. If a current element is z-directed, vector magnetic potential is _________.

58. If vector magnetic potential has only A_z, E_ϕ is _________.

59. Radiation resistance of current element is _________.

60. Radiation resistance of quarter wave monopole is _________.

61. Directional pattern of a short dipole in the horizontal plane is a _________.

62. Directional pattern of a horizontal halfwave centre-fed dipole is _________.

63. Effective length of a dipole is always _________ than the actual length.

64. The directivity in dB of half wave dipole is _________.

65. The directivity in dB of current element is _________.

66. Effective area of a Hertzian dipole operating at 100 MHz is _________.

67. The radiation pattern of a horizontal dipole is of _________.

68. The radiation pattern of vertical dipole is of _________.

69. Vector magnetic potential has the unit of _________.

70. Retarded magnetic potential has the unit of _________.

71. Radiated power is contributed by _________ only.

72. The radiation resistance of quarter wave dipole is _________.

73. The radiation resistance of half wave dipole is _________.

74. The total resistance of an antenna is _________.

75. Power gain of an antenna is _________.

76. The antenna of z_a impedance radiates maximum power when the transmitting line feeding the antenna has an impedance of _________.

77. Directive gain of Hertzian dipole is _________.

78. If the signal level is 1 mW, power gain is

(a) 0 dBm (b) 1 dBm (c) 10^{-3} dBm (d) 10 dBm

79. Whip antenna has a physical length of

(a) $\lambda/4$ (b) $\lambda/2$ (c) $3\lambda/2$ (d) λ

80. For a 300Ω antenna operating with 5 A of current, the radiated power is

(a) 7,500 W (b) 750 W (c) 75 W (d) 1500 W

Answers

1. Yes	2. Yes	3. Yes	4. No	5. No
6. Yes	7. No	8. No	9. No	10. Yes
11. Yes	12. No	13. Yes	14. No	15. Yes
16. Yes	17. Yes	18. Yes	19. Yes	20. No
21. No	22. Yes	23. Yes	24. Yes	25. Yes
26. Yes	27. Yes	28. No	29. No	30. No
31. Yes	32. No	33. Yes	34. No	

35. Watts/unit solid angle 36. Maximum directive gain 37. $\dfrac{w_r}{(w_r + w_l)}$

38. g_p / g_d 39. $\dfrac{\lambda^2}{4\pi} g_d$ 40. $\dfrac{1}{r}$ term 41. $\dfrac{1}{r^2}$ term 42. $\dfrac{1}{r^3}$ term

43. 73Ω 44. 36.5Ω 45. Sinusoidal 46. Constant 47. Triangular

48. 1.5 49. 1.64 50. x-directed 51. $100\left(\frac{l}{\lambda}\right)^2\Omega$ 52. $200\left(\frac{l}{\lambda}\right)^2\Omega$

53. Vertical

54. Power gain of the antenna in dB relative to isotropic antenna

55. Compared to 1 mW 56. Whip antenna 57. z-directed

58. Zero 59. $80\pi^2\left(\frac{dl}{\lambda}\right)^2\Omega$ 60. 36.5Ω 61. Circle

62. Figure of eight 63. Less 64. 2.15 65. 1.64

66. 1.07 m^2 67. Figure of eight shape 68. Dumbell shape

69. Wb/m 70. Wb/m 71. Far-field only 72. 36.5Ω

73. $73\,\Omega$ 74. $R_r + R_l$ 75. The product of efficiency and gain

76. z_a^* 77. 1.5 78. 0 dBm 79. $\lambda/4$ 80. (a)

MULTIPLE CHOICE QUESTIONS

1. Input resistance of an antenna is

(a) R_r (b) R_l (c) $R_r + R_l$ (d) $R_r - R_l$

2. The radiation resistance of a current element is

(a) 36.5Ω (b) 73Ω (c) $80\pi^2\left(\frac{dl}{\lambda}\right)^2$ (d) $80\pi\left(\frac{dl}{\lambda}\right)^2$

3. Horizontal dipole has the directional characteristics of

(a) circle (b) figure of eight (c) four lobes (d) ellipse

4. Retarded vector magnetic potential has the unit of

(a) wb/m (b) V/m (c) V/m^2 (d) wb-sec/m

5. Radiation resistance of half wave dipole is

(a) 73Ω (b) 36Ω (c) $80\pi^2\left(\frac{dl}{\lambda}\right)^2$ (d) 292Ω

6. Effective length of half wave dipole is

(a) $>\lambda/2$ (b) $<\lambda/2$ (c) 0.55λ (d) 0.6λ

7. Far-field consists of

(a) $\frac{1}{r}$ term (b) $\frac{1}{r^2}$ term (c) $\frac{1}{r^3}$ term (d) r term

8. Induction field consists of

(a) $\frac{1}{r^2}$ term (b) $\frac{1}{r}$ term (c) $\frac{1}{r^3}$ term (d) r^2 term

9. Radiated power is proportional to

(a) R_r^2 (b) $\frac{1}{R_r}$ (c) I^2 (d) I

10. The resonant length of a dipole is

(a) $\lambda/2$ (b) 0.6λ (c) 0.4λ (d) 0.35λ

11. The length of Whip antenna is

(a) $\frac{\lambda}{2}$ (b) $\frac{\lambda}{4}$ (c) $\frac{3\lambda}{2}$ (d) λ

12. For a 300Ω antenna operating with 4 A of current, the radiated power is

(a) 7,500 W (b) 4,800 W (c) 480 W (d) 75 W

13. If the signal level is 1 W, the power gain is

(a) 1 dB (b) 0 dB (c) 10 dB (d) 10 dBm

14. The vector magnetic potential of x-directed half wave dipole is

(a) x-directed (b) y-directed (c) z-directed (d) θ-directed

15. The directivity of current element is

(a) 1.5 (b) 1.64 (c) 1.76 (d) 2.15

16. The directivity of half wave dipole is

(a) 1.5 (b) 1.64 (c) 1.76 (d) 2.15

17. If half wave dipole is operating at one wavelength, its effective area is

(a) $0.13\,m^2$ (b) $1.2\,m^2$ (c) $0.012\,m^2$ (d) $120\,m^2$

18. Radiation resistance of dipole of length 0.1λ is

(a) 2Ω (b) 20Ω (c) 200Ω (d) 73Ω

19. Radiation resistance of a monopole of length 0.1λ is

(a) 4Ω (b) 2Ω (c) 8Ω (d) 1Ω

20. If the maximum directive gain of an antenna is 2, its directivity is

(a) 4 (b) 2 (c) 1 (d) 6

Answers

1. (c)	2. (c)	3. (b)	4. (a)	5. (a)
6. (b)	7. (a)	8. (a)	9. (c)	10. (a)
11. (b)	12. (b)	13. (b)	14. (a)	15. (a)
16. (b)	17. (a)	18. (a)	19. (a)	20. (b)

EXERCISE PROBLEMS

1. The field amplitude due to half wave dipole at 10 km is 0.1 V/m. It operates at 100 MHz. Find the dipole length and its radiated power.

2. What is the length of a half wave dipole at frequencies of 10 MHz, 50 MHz and 100 MHz?

3. Find the maximum effective area of an antenna at a frequency of 2 GHz when the directivity is 100.

4. Obtain the gain of an antenna whose area is 12 m^2 and operating at a frequency of 6 GHz.

5. Find the radiated power of an antenna if a current of 10 amperes exists and its radiation resistance is $32.0\,\Omega$.

6. What is the radiation resistance of an antenna if it radiates a power of 120 W and when the current in it is 10 amperes.

7. Find the directivity, efficiency and effective area of an antenna if its $R_r = 80\,\Omega$, $R_l = 10\,\Omega$. The power gain is 10 dB and antenna operates at a frequency of 100 MHz.

8. If the transmitting power is 10 KW, find the power density at distances of 10 km, 50 km, 100 km assuming the radiator is isotropic.

9. If the current element is z-directed, find the far-field components of **H**.

10. Derive an expression for distant θ-field component of **E** for a dipole of length L.

11. (a) Find the current required to radiate a power of 50 W at 60 MHz from 0.1λ Hertzian dipole.
    (b) Determine the radiation resistance of the element.

12. Find the radiation efficiency of a Hertzian dipole of length 0.03λ at a frequency of 100 MHz if the loss resistance is $0.01\,\Omega$.

CHAPTER

9

ADVANCED TOPICS

Electromagnetic Interference is nothing but electromagnetic pollution. It is neither seen nor sensed nor is it audible and hence it is a silent threat.

The main aim of this chapter is to provide the basics of advanced topics related to field theory. They include:

- secondary sources of EM fields
- reciprocity and reaction concepts
- induction and equivalence theorems
- EMI/EMC concepts
- EMI sources and effects of EMI
- biological hazards
- EMC standards, ESD and EMP
- numerical techniques of EMF theory—FDM, FEM, MOM techniques
- solved problems, points/formulae to remember, objective and multiple choice questions and exercise problems.

Do you know?

During thunders a field of about 20 KV/m is produced.

9.1 INTRODUCTION

An elementary treatment of some advanced topics required at the undergraduate level of Engineering is provided in this chapter. Some biological effects of electromagnetic radiation are also presented.

9.2 SECONDARY SOURCES OF ELECTROMAGNETIC FIELDS

In the first and second Maxwell's equations, namely

$$\nabla \times \mathbf{H} = \dot{\mathbf{D}} + \mathbf{J}$$

and
$$\nabla \times \mathbf{E} = -\dot{\mathbf{B}},$$

$\dot{\mathbf{D}}$ is displacement electric current density, $\mathbf{J}$ is conduction current density and $\dot{\mathbf{B}}$ is the magnetic displacement current density. These relations are good enough to solve most of the problems that were considered till now.

However, there are many cases where the knowledge of fictitious magnetic currents and charges is extremely useful. Although the fields are generated by electric current and charge distributions, it is possible to compute the fields from the equivalent distributions of fictitious magnetic currents and charges.

For example, an electric current loop is considered to be equivalent to a magnetic dipole. The EM field generated by a small horizontal electric current loop is identical to that of a vertical magnetic dipole. Similarly, the fields produced by magnetic current loop and electric dipole are the same.

Keeping these facts in mind, the first and second Maxwell's equations are written as:

$$\nabla \times \mathbf{H} = \dot{\mathbf{D}} + \mathbf{J} \tag{9.1}$$

and
$$\nabla \times \mathbf{E} = -\dot{\mathbf{B}} - \mathbf{M} \tag{9.2}$$

where $\mathbf{M}$ is the magnetic conduction current density (V/m^2). Analogy of surface electric current density, $\mathbf{J}_s$ is the surface magnetic current density, $\mathbf{M}_s$. It is evident from Equations (9.1) and (9.2) that there exists full symmetry and duality in Maxwell's equations. These current densities are defined as

$$\mathbf{J}_s \equiv \mathbf{a}_n \times \mathbf{H}$$

and
$$\mathbf{M}_s \equiv \mathbf{E} \times \mathbf{a}_n$$

Maxwell's equations for magnetic currents in the absence of electric currents are given by

$$\nabla \times \mathbf{H}^m = \in \dot{\mathbf{E}}^m + \mathbf{J}$$

$$\nabla \times \mathbf{E}^m = -\mu \dot{\mathbf{H}}^m - \mathbf{M}$$

In the absence of magnetic currents, they are expressed as

$$\nabla \times \mathbf{H}^e = \in \dot{\mathbf{E}}^e + \mathbf{J}$$

$$\nabla \times \mathbf{E}^e = -\mu \dot{\mathbf{H}}^e$$

The superscripts e and m refer to fields due to electric and magnetic currents respectively.

It is interesting to introduce **Vector electric potential**, bringing analogy from magnetic fields.

Vector electric potential, F is defined as

$$\nabla \times \mathbf{F} \equiv -\in \mathbf{E}^m$$

or,

$$\mathbf{F} \equiv \frac{1}{4\pi} \int_{\upsilon} \frac{\in \mathbf{M}(r)}{r} \, d\upsilon$$

9.3 RECIPROCITY IN ELECTROMAGNETIC FIELD THEORY

If the currents $\mathbf{J}^x$, $\mathbf{M}^x$ produce fields $\mathbf{E}^x$, $\mathbf{H}^x$ and if the currents $\mathbf{J}^y$, $\mathbf{M}^y$ produce fields $\mathbf{E}^y$, $\mathbf{H}^y$ in a linear and isotropic medium, then the Maxwell's equations are:

$$\left.\begin{aligned} \nabla \times \mathbf{H}^x &= a\mathbf{E}^x + \mathbf{J}^x \\ \nabla \times \mathbf{E}^x &= -b\mathbf{H}^x - \mathbf{M}^x \\ \nabla \times \mathbf{H}^y &= a\mathbf{E}^y + \mathbf{J}^y \\ \nabla \times \mathbf{E}^y &= -b\mathbf{H}^y - \mathbf{M}^y \end{aligned}\right\} \tag{9.3}$$

where $a = \sigma + j\omega\in$, $b = j\omega\mu$.

Using the properties of isotropic media, from Equation (9.3), we can write

$$\int_S (\mathbf{E}^y \times \mathbf{H}^x - \mathbf{E}^x \times \mathbf{H}^y) . d\mathbf{S} = \int_{\upsilon} (\mathbf{E}^x . \mathbf{J}^y - \mathbf{H}^x . \mathbf{M}^y - \mathbf{E}^y . \mathbf{J}^x + \mathbf{H}^y . \mathbf{M}^x) \, d\upsilon \tag{9.4}$$

If all the sources are contained within a finite volume, the far-fields constitute spherical waves and the surface integral of Equation (9.4) becomes zero. Hence

$$\int_{\upsilon} (\mathbf{E}^x . \mathbf{J}^y - \mathbf{H}^x . \mathbf{M}^y) \, d\upsilon = \int_{\upsilon} (\mathbf{E}^y . \mathbf{J}^x - \mathbf{H}^y . \mathbf{M}^x) d\upsilon \tag{9.5}$$

Equation (9.5) is written as

$$\boxed{\langle x, y\rangle = \langle y, x\rangle} \tag{9.6}$$

where $$\langle x, y\rangle = \int_{\upsilon} (\mathbf{E}^x . \mathbf{J}^y - \mathbf{H}^x . \mathbf{M}^y)\, d\upsilon$$

and $$\langle y, x\rangle = \int_{\upsilon} (\mathbf{E}^y . \mathbf{J}^x - \mathbf{H}^y . \mathbf{M}^x)\, d\upsilon$$

$\langle x, y\rangle$ represents reaction of field y on the source x. Similarly, $\langle y, x\rangle$ represents reaction of x on y.

The principle of reciprocity is an inter-relationship between any two source-field pairs. Equations (9.5) and (9.6) are the expressions of such pairs. It is possible to state that if

$$\langle x, y\rangle = \langle y, x\rangle$$

then the **reciprocity** exists.

9.4 REACTION CONCEPT

If x and y represent two sets of source currents and fields, then according to the principle of reciprocity, we can write the above as

$$\boxed{\langle x, y\rangle = \langle y, x\rangle}$$

Here $\langle x, y\rangle$ is known as 'reaction' of field x on source, y and $\langle y, x\rangle$ is known as 'reaction' of field y on source x.

$\langle x, x\rangle$ represents the reaction of field x on its own source x and it is known as **self-reaction**.

Self reaction is defined as

$$\langle x, x\rangle \equiv -\int_{\upsilon} \mathbf{H}.\mathbf{M} d\upsilon, \text{ (watt)}$$

9.5 INDUCTION AND EQUIVALENCE THEOREMS

Let the region 1 represented by μ_1, σ_1, $\in_1$ contain a system of sources, S_1 and let region 2 represented by $\mu_2, \sigma_2, \in_2$ contain no sources (Fig. 9.1).

S_1 indicates the presence of sources and S_f indicates source free region.

In region 1, the actual fields are $(\mathbf{E}_i + \mathbf{E}_r, \mathbf{H}_i + \mathbf{H}_r)$ and in region 2 the fields are $\mathbf{E}_2$ and $\mathbf{H}_2$.

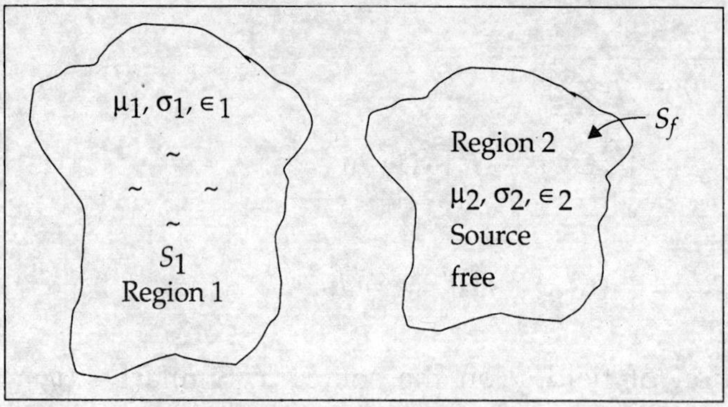

Fig. 9.1 Regions containing sources and source free regions

Induction theorem states that the induced or scattered fields are produced by electric and magnetic current sheets and the current densities are represented by

$$\mathbf{J}_s = \mathbf{a}_n \times \mathbf{H}_i$$

$$\mathbf{M}_s = -\mathbf{a}_n \times \mathbf{E}_i$$

where

$$\mathbf{H}_i = (\mathbf{H}_{t2} - \mathbf{H}_{tr})$$

$$\mathbf{E}_i = (\mathbf{E}_{t2} - \mathbf{E}_{tr})$$

$\mathbf{E}_{t2}$ = tangential component of $\mathbf{E}_2$

$\mathbf{E}_{tr}$ = tangential component of $\mathbf{E}_r$

$\mathbf{H}_{t2}$ = tangential component of $\mathbf{H}_2$

$\mathbf{H}_{tr}$ = tangential component of $\mathbf{H}_r$

$\mathbf{E}_i$ = induced electric field

$\mathbf{H}_i$ = induced magnetic field

$\mathbf{a}_n$ = unit normal vector to the sheet

Equivalence theorem states that current distributions and induced fields are related by

$$\mathbf{J}_s = \mathbf{a}_n \times \mathbf{H}_i$$

$$\mathbf{M}_s = -\mathbf{a}_n \times \mathbf{E}_i$$

where

$$\mathbf{H}_i = \mathbf{H}_{t2} \ [\mathbf{H}_{tr} = 0]$$

$$\mathbf{E}_i = \mathbf{E}_{t2} \ [\mathbf{E}_{tr} = 0]$$

It is clear from this theorem that the **Equivalence Theorem** is a particular case of **Induction Theorem**, that is, when region 1 and region 2 have the same constants, there are no reflected waves. Under these conditions both the theorems are identical.

9.6 ELECTROMAGNETIC INTERFERENCE AND COMPATIBILITY (EMI/EMC)

EMI/EMC Engineering is a unique and specialised subject. It came into prominence with the tremendous advances in technology and proliferation of a variety of electronic instruments and devices which produce electromagnetic emissions. In fact, the world is highly saturated with such emissions from a variety of sources. The effect of such sources is spread throughout the electronic spectrum.

EMI is defined as the undesirable signal which causes unsatisfactory operation of a circuit or device.

EMC is defined as the ability of electronic and communication equipment to be able to operate satisfactorily in the presence of interference and not be a source of interference to the nearby equipment.

Electromagnetic susceptibility (EMS) is the capability of a device to respond to EMI.

Basic types of EMI These are of two types. They are:

(a) **Intra-EMI** EMI is said to be intra-EMI if the functional characteristics of one module within an electronic equipment or system is disturbed due to EMI from another module.

(b) **Inter-EMI** EMI is said to be inter-EMI if the functional characteristics of one equipment is disturbed due to EMI generated by another equipment.

9.7 EMI SOURCES

These are divided mainly into two types:

I. Natural II. Man-made

I. Natural EMI sources are of the following types:

Terrestrial and Extra-terrestrial

1. **Terrestrial sources** These are atmospheric thunderstorms, lightning discharges and precipitation static.

2. **Extra-terrestrial sources** These are sun-disturbed and quiet, cosmic noise and radio stars.

II. Man-made EMI sources are many.

1. **Electric power sources** These consist of:

   (a) Generation equipment, conversion (step up/down) equipment like faulty transformers and faulty insulators.

(b) Transmission equipment like pick-up and re-radiation and faulty insulations.

(c) Distribution equipment like faulty transformers, faulty wiring, faulty insulation, poor grounding, pick-up and re-radiation.

2. **Electronic communication devices** These are:

(a) Police radio

(b) All types of radars

(c) Cellular and mobile communication

(d) Satellite communication

(e) Point-to-point communication

(f) Television and Radio

3. **Machines and Tools** These are:

(a) Industrial machines

- Electric cranes
- Fork-lift trucks
- Milling machines
- Printing presses
- Punch presses and so on

(b) Office/business machines

- Computers
- Cash registers
- Electronic typewriters
- Photocopiers and so on

(c) Welders and heaters

- Arc welders
- RF stabilised welders
- Induction heaters and so on

(d) Transporters

- Elevators
- Escalators
- Conveyer belts and so on

(e) Power tools

- Electric drills and grinders

- Mixers
- Electric hand-saws and so on

(f) Appliances

- Air-conditioners
- Refrigerators
- Microwave ovens
- Vacuum cleaners
- Electric lawn mowers and so on

4. **Ignition systems**

(a) Engines

(b) Vehicles

- Automobiles
- Aircrafts
- Tanks
- Trucks
- Tractors and so on

(c) Tools

- Auxiliary generators
- Lawn mowers
- Portable saws and so on

5. **Consumer and medical electric systems**

(a) Lights

- Fluorescent lamps
- Faulty incandescent lights
- Neon lights
- RF excited gas displays
- Light dimmers and so on

(b) Entertainment

- Home computers
- Cassette players/recorders
- Television receivers

(c) Appliances
- Electric shavers
- Washing machines
- Grinders/mixers
- Vacuum cleaners

(d) Medical equipment
- X-ray machines
- Defibrillators
- Ultrasonic scanners
- Scanners

6. **Nuclear** These are:

(a) Nuclear submarines, ships and nuclear aircraft

(b) Nuclear detonators

(c) Nuclear power stations

The devices which are susceptible to EMI are listed below.

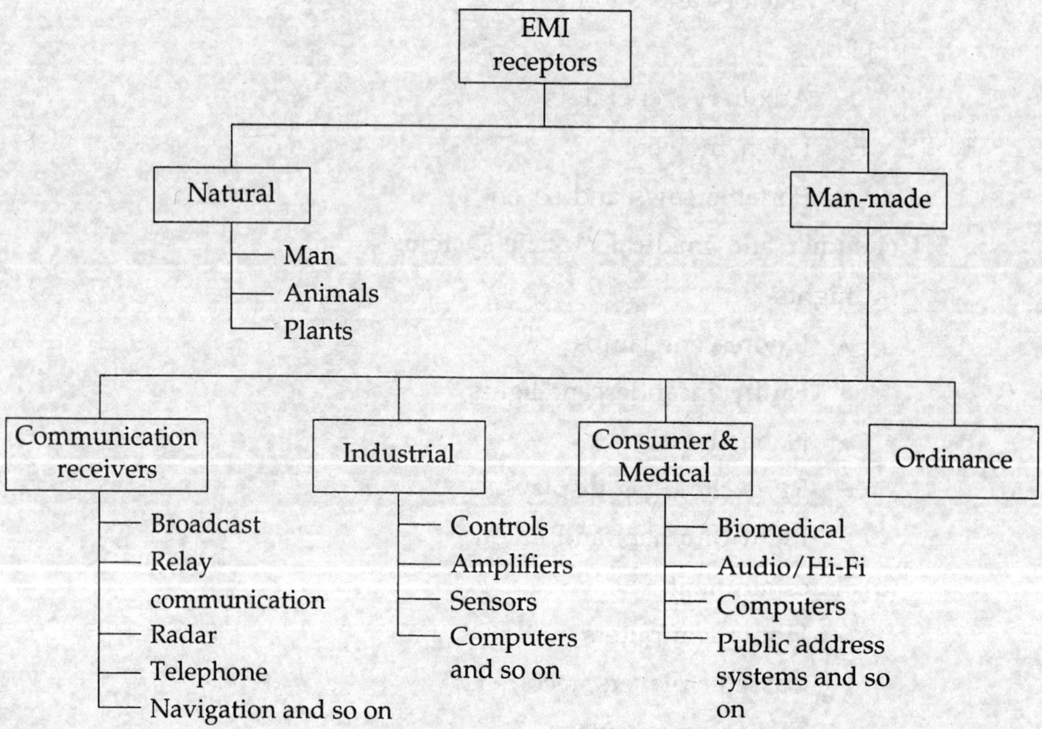

A typical inter-system EMI environment is shown below.

✱ Ship (E, S)

✱ Power lines (E)

✱ TV (E, S)

✱ Aircraft (E, S)

✱ FM (E, S)

✱ Computers (E, S)

✱ Radio (E, S)

✱ Telecommunications (E, S)

✱ Automobiles (E, S)

✱ Police radio (E, S)

✱ Walkie talkie (E, S)

✱ Microwave links (E, S)

✱ Wireless transmitter (E, S)

E represents Emitter

S represents Susceptible

EMI is coupled in two ways:

1. By conduction—Here the EMI is routed along power supply and signal lines.
2. By radiation—Here the EMI is coupled through radiation from wires and antennas. This can be either through near-fields or far-fields.

9.8 EFFECTS OF EMI

EMI is considered to be a silent and unknown threat. On many occasions, intermittent malfunctions or random failures of equipment are experienced by several users. They are attributed to be due to the presence of EMI. In view of this, the threat due to EMI is silent and has proved to be hazardous.

Common effects of EMI

(i) **Annoying effects** Very often, momentary and random disturbances occur in radio and television reception.

(ii) **Disturbing effects** EMI causes unwanted reset and change of status in settings in computers and digital equipment. Malfunctioning of computer keyboards is also noticed.

(iii) **Catastrophic situations** Burning of electronic components, loss of data, change of threshold settings, improper or unwanted operations and sometimes biological hazards occur very often due to EMI.

9.9 METHODS TO ELIMINATE EMI OR DESIGN METHODS FOR EMC

The effective methods to eliminate EMI are:

(i) Shielding
(ii) Grounding
(iii) Bonding
(iv) Filtering
(v) Isolation
(vi) Separation and orientation
(vii) Circuit impedance level control
(viii) Cable design
(ix) Cancellation techniques in frequency or time domain
(x) Proper selection of cables, passive components
(xi) Antenna polarisation control
(xii) Balancing

The first four methods are popular for different applications. A few basics are presented below.

Shielding

The main object of shielding is to restrict radiations to a specified region to prevent it from entering susceptible devices. The quality of shielding is expressed in the form of **shielding effectiveness** of the material. Shielding of materials can be solids, screens and braids. They can be in the form of boxes, partitions, cables and connector shields.

Shielding effectiveness (Se) It is defined as the ratio of incident power to transmitted power.

$$Se = \frac{W_i}{W_t}$$

$$Se\ (\text{dB}) \equiv 10 \log \frac{W_i}{W_t}$$

For electric fields, we have

$$Se\ (\text{dB}) \equiv 20 \log \frac{E_i}{E_t}$$

For magnetic fields,

$$S_e\ (\text{dB}) \equiv 20\log \frac{H_i}{H_t}$$

The shielding effectiveness for electric fields in dB can be written as the sum of three terms, that is,

$$S_e\,(\text{dB}) = R_{\text{dB}} + A_{\text{dB}} + M_{\text{dB}}$$

where R_{dB} = reflected loss

A_{dB} = absorption loss

M_{dB} = multiple re-reflections and transmission losses

Shielding is done by good conducting materials. Some examples of shielding materials are silver, copper, aluminium, gold, brass and bronze.

Reflection loss is very high at low frequencies and for high conductivity materials.

It is expressed as

$$R_{\text{dB}} = 168 + 10\log_{10}\left(\frac{\sigma_r}{\mu_r f}\right)$$

where σ_r = conductivity relative to copper

The absorption loss for the materials $(\mu_r\,\sigma_r >> 1)$ is given by

$$A_{\text{dB}} = 8.6859\, t/\delta$$

where t = thickness of the shield

δ = depth of penetration

The multiple reflection loss is given by

$$M_{\text{dB}} = 20\log_{10}\left|1 - \left(\frac{\eta_0 - \eta}{\eta_0 - \eta}\right)^2 e^{-2t/\delta}\, e^{-j2\beta t}\right|$$

M_{dB} is negligible for shields made of good conductors $(\eta << \eta_0)$. If $t >> \delta$ and $\beta = \alpha = \frac{1}{\delta}$.

$$M_{\text{dB}} \approx 20\log_{10}|1 - e^{-2t/\delta}\, e^{-j2t/\delta}|$$

The effective magnetic shielding materials are the ones with high permeability. Examples: Mumetal, Steel.

Grounding

Grounding provides a conducting path between electronic devices and the ground. Ground is nothing but some reference point. It is a circuit concept.

The ideal ground is characterised by zero potential and impedance.

The types of grounding techniques are:

(a) Floating ground: It isolates circuits from a common ground plane. Sometimes it may be hazardous.

The ground plane is in the form of a wire or a conductive rod.

(b) Single-point grounding: It reduces the effects of facility ground currents. This is used to control EMP energy.

(c) Multiple point grounding: It reduces ground lead lengths.

Bonding

It provides a low-impedance path between two conducting surfaces. It is a part of grounding and represents its physical implementation.

It creates a homogeneous structure for current flow and suppresses the creation of potentials between two metallic parts.

Bonding is useful to protect against the effects of shocks and to protect circuits from current return paths. They reduce potential difference between the devices and carry large faulty currents.

Bonding is of two types:

Direct bonding is made by metal-to-metal between the connected elements. **Indirect bonding** is made by contact using conductive jumpers.

Bonding quality is represented by its DC and AC resistances and also bonding effectiveness.

$$R_{dc} \equiv \frac{l}{\sigma s}$$

$$R_{ac} \equiv \frac{l}{\sigma \delta w}$$

where l = length of the bond, (m)

σ = conductivity, (mho/m)

s = cross-sectional area, (m^2)

δ = depth of penetration, (m)

w = width of the bond, (m)

Bonding effectiveness (Be) It is defined as the difference between induced voltage in the case of an equipment with a bond trap and the induced voltage in the case of an equipment without a bond trap. It is expressed in dB.

Filtering

These are used to filter out conducted EMI. The filtering effectiveness is expressed by Insertion loss (IL). It is defined as

$$IL \equiv 20\log\frac{V_{01}}{V_{02}}$$

where V_{01} = output voltage with filter

V_{02} = output voltage without filter

For low pass inductive filters,

Insertion loss $= 10\log\left(1+\dfrac{\omega L}{2R}\right)\text{dB}$

For low pass capacitive filter,

Insertion loss $= 10\log\left(1+\dfrac{\omega RC}{2}\right)\text{dB}$

where ω = angular frequency $= 2\pi f$

9.10 NEED FOR EMC STANDARDS

EMC standards are required for trouble free co-existence and to ensure satisfactory operation. They are also required to provide compatibility between electrical, electronic, computer, control and other systems. Standards are required as manufacturer-user interaction and user's knowledge on EMI are limited.

They are also required for establishing harmonised standards to reduce international trade barriers and to improve product reliability and life of the product.

9.11 EMC STANDARDS

These are of two types:

Military Standards

These include emission and susceptibility standards. Emission standards specify emission limits in voltage or current, power or field strengths in specified frequency ranges. Susceptibility standards specify conducted spike or radiated field parameters.

Military EMC standards are made in order to ensure system-to-system compatibility in a real time military environment. Equipments are classified based on their deployment environment. In these standards, test procedures are well defined. Military standards are more stringent than civilian standards. Most of the military standards are broadly based on MIL-STD 461 and 462.

Civilian Standards

Civilian EMC standards are applicable for equipments used for commercial, industrial and domestic applications. The emission standards are specified to protect broadcast services from interference. These also take into account the physiological interference effects experienced by human beings.

9.12 ADVANTAGES OF EMC STANDARDS

The advantages are:

1. Compatibility, reliability and maintainability are increased.
2. Design safety margin is provided.
3. The equipment operates satisfactorily in EMI scenario.
4. Product life is increased.
5. Higher profits are possible.

9.13 EMC STANDARDS IN DIFFERENT COUNTRIES

S. no.	*Standard*	*Meaning*	*Country*
1.	CISPR (IEC)	Committee International Special Perturbations Radioelectriques—Europe	International Committee
2.	FCC	Federal Communications Council	USA
3.	SAE	Society of Automobile Engineers	Trade Association Technical Committee
4.	VG	Military Standard	Germany
5.	VDE	Verband Deutscher Electrotecknikev	Germany
6.	ISI	EMI measurements and measuring apparatus	India
7.	DEF STD	59-41 British MIL—STD	UK
8.	GAM-EG-13	France MIL—STD	France
9.	CENELEC	European committee for electro-technical standardisation	Europe
10.	EN	European Norms	Europe

9.14 BIOLOGICAL EFFECTS OF EMI/EMR (ELECTROMAGNETIC INTERFERENCE, ELECTROMAGNETIC RADIATION)

EM waves, light, heat, X-rays and gamma rays are all different forms of electromagnetic radiation. However, they differ in their wavelength. These radiations have hazardous effects on man and material. The effects can be divided into two categories:

1. Thermal Effects
2. Non-thermal Effects

1. **Thermal Effects** The effects of EM radiation whose frequencies range between 0.1 to 100 GHz are given in Table 9.1.

Table 9.1 Effects of EM Radiation

Frequency (GHz)	*Effect*
0.1	Warming of exposed areas.
0.15–1.2	Overheating occurs and causes damage to internal organs.
1.0–3.3	Lens of the eye and kidneys are susceptible to damage when tissues are heated up.
3.3–10	Noticeable skin heating occurs.
10–100	Skin acts as either a reflector or an absorber and hence heating takes place.

The **damaging levels** depend on frequency, ambient temperature, body resistance and weight of individuals.

Exposure over an energy density of **10 mw/cm^2** at any frequency is considered to be not safe.

2. **Non-thermal Effects**

(i) Minor changes in human blood properties take place.
(ii) Buzzing sound is heard upon exposure to EMR.
(iii) Abnormalities of the chromosome structure occurs.
(iv) Movement, orientation and polarisation of protein molecules are noticed.
(v) Epigastric distress, emotional upsets and nausea have been noticed.

Radiation limits in the frequency range of 0.1 to 100 GHz when personnel are exposed to EM radiations are:

(i) The average incident power density should not exceed 10 mw/cm^2 for exposures greater than 30 seconds.

(ii) The average incident energy density should not exceed 300 mJ/cm^2 for intermittent exposures between 3 and 30 seconds.

(iii) According to IEEE, safe power density level is 2 W/m^2.

9.15 ELECTROSTATIC DISCHARGE (ESD)

ESD results from the separation of static charge.

Effects of ESD:

1. High electrostatic field is created by the charge separation prior to the ESD arc.
2. High arc discharge currents are generated.

Methods of Separation of Charge

Rubbing of two types of insulating material causes charge to be separated from one material to the other. The charge separation creates high fields and hence causes a voltage difference between the two materials. This leads to a breakdown of the air and intensive arcs are produced. A direct conduction path will result if one material comes in contact with a conductor. This phenomenon is a familiar one when we walk across a carpet on a dry day and touch a metallic doorknob. When the resulting arc current enters sensitive devices, they are damaged.

9.16 ORIGIN OF ESD EVENT

If two neutral insulators are brought in contact, charge is transferred from one to another. When they are separated, they become charged. At this time, one material is positively charged and the other one is negatively charged. The degree of charge transfer depends on many factors. For example, if Nylon is rubbed against Teflon, electrons will be transferred from Nylon to Teflon. Hence Nylon acquires (+)ve charge and Teflon acquires (–)ve charge.

In fact, touching an insulator with a conductor creates charge separation but the degree is less than that of two insulators.

Grounding a conductor will bleed off the charge but grounding an insulator will not.

ESD wrist straps can be worn to prevent building up of charge.

If a high resistance of the order of 1 MΩ is connected from the installer's wrist to earth ground, any static charge stored on the body's skin is discharged to the ground.

It is interesting to know some facts about the frictionally generated charges.

(i) Sparks with a voltage of about 5 KV are generated when a man walks on a rug on a dry day, touches a metallic body and hits. Nothing happens to the man as the charge is less even when the voltage is high. On the other hand, when the charge is high with less generated voltage, it can be fatal.

(ii) The charges attract dust particles and damage electronic chips and circuits.

(iii) They may cause fire and explosion in inflammable environment.

ESD causes component destruction by:

1. direct conduction
2. secondary arcs or discharges
3. capacitive coupling
4. inductive coupling

The techniques to prevent the ESD effects are:

1. Prevent occurrence of ESD event
2. Prevent coupling
3. Create an inherent immunity to ESD event.

9.17 ELECTROMAGNETIC PULSE (EMP)

Whenever a nuclear detonation takes place, energy is released and this appears as electromagnetic pulse. These pulses affect the operation of electrical and electronic equipment even at great distances. The far-field characteristics of the radiated EMP produced by a single high altitude detonation is found to have significant variation in amplitude, wave shape and propagation characteristics. These depend on weapon yield, location of the observer, the height and location of the burst and orientation of earth's magnetic field.

The EMP amplitude spectrum covers a broad frequency range which extends from low to microwave frequencies.

A typical EMP waveform can be represented by

$$E(t) = \frac{E_m}{a}\,(e^{-\alpha_1 t} - e^{-\alpha_2 t})$$

where t = time in seconds

a = normalisation factor $= e^{-\alpha_1 t_1} - e^{-\alpha_2 t_1}$

t_1 = time to peak value

$E_m = E\,(t_1)$

$$\alpha_1 = 1.5 \times 10^6 \left(\frac{1}{s}\right)$$

$$\alpha_2 = 260 \times 10^6 \left(\frac{1}{s}\right)$$

The Fourier transform of the equation is

$$E\,(j\omega) = \frac{E_m}{a} \frac{(\alpha_2 - \alpha_1)}{(j\omega + \alpha_1)\,(j\omega + \alpha_2)}$$

A typical EMP wave form is shown in Fig. 9.2.

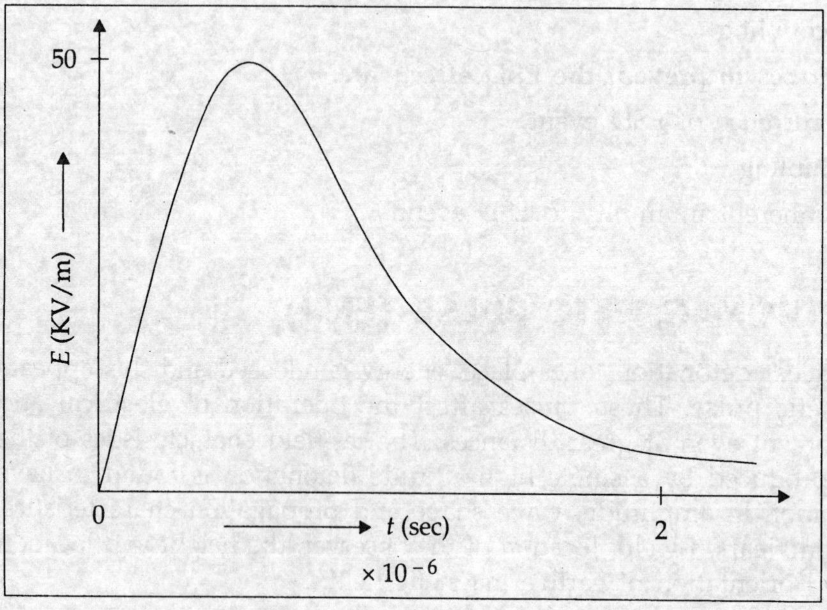

Fig. 9.2 A typical EMP wave form

9.18 NUMERICAL TECHNIQUES FOR THE ANALYSIS OF ELECTROMAGNETIC FIELDS

Most of the engineering problems/models are in the form of differential equations. Systems with one independent variable can be modeled by ordinary differential equations. But systems with two or more independent variables require the use of partial differential equations. It is possible to obtain the solution for some of these equations in closed form. But majority of the large sets of simultaneous differential equations and non-linear ones do not have analytical solutions and hence the application of numerical techniques is required.

The most useful numerical techniques are:

(i) Finite Difference Method (FDM)
(ii) Finite Element Method (FEM)
(iii) Method of Moment (MOM)

In all these methods, discretisation of a continuous region into a finite number of sections is made. Moreover, they require the solution of a set of algebraic equations only. The solution of differential or integral equations is not required. These methods are extremely useful for the design as well as analysis of an electromagnetic system. Problems involving the determination of current distribution in a current element or a dipole and the field distribution in a slot are easily solved by these methods.

For modeling static field systems, Laplace's and Poisson's equations can be used. These require the solution of differential equations. It is possible to obtain analytical solutions when the regions have regular geometrical shapes like triangular, circular, elliptical, rectangular and so on. But for arbitrarily shaped regions, the field solutions are obtained using FDM, FEM and MOM techniques.

9.19 FINITE DIFFERENCE METHOD (FDM)

It is a numerical technique in which the domain is divided into a number of discrete points. It consists of a set of difference equations. FDM is useful to solve the spatial distribution of electromagnetic fields in different media.

Solution by this method is approximate and error can be reduced by taking more number of discrete points in a specified region. For example, Poisson's equation can be solved for $V(x, y, z)$ in any arbitrary region using the boundary conditions. However, the analysis for fields in an arbitrarily shaped region is complex by this method.

9.20 FINITE ELEMENT METHOD (FEM)

FEM is a technique useful to solve the problems containing differential equations numerically. This method is easily applicable to solve the problems involving arbitrarily shaped regions. The method basically consists of:

(i) discretisation of the field region
(ii) derivation of equations for each element
(iii) assembling of all elements in the field region
(iv) solving the set of equations

Solution can be obtained either by **Iterative** method or by **Band matrix** method.

FEM is a better method than FDM and MOM as it is easy to apply to complex regions and it is possible to write universal general purpose programmes to solve different types of problems. At the same time, it is constrained by a few drawbacks. For instance, the preparation of data is complex and involved. The programming part is also involved compared to FDM and MOM.

9.21 METHOD OF MOMENTS (MOM)

It is basically an incremental numerical technique. It is useful to find the integrand from the integral equation. An equation is known as an integral equation if the integrand is an unknown.

This method makes use of equations of unknown potentials or fields in integral form to find out the potential or field distribution in a medium.

An example of integral equation is

$$V(r) = \int_{\upsilon} \frac{\rho_{\upsilon}(r')\, d\upsilon'}{4\pi \epsilon R}$$

where $\rho_{\upsilon}(r')$ is an unknown.

$$r' = (x', y', z'), \quad r = (x, y, z)$$

$$\rho_{\upsilon}(r') = \rho_{\upsilon}(x', y', z')$$

$$V(r) = V(x, y, z)$$

ρ_{υ} is the source of the potential function. r is the distance between the source and potential. ρ_{υ} is an unknown function and V is a known function. The method of moments consists of the following steps:

1. Express the integral equation of the potential function in the form of

   $$F(X) = y$$

   where F is an integral operator and y is a known function.

   X is an unknown output function.

2. Expand X as a linear combination of N terms, that is,

   $$X(x) = a_1 X_1(x) + a_2 X(x) + \ldots + a_N X_N(x)$$

   where a_N is an unknown constant and $X_N(x)$ is a known function and is called a **basis** or **expansion function**.

3. Express F in the form of

   $$\sum_{n=1}^{N} a_n F(X_n) = y$$

This consists of one equation with N unknowns. It is not possible to solve it as such. It is required to have N independent linear equations. Hence, applying boundary conditions at N points, we have N equations.

4. Express the above summation in the form of

$$\sum_{n=1}^{N} a_n \, F(X_n) = y_m$$

$$m = 1, 2, 3, \ldots, N$$

5. Express the above expression in Matrix form, that is,

$$[F_{mn}]\,[a_n] = [y_m]$$

6. Find a_n from

$$[a_n] = [F_{mn}]^{-1}\,[y_m]$$

For the sake of analogy, we can have

$$[a_n] = [I_n]$$

$$[F_{mn}] = [z_{mn}]$$

$$[y_m] = [V_m]$$

$$\boxed{[I_n] = [Z_{mn}]^{-1}\,[V_m]}$$

SOLVED PROBLEMS

Problem 9.1 For a copper shield, find the reflection loss in dB at (a) 1 kHz and (b) 10 MHz.

Solution (a) $f = 1\,\text{kHz}$

For copper $\sigma_r = 1,\ \mu_r = 1$

$$R_{\text{dB}} = 168 + 10\log_{10}\left(\frac{\sigma_r}{\mu_r\, f}\right)$$

$$= 168 + 10\log_{10}\left(\frac{1}{10^3}\right)$$

$$= 168 - 30$$

$$\boxed{R_{\text{dB}} = 138\,\text{dB}}$$

(b) At $f = 10$ MHz

$$R_{dB} = 168 + 10\log_{10}\left(\frac{1}{10^7}\right)$$

$$= 168 - 70$$

$$\boxed{R_{dB} = 98 \text{ dB}}$$

Problem 9.2 Determine reflection loss for a steel shielding at (a) 10 kHz and (b) 10 MHz. For steel, $\mu_r = 1{,}000$, $\sigma_r = 0.1$.

Solution (a)

$$f = 10 \text{ kHz} = 10^4 \text{ Hz}$$

$$\mu_r = 1{,}000$$

$$\sigma_r = 0.1$$

$$R_{dB} = 168 + 10\log_{10}\left(\frac{0.1}{10^3 \times 10^4}\right)$$

$$= 168 + 10\log_{10}(10^{-8})$$

$$\boxed{R_{dB} = 88 \text{ dB}}$$

(b) $f = 10$ MHz

$$R_{dB} = 168 + 10\log_{10}\left(\frac{0.1}{10^3 \times 10^7}\right)$$

$$= 168 - 110$$

$$\boxed{R_{dB} = 58 \text{ dB}}$$

Problem 9.3 For nickel shield, find the absorption loss when $t/\delta = 1$, $t/\delta = 2$, $t/\delta = 4$.

Solution If $t/\delta = 1$

$$A_{dB} = 8.6859\, t/\delta$$

$$= 8.6859 \times 1$$

$$\boxed{A_{dB} = 8.6859 \text{ dB}}$$

If $t/\delta = 2$

$$A_{dB} = 8.6859\, t/\delta$$

$$= 8.6859 \times 2$$

$$\boxed{A_{dB} = 17.3718 \text{ dB}}$$

If $t/\delta = 4$

$$A_{dB} = 8.6859t/\delta$$
$$= 8.6859 \times 4$$
$$\boxed{A_{dB} = 34.7436\,\text{dB}}$$

Problem 9.4 Find multiple reflection loss due to a shielding material for which $\eta << \eta_0$ and $t/\delta = 0.1$.

Solution We have

$$M_{dB} = 20\log_{10} |1 - e^{-2t/\delta}\ e^{-j2t/\delta}|$$
$$= 20\ \log_{10} |1 - e^{-0.2}\ e^{-j0.2}|$$
$$\boxed{M_{dB} = 14.8\,\text{dB}}$$

POINTS/FORMULAE TO REMEMBER

- The General form of second Maxwell's equation is $\nabla \times \mathbf{E} = -\dot{\mathbf{B}} - \mathbf{M}$.
- The surface electric current density is $\mathbf{J}_s = \mathbf{a}_n \times \mathbf{H}$.
- The surface magnetic current density is $\mathbf{M}_s = \mathbf{E} \times \mathbf{a}_n$.
- The vector electric potential is defined as

$$\nabla \times \mathbf{F} \equiv -\epsilon \mathbf{E}^m$$

$$\mathbf{F} \equiv \frac{1}{4\pi} \int_{\upsilon} \frac{\epsilon \mathbf{M}(r)}{r} d\upsilon$$

- Self-reaction is defined as $\langle x, x \rangle = -\int_{\upsilon} \mathbf{H}.\mathbf{M}\, d\upsilon$.
- The principle of reciprocity gives an inter-relationship between any two-source field pairs.
- By reciprocity, $\langle x, y \rangle = \langle y, x \rangle$.
- Most of the devices and systems are sources of EMI and also they are susceptible.
- Systems should be EMC designed for long life.
- EMI sources are natural and man-made.
- Domestic electric appliances are sources of EMI.
- Automobiles are sources of EMI.
- Medical devices and instruments are susceptible to EMI.
- Shielding, grounding, bonding, isolation and filtering are the major EMC design techniques.
- Military and civilian EMC standards are different in EMI permissible limits.
- FEM, FDM and MOM are important numerical techniques.
- ESD is more hazardous during lightning.
- An electric current loop is equivalent to a magnetic dipole.
- The electromagnetic fields produced by a small horizontal electric current loop are identical to that of a vertical magnetic dipole.
- The fields produced by magnetic current ioop and electric dipole are the same.

OBJECTIVE QUESTIONS

1. Electric power lines are sources of EMI. (Yes/No)
2. Computers are susceptible to EMI. (Yes/No)
3. Walkie Talkies radiate and they are susceptible to EMI. (Yes/No)
4. Automobiles are sources of EMI. (Yes/No)
5. Aircraft is not susceptible to EMI. (Yes/No)
6. Submarines are not susceptible to EMI. (Yes/No)
7. Radar is a man-made source of EMI. (Yes/No)
8. Human beings are susceptible to EMR. (Yes/No)
9. Animals are susceptible to EMR. (Yes/No)
10. Medical equipment is susceptible to EMI. (Yes/No)
11. Kitchen mixie is a source of EMI. (Yes/No)
12. Hair drier is a source of EMI. (Yes/No)
13. Television is susceptible to EMI. (Yes/No)
14. Fluorescent lamp is a source of EMI. (Yes/No)
15. Isolation is a method of EMI control. (Yes/No)
16. Grounding is a method of EMC design. (Yes/No)
17. Proper cable design reduces EMI. (Yes/No)
18. Washing machines are sources of EMI. (Yes/No)
19. Light dimmers in automobiles are sources of EMI. (Yes/No)
20. Cell phones are sources of EMI. (Yes/No)
21. FEM is more useful than FDM. (Yes/No)
22. FEM is a numerical technique using a set of algebraic equations. (Yes/No)
23. Numerical evaluation of integrals is more accurate than the numerical evaluation of differential equations. (Yes/No)

24. Reciprocity exists in field theory. (Yes/No)
25. If $\langle x, y\rangle = \langle y, x\rangle$ reciprocity is said to exist. (Yes/No)
26. It is easy to shield electric fields than to shield magnetic fields. (Yes/No)
27. Reflection loss is very large for electric fields and plane waves. (Yes/No)
28. Reflection loss is usually small for low frequency magnetic fields. (Yes/No)
29. Shielding is said to be excellent, if the attenuation is > 90 dB. (Yes/No)
30. The unit of magnetic conduction density is _________.
31. The unit of surface magnetic current density is _________.
32. Surface magnetic current density is defined as _________.
33. Vector electric potential, **F** is defined by _________.
34. The unit of vector electric potential is _________.
35. The unit of reaction is _________.
36. Self-reaction is represented by _________.
37. Reciprocity in field theory is said to exist if _________.
38. EMI means _________.
39. EMC represents _________.
40. EMS means _________.
41. EMI is mainly from _________.
42. EMI enters the system by _________.
43. EMI can be controlled by _________.
44. EMI from antennas can be controlled by _________.
45. Filtering is a technique suitable to control _________.
46. Two types of EMC standards are _________.
47. FCC means _________.
48. CISPR represents _________.
49. The main advantage of EMC standards is that _________.

50. VDE is __________ standard.

51. The main effects of EMI/EMR are __________.

52. High EMR susceptible organs of human beings are __________.

53. ESD means __________.

54. ESD originates due to __________.

55. The safe EMR exposure limit is __________.

56. The effect of ESD is that __________.

57. ESD causes __________.

58. EMP means __________.

59. The damaging levels due to EMR in human beings depend on __________.

60. Two major sources of EMI are __________.

61. FDM means __________.

62. FEM represents __________.

63. MOM means __________.

64. The unit of self-reaction is __________.

Answers

1. Yes	2. Yes	3. Yes	4. Yes	5. No
6. No	7. Yes	8. Yes	9. Yes	10. Yes
11. Yes	12. Yes	13. Yes	14. Yes	15. Yes
16. Yes	17. Yes	18. Yes	19. Yes	20. Yes
21. Yes	22. Yes	23. Yes	24. Yes	25. Yes
26. Yes	27. Yes	28. Yes	29. Yes	30. V/m^2

31. V/m 32. $\mathbf{M}_s = \mathbf{E} \times \mathbf{a}_n$ 33. $\nabla \times \mathbf{F} \equiv -\in \mathbf{E}^m$

34. Farad-Volt/m 35. Watts 36. $\langle x, x \rangle$

37. $\langle x, y \rangle = \langle y, x \rangle$ 38. Electromagnetic interference

39. Electromagnetic compatibility
40. Electromagnetic susceptibility
41. Man-made and natural sources
42. Conduction and radiation
43. Shielding
44. Polarisation control
45. Conducted EMI
46. Military and civilian
47. Federal Communications Council
48. Committee International Special Perturbations Radioelectriques
49. Product life is increased
50. German EMC standard
51. Thermal and non-thermal
52. Eyes and kidneys
53. Electrostatic discharge
54. Separation of static charges
55. 10 mW/cm^2 for exposures greater than 30 seconds
56. Intense arc discharge current takes place
57. Component destruction
58. Electromagnetic pulse
59. Frequency and ambient temperature
60. Nuclear detonations and lightning
61. Finite difference method
62. Finite element method
63. Method of moment
64. Watt

MULTIPLE CHOICE QUESTIONS

1. Maxwell's equation is
   (a) $\nabla \times \mathbf{E} = -\dot{\mathbf{B}} + \mathbf{J}$
   (b) $\nabla \times \mathbf{E} = -\dot{\mathbf{B}} - \mathbf{M}$
   (c) $\nabla . \mathbf{E} = -\dot{\mathbf{B}}$
   (d) $\nabla . \mathbf{D} = -\mathbf{B}$

2. The unit of magnetic conduction current density is
   (a) V/m (b) V/m^2 (c) A/m^2 (d) A/m

3. The unit of surface magnetic current density is
   (a) A/m (b) V/m (c) A/m^2 (d) A/m^3

4. According to reciprocity in EMF theory
   (a) $\langle x, y \rangle = \langle y, x \rangle$
   (b) $\langle x, y \rangle \neq \langle y, x \rangle$
   (c) $\langle x, y \rangle = \langle x, z \rangle$
   (d) $\langle z, x \rangle = \langle y, z \rangle$

5. Self-reaction is defined by
   (a) $\langle a, a \rangle = \int_{\upsilon} \mathbf{H}.\mathbf{M}\, d\upsilon$
   (b) $\langle a, a \rangle = -\int_{\upsilon} \mathbf{H}.\mathbf{M}\, d\upsilon$
   (c) $\langle a, a \rangle = \int_{\upsilon} \mathbf{H}\mathbf{M}\, d\upsilon$
   (d) $\langle a, a \rangle = \int_{\upsilon} \mathbf{M}.\mathbf{H}\, d\upsilon$

6. The unit of self-reaction is
   (a) watts (b) $watts/m^2$ (c) volt-amp (d) $amper^2$

7. The surface electric conduction current density is
   (a) $\mathbf{J}_s = \mathbf{a}_n \times \mathbf{H}$ (b) $\mathbf{J}_s = \mathbf{H} \times \mathbf{a}_n$ (c) $\mathbf{J}_s = \mathbf{a}_n . \mathbf{H}$ (d) $\mathbf{J}_s = \mathbf{H}$

8. The surface magnetic conduction density is
   (a) $\mathbf{M} = \mathbf{a}_n \times \mathbf{E}$ (b) $\mathbf{M} = -\mathbf{a}_n \times \mathbf{E}$ (c) $\mathbf{M} = \mathbf{E}.\mathbf{a}_n$ (d) $\mathbf{M} = \mathbf{E}$

9. The system which is not susceptible to EMI is
   (a) transmission lines
   (b) computer keyboard
   (c) television
   (d) aircraft

10. Radar is
    (a) conventional EMI source
    (b) natural EMI source
    (c) not an EMI source
    (d) unconventional EMI source

11. Lightning is
    (a) a source of EMI
    (b) a natural source of EMI
    (c) a man-made source of EMI
    (d) is not a source of EMI

12. EMI occurs in a system due to antenna by
    (a) conduction
    (b) radiation
    (c) filtering
    (d) isolation

13. EMC method is
    (a) shielding
    (b) conduction
    (c) radiation
    (d) non-conduction

14. American EMC standard name is
    (a) FCC
    (b) VDE
    (c) ISI
    (d) VG

15. Bio-effects are more prominent in the frequency range of
    (a) HF
    (b) UHF
    (c) Microwave
    (d) LF

16. EMR effects are not prominent in
    (a) human beings
    (b) semiconductors
    (c) conductors
    (d) insulators

17. FEM is
    (a) empirical method
    (b) numerical method
    (c) analytical method
    (d) none of these

18. EMI is more hazardous in
    (a) television
    (b) radars
    (c) spacecrafts
    (d) trains

19. Cell phones create
    (a) conducted EMI
    (b) radiated EMI
    (c) filtering
    (d) shielding

20. VDE is EMC standard in
    (a) America
    (b) India
    (c) Japan
    (d) Germany

Answers

1. (b)	2. (b)	3. (b)	4. (a)	5. (b)
6. (a)	7. (a)	8. (b)	9. (a)	10. (a)
11. (a)	12. (b)	13. (a)	14. (a)	15. (c)
16. (d)	17. (b)	18. (c)	19. (b)	20. (d)

EXERCISE PROBLEMS

1. For silver shield, find the reflection loss in dB at

   (a) 1 kHz (b) 10 MHz (c) 100 MHz

   $\mu_r = 1$, $\sigma_r = 1.05$ for silver.

2. Find the reflection loss for aluminium shielding at

   (a) 100 kHz (b) 10 MHz

   For aluminium, $\mu_r = 1$, $\sigma_r = 0.61$.

3. Determine absorption loss when $t/\delta = 1.3$, $t/\delta = 2.5$, $t/\delta = 3$ for steel shield.

4. Find the multiple loss due to a shielding material for which $\eta = 10\Omega$ and $t/\delta = 0.2$, $t/\delta = 0.4$.

OBJECTIVE QUESTIONS AND ANSWERS

An objective-type examination improves sharpness, prudent judgement, depth of knowledge, overall understanding, logic and concepts of the subject. A subjective-type examination improves communicaion and explanatory skills of the subject.

1. If Coulomb's force on 2 C due to 5 C is $10\mathbf{a}_x$ (N), the force on 5 C due to 2 C is
(a) $-10\mathbf{a}_x$ (N) (b) $10\mathbf{a}_x$ (N) (c) $10\mathbf{a}_y$ (N) (d) $-10\mathbf{a}_y$ (N)

2. Gradient of a potential and equipotential surface are
(a) orthogonal to each other (b) in the same direction
(c) out of phase (d) at 45° with one another

3. The work in moving a charge between two points depends on
(a) the path (b) Q, **E** and the path
(c) Q and **E** only (d) Q, **E** and end points

4. Unit of potential is
(a) Joules (b) N/C
(c) N/Joules (d) Joules/C

5. Field due to infinitely long line charge along z-axis varies with
(a) ϕ (b) z
(c) ρ (d) both ϕ and z

6. When the surface density is in y-z plane, field varies with
(a) y (b) z
(c) x (d) both y and z

7. Under static conditions, ρ_v inside an ideal conductor is
(a) 0 (b) ∞ (c) high (d) finite

8. Under static conditions, **E** inside an ideal conductor is
(a) 0 (b) ∞ (c) very high (d) finite

9. Under static conditions, **H** inside an ideal conductor is
(a) 0 (b) ∞ (c) very high (d) finite

10. Boundary condition on $\mathbf{E}_{tan}$ at an ideal conductor/free space interface is
(a) 0 (b) ∞ (c) very high (d) finite

11. In free space, for static fields,
(a) $\nabla \times \mathbf{E} = 0$ (b) $\nabla \times \mathbf{E} = -\mathbf{B}$ (c) $\nabla \times \mathbf{E} = -\mathbf{H}$ (d) $\nabla \times \mathbf{E} = \infty$

12. Which of the following is correct?
(a) $\nabla . \mathbf{E} = \rho_v$ (b) $\nabla . \mathbf{E} = \rho_v / \epsilon_0$ (c) $\nabla . \mathbf{E} = -\rho_v$ (d) $\nabla . \mathbf{E} = \epsilon_0 \rho_v$

13. For steady charge density

(a) $\nabla.J=0$ (b) $\nabla.J=\dot{\rho}_\upsilon$ (c) $\nabla.J=-\dot{\rho}_\upsilon$ (d) $\nabla.J=\epsilon E$

14. If the number of magnetic lines of force that cut across a conductor increases, the amount of induced voltage is _________.

15. If the number of turns in a coil is high, the induced voltage is _________.

16. An induced voltage exists only when the flux is changing. (Yes/No)

17. The flux of 1 mwb increasing to 1.01 mwb in 1.0 second produces a flux change $\frac{d\phi}{dt}$ equal to

(a) 0.01 mwb/s (b) 1.01 mwb/s

(c) 0.1 mwb/s (d) 1μ wb/s

18. Faraday's law is useful to find the amount of induced voltage. (Yes/No)

19. Lines of force of two magnetic fields in the same direction assist each other to provide a stronger magnetic field. (Yes/No)

20. If the magnetic flux of 8 mwb changes to 10 mwb in 1.0 second, the rate of flux is

(a) 1.0 mwb/sec (b) 2 mwb/sec

(c) 0 mwb/sec (d) 18 mwb/sec

21. If the flux changes from 8 wb to 10 wb in 2 seconds, the rate of change of flux is

(a) 2 wb/sec (b) 1 wb/sec (c) 0 wb/sec (d) 1 wb/sec

22. When a magnetic flux cuts across 200 turns at the rate of 2 wb/s, the induced voltage is

(a) 400 V (b) 100 V (c) 600 V (d) 0 V

23. When flux changes at a constant rate, the induced voltage is

(a) constant (b) 0 (c) ∞ (d) $-\infty$

24. The S.I unit of magnetic flux is

(a) Weber (b) Coulomb (c) Tesla (d) Gauss

25. If a magnetic flux of 4μ wb passes through an area of 5×10^{-4} m^2, the flux density is

(a) 8 mT (b) 80 mT (c) 0.8 mT (d) 0.08 mT

26. The divergence of the curl of a magnetic field is

(a) zero (b) ∞ (c) a vector (d) finite

27. The curl of a gradient of scalar magnetic potential is
(a) ∞ (b) a scalar (c) zero (d) ampere

28. Unit of vector magnetic potential is
(a) wb/m (b) Tesla (c) Volts (d) Ampere

29. Unit of scalar magnetic potential is
(a) Ampere (b) Tesla (c) Volt (d) wb/m

30. Unit of permeability of a medium is
(a) Henry (b) Henry/m (c) Farad/m (d) Weber

31. Unit of magnetic susceptibility is
(a) no units (b) Henry/m (c) wb/m (d) wb

32. The direction of **H** is the same as that of
(a) **B** (b) **A** (c) **E** (d) $\dot{\mathbf{D}}$

33. Unit of ∇^2 is
(a) 1/m (b) $1/m^2$ (c) none (d) A-m

34. ∇ is
(a) a scalar (b) a constant
(c) a vector (d) integral operator

35. Ampere's law is applicable for
(a) open path only (b) closed path only
(c) either open or closed path (d) square path only

36. The tangential component of a magnetic field is continuous across any discontinuity. (Yes/No)

37. The normal component of magnetic flux density is continuous. (Yes/No)

38. Force on a charge at rest due to a magnetic field is
(a) zero (b) a constant (c) ∞ (d) finite

39. Scalar magnetic potential in a region where **J** is present, is
(a) zero (b) ∞ (c) $\nabla \times \mathbf{H}$ (d) finite

40. Scalar magnetic potential exists when the field is produced by a current element. (Yes/No)

41. Unit of torque is
(a) the same as that of force (b) Newton-m^2
(c) Newton-m (d) Newton/m

42. Magnetic field lines are closed loops. (Yes/No)

43. When the current in a current element is upwards, the direction of the magnetic field is clockwise. (Yes/No)

44. If the current in a current element is downwards, the direction of the magnetic field is along $\mathbf{a}_\phi$. (Yes/No)

45. The direction of magnetisation is
(a) that of **H** (b) that of **A**
(c) that of **E** (d) that of **D**

46. $M_{12} = M_{21}$ (Yes/No)

47. The units of self-inductance and mutual-inductance are the same. (Yes/No)

48. Energy stored in an inductor is
(a) LI^2 (b) $1/2LI^2$ (c) $1/2LV^2$ (d) $1/2VI^2$

49. Unit of magnetic dipole moment is
(a) C-m (b) Amp-m (c) Amp/m (d) Amp-m^2

50. Magnetic susceptibility of free space is
(a) zero (b) 1 (c) μ_r (d) μ_o

51. Magnetisation satisfies the relation
(a) $\nabla \times \mathbf{M} = J_b$ (b) $\nabla \times \mathbf{M} = \mathbf{J}$
(c) $\nabla \times \mathbf{M} = 0$ (d) $\nabla \times \mathbf{M} = I_b$

52. The self-inductance of two coils are 4 H and 9 H. If the coefficient of coupling is 0.5, the mutual-inductance between the two coils is
(a) 12 H (b) 3 H (c) 6 mH (d) 10 H

53. Magnetic field in a perfect conductor is
(a) high (b) zero
(c) moderate (d) finite

54. The net magnetic flux emerging from any closed surface is
(a) zero (b) constant (c) unity (d) finite

55. The force on a charge due to magnetic field is

(a) QVB (b) $Q\,\mathbf{V}\,.\,\mathbf{H}$ (c) $Q\,(\mathbf{V} \times \mathbf{B})$ (d) zero

56. A conductor 1 m long carries a current of 5 mA and is at an angle of 30° with $B = 1.5$ wb/m^2. The magnitude of the force is

(a) 7.5 mN (b) 5 mN (c) 3.75 mN (d) 7.5 N

57. The first Maxwell's equation in free space is

(a) $\nabla\times\mathbf{H}=\mathbf{D}+\mathbf{J}$ (b) $\nabla\times\mathbf{H}=\dot{\mathbf{D}}$
(c) $\nabla\times\mathbf{H}=0$ (d) $\nabla\times\mathbf{H}=\mathbf{J}$

58. Identify which of the following waves do not exist in hollow waveguides:

(a) TE (b) TM (c) TE and TM (d) TEM

59. Poynting vector is given by

(a) $\mathbf{E}\times\mathbf{H}$ (b) $\mathbf{E}.\mathbf{H}$ (c) $\mathbf{H}\times\mathbf{E}$ (d) $\mathbf{H}.\mathbf{E}$

60. Poynting vector gives

(a) rate of energy flow (b) direction of polarisation
(c) electric field (d) magnetic field

61. $\mathbf{E}.\mathbf{H}$ of a uniform plane wave is

(a) EH (b) 0 (c) ηE^2 (d) ηH^2

62. Identify the scalar quantity:

(a) V (b) E (c) H (d) A

63. The minimum value of VSWR is

(a) 0 (b) 1 (c) -1 (d) 0.1

64. VSWR is

(a) V_{max}/V_{min} (b) V_{min}/V_{max} (c) $V_{max}\,V_{min}$ (d) $V_{max}+V_{min}$

65. Poisson's equation is

(a) $\nabla^2 V=-\rho/\epsilon_0$ (b) $\nabla^2 V=\rho/\epsilon_0$
(c) $\nabla^2 V=0$ (d) $\nabla^2 V=J/\epsilon_0$

66. In conductors,

(a) $\nabla\times\mathbf{D}=0$ (b) $\nabla\times\mathbf{D}=\rho$
(c) $\nabla\times\mathbf{D}=\mathbf{J}_s$ (d) $\nabla\times\mathbf{E}=\mathbf{J}_s$

67. The unit of magnetic current density is

(a) $\frac{V}{m^2}$ (b) $\frac{V}{m}$ (c) $\frac{A}{m^2}$ (d) $\frac{A}{m}$

68. If a dielectric material is placed in an electric field, the field intensity

(a) increases (b) does not change
(c) becomes zero (d) decreases

69. For static fields

(a) $\nabla \times \mathbf{H} = \mathbf{D} + \mathbf{J}$ (b) $\nabla \times \mathbf{H} = \mathbf{J}$
(c) $\nabla \times \mathbf{H} = 0$ (d) $\nabla \times \mathbf{H} = \mathbf{D}$

70. Two waves are said to be out of phase if their phase difference is

(a) 180° (b) 360° (c) 90° (d) 270°

71. Absolute permeability of free space is

(a) $4\pi \times 10^{-7}$ A/m (b) $4\pi \times 10^{-7}$ H/m
(c) $4\pi \times 10^{-7}$ F/m (d) $4\pi \times 10^{-7}$ H/m^2

72. For a static magnetic field

(a) $\nabla \times \mathbf{B} = \rho$ (b) $\nabla \times \mathbf{B} = \mu \mathbf{J}$
(c) $\nabla . \mathbf{B} = \mu_0 \mathbf{J}$ (d) $\nabla \times \mathbf{B} = 0$

73. The electric field in free space is

(a) $\frac{\mathbf{D}}{\epsilon_0}$ (b) $\frac{\mathbf{D}}{\mu_0}$ (c) $\epsilon_0 \mathbf{D}$ (d) $\frac{\sigma}{\epsilon_0}$

74. For a uniform plane wave in the x-direction

(a) $E_x = 0$ (b) $H_x = 0$
(c) $E_x = 0$ and $H_x = 0$ (d) $E_y = 0$

75. A transmission line whose $Z_0 = 75\Omega$ is terminated by 75Ω. Its input impedance is

(a) 75Ω (b) 150Ω
(c) 375Ω (d) 300Ω

76. Displacement current density is

(a) $\mathbf{D}$ (b) $\mathbf{J}$
(c) $\partial \mathbf{D} / \partial t$ (d) $\partial \mathbf{J} / \partial t$

77. If Γ is the coefficient of reflection, VSWR is given by

(a) $\frac{1-|\Gamma|}{1+|\Gamma|}$ (b) $\frac{1+|\Gamma|}{1-|\Gamma|}$ (c) $\frac{1+\Gamma}{1-\Gamma}$ (d) $\frac{1+|\Gamma|^2}{1-|\Gamma|^2}$

78. Depth of penetration in free space is

(a) α (b) $1/\alpha$ (c) 0 (d) infinity

79. When an EM wave is incident on a dielectric, it is

(a) fully transmitted
(b) fully reflected
(c) partially transmitted and partially reflected
(d) none of these

80. In circular polarisation of EM wave

(a) $E_x = E_y$ (b) $E_x < E_y$ (c) $E_x > E_y$ (d) $E_x \neq E_y$

81. Cut-off wavelength for dominant mode in a rectangular waveguide is

(a) 1 (b) 0 (c) very high (d) $2a$

82. If the dimensions of the narrow and broad walls of a waveguide are 3 cm and 4.5 cm, cut-off wavelength for the dominant mode is

(a) 6 cm (b) 9.0 cm
(c) 13.5 cm (d) 1.5 cm

83. Magnetic field due to an infinitely long current element is

(a) $I/4\pi\rho$ (b) $I/2\pi\rho$
(c) $I^2/2\pi\rho$ (d) $I^2/4\pi\rho$

84. Tesla is the unit of

(a) magnetic flux density (b) magnetic flux
(c) magnetisation (d) magnetic susceptibility

85. If a line is terminated in an open circuit, the VSWR is

(a) 0 (b) 1 (c) ∞ (d) -1

86. Complex Poynting vector, **P** is

(a) $\mathbf{P} = \mathbf{E} \times \mathbf{H}^*$ (b) $\mathbf{P} = \mathbf{E} \times \mathbf{H}^*$
(c) $\mathbf{P} = \frac{1}{2}\mathbf{E} \times \mathbf{H}^*$ (d) $\mathbf{P} = \mathbf{H} \times \mathbf{E}^*$

87. At the line surface of a dielectric

(a) $\mathbf{D}_{n1} = \mathbf{D}_{n2}$ (b) $\mathbf{D}_{n1} - \mathbf{D}_{n2} = \rho_s$

(c) $\mathbf{D}_{n1} - \mathbf{D}_{n2} = \mathbf{J}_s$ (d) $\mathbf{D}_{n1} - \mathbf{D}_{n2} = \frac{\rho_s}{\in}$

88. Brewster angle is the

(a) angle of incidence at which no reflection occurs
(b) angle of reflection
(c) angle of transmission
(d) angle of refraction

89. Distortionless condition for a transmission line is

(a) $LG = RC$ (b) $LR = GC$ (c) $GR = LC$ (d) $LC = Q$

90. Intrinsic impedance of free space is

(a) $120\pi\,\Omega$ (b) 300Ω (c) 75Ω (d) 73Ω

91. The reflection coefficient is generally

(a) complex (b) scalar (c) real (d) imaginary

92. The velocity of an EM wave in a medium whose $\in_r = 2$, $\mu_r = 2$ is

(a) 3×10^8 m/s (b) 3×10^8 cm/s
(c) 1.5×10^8 m/s (d) 1.5×10^8 cm/s

93. Time varying electric field is

(a) $\mathbf{E} = -\nabla V$ (b) $\mathbf{E} = -\nabla V - \dot{\mathbf{A}}$
(c) $\mathbf{E} = -\nabla V - \mathbf{B}$ (d) $\mathbf{E} = -\nabla V - \mathbf{D}$

94. The polarisation of dielectric material results in the

(a) creation of electrons (b) creation of electric dipoles
(c) creation of eddy currents (d) creation of magnetic dipoles

95. Any electric and magnetic field, irrespective of the fields of EM waves, are

(a) perpendicular to each other (b) parallel to each other
(c) may have any direction (d) are in the same direction

96. Uniform plane wave is

(a) longitudinal in nature

(b) transverse in nature
(c) neither transverse nor longitudinal
(d) vertical

97. The surface impedance of a conductor is nothing but
(a) Z_0 (b) reactive impedance
(c) load impedance (d) $\mathbf{E}_t / \mathbf{J}_s$

98. In a homogeneous medium, the direction of **E** is the same as
(a) **D** (b) **B** (c) **H** (d) **P**

99. Magnetic susceptibility of a medium is
(a) $\frac{M}{H}$ (b) $\frac{B}{H}$ (c) $\frac{H}{B}$ (d) $\frac{H}{M}$

100. Polarisation and direction of propagation of EM wave are one and the same. (Yes/No)

101. The direction of propagation of EM wave is obtained from
(a) $\mathbf{E} \times \mathbf{H}$ (b) $\mathbf{E} . \mathbf{H}$ (c) **E** (d) **H**

102. When an EM wave undergoes reflections while propagation, its group velocity v is
(a) greater than free space velocity v_0 (b) greater than phase velocity
(c) equal to v_0 (d) less than v_0

103. A medium is isotropic if
(a) $\in = 0$ (b) $\in = 1$
(c) $\in =$ scalar constant (d) $\in = \infty$

104. A hollow rectangular waveguide acts as a
(a) high pass filter (b) low pass filter
(c) band pass filter (d) low frequency radiator

105. Equation of continuity is
(a) $\nabla . \mathbf{J} = \dot{\rho}_v$ (b) $\nabla \times \mathbf{J} = -\dot{\rho}_v$
(c) $\nabla . \mathbf{J} = -\dot{\rho}_v$ (d) $\nabla \times \mathbf{J} = \rho_v$

106. The electric field inside a conducting sphere is
(a) uniform (b) zero
(c) maximum (d) minimum

107. If VSWR is 2, magnitude of the reflection coefficient is

(a) 1/2 (b) 1/3 (c) 1 (d) –1

108. The velocity of EM wave in a conductor is

(a) very high
(b) 3×10^8 m/s
(c) greater than 3×10^8 m/s
(d) very low

109. The velocity of propagation in a lossless transmission line is

(a) $\frac{1}{\sqrt{LC}}$ (b) $\frac{1}{\sqrt{\mu_0 \in_0}}$ (c) 3×10^8 m/s (d) $\sqrt{\frac{L}{C}}$

110. The characteristic impedance of a lossless transmission line is

(a) $\sqrt{\frac{L}{C}}$ (b) $\sqrt{LC}$ (c) $\frac{1}{\sqrt{LC}}$ (d) $\sqrt{\frac{C}{L}}$

111. The reflection coefficient, when a wave is incident normally on a perfect conductor, is

(a) 0
(b) 1
(c) greater than 1
(d) ∞

112. Inductance increases with the permeability of the core. (Yes/No)

113. Inductance decreases with length for the same number of turns. (Yes/No)

114. The current in an inductor changes from 10 to 16 mA in 2 seconds. The rate of change of current is

(a) 3 mA/s (b) 6 A/s (c) 6 A (d) 3 A

115. The current in an inductor changes by 100 mA in 2μ sec. The rate of change of current is

(a) 50000 A/s
(b) 100 A/s
(c) 50 A/s
(d) 50 mA/s

116. The inductance of a coil which induces 20 V when the rate of change of current in 4 A/s is

(a) 5 H (b) 10 H (c) 80 H (d) 16 H

117. The induced voltage across a 2 H inductance produced by a current charge of 12 A/s is

(a) 24 V (b) 48 V (c) 1/3 V (d) 8 V

118. A coil L_1 produces 100μ wb of magnetic flux. Out of this, 50μ wb is linked with a second coil, L_2. Then the coefficient of coupling is

(a) 0.50 (b) 2 (c) 50 (d) 4/3

119. If two coils of $L_1 = L_2 = 400\,\text{mH}$ have coefficient of coupling of 0.2, the mutual coupling is

(a) 80 mH (b) 40 m (c) 800 mH (d) 20 mH

120. In an inductance of 10 H with a current 3 A, energy stored in the inductance is

(a) 45 J (b) 30 J (c) 300 J (d) 3.3 J

121. If magnitude of magnetic flux density in free space is $4\pi \times 10^{-7}$ wb/m^2, the magnetic field strength is

(a) 1 A/m (b) $4\pi \times 10^{-7}$ A/m
(c) 1 wb/m (d) $4\pi \times 10^{-7}$ wb/m

122. The unit of μ_0 is

(a) H/m (b) Henry (c) F/m (d) H/m^2

123. The unit of $\in_0$ is

(a) Farad (b) F/m (c) F/m^2 (d) H/m

124. Unit of reluctance is

(a) Henry (b) Henry/m (c) Weber (d) 1/Henry

125. J in free space is

(a) 0 (b) ∞ (c) E/σ (d) none

126. One Neper is

(a) greater than 1 dB (b) less than 1 dB
(c) equal to 1 dB (d) 4.3 dB

127. For a good conductor, the depth of penetration is

(a) β (b) α (c) $1/\beta$ (d) ∞

128. At low frequencies, Earth is a

(a) good conductor (b) bad conductor
(c) excellent conductor (d) a good capacitor

129. Unit of Poynting vector is

(a) Watts (b) Watts/m^2 (c) Volt-amp (d) Joules

130. If reflection coefficient is complex, VSWR is

(a) scalar (b) complex (c) zero (d) ∞

131. If Z_L is purely resistive and $Z_0 > Z_L$, VSWR is

(a) $\frac{Z_0}{Z_L}$ (b) $\frac{Z_L}{Z_0}$ (c) 0 (d) 1

132. Z_0 is defined as

(a) $\frac{V_r}{I_r}$ (b) $\frac{V_{max}}{I_{min}}$ (c) $\frac{V_{min}}{I_{max}}$ (d) $\frac{V_i}{I_i}$

133. The static electric field at the surface of a conductor is directed normal to that surface everywhere. (Yes/No)

134. The conductor surface is an equipotential surface. (Yes/No)

135. Electric field is conservative. (Yes/No)

136. Magnetic field is conservative. (Yes/No)

137. Charge density within a conductor is

(a) zero (b) ∞ (c) moderate (d) high

138. The surface charge density resides

(a) on the exterior surface (b) on the interior of the surface
(c) on the middle of the conductor (d) in the interior depth

139. In static conditions, the current flow in a conductor is

(a) zero (b) ∞ (c) moderate (d) low

140. The current density in silver $(\sigma = 61.7 \times 10^6 \ \Omega/m)$ when the electric field is 1.0 V/m is

(a) 61.7×10^6 A/m^2 (b) 6.17×10^6 A/m^2
(c) 617×10^6 A/m^2 (d) 0.617×10^6 A/m^2

141. The charge stored in a 40 μF capacitor with 50 V across it is

(a) 2.0 mc (b) 20 mc (c) 200 mc (d) 0.2 mc

142. The charge stored in a capacitor when a constant current of 2 μA flows for 20 seconds is

(a) 40 μc (b) 10 μc (c) 400 μc (d) 200 μc

143. The capacitance of a capacitor which is charged to $40\mu c$ when 20 V is applied is

(a) $2\mu F$ (b) $800\mu F$ (c) $20\mu F$ (d) $200\mu F$

144. The voltages across a capacitor of $10\mu F$ when a constant current of 5 mA flows for 1 sec is

(a) 50 V (b) 500 V (c) 5 V (d) 0.5 V

145. If free space is replaced by a dielectric material between the plates of a capacitor, the stored energy in the capacitor

(a) decreases (b) increases

(c) remains the same (d) zero

146. Unit of energy density in an electrostatic field is

(a) Joule/m^3 (b) Joule/m^2 (c) Joules (d) Joules/m

147. If the electric susceptibility of a medium is 3, the permeability is

(a) 4 F/m (b) 35.416×10^{-12} F/m

(c) 2 F/m (d) 8 F/m

148. Energy stored in a capacitor is

(a) proportional to $\in_r$ (b) inversely proportional to $\in_r$

(c) independent of $\in_r$ (d) proportional to $1/\in_r$

149. Magnetic torque on a loop is

(a) a vector (b) a scalar (c) a constant (d) zero

150. Magnetic dipole is

(a) a pair of charges (b) a current loop

(c) a conductor (d) an electric dipole

151. Reluctance is

(a) the reciprocal of resistance (b) the reciprocal of inductance

(c) the reciprocal of capacitance (d) the reciprocal of permeance

152. The unit of magnetic charge is

(a) Coulomb (b) Ampere

(c) Ampere-metre square (d) Ampere-metre

153. If the flux passing through a cube is equal to $10\mu c$, the total charge enclosed by the cube defined by $0 \le x \le 2$, $0 \le y \le 2$ and $0 \le z \le 2$ is

(a) $10\mu c$ (b) $6\mu c$ (c) $10/3\mu c$ (d) $80\mu c$

154. Equation $\nabla.(-\epsilon \nabla V) = \rho_v$ is Poisson's equation. (Yes/No)

155. The electric susceptibility of air is

(a) 0 (b) 2 (c) 1 (d) 4

156. An electrostatic field cannot maintain a steady current in a closed circuit. (Yes/No)

157. In electrostatics, voltage and potential difference are equivalent. (Yes/No)

158. Except in electronics, voltage and potential difference are not equivalent. (Yes/No)

159. There are no isolated magnetic poles. (Yes/No)

160. There are no isolated magnetic charges. (Yes/No)

161. The lines of magnetic flux are continuous. (Yes/No)

162. The electric flux lines are continuous. (Yes/No)

163. $\frac{\partial \mathbf{B}}{\partial t}$ represents

(a) magnetic current (b) magnetic current density

(c) magnetic flux density (d) magnetic flux

164. The electromagnetic field inside an ideal conductor is

(a) zero (b) ∞

(c) maximum (d) minimum

165. Time varying field can exist in a conductor ($\sigma < \infty$). (Yes/No)

166. Surface charge density, ρ_s and $\mathbf{J}_s$ can exist on the surface of a perfect conductor. (Yes/No)

167. At the interface between two dielectrics, $\mathbf{J}_s$ is

(a) zero (b) 1 (c) ∞ (d) high

168. J_s can exist at the boundary between a conductor and a perfect dielectric. (Yes/No)

169. ρ_s can exist at the boundary between a conductor and a perfect dielectric. (Yes/No)

170. A field can exist if it satisfies

(a) Gauss's law (b) Faraday's law
(c) Coulomb's law (d) all Maxwell's equations

171. Source free region means

(a) $\sigma = 0$ (b) $\rho_v = 0$
(c) $\mathbf{J} = 0$ (d) $\rho_v = 0, \sigma = 0, J = 0$

172. If σ = 2.0 mho/m, E = 10.0 V/m, the conduction current density is

(a) 5.0 A/m^2 (b) 20.0 A/m^2
(c) 40.0 A/m^2 (d) 20 A

173. Maxwell's equations give the relations between

(a) different fields (b) different sources
(c) different boundary conditions (d) different potentials

174. Boundary condition on **J** is

(a) $\mathbf{a}_n \times (\mathbf{J}_1 - \mathbf{J}_2) = 0$ (b) $\mathbf{a}_n . (\mathbf{J}_1 - \mathbf{J}_2) = 0$
(c) $\mathbf{J}_1 = \mathbf{J}_2$ (d) $(\mathbf{J}_1 - \mathbf{J}_2) \times \mathbf{a}_n = 0$

175. Boundary condition on **E** is

(a) $\mathbf{a}_n \times (\mathbf{E}_1 - \mathbf{E}_2) = 0$ (b) $\mathbf{a}_n . (\mathbf{E}_1 - \mathbf{E}_2) = 0$
(c) $\mathbf{E}_1 = \mathbf{E}_2$ (d) $\mathbf{E}_{t1} - \mathbf{E}_{t2} = \rho_s$

176. Boundary condition on **H** is

(a) $\mathbf{a}_n \times (\mathbf{H}_1 - \mathbf{H}_2) = \mathbf{J}_s$ (b) $\mathbf{a}_n . (\mathbf{H}_1 - \mathbf{H}_2) = 0$
(c) $\mathbf{a}_n \times (\mathbf{H}_1 - \mathbf{H}_2) = \mathbf{J}_s$ (d) $\mathbf{a}_n . (\mathbf{H}_1 - \mathbf{H}_2) = 0$

177. Velocity of the EM wave is

(a) inversely proportional to β (b) inversely proportional to α
(c) directly proportional to β (d) directly proportional to α

178. Velocity of the wave in an ideal conductor is

(a) zero (b) very large
(c) moderate (d) small

179. If a wave in free space has $E = 2$ V/m, H is

(a) $\frac{1}{60\pi}$ A/m (b) 60π A/m
(c) 120π A/m (d) 240π A/m

180. If wet soil has $\sigma = 10^{-2}\,\Omega/\text{m}, \epsilon_r = 15, \mu_r = 1, f = 60\,\text{Hz}$, it is a

(a) good conductor (b) good dielectric
(c) semiconductor (d) magnetic medium

181. If wet soil has $\sigma = 10^{-2}\,\text{V/m}, \epsilon_r = 15, \mu_r = 1$ at 10 GHz, it is a

(a) good conductor (b) good dielectric
(c) semiconductor (d) semi dielectric

182. The cosine of the angle between two vectors is

(a) sum of the products of their direction cosines
(b) difference of the products of the direction cosines of the two vectors
(c) product of the products of the direction cosines of the two vectors
(d) division of the products of the direction cosines of the two vectors

183. Equiphase surfaces are

(a) planes (b) only lines (c) only cones (d) only circles

184. The wavelength of a wave travelling along the wave normal to the plane lying in y-z plane is

(a) greater than that along x-direction
(b) smaller than that along x-direction
(c) equal to that along x-direction
(d) zero

185. The phase velocity v_p of a wave travelling along the wave normal to the plane lying in y-z plane is

(a) smaller than v_p in x-direction (b) greater than v_p in x-direction
(c) equal to v_p in x-direction (d) infinity

186. Electric flux density and field are related by

(a) $\mathbf{D} = \epsilon\,\mathbf{E}$ (b) $\mathbf{D} = \dfrac{\mathbf{E}}{\epsilon}$ (c) $\mathbf{D} = \mu\,\mathbf{E}$ (d) $\mathbf{D} = \epsilon_r\,\mathbf{E}$

187. Magnetic flux flowing through a closed surface is nothing but the charge enclosed. (Yes/No)

188. The electric field intensity E at a point (1, 2, 2) due to (1/9) nc located at (0, 0, 0) is

(a) 33 V/m (b) 0.333 V/m
(c) 0.33 V/m (d) zero

189. If **E** is a vector, then $\nabla.\nabla\times\mathbf{E}$ is

(a) 0 (b) 1 (c) does not exist (d) infinity

190. Maxwell's equation, $\nabla\times\mathbf{B}=0$ is due to

(a) $\mathbf{B}=\mu\mathbf{H}$ (b) $\mathbf{B}=\frac{\mathbf{H}}{\mu}$

(c) non-existence of a monopole (d) $\mathbf{B}=\mathbf{H}$

191. For charge free regions,

(a) $\nabla.\mathbf{D}=0$ (b) $\nabla.\mathbf{D}=\rho_v$

(c) $\nabla.\mathbf{D}=\rho_v/\in$ (d) $\nabla.\mathbf{D}=\in\rho_v$

192. A static electric field cannot exist in the absence of

(a) **H** (b) **B** (c) Q (d) **M**

193. Velocity of EM wave in free space is

(a) independent of f (b) increases with increase in f

(c) decreases with increase in f (d) increases with f^2

194. Divergence theorem is applicable for

(a) static fields only

(b) time varying fields only

(c) both static and time varying fields

(d) electric fields only

195. The direction of propagation of EM wave is given by

(a) the direction of **E** (b) the direction of **H**

(c) the direction of **E** × **H** (d) the direction of **E** . **H**

196. For uniform plane wave propagating in z-direction

(a) $E_x=0$ (b) $H_x=0$

(c) $E_y=0,\ H_y=0$ (d) $E_z=0,\ H_z=0$

197. For free space

(a) $\sigma=\infty$ (b) $\sigma=0$ (c) $J\neq 0$ (d) $\rho_v=\infty$

198. 1 dB is

(a) $\log_{10}\frac{P_1}{P_2}$ (b) $10\log_{10}\frac{P_1}{P_2}$ (c) $\log_e\frac{P_1}{P_2}$ (d) $10\log_{10}\frac{V_1}{V_2}$

199. Velocity of propagation of EM wave is

(a) $\sqrt{\frac{\epsilon_0}{\mu_0}}$ (b) $\frac{\mu_0}{\epsilon_0}$ (c) $\frac{1}{\sqrt{\mu_0 \epsilon_0}}$ (d) $\frac{\epsilon_0}{\mu_0}$

200. The electric field for time varying potentials

(a) $\mathbf{E} = -\nabla V$ (b) $\mathbf{E} = -\nabla V - \dot{\mathbf{A}}$
(c) $\mathbf{E} = \nabla V$ (d) $\mathbf{E} = -\nabla V + \mathbf{A}$

201. The ratio of displacement and conduction current densities is

(a) $\frac{\epsilon}{\sigma}$ (b) $\frac{\sigma}{\epsilon}$ (c) $\sigma.\epsilon$ (d) $\frac{\omega\epsilon}{\sigma}$

202. The depth penetration of a wave in a dielectric increases with increasing

(a) σ (b) μ (c) ϵ (d) λ

203. The cut-off frequency of dominant mode in a waveguide is

(a) lowest (b) highest (c) zero (d) ∞

204. The dominant mode in a waveguide is characterised by

(a) highest cut-off wavelength (b) lowest cut-off wavelength
(c) zero cut-off wavelength (d) highest cut-off frequency

205. The intrinsic impedence of the medium whose $\sigma = 0, \epsilon_r = 9, \mu_r = 1$ is

(a) $40\pi\Omega$ (b) 9Ω (c) $120\pi\Omega$ (d) $60\pi\Omega$

206. For time varying EM fields

(a) $\nabla \times \mathbf{H} = \mathbf{J}$ (b) $\nabla \times \mathbf{H} = \dot{\mathbf{D}} + \mathbf{J}$
(c) $\nabla \times \mathbf{E} = 0$ (d) $\nabla \times \mathbf{H} = 0$

207. The wavelength of a wave with a propagation constant $= 0.1\pi + j0.2\pi$ is

(a) 10 m (b) 20 m (c) 30 m (d) 25 m

208. The characteristic impedance Z_0 of a lossless transmission line is

(a) $\sqrt{LC}$ (b) $\sqrt{\frac{L}{C}}$ (c) $\sqrt{\frac{C}{L}}$ (d) $\frac{1}{\sqrt{LC}}$

209. VSWR $= 1$ is obtained when

(a) $Z_L = 0$ (b) $Z_L = \infty$
(c) $Z_L = Z_0$ (d) Z_L is reactive

210. Relaxation time for good conductors is

(a) short (b) long (c) zero (d) ∞

211. Relaxation time for good dielectrics is

(a) short (b) long (c) zero (d) 10 ms

212. If $\sigma = 5.8 \times 10^7$ mho/m, $\in_r = 1$ for copper, relaxation time is

(a) 1.53×10^{-19} sec (b) 5.8×10^{-7} sec

(c) 0.1724×10^{-7} sec (d) 1.53 ms

213. For fused quartz, $\sigma = 10^{-17}$ mho/m, $\in_r = 5.0$, the relaxation time is

(a) 51.2 days (b) 5.12 days (c) 0.512 days (d) 0.0512 days

214. Within a conductor, ρ_v is

(a) 0 (b) ∞ (c) moderate (d) ϕ

215. The potential difference between any two points in a conductor is

(a) zero (b) ∞ (c) very high (d) 10 V

216. Electrostatic screening is obtained by

(a) a conductor (b) a dielectric

(c) a magnetic material (d) a semiconductor

217. A wave is totally reflected when

(a) the angle of incidence (θ_i) is large (b) θ_i is small

(c) $\theta_i = 0$ (d) $\theta_i = \theta_t$

218. A wave is totally reflected when

(a) medium 1 is denser than medium 2

(b) medium 1 is less denser than medium 2

(c) $\in_1 = \in_2$

(d) $\theta_i = \theta_t$

219. The surface resistance of a flat conductor is equal to its

(a) DC resistance of δ thickness (b) AC resistance of δ thickness

(c) zero (d) ∞

220. For a lossless medium with $\in_r = 10$, $\mu_r = 5$, impedance is

(a) $266\,\Omega$ (b) $26.6\,\Omega$ (c) $2660\,\Omega$ (d) $377\,\Omega$

221. If the magnitude of **H** for a plane wave in free space is 2.0 m A/m, the magnitude of **E** is

(a) 753.4 mV/m (b) 7.534 mV/m

(c) 75.34 mV/m (d) 188.5 mV/m

222. If distilled water has $\sigma = 0, \epsilon_r = 81$ and $\mu_r = 0$, its impedance is

(a) 418 Ω (b) 41.8 Ω (c) 9 Ω (d) 81 Ω

223. A medium has $\mu_r = 1$ and $\epsilon_r = 25$. The phase velocity of the wave in this medium is

(a) 0.6×10^8 m/sec (b) 6×10^8 m/sec

(c) 0.06×10^8 m/sec (d) 0.6×10^8 cm/sec

224. The magnitude of **H** of a plane wave in a medium is 5 A/m. The medium constants are $\epsilon_r = 4$, $\mu_r = 1$. The average power flow is

(a) 2354 w/m^2 (b) 23.54 w/m^2

(c) 235.4 w/m^2 (d) 2.354 w/m^2

225. For a plane travelling wave if the electric energy density is 10 mJ/m^3, the magnetic energy density is

(a) 10 mJ/m^3 (b) 1.0 mJ/m^3 (c) zero (d) 100 mJ/m^3

226. The electric field, **E** is due to a fixed charge is conservative because

(a) $\nabla \times \mathbf{E} = 0$ (b) $\nabla \times \mathbf{E} = -\mathbf{B}$ (c) $\nabla . \mathbf{D} = 0$ (d) $\nabla . \mathbf{B} = 0$

227. **H** field is

(a) rotational (b) irrotational

(c) transverse (d) longitudinal

228. *H* field in a current-carrying region is not conservative because

(a) $\nabla \times \mathbf{H} = \mathbf{J}$ (b) $\nabla \times \mathbf{H} = 0$ (c) $\nabla . \mathbf{H} = 0$ (d) $\nabla . \mathbf{D} = 0$

229. A magnetic circuit is a __________ followed by the flux in a magnetic material.

230. Leakage flux, produced by the coil completes its path through the medium surrounding the magnetic circuit. (Yes/No)

231. Magnetic circuits are part __________.

232. The magnetic flux density in a magnetic material is

(a) uniform (b) non-uniform

(c) zero (d) ∞

233. In series magnetic circuits the magnetic flux in the magnetic material is
   (a) equal to the magnetic flux in air gap
   (b) not equal to the magnetic flux in air gap
   (c) zero
   (d) ∞

234. The fringing of magnetic flux means
   (a) spreading of magnetic flux in the air gap
   (b) flux is zero
   (c) flux is infinity
   (d) flux is moderate

235. One application of electrostatic field is _________.

236. One application of magnetostatic field is _________.

237. Electric field on free charged particle
   (a) increases kinetic energy of the particle
   (b) decreases kinetic energy of the particle
   (c) makes kinetic energy zero
   (d) makes it immobile

238. Induced electric field is
   (a) conservative
   (b) non-conservative
   (c) zero
   (d) infinity

239. Poynting vector is
   (a) $\mathbf{P} = \mathbf{E}.\mathbf{H}$
   (b) $\mathbf{P} = \mathbf{E} \times \mathbf{H}$
   (c) $\mathbf{P} = \mathbf{E} \times \mathbf{H}^*$
   (d) $\mathbf{P} = \frac{1}{\sqrt{2}}\mathbf{E} \times \mathbf{H}^*$

240. **B** is said to be linear if
   (a) **B** and **H** are parallel
   (b) **B** and **H** are perpendicular
   (c) **E** and **H** are parallel
   (d) **E** and **H** are perpendicular

241. When a plane wave travels in a dielectric medium, the average electric energy density and average magnetic density are
   (a) equal
   (b) zero
   (c) unequal
   (d) ∞

242. In a dispersive medium
(a) signal is distorted
(b) signal is not distorted
(c) $\mathbf{E} = \mathbf{D}$
(d) $\mathbf{D} = \mathbf{H}$

243. An example of a dispersive medium is
(a) conducting medium
(b) magnetic medium
(c) magnetic material
(d) non-magnetic material

244. Brewster angle is an angle of incidence at which there exists
(a) full reflection
(b) no reflection
(c) both reflection and transmission
(d) only transmission

245. If potential is anti-symmetric, the electric field is
(a) symmetric
(b) anti-symmetric
(c) not present
(d) not related to potential

246. If a line charge is along the axis of a cylinder, the flux passes through
(a) the top surface
(b) the bottom surface
(c) both the top and bottom
(d) the curved surface

247. If the line charge is along the axis of a cylinder, the Guassian surface is
(a) top surface
(b) bottom surface
(c) curved surface
(d) not present

248. If a line charge of $\rho_L = 10\,\text{PC/mm}$ is distributed along the z-axis from $-\infty$ to ∞, the electric field on the curved surface of the cylinder whose radius is 10 cm is
(a) 1.8 V/m
(b) 18 V/m
(c) 180 V/m
(d) 1800 V/m

249. If $\mathbf{D} = 10\mathbf{a}_x\ \text{C/m}^2$, find the flux crossing 1-m^2 area that is normal to the axis at $x = 3$ m.
(a) 10 C
(b) 30 C
(c) 1.0 C
(d) 0.3 C

250. $\nabla.\mathbf{E}$ for the field of a uniform sheet charge is
(a) zero
(b) ∞
(c) 10
(d) $-\infty$

251. If $\mathbf{E} = 10r\,\mathbf{a}_\phi + 5\mathbf{a}_z$, $\nabla.\mathbf{E}$ is
(a) ∞
(b) zero
(c) $-\infty$
(d) 140 V/m^2

252. If the field at a point due to a pair of charges at $\pm d$ on the z-axis is

$$\mathbf{E} = \frac{Qd}{4\pi\epsilon_0 r^3}(2\cos\theta\,\mathbf{a}_r + \sin\theta\,\mathbf{a}_\theta),$$

the ϕ-compnent of **E** is

(a) zero (b) $\frac{Qd}{2\pi\epsilon_0 r^3}$ (c) $\frac{Qd}{4\pi\epsilon_0 r^3}$ (d) ∞

253. Two infinite sheets with a charge density ρ_s on each are located at $x = \pm 2.0$ m. The field E in $-1 < x < 1$ is

(a) ∞ (b) $\frac{\rho_s}{\epsilon_0}$ (c) zero (d) $-\frac{\rho_s}{\epsilon_0}$

254. If E = 10 V/m in free space, D is

(a) zero (b) $10\epsilon_0$ C/m (c) $\frac{10}{\epsilon_0}$ C/m^2 (d) $10\epsilon_0$ C/m^2

255. Conduction takes place when
(a) there is a direct electrical connection
(b) there is free space
(c) the circuits are isolated
(d) the circuits are widely separated

256. Induction taken place when
(a) there is direct electrical connection
(b) there are more circuits
(c) there is magnetic coupling
(d) there is electric coupling

257. An ideal transformer has
(a) infinite permeability (b) zero permeability
(c) $\mu = \mu_0$ (d) $\epsilon = \epsilon_0$

258. An ideal transformer has
(a) winding resistance of zero
(b) winding resistance of ∞
(c) winding resistance of moderate value
(d) high capacitance value

259. An auto transformer is
(a) an ordinary transformer
(b) a step-up transformer
(c) a step-down transformer
(d) one in which two windings of a transformer are also interconnected electrically

260. An EM wave in a hollow rectangular waveguide is characterised by
(a) TEM wave (b) only TE wave
(c) only TM wave (d) both TE and TM waves

261. Magnetic susceptibility has units of $\in$. (Yes/No)

262. Amperian is the current due to free electrons. (Yes/No)

263. $\oint \mathbf{B}.d\mathbf{s} = 0$ is always true. (Yes/No)

264. Magnetisation has units of **H**. (Yes/No)

265. Mutual inductance, *M* depends on
(a) flux linkage (b) current
(c) number of turns (d) μ and geometrical path

266. Magnetic dipole means
(a) two magnetic charges
(b) a current loop
(c) two electric charges in magnetic field
(d) electric dipole

267. Magnetic dipole moment has
(a) Weber as unit (b) Tesla as unit
(c) Amp-m^2 as unit (d) no units

268. Magnetic current density has the unit of
(a) V/m^2 (b) V/m
(c) A/m (d) A/m^2

269. If an isolated semiconductor is placed in an electric field, the motion of free electrons produce an electric field that cancels the external applied field. (Yes/No)

270. A semiconductor and conductor behave in the same manner when they are subjected to electric field. (Yes/No)

271. Dielectric strength of a material indicates the maximum **E** field before break down. (Yes/No)

272. Polarisation results in bound charge distribution. (Yes/No)

273. The electric field just above a conductor is always
(a) normal to the surface (b) tangential to the surface
(c) zero (d) ∞

274. The normal component of **D** in a dielectric medium just above the surface of a conductor is
(a) equal to the surface charge density
(b) equal to the tangential component
(c) zero
(d) infinity

275. The normal components of **D** are
(a) continuous across a dielectric boundary
(b) discontinuous across a dielectric boundary
(c) zero
(d) ∞

276. The number of lines of force from a point charge of 20 C is
(a) 10 (b) 20
(c) 30 (d) 40

277. Potential difference is equal to the change in potential energy per unit charge in the limit $Q \to 0$. (Yes/No)

278. In a charge free region, $E_x = 2x$, $E_y = 2y$ and E_z is
(a) $-4z$ (b) $4z$ (c) $2z$ (d) 4

279. An electric field given by $\mathbf{E} = \mathbf{a}_x + 2\mathbf{a}_y + 2\mathbf{a}_z$ V/m is uniform. (Yes/No)

280. The potential distribution within a conducting medium satisfies Laplace's equation as long as the medium is homogeneous and the current distribution is time invariant. (Yes/No)

281. Relaxation time is given by
(a) $\frac{\epsilon}{\sigma}$ (b) $\frac{\sigma}{\epsilon}$ (c) $\frac{\mu}{\epsilon}$ (d) $\sqrt{\frac{\epsilon}{\sigma}}$

282. The relaxation time for pure water is
(a) 30 ns (b) 40 ns (c) 50 ns (d) 4 s

283. The relaxation time for amber is about
(a) 40 min (b) 70 min (c) 70 sec (d) 70 hrs

284. Relaxation time for copper is
(a) 1.52×10^{-19} sec (b) 1.52 sec
(c) 1.52 minutes (d) 1.52×10^{-19} minutes

285. The permeability of all non-magnetic materials is
(a) the same as that of free space (b) zero
(c) ∞ (d) very high

286. Does a charge at rest establish a magnetic field? (Yes/No)

287. Does a charge in motion establish an electrostatic field? (Yes/No)

288. Is $\phi = \oint_S \mathbf{B}.d\mathbf{S}$ true? (Yes/No)

289. For static fields, $\int (\mathbf{E}\times\mathbf{H}).d\mathbf{s} = -\int_v \mathbf{J}.\mathbf{E}\,dv$ is true. (Yes/No)

290. Phase velocity and the velocity of wave propagation in free space are one and the same. (Yes/No)

291. TEM does not exist in a hollow rectangular waveguide but exists in waveguides of other shapes. (Yes/No)

292. If a waveguide for *s*-band system has a = 7 cm, b = 3 cm and TE_{10} propagates in *z*-direction, then the cut-off frequency for TE_{10} is
(a) 2.142 GHz (b) 21.42 GHz
(c) 0.2142 GHz (d) 2.142 MHz

293. Cut-off frequency of TE_{10} mode for 2.286 cm × 1.016 cm waveguide is
(a) 6.562 GHz (b) 0.6562 GHz
(c) 65.62 GHz (d) 6.562 MHz

294. For a square waveguide, $TE_{10} = TE_{01}$. (Yes/No)

295. *R*, *L*, *C* and *G* of a transmission line are functions of frequency. (Yes/No)

296. The shunt conductance, *G* of a transmission line is equal to $1/R$ where *R* is the series resistance. (Yes/No)

297. *R*, *L*, *C* and *G* of a transmission line are known as secondary constants. (Yes/No)

298. The units of *R*, *L*, *C* and *G* of a transmission line are Ω, *H*, *F*, mho. (Yes/No)

299. *R*, *L*, *G*, *C* of a transmission line are called lumped constants. (Yes/No)

300. The term long line means

(a) $\frac{\lambda}{4}$ length or more (b) ∞

(c) 1,000 km (d) 0.1λ length

301. Short line means its length is much less than $\frac{\lambda}{4}$. (Yes/No)

302. SWR is measured by

(a) reflectometer (b) voltmeter

(c) ammeter (d) power meter

303. Z_0 of a transmission line is independent of the length of the line. (Yes/No)

304. Copper loss in transmission line is more if Z_0 is small. (Yes/No)

305. Copper loss increases as the transmission line ages. (Yes/No)

306. Dielectric loss in a transmission line increases with increase in frequency. (Yes/No)

307. Radiation losses are more when the spacing between transmission lines is more. (Yes/No)

308. Radiation losses in coaxial cable are smaller than those of parallel-wire lines. (Yes/No)

309. Radiation losses become small if the frequency is increased. (Yes/No)

310. Matched load means $Z_L = Z_0$ (Yes/No)

311. Crystallisation reduces the copper loses in transmission lines. (Yes/No)

312. Skin effect reduces the copper loses in transmission lines. (Yes/No)

313. For air, dielectric losses are very high. (Yes/No)

314. The *r* and *x* circles in Smith chart are orthogonal to each other. (Yes/No)

315. The perimeter of the outer rim of the Smith chart is of $\frac{\lambda}{4}$ length. (Yes/No)

316. VSWR is given by $\frac{V_{max}}{V_{min}}$. (Yes/No)

317. If $R_L > Z_0$, VSWR is given by R_L/Z_0. (Yes/No)

318. If $Z_L = 100 + j200$ and $Z_0 = 50\Omega$, the normalised impedance is
(a) $1 + j2$
(b) $2 + j4$
(c) $\sqrt{20}$
(d) 6

319. VSWR has a range of $1 \leq \text{VSWR} < \infty$. (Yes/No)

320. VSWR has a range of $0 \leq \text{VSWR} \leq \infty$. (Yes/No)

321. VSWR has a range of $-\infty \leq \text{VSWR} \leq \infty$. (Yes/No)

322. The input current in a matched line is 50 mA and the load current is 1 mA. If line is 1 km long, attenuation in Nepers is
(a) 3.9 NP
(b) 39 NP
(c) – 39 NP
(d) – 39 mNP

323. At low frequencies, Z_0 of a transmission line is
(a) $\sqrt{\frac{R}{G}}\ \Omega$
(b) $\sqrt{\frac{G}{R}}$
(c) $\sqrt{\frac{L}{C}}$
(d) $\sqrt{\frac{C}{L}}$

324. At high frequencies Z_0 of a transmission line is
(a) $\sqrt{\frac{R}{G}}\ \Omega$
(b) $\sqrt{\frac{G}{R}}$
(c) $\sqrt{\frac{L}{C}}$
(d) $\sqrt{\frac{C}{L}}$

325. Unit of phase constant is
(a) rad/m
(b) m
(c) degrees
(d) rad/m^2

326. At LF and VLF, polarisation often used is
(a) vertical
(b) horizontal
(c) theta
(d) elliptical

327. dB_i means power gain of the antenna in dB relative
(a) dipole (b) isotropic antenna
(c) dish (d) horn

328. dB_m means power gain compared to
(a) 1 W (b) 1 μW (c) 1 mW (d) 1 MW

329. If the signal level is 1 mW, power gain is
(a) 0 dBm (b) 1 dBm
(c) 10^{-3} dBm (d) 10 dBm

330. Whip antenna has a physical length of
(a) $\frac{\lambda}{4}$ (b) $\frac{\lambda}{2}$
(c) $\frac{3\lambda}{2}$ (d) λ

331. For a 300Ω antenna operating with 5 A of current, the radiated power is
(a) 7500 W (b) 750 W
(c) 75 W (d) 7500 mW

332. Effective area of antenna is a function frequency. (Yes/No)

333. Antenna used in mobile communications is
(a) whip antenna (b) dipole
(c) dish (d) horn

334. Half-power beamwidth of a dish antenna is
(a) $70\lambda/D$ (b) $70D/\lambda$
(c) $7D/\lambda$ (d) $7\lambda/D$

335. If a current element is *z*-directed, vector magnetic potential is
(a) *x*-directed (b) *y*-directed
(c) θ-directed (d) *z*-directed

336. If vector magnetic potential has only A_z, E_ϕ is ________.

337. Radiation resistance of a current element is

(a) $80\pi^2\left(\frac{dl}{\lambda}\right)^2 \Omega$ (b) 80Ω (c) $80\pi^2\,\Omega$ (d) $80\left(\frac{dl}{\lambda}\right)^2$

338. Radiation resistance of quarter wave monopole is ________.

339. Directional pattern of a short dipole in the horizontal plane is a ________.

340. Directional pattern of a horizontal half wave centre fed dipole is ________.

341. Effective length of a dipole is always ________ than the actual length.

342. The directivity of half wave dipole is ________.

343. The directivity of current element is ________.

344. Effective area of a Hertzian dipole operating at 100 kHz is ________.

345. FCC means
   (a) Federal Communication Council   (b) Foreign Communication Council
   (c) Fixed Communication Council   (d) France EC

346. VDE is EMC standard of
   (a) USA   (b) India   (c) France   (d) Germany

347. Grounding a conductor will ________ the charge.

348. EMP is created by ________.

349. Shielding is one of the popular methods for ________.

350. The average incident radiation power density should not exceed ________ for human exposures greater than 30 seconds.

351. The unit of self-reaction is ________.

Answers

1. (a)	2. (a)	3. (d)	4. (d)	5. (c)
6. (c)	7. (a)	8. (a)	9. (a)	10. (a)
11. (a)	12. (b)	13. (a)	14. Higher	15. High
16. Yes	17. (a)	18. Yes	19. Yes	20. (b)
21. (b)	22. (a)	23. (a)	24. (a)	25. (a)
26. (a)	27. (c)	28. (a)	29. (a)	30. (b)
31. (a)	32. (a)	33. (b)	34. (c)	35. (b)

36. No	37. Yes	38. (a)	39. (a)	40. No
41. (c)	42. Yes	43. No	44. No	45. (a)
46. Yes	47. Yes	48. (b)	49. (d)	50. (a)
51. (a)	52. (b)	53. (b)	54. (a)	55. (c)
56. (c)	57. (b)	58. (d)	59. (a)	60. (a)
61. (b)	62. (a)	63. (b)	64. (a)	65. (a)
66. (a)	67. (a)	68. (c)	69. (b)	70. (a)
71. (b)	72. (b)	73. (a)	74. (c)	75. (a)
76. (c)	77. (b)	78. (d)	79. (c)	80. (a)
81. (d)	82. (b)	83. (b)	84. (a)	85. (c)
86. (c)	87. (a)	88. (a)	89. (a)	90. (a)
91. (a)	92. (c)	93. (b)	94. (b)	95. (c)
96. (b)	97. (d)	98. (a)	99. (a)	100. No
101. (a)	102. (d)	103. (c)	104. (a)	105. (c)
106. (b)	107. (b)	108. (d)	109. (a)	110. (a)
111. (b)	112. Yes	113. Yes	114. (a)	115. (a)
116. (a)	117. (a)	118. (a)	119. (a)	120. (a)
121. (a)	122. (a)	123. (b)	124. (d)	125. (a)
126. (a)	127. (c)	128. (b)	129. (b)	130. (a)
131. (a)	132. (d)	133. Yes	134. Yes	135. Yes
136. No	137. (a)	138. (a)	139. (a)	140. (a)
141. (a)	142. (a)	143. (a)	144. (b)	145. (b)
146. (a)	147. (a)	148. (a)	149. (a)	150. (b)
151. (d)	152. (d)	153. (a)	154. Yes	155. (a)
156. Yes	157. Yes	158. Yes	159. Yes	160. Yes
161. Yes	162. No	163. (b)	164. (a)	165. Yes
166. Yes	167. (a)	168. No	169. Yes	170. (d)
171. (d)	172. (b)	173. (a)	174. (b)	175. (a)
176. (a)	177. (a)	178. (a)	179. (a)	180. (a)
181. (b)	182. (a)	183. (a)	184. (b)	185. (a)
186. (a)	187. No	188. (c)	189. (a)	190. (c)
191. (a)	192. (c)	193. (a)	194. (c)	195. (c)
196. (d)	197. (b)	198. (b)	199. (c)	200. (b)

201. (d)	202. (d)	203. (a)	204. (a)	205. (a)
206. (b)	207. (a)	208. (b)	209. (c)	210. (a)
211. (b)	212. (a)	213. (a)	214. (a)	215. (a)
216. (a)	217. (a)	218. (a)	219. (a)	220. (a)
221. (a)	222. (b)	223. (a)	224. (a)	225. (a)
226. (a)	227. (a)	228. (a)	229. Closed path	
230. Yes	231. Rotating machines, relays			232. (a)
233. (a)	234. (a)	235. (d) Deflect the electrons in CRT		
236. Magnetic separator		237. (a)	238. (b)	239. (b)
240. (a)	241. (a)	242. (a)	243. (a)	244. (b)
245. (a)	246. (d)	247. (c)	248. (a)	249. (a)
250. (a)	251. (b)	252. (a)	253. (c)	254. (d)
255. (a)	256. (c)	257. (a)	258. (a)	259. (d)
260. (d)	261. No	262. No	263. Yes	264. Yes
265. (a)	266. (b)	267. (c)	268. (a)	269. Yes
270. Yes	271. Yes	272. Yes	273. (a)	274. (a)
275. (a)	276. (b)	277. Yes	278. (a)	279. Yes
280. Yes	281. (a)	282. (b)	283. (b)	284. (a)
285. (a)	286. No	287. No	288. Yes	289. Yes
290. Yes	291. No	292. (a)	293. (a)	294. Yes
295. Yes	296. No	297. No	298. No	299. No
300. (a)	301. Yes	302. (a)	303. Yes	304. Yes
305. Yes	306. Yes	307. Yes	308. Yes	309. No
310. Yes	311. No	312. No	313. No	314. Yes
315. No	316. Yes	317. Yes	318. (b)	319. Yes
320. No	321. No	322. (a)	323. (a)	324. (c)
325. (a)	326. (a)	327. (b)	328. (c)	329. (a)
330. (a)	331. (a)	332. Yes	333. (a)	334. (a)
335. (d)	336. Zero	337. (a)	338. 36.5 Ω	339. Circle
340. Figure of eight		341. Less	342. 1.64	343. 1.5
344. 1.07 m^2	345. (a)	346. (d)	347. Bleed off	
348. Nuclear detonations		349. EMC design		
350. 10 mW/cm^2		351. Watt		

BIBLIOGRAPHY

- Administer, Joseph A. (1993). *Schaum's Outline of Theory and Problems of Electromagnetics*. New Delhi: McGraw-Hill.
- Carstensen, Russell V. (1981). *EMI Control in Boats and Ships*. USA: Don White Consultants, Inc.
- Crowley, J. M. (1986). *Fundamentals of Applied Electrostatics*. New York: Wiley.
- Guru, Bhagsingh and Hüseyin R. Hüziroğlu (2001). *Electromagnetic Field Theory Fundamentals*. Singapore: Thomson Asia Pvt. Ltd.
- Hayt, Jr. W. H. (1989). *Engineering Electromagnetics*. New Delhi: Tata McGraw-Hill.
- Jordan, E. C. and K. G. Balmain (1968). *Electromagnetic Waves and Radiating Systems*, New Jersey, US: Prentice-Hall.
- King, David K. (1999). *Fields and Wave Electromagnetics*. New York: Pearson Education.
- Narayana Rao, N. (1972). *Basic Electromagnetics with Applications*. New Jersey, US: Prentice-Hall.
- Nasar, Syed A. (1992). *Electromagnetics, Schaum's Series*. New York: McGraw-Hill.
- Paul, R. C. (1992). *Electromagnetic Compatibility*. New York: John Wiley & Sons.
- Plonsey, R. and R. E. Collin (1982). *Principles and Applications of Electromagnetic Fields*. New York: McGraw-Hill.
- Popvic, B. D. (1971). *Introductory Engineering Electromagnetics*. Reading Mass: Addison Wesley.
- Raju, G. S. N. (2005). *Antennas and Wave Propagation*, New Delhi: Pearson Education.
- Ramo, S., J. R. Whinnery and T. Van Duzer (1984). *Fields and Waves in Communication Electronics*. New York: Wiley.
- Sadiku, Mathew N. O. (2001). *Elements of Electromagnetics*. London: Oxford University Press.
- Sander, K. F. and G. A. L. Read (1986). *Transmission and Propagation of EM Waves*. England: Cambridge University Press.
- Seshadri, S. R. (1971). *Fundamentals of Transmission Lines and Electromagnetic Fields*. Reading, Mass: Addison Wesley.
- Shen, L. C. and J. A. Kong (1987). *Applied Electromagnetism*. Boston, Mass: PWS Engineering.
- Zahn, M. (1979). *Electromagnetic Field Theory*. New York: John Wiley.

INDEX

W